Methods in Enzymology

Volume XXXVI

HORMONE ACTION

Part A

Steroid Hormones

METHODS IN ENZYMOLOGY

EDITORS-IN-CHIEF

Sidney P. Colowick   Nathan O. Kaplan

Methods in Enzymology

Volume XXXVI

Hormone Action

Part A

Steroid Hormones

EDITED BY

Bert W. O'Malley

DEPARTMENT OF CELL BIOLOGY
BAYLOR COLLEGE OF MEDICINE
TEXAS MEDICAL CENTER
HOUSTON, TEXAS

Joel G. Hardman

DEPARTMENT OF PHYSIOLOGY
VANDERBILT UNIVERSITY SCHOOL OF MEDICINE
NASHVILLE, TENNESSEE

1975

ACADEMIC PRESS New York San Francisco London
A Subsidiary of Harcourt Brace Jovanovich, Publishers

ACADEMIC PRESS, INC.
111 Fifth Avenue, New York, New York 10003

United Kingdom Edition published by
ACADEMIC PRESS, INC. (LONDON) LTD.
24/28 Oval Road, London NW1

Library of Congress Cataloging in Publication Data

O'Malley, Bert W
Hormones and cyclic nucleotides.

(Methods in enzymology ; v. 36-)
Includes bibliographical references and indexes.
CONTENTS: pt. A. Steroid hormones.–pt. C. Cyclic nucleotides.–pt. D. Isolated cells, tissues, and organ systems.–pt. E. Nuclear structure and function.
1. Enzymes. 2. Hormones. 3. Cyclic nucleotides. I. Hardman, Joel G., joint author. II. Title. III. Series: Methods in enzymology ; v. 36 [etc.]
[DNLM: 1. Cell nucleus. 2. Hormones. 3. Nucleotides, Cyclic. W1 ME9615K v. 40 / QH595 H812]
QP601.C733 vol. 36, etc. [QP601] 574.1'925'08s
[574.1'92] 74-10710
ISBN 0–12–181936–1 (v. 36) (pt. A)

PRINTED IN THE UNITED STATES OF AMERICA

Table of Contents

Section I. Hormone-Binding Proteins and Assays for Steroid Hormones

Section II. Serum-Binding Proteins for Steroid and Thyroid Hormones

Section III. Cytoplasmic Receptors for Steroid Hormones

Section IV. Nuclear Receptors for Steroid Hormones

Section V. Purification of Receptor for Steroid Hormones

Section VI. Steroid Hormone Effects on Biochemical Processes

Section VII. Isolation of Biologically Active Metabolites of Steroid and Thyroid Hormones

Contributors to Volume XXXVI

Article numbers are in parentheses following the names of contributors. Affiliations listed are current.

A. M. AKBAR (2), *Department of Endocrinology, Cook County General Hospital, Chicago, Illinois*

PIERRE ALLOUGH (11), *Unité de Recherches sur le Métabolisme Moléculaire de l'Institut National de la Santé et de la Recherche Médicale, Départment de Chimie Biologique, Université Paris-Sud, Bicêtre, France*

J. N. ANDERSON (25), *Department of Biological Sciences, Stanford University, Stanford, California*

FERNANDO ARIAS (33), *Departments of Obstetrics, Gynecology, and Biological Chemistry, Washington University School of Medicine, St. Louis, Missouri*

MICHEL ATGER (11). *Unité de Recherches sur le Métabolisme Moléculaire de l'Institut National de la Santé et de la Recherche Médicale, Département de Chimie Biologique, Université Paris-Sud, Bicêtre, France*

C. WAYNE BARDIN (6), *Departments of Medicine, Physiology and Comparative Medicine, Pennsylvania State University, Hershey, Pennsylvania*

E. E. BAULIEU (29), *Unité de Recherches sur le Métabolisme Moléculaire et la Physio-Pathologie des Stéroides de l'Institut National de la Santé et de la Recherche Médicale, Département de Chimie Biologique, Université Paris-Sud, Bicêtre, France*

JOHN D. BAXTER (19, 20), *Metabolic* Re*search Unit and Department of Medicine, University of California, San Francisco, California*

M. M. BOUTON (29), *Roussel-UCLAF, Romainville, France*

H. LEON BRADLOW (9, 44), *The Institute for Steroid Research, Montefiore Hospital and Medical Center, Bronx, New York*

P. I. BRECHER (23), *Department of Medicine, Boston University School of Medicine, Boston, Massachusetts*

FRANCESCO BRESCIANI (30), *Institute of General Pathology, 1st Medical Faculty, University of Naples, Naples, Italy*

RICHARD E. BULLER (27), *Department of Cell Biology, Baylor College of Medicine, Houston, Texas*

LESLIE P. BULLOCK (6), *Department of Medicine, Division of Endocrinology, Milton S. Hershey Medical Center, Hershey, Pennsylvania*

ELIZABETH P. BURROWS (34), *Department of Chemistry, Vanderbilt University, Nashville, Tennessee*

ROBERT M. BURTON (8), *Department of Obstetrics and Gynecology, University of Louisville School of Medicine, Louisville, Kentucky*

EVANGELINA CASTAÑEDA (4a), *The Ben May Laboratory for Cancer Research, The University of Chicago, Chicago, Illinois*

M. G. CATELLI (29), *Unité de Recherches sur le Métabolisme Moléculaire et la Physio-Pathologie des Stéroides de l'Institut National de la Santé et de la Recherche Médicale, Département de Chimie Biologique, Université Paris-Sud, Bicêtre, France*

RALPH R. CAVALIERI (12), *Department of Medicine, University of California and Veterans Administration Hospital, San Francisco, California*

GARY C. CHAMNESS (16), *Department of Medicine, University of Texas Health Science Center, San Antonio, Texas*

J. H. CLARK (25), *Department of Cell Biology, Baylor College of Medicine, Houston, Texas*

H. F. DELUCA (46), *Department of Biochemistry, University of Wisconsin at Madison, Madison, Wisconsin*

EUGENE R. DESOMBRE (23, 31), *The Ben May Laboratory for Cancer Research, The University of Chicago, Chicago, Illinois*

I. S. EDELMAN (26), *Cardiovascular Research Institute and the Departments of Medicine and of Biochemistry and Biophysics, University of California School of Medicine, San Francisco, California*

H. A. FELDMAN (1), *Division of Engineering and Applied Physics, Harvard University, Cambridge, Massachusetts*

DAVID K. FUKUSHIMA (44), *Institute for Steroid Research, Montefiore Hospital and Medical Center, Bronx, New York*

STANLEY R. GLASSER (39, 41), *Department of Cell Biology, Baylor College of Medicine, Houston, Texas*

THOMAS A. GORELL (31), *The Ben May Laboratory for Cancer Research, The University of Chicago, Chicago, Illinois*

JACK GORSKI (15, 24, 38), *Departments of Biochemistry and Meat and Animal Science, University of Wisconsin at Madison, Madison, Wisconsin*

ERLIO GURPIDE (7), *Departments of Biochemistry and Obstetrics and Gynecology, Mount Sinai School of Medicine, New York, New York*

GEORGE B. HARDING (8), *Department of Biochemistry,. University of Louisville School of Medicine, Louisville, Kentucky*

ROSS G. HARDISON (34), *Department of Biochemistry, University of Iowa, Iowa City, Iowa*

STEPHEN J. HIGGINS (20), *Androgen Physiology Department, Imperial Cancer Research Fund Laboratories, London, England*

MICHAEL F. HOLICK (46), *Department of Biochemistry, University of Wisconsin at Madison, Madison, Wisconsin*

SIDNEY H. INGBAR (12), *Department of Medicine, University of California and Veterans Administration Hospital, San Francisco, California*

R. IRVING (32), *Androgen Physiology Department, Imperial Cancer Research Fund Laboratories, London, England*

E. V. JENSEN (23), *The Ben May Laboratory for Cancer Research, The University of Chicago, Chicago, Illinois*

PETER LETH JØRGENSEN (36), *Institute of Physiology, University of Aarhus, Denmark*

BENITA S. KATZENELLENBOGEN (38), *Department of Physiology and Biophysics, and School of Basic Medical Sciences, University of Illinois, Urbana, Illinois*

MARVIN A. KIRSCHNER (5), *Newark Beth Israel Medical Center, New Jersey Medical School, Newark, New Jersey*

S. KORENMAN (4), *Departments of Internal Medicine and Biochemistry, College of Medicine, University of Iowa, Iowa City, Iowa*

TEHMING LIANG (28), *The Ben May Laboratory for Cancer Research, The University of Chicago, Chicago, Illinois*

SHUTSUNG LIAO (4a, 28), *The Ben May Laboratory for Cancer Research and the Department of Biochemistry, The University of Chicago, Chicago, Illinois*

WILLIAM L. MCGUIRE (16, 21), *Department of Medicine, University of Texas Health Science Center, San Antonio, Texas*

W. I. P. MAINWARING (32), *Androgen Physiology Department, Imperial Cancer Research Fund Laboratories, London, England*

D. MARVER (26), *Cardiovascular Research Institute, University of California School of Medicine, San Francisco, California*

EDWIN MILGROM (11), *Unité de Recherches sur le Métabolisme Moléculaire de l'Institut National de la Santé et de la Recherche Médicale, Département de Chimie Biologique, Université Paris-Sud, Bicêtre, France*

RONALD J. MOORE (40), *Department of Internal Medicine, University of Texas Southwestern Medical School, Dallas, Texas*

ALLAN MUNCK (22, 35), *Department of Physiology, Dartmouth Medical School, Hanover, New Hampshire*

Jon R. Neergaard (34), *Department of Chemistry, Vanderbilt University, Nashville, Tennessee*

T. M. Nett (2), *Department of Physiology and Biophysics, Colorado State University, Fort Collins, Colorado*

G. D. Niswender (2), *Department of Physiology and Biophysics, Colorado State University, Fort Collins, Colorado*

Ernesto Nola (30), *Institute of General Pathology, 1st Medical Faculty, University of Naples, Naples, Italy*

M. Numata (23), *Department of Obstetrics and Gynecology, Tokyo Medical and Dental University, Tokyo, Japan*

Jack H. Oppenheimer (47), *Division of Endocrinology, Albert Einstein College of Medicine, Montefiore Hospital and Medical Center, Bronx, New York*

E. J. Peck, Jr. (25), *Department of Cell Biology, Baylor College of Medicine, Houston, Texas*

Donald Prough (6), *Milton S. Hershey Medical Center, Hershey, Pennsylvania*

Giovanni Alfredo Puca (30), *Institute of General Pathology, 1st Medical Faculty, University of Naples, Naples, Italy*

Leyla C. Ramirez (45), *Veterans Administration Hospital, Bronx, New York*

C. Raynaud-Jammet (29), *Unité de Recherches sur le Métabolisme Moléculaire et la Physio-Pathologie des Stéroides de l'Institut National de la Santé et de la Recherche Médicale, Département de Chimie Biologique, Université Paris-Sud, Bicêtre, France*

D. Rodbard (1), *Reproduction Research Branch, National Institute of Child Health and Human Development, National Institutes of Health, Bethesda, Maryland*

William Rosner (9, 10), *Department of Medicine, Roosevelt Hospital, and the College of Physicians and Surgeons, Columbia University, New York, New York*

Guy G. Rousseau (19, 20), *Département de Physiologie, Université de Louvain, Louvain, Belgium*

Daniel V. Santi (19), *Department of Biochemistry and Biophysics, University of California, San Francisco, California*

Madhabananda Sar (13), *Laboratories for Reproductive Biology, University of North Carolina, Chapel Hill, North Carolina*

William T. Schrader (17, 27), *Department of Cell Biology, Baylor College of Medicine, Houston, Texas*

Geoffrey W. G. Sharp (37), *Massachusetts General Hospital and Harvard Medical School, Boston, Massachusetts*

Merry R. Sherman (14, 18), *Memorial Sloan-Kettering Cancer Center, New York, New York.*

Vincenzo Sica (30), *Institute of General Pathology, 1st Medical Faculty, University of Naples, Naples, Italy*

Pentti K. Siiteri (42), *Department of Obstetrics and Gynecology, University of California, San Francisco, San Francisco, California*

Howard E. Smith (34), *Department of Chemistry, and Center for Population Research and Studies in Reproductive Biology, Vanderbilt University, Nashville, Tennessee*

Rex Smith (10), *Department of Medicine, Roosevelt Hospital, and College of Physicians and Surgeons, Columbia University, New York, New York*

Roy G. Smith (34), *Department of Cell Biology, Baylor College of Medicine, Houston, Texas*

S. Smith (23), *The Ben May Laboratory for Cancer Research, The University of Chicago, Chicago, Illinois*

Susan H. Socher (27), *Department of Cell Biology, Baylor College of Medicine, Houston, Texas*

George M. Stancel (15), *Program in Pharmacology, University of Texas Medical School, Texas Medical Center, Houston, Texas*

Charles A. Strott (3, 43), *Reproduction Research Branch, National Institute of Child Health and Human Development, National Institutes of Health, Bethesda, Maryland*

Walter E. Stumpf (13), *Departments of Anatomy and Pharmacology, Laboratories for Reproductive Biology, University of North Carolina, Chapel Hill, North Carolina*

Martin I. Surks (47), *Division of Endocrinology, Albert Einstein College of Medicine, Montefiore Hospital and Medical Center, Bronx, New York*

Frederick Sweet (33), *Departments of Obstetrics, Gynecology, and Biological Sciences, Washington University School of Medicine, St. Louis, Missouri*

David O. Toft (14), *Department of Endocrine Research, Mayo Clinic, Rochester, Minnesota*

Stanley Ulick (45), *Veterans Administration Hospital and Department of Medicine, Mount Sinai School of Medicine, Bronx, New York*

James C. Warren (33), *Departments of Obstetrics, Gynecology, and Biological Chemistry, Washington University School of Medicine, St. Louis, Missouri*

Ulrich Westphal (8), *Department of Biochemistry, University of Louisville School of Medicine, Louisville, Kentucky*

David Williams (24), *Department of Biochemistry, University of California Medical Center, San Francisco, California*

Jean D. Wilson (40), *Department of Internal Medicine, The University of Texas, Southwestern Medical School, Dallas, Texas*

Charles Wira (22), *Department of Physiology, Dartmouth Medical School, Hanover, New Hampshire*

Lawrence Zyskowski (35), *Department of Physiology, Dartmouth Medical School, Hanover, New Hampshire*

Preface

The creation of a series of volumes dealing with methodological aspects of hormone action has been a substantial undertaking. The large investment of time required for this project on the part of both the editors and the contributors appears to be justified because of the rapidly increasing number of investigators in the field of hormone action. A rough estimation gleaned from journal articles and programs of national meetings leads us to the striking conclusion that an approximate sixfold expansion of this field has occurred over the past eight years. For this reason we have attempted to select a representative sample of the basic methods employed in this field and present them in this series of volumes for "Methods in Enzymology." The volumes have been arbitrarily subdivided into five major categories, the first three of which deal with techniques employed primarily in studies on steroid hormones, peptide hormones, and cyclic nucleotides. An additional volume deals with isolated cell, tissue, and organ systems used for studies of all hormones, and a final volume contains contributions in the areas of nuclear structure and function in addition to various other techniques covering a wide variety of topics.

The progress made by the investigators working on steroid hormone action in the early 1960's appears to have formed much of the basis for development of the "hormone action" field. Preparation of radioactive tracers which could be used as probes for intracellular hormone action and the realization that steroid hormones control nucleic acid and protein synthesis in target cells provided the foundation for the growth of interest in this area.

This volume contains a compilation of techniques, and as such is intended to provide a methodological reference for studies of steroid hormone action. Thyroid hormones are also included in this volume because of their apparent similarities in regard to molecular mechanism of action. Considerations are provided concerning the manner in which hormones circulate in the blood stream and are eventually sequestered in target cells. This leads to description of their interactions with cytoplasmic and nuclear receptors as well as presentation of techniques for purification of the steroid hormone receptor molecules. Methods dealing with the biochemical and biological processes stimulated by steroid hormones are included in addition to methods for assessing hormone metabolism. Although a good deal of work and interest in this field has been at the level of gene transcription and nucleic acid synthesis, such methods have been and will continue to be dealt with in other volumes of this series. Although some overlap exists between contributions to this volume, it is our opinion that investigators should have a choice of laboratory approaches in cases where some controversy exists.

Omissions have inevitably occurred—some because potential authors were overcommitted, some because of editorial oversight, some because of the timing of new developments relative to the publication deadline. Some apparent omissions have been covered in previous volumes of "Methods in Enzymology."

We thank Drs. S. P. Colowick and N. O. Kaplan who originated the idea for and encouraged the compilation of this volume. We thank the staff of Academic Press for their help and advice. We especially thank the contributing authors for their patience and full cooperation and for carrying out the research that made this volume possible.

BERT W. O'MALLEY
JOEL G. HARDMAN

METHODS IN ENZYMOLOGY

EDITED BY

Sidney P. Colowick and Nathan O. Kaplan

VANDERBILT UNIVERSITY
SCHOOL OF MEDICINE
NASHVILLE, TENNESSEE

DEPARTMENT OF CHEMISTRY
UNIVERSITY OF CALIFORNIA
AT SAN DIEGO
LA JOLLA, CALIFORNIA

I. Preparation and Assay of Enzymes
II. Preparation and Assay of Enzymes
III. Preparation and Assay of Substrates
IV. Special Techniques for the Enzymologist
V. Preparation and Assay of Enzymes
VI. Preparation and Assay of Enzymes (*Continued*)
Preparation and Assay of Substrates
Special Techniques
VII. Cumulative Subject Index

METHODS IN ENZYMOLOGY

EDITORS-IN-CHIEF

Sidney P. Colowick   Nathan O. Kaplan

Volume XX. Nucleic Acids and Protein Synthesis (Part C)
Edited by Kivie Moldave and Lawrence Grossman

Volume XXI. Nucleic Acids (Part D)
Edited by Lawrence Grossman and Kivie Moldave

Volume XXII. Enzyme Purification and Related Techniques
Edited by William B. Jakoby

Volume XXIII. Photosynthesis (Part A)
Edited by Anthony San Pietro

Volume XXIV. Photosynthesis and Nitrogen Fixation (Part B)
Edited by Anthony San Pietro

Volume XXV. Enzyme Structure (Part B)
Edited by C. H. W. Hirs and Serge N. Timasheff

Volume XXVI. Enzyme Structure (Part C)
Edited by C. H. W. Hirs and Serge N. Timasheff

Volume XXVII. Enzyme Structure (Part D)
Edited by C. H. W. Hirs and Serge N. Timasheff

Volume XXVIII. Complex Carbohydrates (Part B)
Edited by Victor Ginsburg

Volume XXIX. Nucleic Acids and Protein Synthesis (Part E)
Edited by Lawrence Grossman and Kivie Moldave

Volume XXX. Nucleic Acids and Protein Synthesis (Part F)
Edited by Kivie Moldave and Lawrence Grossman

Volume XXXI. Biomembranes (Part A)
Edited by Sidney Fleischer and Lester Packer

Volume XXXII. Biomembranes (Part B)
Edited by Sidney Fleischer and Lester Packer

Volume XXXIII. Cumulative Subject Index Volumes I–XXX
Edited by Martha G. Dennis and Edward A. Dennis

Methods in Enzymology

Volume XXXVI

HORMONE ACTION

Part A

Steroid Hormones

Section I

Hormone-Binding Proteins and Assays for Steroid Hormones

[1] Theory of Protein–Ligand Interaction

By D. RODBARD and H. A. FELDMAN

The concentration of "free" (unbound) hormones (e.g., steroids, thyroxine) and drugs in plasma is "buffered" by their interactions with plasma proteins. The initial step in the action of both steroid and protein hormones is binding to a receptor. The hormone–receptor complex may, in turn, bind to other proteins and/or nucleic acids. Thus, the theory of protein–ligand interaction is central to evaluation of the mechanism of hormone action. It appears that the theory of protein–ligand interaction previously developed in connection with the enzyme–substrate, antigen–antibody, drug–receptor, DNA–RNA hybridization, and oxygen–hemoglobin reactions will suffice for description of most hormone–protein interactions (see Rodbard[1] for recent review).

The basic problem is to describe quantitatively the ligand–protein interaction in terms of a biochemical (and the corresponding mathematical) model. We wish to estimate the affinity constant (or constants), molar concentration of binding sites (binding capacities), or, alternatively, the number of binding sites per molecule of protein and the reaction rate constants. Further, we seek to develop criteria to select the best from among a series of closely related models.

We shall now present a review of the more important, popular, simple models and the corresponding equations which are most frequently needed. We begin with very simple models and progress to consider relatively more complex (and realistic) ones.

Equilibrium Models

Two Parameter. The most elementary model of protein–ligand interaction is to assume that both species are homogeneous and univalent.

$$P + Q \rightleftharpoons PQ \tag{1}$$

This reaction is described by an association rate constant (k), a dissociation rate constant (k' or k_{-1}), and the equilibrium constant of association.

$$K = k/k' = \frac{[PQ]}{[P][Q]} \tag{2a}$$

[1] D. Rodbard, *in* "Receptors for Reproductive Hormones" (B. W. O'Malley and A. R. Means, eds.), p. 289–326. Plenum, New York, 1973.

Combining the expression for K with the expressions for conservation of mass for P and Q,

$$\begin{aligned} p &= [P] + [PQ] \\ q &= [Q] + [PQ] \end{aligned} \tag{2b}$$

and the definition of the bound-to-free ratio,

$$R \equiv B/F \equiv [PQ]/[P] \tag{2c}$$

one obtains the Scatchard plot[2] which is analogous to the Eadie[3] or the Hofstee[4] plot of enzymology.

$$R \equiv B/F = K[Q] = K(q - B) \tag{3}$$

In this simple, ideal case, a plot of $R = B/F$ vs $B = [PQ]$ produces a straight line with slope $-K$ and an intercept on the bound-axis equal to q. [Our use of the term Scatchard plot is analogous but not identical with the original method of Scatchard, which involves dividing both sides of Eq. (3) by the molar concentration of the binding protein. We adapt this approach, since in many circumstances the molarity of the binding protein is unknown. Thus, our expression for "bound," B, has dimensions of concentration, rather than being a dimensionless number (of binding sites filled per molecule). Note that the $\bar{\nu}$ of Scatchard corresponds to $[PQ]/q$ in our nomenclature, for the case of the univalent binding protein. The c of Scatchard corresponds to our Free $= F = [P]$. In the case of a multivalent binding protein, our q represents the total concentration of sites and not the molar concentration of the binding protein.]

However, when interested in the relationship between B/F (or F/B or B/T) and the *free* ligand concentration, $F = [P]$, one uses[5]

$$R = B/F = \frac{Kq}{1 + KF} \tag{4}$$

The relations between B/F, F/B, or B/T and *total* ligand concentration, Total $= T = p$, are given by[6]

$$R^2 + R(1 + Kp - Kq) - Kq = 0 \tag{5a}$$

[2] G. Scatchard, *Ann. N.Y. Acad. Sci.* **51,** 660 (1949).

[3] G. S. Eadie, *J. Biol. Chem.* **146,** 85 (1942).

[4] B. H. J. Hofstee, *Science* **116,** 329 (1952).

[5] J. T. Edsall and J. Wyman, "Biophysical Chemistry," Vol. 1, p. 610ff. Academic Press, New York, 1958.

[6] R. P. Ekins, G. B. Newman, and J. L. H. O'Riordan, *in* "Radioisotopes in Medicine: In Vitro Studies" (R. L. Hayes, F. A. Goswitz, and B. E. P. Murphy, eds.), p. 59. U.S. At. Energy Comm., Oak Ridge, Tennessee, 1968.

$$(F/B)^2 + (F/B)\left(1 - \frac{p}{q} - \frac{1}{Kq}\right) - \frac{1}{Kq} = 0 \qquad (5b)$$

$$(B/T)^2 - (B/T)\left(1 + \frac{q}{p} + \frac{1}{Kp}\right) + \frac{q}{p} = 0 \qquad (5c)$$

Usually, Eq. (5a) is the best form to use, since it is well behaved as either p or q approach zero. After calculating B/F, one can then calculate F/B, B/T, etc., using the convenient forms

$$F/B = [P]/[PQ] = \frac{1}{B/F} = \frac{1}{R} \qquad (6a)$$

$$B/T = \frac{[PQ]}{p} = \frac{B/F}{1 + B/F} \qquad (6b)$$

$$[PQ] = p[B/T] \qquad (6c)$$

$$[P] = p - [PQ] = \frac{p}{1 + R} \qquad (6d)$$

$$[Q] = q - [PQ] \qquad (6e)$$

Thus, it is possible to calculate the concentrations of all three species present ($[P]$, $[Q]$, $[PQ]$) when given K, p, and q.

"One on Two" or Four Parameter Model. When dealing with two distinct and independent, noninteracting species of binding sites, e.g.,

$$\begin{aligned} P_1 + Q_1 &\rightleftharpoons P_1Q_1 \\ P_1 + Q_2 &\rightleftharpoons P_1Q_2 \end{aligned} \qquad (7)$$

with two distinct affinity constants, K_{11}, K_{12}, the Scatchard plot of B/F vs. bound, where

$$\begin{aligned} B/F &= \frac{[P_1Q_1] + [P_1Q_2]}{[P_1]} \\ \text{Bound} = B &= [P_1Q_1] + [P_1Q_2] \end{aligned} \qquad (8)$$

is now a hyperbola, given by[7]

$$\begin{aligned} R = \tfrac{1}{2}(K_{11}(q_1 - B) + K_{12}(q_2 - B) \\ + \{[K_{11}(q_1 - B) - K_{12}(q_2 - B)]^2 + 4K_{11}K_{12}q_1q_2\}^{1/2}) \end{aligned} \qquad (9)$$

Graphical methods for fitting this hyperbola are discussed by Berson and Yalow[8] and Feldman.[7]

[7] H. A. Feldman, *Anal. Biochem.* **48**, 317 (1972).
[8] S. A. Berson and R. S. Yalow, *J. Clin. Invest.* **38**, 1996 (1959).

The B/F ratio is related to the free ligand concentration as[5]

$$B/F = \frac{K_{11}q_1}{1 + K_{11}F} + \frac{K_{12}q_2}{1 + K_{12}F} \tag{10}$$

It is possible to calculate B/F "parametrically" as a function of total ligand concentration, using either Eq. (9) or Eq. (10). Procedure: Specify amount bound; calculate B/F; then calculate total from bound and B/F. Alternatively, specify free, calculate B/F, and then calculate total:

$$B/F = fn(B) \qquad \text{[Eq. (9)]}$$

or

$$B/F = fn(F) \qquad \text{[Eq. (10)]}$$

$$p = B\left(\frac{1 + R}{R}\right) \qquad p = F(1 + R). \tag{11}$$

Three Parameter Model. In many cases, the second class of sites may appear to be present with unlimited, nonsaturable, or "infinite" binding capacity but "zero" affinity, thus producing a horizontal asymptote on the Scatchard plot, so that $K_{12}q_2$ remains finite (designated K_3N_3 by Baulieu and Raynaud[9] or K_3 by Rodbard[1]) whereas terms involving K_{12} or $1/q_2$ vanish. Then

$$B/F = \frac{K_{11}q_1}{1 + K_{11}F} + K_3 \tag{12}$$

and the corresponding Scatchard plot is described by[10]

$$R = \tfrac{1}{2}(K_{11}(q_1 - B) - K_3 + \{(K_{11}B - K_{11}q_1 - K_3)^2 + 4K_{11}K_3B\}^{1/2}) \tag{13}$$

The "dose response curve" analogous to Eq. (5a) becomes[10]

$$R^2 + R(1 + K_{11}p - K_{11}q_1 - K_3) - K_{11}q_1 - K_3(1 + K_{11}p) = 0 \tag{14}$$

Five Parameter Model. If we are dealing with two distinct classes or orders of saturable sites, and, in addition, there is a third class of sites with infinite capacity but zero affinity (as in the three parameter case), then[9]

$$R = \frac{K_{11}q_1}{1 + K_{11}F} + \frac{K_{12}q_2}{1 + K_{12}F} + K_3 \tag{15}$$

when all three classes of sites are noninteracting. The importance of this model [especially Eq. (15)] and the importance of the K_3 term was pointed

[9] E. E. Baulieu and J.-P. Raynaud, *Eur. J. Biochem.* **13**, 293 (1970).

[10] R. E. Bertino, personal communication.

out by Baulieu and Raynaud.[9] Without this term (i.e., in the two and especially the four parameter models), many parameter-fitting programs will frequently fail to converge, as $K_{12} \rightarrow 0$ and $q_2 \rightarrow \infty$. In this case, the Scatchard plot is given by a cubic of R in terms of B or a quadratic of B in terms of R.[11]

$$R^3 + R^2[K_{11}(B - q_1) + K_{12}(B - q_2) - K_3] + R[K_{11}K_{12}B(B - q_1q_2) - K_3B(K_{11} + K_{12})] - K_{11}K_{12}K_{13}B^2 = 0 \tag{16}$$

The above expressions are useful in plotting and curve-fitting for the two, three, four, and five parameter cases.

One Ligand – m Classes of Binding Sites. The next obvious generalization is to consider any arbitrary number of classes or orders of binding sites.[5,6] Here, the expressions defining the affinity constants and the statements for conservation of mass may be combined into one equation, similar to Eqs. (4), (10), and (15):

$$B/F = \frac{\sum_{j=1}^{m} [P_1Q_j]}{[P_1]} = \sum_{j=1}^{m} \frac{K_{1j}q_j}{1 + K_{1j}F} \tag{17}$$

(again, assuming no interaction between sites).

These equations may be used to calculate B/F and, in turn, the complete chemical composition of a reaction system of equilibrium, when p, all m q_j values, and all m K_{1j} values are known.

Two Ligands – One Binding Site. When dealing with two different ligands (e.g., labeled and unlabeled species, drug and antagonist, two closely related congeners) reacting with the same binding site, then the bound-to-free ratio for one of the ligands (e.g., P_1) is related to the amount of that ligand bound to one of the binding proteins (B_1) as Eq. (18) of Feldman.[7]

$$R_1 = \frac{K_{11}}{2K_{21}}(-1 - K_{21}(B_1 + p_2 - q_1) + \{[1 - K_{21}(B_1 + p_2 - q_1)]^2 + 4K_{21}p_2\}^{1/2}) \tag{18}$$

The amount bound, $B_1 = [P_1Q_1]$, does not represent the total concentration of all binding sites filled. This plot differs from the "apparent" Scatchard plot which one would erroneously obtain by plotting R_1 versus

[11] G. R. Frazier, personal communication.

the amount presumed bound if P_1 and P_2 were assumed to have the same bound-to-free ratio.

$$\text{Apparent bound} = (p_1 + p_2)\left(\frac{R_1}{1 + R_1}\right) \tag{19}$$

R_1 may be expressed in terms of total ligand concentration as [Eq. (31) of Ekins[6] in slightly modified nomenclature]

$$R_1^2 + R_1(1 + K_{11}p_1 - K_{11}q_1) - K_{11}q_1 + \frac{p_2K_{21}(R_1 + 1)R_1}{(K_{21}/K_{11})R_1 + 1} = 0 \tag{20}$$

The "n × m" Case. When dealing with n ligands, each reacting simultaneously with m classes of binding sites, we have $n \times m$ reactions of the type

$$P_i + Q_j \rightleftharpoons P_iQ_j \tag{21}$$

This set of reactions is described by the set of $(nm + n + m)$ simultaneous equations[12]

$$K_{ij} = \frac{[P_iQ_j]}{[P_i][Q_j]}, \tag{22a}$$

$$p_i = [P_i] + \sum_{j=1}^{m} [P_iQ_j], \tag{22b}$$

$$q_j = [Q_j] + \sum_{i=1}^{n} [P_iQ_j] \tag{22c}$$

These may be condensed to n equations,[12] which are quite similar to Eqs. (5), (10), (15), and (17).

$$R_i = \frac{p_i}{[P_i]} - 1 = \sum_{j=1}^{m} \frac{K_{ij}q_j}{1 + \sum_{a=1}^{n} K_{aj}[P_a]} \tag{23}$$

By virtue of symmetry of P and Q in Eq. (21), we can use either n or m simultaneous equations, whichever is smaller (either P or Q may be regarded as the ligand). This is quite important computationally: e.g., for a 6×2 system, we need to invert a 2×2 matrix instead of a 20×20 matrix!

[12] H. Feldman, D. Rodbard, and D. Levine, *Anal. Biochem.* **45**, 530 (1972).

This family of equations can be solved to any desired precision by numerical methods (e.g., the Newton-Raphson method) to provide a complete description of all $(nm + n + m)$ chemical moieties present when the $n \times m$ K values and the total concentrations (p's, q's) are given.[12] Conversely, given an appropriate set of data, one can calculate the parameters (K_{ij}'s, q_j's) which best describe the system for any given model (n and m).[7] Computer programs for this purpose are available on request from the authors.[7,12a]

Multivalency and Stepwise Binding Reactions at Equilibrium

Until now, we have assumed that all "classes" or "orders" of sites are available and presumably reacting simultaneously, and that all species were univalent. Another general class of models, involving a single type of ligand but n classes of binding sites, is the series of sequential or stepwise reactions[5,13]:

$$P \rightleftharpoons QP \rightleftharpoons Q_2P \rightleftharpoons \cdot \cdot \cdot \cdot \rightleftharpoons Q_{n-1}P \rightleftharpoons Q_nP \tag{24a}$$

(The free or unbound Q's are not shown for simplicity.)

Here we change our nomenclature: P represents the multi- (n-) valent binding protein; Q represents the ligand, and corresponds to L of Edsall and Wyman[5] or A of Fletcher *et al.*[13]; the subscript indicates the stoichiometry and class of binding site and not the type of ligand. Then

$$K_1 = \frac{[QP]}{[Q][P]}$$

$$K_2 = \frac{[Q_2P]}{[QP][P]}$$

$$\cdot \qquad \cdot$$

$$\cdot \qquad \cdot$$

$$\cdot \qquad \cdot$$

$$K_n = \frac{[Q_nP]}{[Q_{n-1}P][P]} \tag{24b}$$

represent the macroscopic, stepwise equilibrium constants. This model includes the "$1 \times m$" model [Eq. (17) above] at equilibrium, but the two models lead to different predictions in terms of the kinetics of the

[12a] D. Rodbard and G. R. Frazier, "Radioimmunoassay Data Processing, 2nd ed.," PB217367. Nat. Tech. Infor. Serv., U.S. Dept. of Commerce, Springfield, Virginia, 1973.

[13] J. E. Fletcher, A. A. Spector, and J. D. Ashbrook, *Biochemistry* 9, 4580 (1970).

transient state of the reaction. This model provides one of the simplest though still a completely general model of cooperative binding phenomena and/or "allosteric" effects. No assumptions are made concerning the changes in conformational state of the protein due to binding of ligand. The properties of this model and its interconvertibility with the model of Scatchard and others have been discussed.[13–15] The relationship of this model to those of Monod-Changeau-Wyman[16] and of Koshland[17] and Frieden is considered in Magar *et al.*[17a] The ratio of bound ligand to total macromolecule, i.e., the average number of ligand molecules bound to each molecule of binding protein, is given by[5,13]

$$\bar{\nu} = \frac{K_1[Q] + 2K_1K_2[Q]^2 + \cdots + nK_1K_2 \cdots K_n[Q]^n}{(1 + K_1[Q] + K_1K_2[Q]^2 + \cdots + K_1K_2 \cdots K_n[Q]^n)} = \frac{\sum_{i=1}^{n} i[Q]^i \prod_{j=1}^{i} K_j}{1 + \sum_{i=1}^{n} [Q]^i \prod_{j=1}^{i} K_j} \tag{25}$$

The bound-to-free ratio for *ligand Q* is

$$R_q = \frac{\bar{\nu}}{[Q]} \cdot p \tag{26}$$

where p is the molarity of the binding protein in this case and not the total number of binding sites. For discussion of particular cooperative models, one should see Fletcher *et al.*,[14] Monod *et al.*,[16] and Koshland *et al.*[17] However, the stepwise model should be adequate for most purposes.[13,14,17a]

Sips and Hill Plot

The preceding models are based on the concept of the presence of discrete classes or orders of sites. Another approach is to regard the

[14] J. E. Fletcher and J. D. Ashbrook, *Ann. N.Y. Acad. Sci.* **226,** 69 (1973).

[15] R. I. Shrager, "MODELAIDE: A Computer Graphics Program for the Evaluation of Mathematical Models," Tech. Rep. No. 5. Div. Comput. Res. Technol., National Institutes of Health, U.S. Dept. of Health, Education and Welfare, Washington, D.C., 1970.

[16] J. Monod, J. Wyman, and J.-P. Changeux, *J. Mol. Biol.* **12,** 88 (1965).

[17] D. E. Koshland, G. Némethy, and D. Filmer, *Biochemistry* **5,** 365 (1966).

[17a] M. E. Magar, "Data Analysis in Biochemistry and Biophysics," Chapter 13, p. 411ff. Academic Press, New York, 1972; K. Kirschner, E. Gallego, I. Schuster, and D. Goodall, *J. Mol. Biol.* **58,** 29 (1971); K. Kirschner, *Curr. Top. Cell. Regul.* **4,** 167 (1971).

affinity of the ligand for the receptor(s) or the protein as a continuously distributed variable, e.g., with a Gaussian distribution for $\log(K)$, i.e., a log-normal distribution for K. The behavior of these systems may be calculated directly, as by use of the above models [e.g., Eq. (17) for the "$1 \times m$" case] to approximate any desired frequency distribution. When $\log(K)$ is normally distributed, then $\sigma_{\text{eog}(K)}$ becomes a measure of "heterogeneity" and K_0 represents the median and modal K value.[18]

In lieu of the use of the log-normal distribution, which leads to some unwieldly results, one can utilize the Sips distribution.[19] Then the binding equations may be put in the form

$$\frac{1}{B} = \frac{1}{K^a q[P]^a} + \frac{1}{q} \tag{27}$$

(We return to use of Q as the binding protein, P as the ligand.)

If $a = 1$, this is equivalent to the Lineweaver-Burk equation.[20] In the original application of the Sips distribution to hapten–antibody reactions, Nisonoff and Pressman[21] recommended the use of a plot of

$$\frac{1}{B} \text{ vs. } \frac{1}{F^a}$$

for various values of a. (When $a = 1$, there is no apparent heterogeneity. The lower the value of a, the greater the apparent degree of heterogeneity of the K values of the binding sites.)

However, most workers prefer to rearrange this equation in the form[5]

$$\text{logit}\left(\frac{B}{q}\right) = \log\left(\frac{B}{q - B}\right) = a \log(K) + a \log[P] \tag{28a}$$

or

$$B = \frac{q}{1 + \left(\dfrac{1}{K[P]}\right)^a} \tag{28b}$$

Thus, there is a linear relationship between the log of the bound-to-free ratio for the antibody (or receptors or binding protein), which corresponds to the logit of the fraction of antibody sites filled, and the log of the concentration of free (not total) ligand.

A value of $a < 1$ here corresponds to a Scatchard plot which is concave (multiple of orders of binding sites, and/or "negative cooperativity").

[18] L. Pauling, D. Pressman, and A. L. Grossberg, *J. Amer. Chem. Soc.* **66**, 784 (1944); F. Karush, *ibid.* **72**, 2705 (1950).

[19] R. Sips, *J. Chem. Phys.* **16**, 490 (1948).

[20] H. Lineweaver and D. Burk, *J. Amer. Chem. Soc.* **56**, 65 (1934).

[21] A. Nishonoff and D. Pressman, *J. Immunol.* **81**, 126 (1958).

A value of $a > 1$ (usually in the presence of nonlinearity of the Hill or Sips plot) is suggestive of "positive" cooperativity. Note that the Sips plot of immunology, as described here, and the Hill plot of enzymology are directly analogous aside from the change in nomenclature, if we assume that "velocity" is analogous to concentration of bound complex B, and maximal velocity is analogous to binding capacity q.

Kinetic Models

The time course of the reaction shown in Eq. (1) can be described by

$$\frac{d[PQ]}{dt} = k[P][Q] - k'[PQ]$$
$$\frac{d[P]}{dt} = \frac{d[Q]}{dt} = -\frac{d[PQ]}{dt} \tag{29}$$

The analytical solution to these equations is given by Vassent and Jard[22] and by Rodbard and Weiss.[23]

Further, if a fraction of the ligand is "labeled" (e.g., with radioisotope) without affecting the rate constants k or k', the analytical solution for the concentrations of all species, viz., $[P]$, $[P^*]$, $[PQ]$, $[P^*Q]$, and $[Q]$, as a function of time is available.[22,23] Note: when multiple cross-reacting or multivalent species are present, then we can use numerical integration to describe the time course of the reactions.[24,25]

Methods for estimation of the association and dissociation rate constants are discussed elsewhere (e.g., see Rodbard[1]). When there are multiple classes of sites, but only a single species of ligand, one can estimate the various k' values by permitting the ligand and binding protein(s) to react (nearly) to equilibrium, and then remove all the free ligand. The amount of ligand bound as a function of time is a linear combination of negative exponential terms.

The multiple orders of k' values can be estimated as the parameters of the equation

$$\text{Bound} = \sum_{j=1}^{m} A_j e^{-k'_j t} \tag{30}$$

where A_j corresponds to $[P_1Q_j]_0$, and m itself is unknown.

[22] G. Vassent and S. Jard, *C. R. Acad. Sci.* **272,** 880 (1971).
[23] D. Rodbard and G. H. Weiss, *Anal. Biochem.* **52,** 10 (1973).
[24] D. Rodbard, H. J. Ruder, J. Vaitukaitis, and H. S. Jacobs, *J. Clin. Endocrinol. Metab.* **33,** 343 (1971)
[25] K. Brown-Grant, R. D. Brennan, and F. E. Yates, *J. Clin. Endocrinol. Metab.* **30,** 733 (1970).

As with the equilibrium models, it is essential to try the "next-most-complicated" and the "next-most-simple" models, involving one or two additional or fewer parameters. Then one should use statistical methods to evaluate which model appears to give the best "fit." The estimation of k and k' (together with q) for even a single reaction is a challenging problem in terms of statistical curve-fitting. For homogeneous systems, one can perform the transient-state and the equilibrium curve-fitting simultaneously.[1] In the presence of multiple classes of sites, precise estimation of the various k and k' values often becomes formidable. Then it becomes necessary to purify the protein (receptors, etc.) to "homogeneity" (by binding criteria) to obtain accurate estimates of rate constant. The value of the ratio k/k' obtained by kinetic analysis should be compared with the K value from analysis of equilibrium models.

Only by analysis of the time course of the reaction(s) can we hope to distinguish between the simultaneous reaction model and the sequential or stepwise model, particularly in the case of negative cooperativity. Likewise, kinetic data will be needed to discriminate between the various models proposed for allosteric effects. However, the models are quite similar, both algebraically and conceptually. Thus, it is almost impossible for data of the highest quality to make this discrimination. Also, we must remember that kinetic data alone (including both equilibrium or steady state and non-steady state) cannot be used as "proof" of reaction schemes. However, it can provide proof of a negative variety in the ability to exclude certain models.

Notes on Parameter-Fitting

Recent reviews of the problems and pitfalls or parameter-fitting for various models are available.[1,13,14,26] Here we only comment as follows.

(1) Parameter-fitting can and should be performed by general nonlinear curve-fitting routines and packages now available at nearly every major computer center (e.g., see Shrager[15] and others[27–29]). Alternatively, special purpose programs and packages are available (e.g., Rodbard and

[26] J. G. Reich, G. Wangerman, M. Falck, and K. Rohde, *Eur. J. Biochem.* **26,** 368 (1972)

[27] G. D. Knott and D. K. Reece, "Proceedings, On-Line '72 Symposium." Vol. 1, p. 497. Brunel University, Uxbridge, Middlesex, England, 1972.

[28] M. Berman and M. P. Weiss, "Users Manual for SAAM (Simulation, Analysis and Modeling)," Version SAAH25. Public Health Service, National Institutes of Health, U.S. Dept. of Health, Education and Welfare, NIAMD, 1971; M. Berman, E. Shahn, and M. F. Weiss, *Biophys. J.* **2,** 275 (1962).

[29] P. F. Sampson, "BMDX85." UCLA Computer Center, Los Angeles, California, 1970; R. I. Jennrich and P. F. Sampson, *Technometrics* **10,** 63 (1968).

Frazier[12a] and Reich *et al.*[26]). This is distinctly preferable to the use of simple unweighted linear regression, which is commonly incorrectly applied to the linearized forms of the binding equations, such as the Scatchard, Lineweaver-Burk, Hill, Sips plots, etc.[1] This is to avoid (or reduce) the problems of nonuniformity of variance and of correlation of errors in the dependent and independent variables, which are usually introduced by these linearizing transforms. Whenever possible, the curve-fitting should be performed in terms of the variables which were actually measured directly. The subtraction of "blanks" and nonspecific counts prior to curve-fitting may introduce systematic errors, which may lead to spurious, fallacious interpretation in terms of "cooperativity" and change the overall behavior of the system. [The Newton-Raphson, conjugate-gradient, and Marquardt-Levenberg methods have all been used successfully, although the latter is probably best in general in terms of stability and speed of convergence. Also, the Nelder-Mead method (and minor modifications thereof) have been used successfully.]

(2) After the curve-fitting procedure, the goodness-of-fit should be evaluated both graphically and by appropriate statistical methods (e.g., examination of number of residual sign changes, a plot of the residuals versus the position on the curve). The curves should be plotted in terms of the variables which were actually measured. In addition, they may be plotted in one or more of the coordinate systems which provide linearization (e.g., the Scatchard plot), together with a plot of the individual "components."[1]

(3) Almost without exception, it is necessary to use a weighted regression or curve-fitting procedure. If an unweighted procedure is used, the onus of proof is on the investigator to show that there is uniformity of variance. The weighting function is the reciprocal of the variance of the dependent variable.

$$w = \frac{1}{\sigma_y^{2}}$$

In general, one should measure σ_y^2 at several points on the curve, and then describe σ_y^2 as a smooth (e.g., quadratic) function of y (or of x).

(4) It is necessary that the independent variable be error-free, or that errors in the dependent and independent variables are independent.

(5) Confidence limits should be stated for all parameters (K's, etc.), even when these are approximate. Monte-Carlo techniques can be used to estimate the errors and the correlation between the errors of the parameters. An empirical estimate of between-experiment error for the parameters should also be obtained.

(6) Before selecting any model, it is necessary to compare its perfor-

mance with the "next-most-simple" and the "next-most-complicated" models, involving one (or two) less or one (or two) more parameters.

A Practical Guide

(1) Study time course of binding of ligand to protein at several widely spaced concentrations of ligand and receptor. Verify results by use of at least two different methods to separate bound and free ligand.[30] Estimate time necessary to achieve equilibrium.

(2) Perform dissociation studies: estimate k' (or several k' values).

(3) Perform "equilibrium" binding studies (measure B/F as fraction of total). Construct a Scatchard plot. Estimate K and q values and compare various models.

(4) Using "q" value(s) from the equilibrium studies, recalculate the association and dissociation rate constant(s) using data of steps 1 and 2.

(5) If the Scatchard plot appears linear (on testing by formal statistical analysis), one can use the data of step 1 and Eq. (29) to estimate k, k', and q simultaneously.

(6) Construct joint confidence limits for parameters, e.g., K_{11} and q_1, by Monte-Carlo methods or by studying the magnitude of the residual variance for various combinations of parameters.

(7) The Sips or Hill plot should be constructed. This is quite useful for detecting cooperativity and heterogeneity. If cooperativity is present by this criterion (or if the Scatchard plot is convex), then proceed with the stepwise model or one of the allosteric models.

(8) Repeat the binding studies at different temperatures. Relate k values to temperature by the Arrhenius equation.

(9) Test stability of reagents; metabolic degradation of either ligand or receptor (as by proteolytic enzymes) can have disastrous effects on the apparent K and k values. Further, no satisfactory mathematical model is available to describe these effects.

Summary and Conclusions

The theory of ligand–protein interaction has been extensively developed. This permits description of both the equilibrium and the nonequilibrium behavior of reaction systems involving any number of ligands and any number of orders or classes of "proteins" or receptor sites), with or without "interaction" between these sites. However, in practice, it is difficult to obtain data with sufficient precision, freedom from system-

[30] D. Rodbard and K. J. Catt, *J. Steroid Biochem.* 3, 255 (1972).

atic errors, and number of observations to permit (a) calculation of parameters and testing goodness-of-fit for a given model and (b) selection among competing models. Appropriate methods of statistical analysis must be applied to extract maximal information from the data of binding studies.

Acknowledgments

R. E. Bertino provided Eqs. (13) and (14). R. C. Rodgers provided Eq. (5c). G. R. Frazier provided Eq. (16). J. E. Fletcher, J. D. Ashbrook, and H. A. Saroff provided stimulating discussions and valuable critical reviews of the manuscript.

[2] Use of Specific Antibodies for Quantification of Steroid Hormones

By G. D. NISWENDER, A. M. AKBAR, and T. M. NETT

Since the first reports in 1969 of the use of antibodies for the quantification of testosterone[1] and estradiol,[2,3] radioimmunoassays have been developed for nearly every major steroid hormone, and the procedures have been described in detail.[4,5] The broad application of this procedure is based on the unique potential for specificity with antibodies and the lack of specific binding proteins for a number of physiologically important steroid hormones. The ability of nonradioactive antigen to compete with radioactive antigen for the available binding sites on the antibody molecule is the theoretical basis of radioimmunoassay. If the number of antibody binding sites and the amount of radioactive antigen are kept constant, then the amount of radioactivity associated with the antibody is a quantitative function of the mass of nonradioactive antigen present in the assay tube. The theoretical aspects of radioimmunoassays have been discussed in detail in several publications.[4-7]

[1] G. D. Niswender and A. R. Midgley, Jr., *Proc. Endocrine Soc., 51st* Abstract No. 22 (1969).

[2] G. E. Abraham, *J. Clin. Endocrinol. Metab.* **29**, 866 (1969).

[3] A. R. Midgley, Jr., G. D. Niswender, and S. Ram, *Steriods* **13**, 731 (1969).

[4] F. G. Péron and B. V. Caldwell, eds., "Immunologic Methods in Steroid Determination." Appleton, New York, 1970.

[5] E. Diczfalusy, *Acta Endocrinol.* (*Copenhagen*), *Suppl.* **147** (1970).

[6] E. Diczfalusy, *Acta Endocrinol.* (*Copenhagen*), *Suppl.* **142** (1969).

[7] W. D. Odell and W. H. Daughaday, eds., "Principles of Competitive Protein Binding Assays." Lippincott, Philadelphia, Pennsylvania, 1971.

Development of Antisera

Methods for Conjugation of Steroid Molecules to Protein

In general, steroid hormones, by themselves, are nonantigenic but are effective in eliciting antibody formation when used as haptens covalently linked to large protein molecules. The same general approach has been used to develop antibodies to a variety of small molecules including nucleotides,[8,9] vasopressin,[10] digitoxin,[11] penicillin,[12] morphine,[13] angiotensin,[14] thyrotropin-releasing hormone,[15] and gonadotropin-releasing hormone.[16]

Derivative Formation. The most widely used procedures for covalently linking steroid molecules to proteins have been those originally described by Erlanger *et al.*[17] These procedures involve the formation of a derivative of the steroid which terminates in a reactive group such as carboxylic acid. When the steroid molecule is to be conjugated through a ketone substituent an oxime derivative is usually prepared, whereas if conjugation is to be done through a hydroxyl group either a hemisuccinate or a chlorocarbonate derivative is made. Although the exact procedures for the formation of steroid derivatives depend on the individual steroid molecule of interest, the following general discussion should be useful to those who wish to develop specific antisera to steroid hormones.

There are several considerations for making oxime derivatives of steroids. The usual procedure is to reflux *O*-(carboxymethyl) hydroxylamine with the steroid in alkalinized ethanol for 3 hours. The reaction product is collected by acidification with HCl to pH 2.0 and extracted with di-

[8] B. F. Erlanger, D. Senitzer, O. J. Miller, and S. M. Beiser, *Acta Endocrinol. (Copenhagen), Suppl.* **168** (1972).

[9] A. L. Steiner, C. W. Parker, and D. M. Kipnis, *Biochem. Psychopharmacol.* **3,** 89 (1970).

[10] M. Miller and A. M. Moses, *Endocrinology* **84,** 557 (1969).

[11] G. C. Oliver, Jr., B. M. Parker, D. L. Brasfield, and C. W. Parker, *J. Clin. Invest.* **47,** 1035 (1968).

[12] C. W. Parker, *in* "Methods in Immunology and Immunochemistry" (C. A. Williams and M. W. Chase, eds.), p. 133. Academic Press, New York, 1967.

[13] B. Berkowitz and S. Spector, *Science* **178,** 1290 (1972).

[14] T. L. Goodfriend, *in* "Methods in Investigative and Diagnostic Endocrinology" (S. A. Berson and R. S. Yalow, eds.). In press.

[15] R. M. Bassiri and R. D. Utiger, *Endocrinology* **90,** 722 (1972).

[16] T. M. Nett, A. M. Akbar, G. D. Niswender, M. T. Hedlund, and W. F. White, *J. Clin. Endocrinol. Metab.* **36,** 880 (1973).

[17] B. F. Erlanger, F. Borek, S. M. Beiser, and S. Lieberman, *J. Biol. Chem.* **228,** 713 (1958).

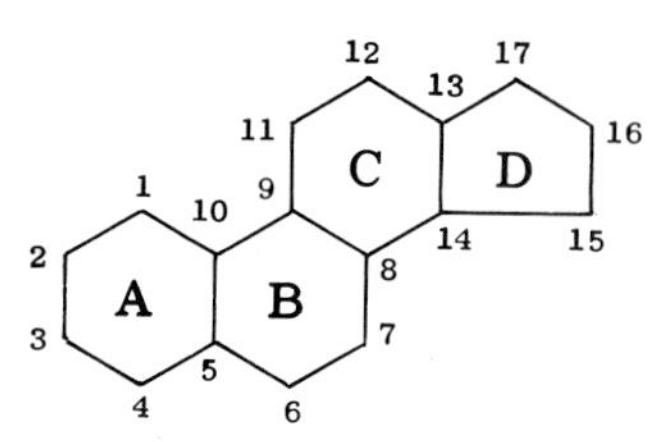

FIG. 1. The numbering sequence for the 17 carbon atoms and the letter designations for the four rings in the cyclopentano-perhydro-phenanthrene nucleus which is the basic structure of all steroid hormones.

ethylether for crystallization. If the steroid contains a single ketone substituent (i.e., testosterone) then conditions should be used which optimize the yield of derivative. Ratios of 4 to 6 moles of *O*-(carboxymethyl) hydroxylamine per mole of steroid optimize yield. However, other considerations become important if there is more than one ketone substituent present on the steroid molecule (i.e., progesterone). One can derivatize a related steroid which has a single ketone group (i.e., 20β-hydroxypregn-4-en-3-one or pregnenolone) using conditions which optimize yields and then chemically convert this molecule to the desired steroid derivative. It is also possible to alter the reaction conditions to favor formation of the oxime at the desired position and then chemically isolate the desired steroid derivative. For example, again in the case of progesterone, a ratio of 0.5 mmoles *O*-(carboxymethyl) hydroxylamine per millimole of progesterone favors oxime formation at the C-3 (Fig. 1) and this product can be isolated chromatographically.

Hemisuccinate derivatives can be made at hydroxyl substituents by reacting 3 to 5 mmoles of succinic anhydride per millimole steroid refluxed with pyridine for 4 to 5 hours. The solution is taken to dryness under reduced pressure, the residue dissolved in ethyl acetate, washed with water, and extracted with 5% NaOH. This alkaline extract is washed with ethyl acetate and acidified with HCl to pH 2.0. The derivative precipitates and should be washed and recrystallized from appropriate solvents. In the case where more than one hydroxyl substituent is present on the steroid molecule it is usually possible to adjust the reaction conditions to take advantage of the varying reactivities of different hydroxyl groups. For example, the phenolic hydroxyl at C-3 in estradiol-17β is more reactive than the secondary alcohol at C-17. To make estradiol-3-monosuccinate a ratio of 1 mmole estradiol to 3 mmole succinic anhydride in pyridine should be reacted for 5 minutes at 25°. Under these very mild conditions the yield will be 55 to 80% of the estradiol-3-monosuccinate.[18] The reaction product can be purified by thin-layer chromatography. If the reaction is continued for 3 days, only the 3,17-disuccinate is formed and selective hydrolysis of the C-3 ester may be performed by adding 0.5 g of sodium carbonate to 10 ml of a methanol solution containing 500 mg of the estradiol-3,17-disuccinate. The reaction should be carried out at room temperature for 18 to 24 hours. This reaction product should also be purified by thin-layer chromatography.

To prepare a chlorocarbonate derivative at a hydroxyl position on a steroid, the method of Miescher *et al.*[19] is used. One millimole of steroid is dissolved in 10 ml anhydrous chloroform, cooled to 0°, and treated with excess (5 to 10 mmoles) of liquid phosgene for 4 hours with cooling. At room temperature, the excess methanol and phosgene are evaporated under reduced pressure, the oily residue is azeotroped with dry benzene, and the product is crystallized from acetone–hexane.

It is very important that all steroid derivatives be extensively characterized by melting point, nuclear magnetic resonance, infrared spectroscopy, ultraviolet spectrophotometry, or combinations of these procedures to determine if the steroid derivative has the intended structure and to assess its purity. It seems likely that in many cases the failure to produce antibodies to steroid hormones is due to inadequate chemical procedures rather than to a failure of the immunized animal to respond.

The structure of steroid hormones differs only in the type and location of functional groups attached to the basic cyclopentanophenanthrene nucleus (Fig. 1). Hence, in order to obtain antibodies with maximal specificity one must take full advantage of the structural differences which exist between the steroid of interest and related compounds. This may be done by conjugating the steroid molecule to protein through a site distant to the functional groups of interest. For example, if one wishes to discriminate between testosterone and androstenedione the testosterone molecule should be conjugated to protein at the 3, 6 or 11 position (Fig. 1) but not at the 17 position since this is the site at which the only difference between these molecules exists.

To illustrate this point more clearly we have included the data in Table I depicting the relative activities of 17 representative steroids when antibodies were developed to progesterone conjugated to protein in the A (C—3), B (C—6), C (C—11), or D (C—20) ring of the molecule. Immunization with progesterone conjugated at C—20 resulted in the forma-

[18] G. E. Abraham and P. K. Grover, *in* "Principles of Competitive Protein-Binding Assays" (W. D. Odell and W. H. Daughaday, eds.), p. 134. Lippincott, Philadelphia, Pennsylvania, 1971.

[19] K. Miescher, H. Kagi, C. Scholz, A. Wettstein, and E. Tschzep, *Biochem. Z.* **294**, 39 (1937).

tion of antibodies which could not distinguish between progesterone and other 3-keto, Δ^4 steroids whose structures are similar in the A, B, C, and D rings (i.e., 17α-hydroxyprogesterone, testosterone, deoxycorticosterone, etc.). Immunization with progesterone-3-BSA resulted in formation of antibodies which easily distinguished between progesterone and steroids with different functional groups attached to the C and D rings. However, this antiserum was less able to distinguish between progesterone and steroids where the only differences in structure occur in the A or B rings (i.e., pregnenolone, 5α-pregnane-3,20-dione). Antibodies developed against progesterone-11α-BSA were quite specific for progesterone and easily detected differences in the A, B, or D ring of the molecule. In fact, this antibody could also distinguish between 11α-hydroxyprogesterone and 11β-hydroxyprogesterone. Antibodies developed against progesterone-6β-BSA easily detected differences in the A, C, or D ring of the molecule but, as expected, were less able to distinguish changes in the B ring (i.e., 5α-pregnane-3,20-dione). Similar data suggests that antibodies produced against testosterone and estradiol-17β conjugated through the 6 or 11 positions are much more specific than antibodies obtained with these steroids conjugated at the 3 or 17 positions (Table I).

From the preceding discussion it should be obvious that it is possible to develop an antibody to perform a particular analytical task. For example, if it is important to distinguish between progesterone and 17α-hydroxyprogesterone the antibody should be developed against progesterone conjugated to protein at the 3, 6 or 11 positions but not at the 20 position. However, if one wishes to quantify progesterone, testosterone, deoxycorticosterone, 17α-hydroxyprogesterone or pregn-4-en-20β-ol-3-one following chromatographic purification of the sample, then the antibody against progesterone-20-BSA has the advantage because it will quantify all these steroids with equal sensitivity. The trend has been to develop antisera which are highly specific for a given steroid so that many of the expensive and tedious procedures for preparation of the sample can be eliminated. For example, it is now possible to quantify testosterone,[20] progesterone,[21] and estradiol-17β[22] without chromatographic purification of the sample. Similar procedures have been or are being developed for most of the other major steroid hormones.

The complexity of the chemistry involved in making the appropriate derivative is increased considerably when the steroid is to be conjugated

[20] A. A. A. Ismail, G. D. Niswender, and A. R. Midgley, Jr., *J. Clin. Endocrinol. Metab.* **34**, 177 (1972).

[21] G. D. Niswender, *Steroids* (in press).

[22] B. G. England, G. D. Niswender, and A. R. Midgley, Jr., *J. Clin. Endocrinol. Metab.* **38**, 42 (1974).

TABLE I
RELATIVE ACTIVITY[a] OF SELECTED STEROIDS IN FOUR DIFFERENT PROGESTERONE RADIOIMMUNOASSAYS

	Progesterone-20	Progesterone-3	Progesterone-11α	Progesterone-6β
Progesterone	1.00	1.00	1.00	1.00
Pregnenolone	.011	.133	.005	.034
11α-hydroxyprogesterone	.071	.134	.345	.0092
11β-hydroxyprogesterone	.02	.08	.08	.015
17α-hydroxyprogesterone	.981	.046	.0122	.0032
5α-pregnane-3,20-dione	.46	.61	.13	.30
Δ^4-pregnen-20β-ol-3-one	.965	.003	.0011	.0012
Δ^4-pregnen-20α-ol-3-one	.336	.001	.0014	.0016
5β-pregnane-3α,20α-diol	.003	<.0001	<.0001	.0004
5α-pregnane-3β,20β-diol	—	—	<.0001	.0001
Deoxycorticosterone	.965	.016	.007	.0017
Corticosterone	.004	.004	.002	.00004
Aldosterone	.071	.0004	.0001	—
Cortisone	.056	.0002	.0003	<.00001
Hydrocortisone	.004	.0002	<.0001	<.00001
Testosterone	.952	.0005	.0004	.00003
Estradiol-17β	.0004	<.0001	<.0001	<.00001

[a] Based on at least four different dose levels in duplicate for each steroid tested.

through the 6 or 11 position. However, in some cases, the necessary starting material such as 6-ketoestradiol or 11α-hydroxyprogesterone can be obtained commercially.

Conjugations of Steroid to Protein. A number of different protein molecules have been used in the formation of hapten immunogens including bovine, human and rabbit serum albumin, and thyroglobulin. Successful antibody formation has been reported with all these proteins, and there is no good evidence that any one is better than the others. Bovine serum albumin (BSA), the protein first used by Erlanger *et al.*,[17] has been the one most frequently used.

There are several methods for conjugation of steroid derivatives to protein with the most commonly used procedures being the mixed anhydride reaction or carbodiimide condensation. Either of these procedures attaches the steroid to the ϵ amino group of the lysine residue in the protein molecule.

Although absolute quantities may vary, depending on the availability of derivatives, in general, in the mixed anhydride procedure 1 mmole of either the oxime or hemisuccinate steroid derivative and 2 mmoles of tri-*n*-butylamine are dissolved in 15 ml dioxane. The solution is cooled and treated with 1 mmole isobutylchlorocarbonate. The reaction is allowed to proceed at 4° for 30 minutes and the mixture is then added to a cooled, stirred solution of 0.02 mM BSA in 30 ml water, 20 ml of dioxane, and 1.17 ml of N NaOH. Stirring and cooling is continued for 4 hours. The pH should remain at 8.0 throughout the reaction. The solution is dialyzed against running water overnight and brought to pH 4.6 with N HCl. The resulting precipitate is collected by centrifugation and washed twice with cold water. The precipitate is then solubilized with 5% NaH CO_3, dialyzed overnight, and lyophilized.

The technique for conjugation of steroids using carbodiimide condensation has been described in detail in other publications[11,18] and will not be discussed here. Likewise the details of the Shotten-Baumann reaction used to conjugate chlorocarbonate derivatives to protein have been described elsewhere.[23]

It is essential that the final steroid—protein conjugate be extensively characterized to determine the average number of steroid molecules per protein molecule. This can be accomplished by UV analysis[17] of the conjugate or by the use of trace quantities of radioactive steroid in producing the derivative and determination of the radioactivity in the final product. This ratio is important, since it appears to influence the efficacy of the conjugate for antibody production. We have always been successful in obtaining usable antibody titers when the steroid-to-protein ratio was 20 or greater but have had little success with ratios of 10 or less.[24] It seems likely that in many cases when investigators have failed to elicit antibody formation it was due to an inadequate concentration of steroid in the conjugate.

Immunization. In recent years numerous methods of immunization have been described, with different investigators advocating their favorite route, species, adjuvant, dose, site, and frequency of injection. The immunization scheme we have used successfully to elicit the formation of antibodies to more than 30 different steroid–protein conjugates is described below.[25]

[23] B. F. Erlanger, S. M. Beiser, F. Borek, F. Edel, and S. Lieberman, *in* "Methods in Immunology and Immunochemistry" (C. A. Williams and M. W. Chase, eds.), p. 144. Academic Press, New York, 1967.

[24] G. D. Niswender and A. R. Midgley, Jr., *in* "Immunologic Methods in Steroid Determination" (F. G. Péron and B. V. Caldwell, eds.), p. 149. Appleton, New York, 1970.

[25] A. R. Midgley, Jr., G. D. Niswender, V. L. Gay, and L. E. Reichert, Jr., *Recent Progr. Horm. Res.* **26,** 235 (1970).

The steroid–protein conjugate is dissolved or suspended in saline at a concentration of 1 mg/ml. An equal volume of complete Freund's adjuvant is placed in the bottom of a 20-ml glass syringe; the steroid–protein conjugate is added in small aliquots, each aliquot being thoroughly mixed on a Virtis homogenizer between additions. The mixture is finally emulsified by mixing for 30 seconds to 1 minute at 80,000 rpm. During this period the syringe is kept in an ice bath to prevent heating of the adjuvant. "Methods in Immunology"[26] is an excellent reference for details of preparing and testing adjuvant.

The emulsified adjuvant is injected into the footpads of rabbits, 50 to 100 μl per site; the remainder of the adjuvant is injected at multiple subcutaneous sites. Alternatively, the adjuvant may be injected intradermally into at least 50 sites in rabbits[27] or sheep, or into multiple subcutaneous sites in sheep. Booster immunizations, emulsified in incomplete Freund's adjuvant just prior to injection, are administered at 3- to 4-week intervals. Each animal receives 0.05 to 6.25 mg of conjugate per immunization. The animals are bled at weekly intervals beginning 3 weeks following the initial immunization. The rabbits are bled under a slight vacuum following incision of the marginal ear vein. Sheep are bled by jugular puncture. It is necessary to store serum from each bleeding of each animal separately since striking changes in titer, affinity, and specificity may occur between bleedings particularly following additional immunizations.

In recent years considerable controversy has arisen concerning the route of administration of the adjuvant and the species most suitable for production of steroid antibodies. Therefore, we have compared the response in rabbits immunized by footpad injections to that in rabbits immunized by intradermal injections. In addition, we have compared the response in rabbits immunized intradermally to the response in sheep following a similar immunization. As can be seen in Fig. 2, there was no difference in the resulting antibody titer following different routes of administration or when rabbits were compared to sheep. In general, the antiserum titer in sheep was more variable than that observed in rabbits.

Radioactive Hormone

To quantify the hormone present in an assay tube, it is necessary to determine the amount of hormone bound to antibody. This can best

[26] D. H. Campbell, J. S. Garvey, N. E. Cremer, and D. H. Sussdorf, "Methods in Immunology." Benjamin, New York, 1963.

[27] J. Vaitukaitis, J. B. Robbins, E. Neischlag, and G. T. Ross. *J. Clin. Endocrinol. Metab.* **33**, 988 (1971).

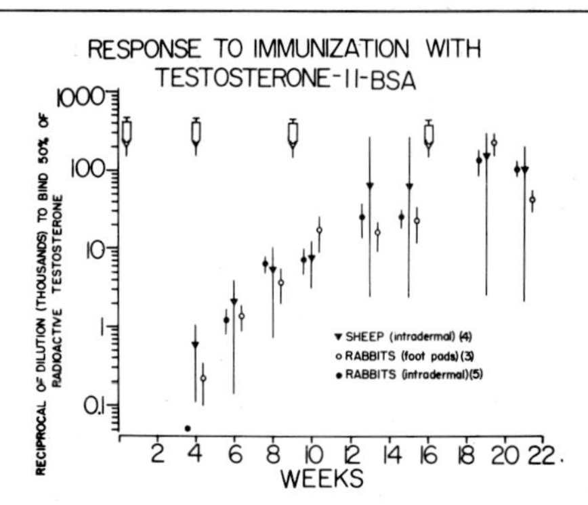

FIG. 2. Antibody titer (mean ± 1 S.E.) following intradermal immunization of sheep (4 mg per immunization), intradermal immunization of rabbits (1 mg per immunization), and foot pad immunization of rabbits (1 mg per immunization) with testosterone-11-BSA. The syringes at the top of the figure depict the times at which immunizing doses were administered.

be done by the use of a radioisotopically labeled steroid. High specific activity, tritiated steroid hormones have usually been used for this purpose. However, we have developed a procedure which allows the use of a radioiodinated form of steroids and has several advantages. A higher specific activity may be attained with radioiodine which permits the use of a considerably less mass of radioactive antigen and results in a more sensitive assay system if the affinity of the antibody is not limiting. The higher specific activity of the radioiodinated steroid also allows the addition of considerably more counts per minute (i.e., 60,000) thereby reducing the counting time necessary to obtain reliable counting statistics. The use of radioiodinated steroid hormones results in easier and less expensive preparation of the sample for counting and also simplifies the use of tritiated steroids to monitor procedural losses.

Although the specific activity of ^{131}I is approximately 7 times higher than that for ^{125}I the isotopic abundance of ^{131}I is rarely above 13 to 18%, whereas ^{125}I is available in almost isotopically pure form. Therefore, based on the resultant specific activity of radioiodinated steroids there is no particular advantage for either isotope. However, ^{125}I has a slower decay rate which reduces the number of radioiodinations necessary and is usually counted with a higher efficiency.

With the exception of steroids possessing a phenolic A ring, it is not possible to radioiodinate directly the cyclopentanophenanthrene nucleus or its substituents. Though phenolic steroids can be radioiodinated directly at the 2 and/or 4 position, this seems to alter the configuration of the molecule and affects the binding to the antibody.[3] This is not surprising since the physical dimensions of an iodine atom approximate those of the complete phenolic A ring.[28]

If the steroid is conjugated to a tyrosine-containing protein, for example RSA,[23] it is possible to radioiodinate the protein without affecting the ability of the steroid to bind to antibody. However, it has not been possible to insure that only one steroid molecule is attached to each albumin molecule. In fact, the steroid-to-protein ratio usually obtained is at least 20. If the albumin conjugates are radioiodinated for use in the radioimmunoassay then each steroid molecule on the albumin has the potential for binding to antibody. To accomplish separation of free radioactive antigen from that bound to antibody it is necessary that every antibody molecule bound to the steroid–albumin conjugate be displaced, which requires a large number of nonradioactive steroid molecules. Assays conducted under these conditions do not produce inhibition curves with adequate slope and result in imprecise estimates of the steroid being measured. To insure a preparation with one steroid per radioiodinated protein, the steroid derivatives were conjugated to the methyl ester of tyrosine (TME), making the phenolic tyrosine ring available for radioiodination.[24]

For the radioiodination of steroid–TME conjugates we have used the following procedures. Two and one-half micrograms of steroid–TME in 2.5 μl of methanol are added to a 1-ml serum bottle (Fisher Scientific) and 50 μl of 0.5 *M* phosphate buffer (pH 7.5) are added. One millicurie of high specific activity $Na^{125}I$ is added. The vial is stoppered and the contents are mixed gently. Thirty micrograms of chloramine-T in 15 μl of 0.05 *M* phosphate buffer (pH 7.5) are added and the reaction mixture is gently mixed by finger tapping for exactly 2 minutes and 60 μg of sodium metabisulfite in 30 μl of 0.05 *M* phosphate buffer are added to terminate the reaction.

Steroid–TME–^{125}I is separated from free iodine either on a 0.6×20 cm column of Sephadex G-25 (progesterone and testosterone) or by electrophoresis on polyacrylamide gel.[25] One-milliliter fractions are collected from the columns. Free radioiodine elutes in fractions 9 through 12, clearly separated from that of labeled steroid which elutes as a single peak in fractions 15 through 18. Electrophoretic separations are accomplished on 7.5% gel, 0.5×6.5 cm columns at 4 mA per gel for 1 to 3 hours with 0.011 *M* boric acid, 0.0332 *M* disodium EDTA, and 0.007 *M* Tris buffer at pH 8.0. After electrophoresis the positions of the radioactive hormones are located by autoradiography with Kodak no-screen

[28] G. E. Abraham and W. D. Odell, *in* "Immunologic Methods in Steroid Determination" (F. G. Péron and B. V. Caldwell, eds.), p. 87. Appleton, New York, 1970.

X-ray film. The regions of interest are dissected, the gel segments crushed between two slides, and the radioiodinated hormone is eluted into 4 to 5 ml assay buffer by standing overnight at 4°. This concentrated radioiodinated steroid hormone is further diluted to 40,000 to 60,000 cpm/100 μl prior to addition to radioimmunoassay tubes.

Development of Steroid Radioimmunoassays

Characterization of Antiserum. Before an antiserum can be used for development of a radioimmunoassay it must be carefully characterized. The most important properties to be examined are titer, affinity, and specificity.

Titer is defined as the number of antibody molecules present in a given quantity of serum and is reflected by the final dilution of antiserum used in the assay. This is established by measuring the ability of varying dilutions of antiserum to bind radioactive hormone. In general, the dilution of antiserum which binds 30 to 50% of the radioactive antigen is used for radioimmunoassay since these conditions optimize the ratio of sensitivity to precision.[29]

Affinity is the strength with which antibody binds to antigen and is usually expressed as an association constant. Sensitivity of a radioimmunoassay system is dependent on the affinity of the antibody and the specific activity of the radioactive antigen. Sensitivity can be defined as the least amount of steroid hormone that can be reliably distinguished from no hormone.[30]

Specificity of an antiserum is the freedom from interference by substances other than the steroids intended to be measured and is the characteristic which ultimately determines its usefulness. As discussed previously, the specificity of antisera against steroids is dependent on the site through which the steroid molecule is conjugated to protein. A convenient method for assessing the specificity of antiserum is to determine the ability of related steroidal compounds to inhibit the binding of the radioactive steroid to antibody. However, this is not the ultimate test of specificity since the number of steroids tested is usually limited. The best test of specificity is close agreement between estimates made by radioimmunoassay and estimates made by totally different, well-established procedures such as gas–liquid chromatography or double isotope derivative techniques. This agreement should exist in a reasonable number of samples obtained from subjects in a variety of physiological states. Al-

[29] R. Ekins and B. Newman, *Acta Endocrinol. (Copenhagen), Suppl.* **147,** (1970).

[30] A. R. Midgley, Jr., G. D. Niswender, and R. W. Rebar, *Acta Endocrinol. (Copenhagen), Suppl.* **142,** 163 (1969).

ternatively, when double isotope derivative or gas–liquid chromatographic procedures are not available, it is possible to compare radioimmunoassay estimates before and after chromatographic purification of the sample. It should be obvious that this procedure is only as reliable as the purification technique employed.

Steroid Radioimmunoassay

General Considerations

A major consideration when developing a radioimmunoassay is the time allowed for reaction between the antibody, the sample or standard, and the radioactive antigen. Equally important is the selection of a suitable method for the separation of free steroid from that bound to antibody. A variety of reaction times varying from 30 minutes[31] to 48 hours[24] have been used. The reaction time of an antiserum may be determined by allowing the radioactive antigen to incubate with antibody for periods varying from a few minutes to several hours and determining the minimum time necessary to reach equilibrium. Considerable variation exists among different antibodies, hence, this characteristic should be determined for each individual antibody. Likewise, the buffer used for the radioimmunoassay may have a pronounced effect on the time necessary to achieve equilibrium. In most assay systems, protein (i.e., gelatin and albumin) is added to the assay buffer to minimize nonspecific adsorption of the antibody and antigen to glass. Albumin has a high affinity for steroids and hence greatly increases time needed to reach equilibrium.

The use of nonequilibrium assay conditions, i.e., the reaction of antibody and sample or standard for a finite time before the radioactive hormone is added, has been shown to increase the sensitivity of radioimmunoassays for glycoprotein hormones.[25] However, only small increases in the sensitivity of radioimmunoassays for steroid hormones have been reported,[25,28] probably due to the relatively short reaction times.

A host of techniques have been used to separate free radioactive steroid from that which is bound to antibody. The ideal method should provide a clean separation of these components and be unaffected by serum or other nonspecific substances in the reaction mixture. In addition, the method should be rapid, simple, reproducible, and inexpensive. The methods of separation most commonly employed are (1) solid-phase antibodies, (2) solid-phase adsorption of antigen, (3) chemical precipitation of antigen–antibody complexes, and (4) immunoprecipitation of antigen–antibody complexes.

[31] D. M. Hendricks, J. F. Dickey, and J. R. Hill, *Endocrinology* **89**, **1350** **(1971)**.

Abraham[2] developed a radioimmunoassay for estradiol-17β using antibody-coated polystyrene tubes. In this procedure the sample or standard and the radioactive hormone are allowed to react in the coated tube and the reaction is terminated by decanting into a liquid scintillation vial for counting. This procedure is very simple, but is wasteful of antibody and suffers from a lack of precision and poor reproducibility. Mikhail *et al.*[32] used a suspension of polymerized antiserum, which could be centrifuged directly to effect separation of free radioactive hormone from that which is bound to antibody for the radioimmunoassay of estradiol and estrone. This assay procedure was also simple, but wasteful of antibody.

Adsorption of the unbound steroid to insoluble particles such as charcoal, Florisil, Fuller's earth, talc, and similar compounds has also been used to separate free steroid from that bound to antibody. Of these substances, charcoal has been the most widely used; hence, our discussion will be limited to the use of this method. The advantages of charcoal are its simplicity and speed. At the completion of the antigen–antibody reaction, 500 μl of a solution of 1.0% Dextran-80 and 0.05% charcoal (Norite A) in PBS is added to each tube and thoroughly mixed. Many other laboratories which routinely use the charcoal separation procedure use 500 μl of a solution which contains 0.025% dextran and 0.25% charcoal. The charcoal binds the free steroid almost instantaneously and it is for this reason that relatively short incubation times (usually 10 minutes to 1 hour at 4°) may be employed. At the end of the incubation the free steroid hormone adsorbed to charcoal is precipitated by centrifugation for 5 minutes at 3000 rpm after which the supernatant is decanted for counting.

Several disadvantages are inherent in this method of separation. The charcoal has some affinity for steroid–antibody complexes, resulting in removal of some of the complexes from the reaction mixture. Pretreatment of the charcoal with dextrans has been reported to increase the selectivity of the charcoal particles and helps to alleviate this problem.[33] In addition, a great excess of charcoal is generally used to effect rapid separation which necessitates critical timing of the reaction so that the contents of each tube receive the same exposure to adsorbent. Unfortunately, the charcoal has a greater affinity for steroid hormones than do some antibodies. Since an excess of adsorbent is used, considerable

[32] C. Mikhail, H. W. Chung, M. Ferin, and R. L. Vande Wiele, *in* "Immunologic Methods in Steroid Determination" (F. G. Péron and B. V. Caldwell, eds.), p. 113. Appleton, New York, 1970.

[33] V. Herbert, K. S. Lau, C. W. Gottlieb, and S. J. Bleicher, *J. Clin. Endocrinol. Metab.* **25**, 1375 (1965).

radioactive hormone may be "stripped" from the antibody during an extended reaction. Temperature is also critical in the charcoal method of separation, since it influences the dissociation of antigen–antibody complexes. Most investigators have effected separation at 4° when using charcoal.

Chemical means to precipitate antigen–antibody complexes have also been used to separate free radioactive steroid from that which is bound to antibody. Mayes *et al.*[34] have used 50% saturated ammonium sulfate to precipitate steroid–antibody complexes, leaving the free steroid in solution. Polyethylene glycol has also been used very successfully to precipitate a variety of antigen–antibody complexes.[35] Both of these procedures are very rapid and easy to perform.

The precipitation of soluble steroid–antibody complexes by the addition of a second antibody (prepared against the first antibody) to yield insoluble immune complexes has proved very successful. In order to insure the formation of a visible precipitate, nonimmune serum from the species in which the steroid antibody was produced is added to the reaction mixture in a final dilution of 1:50 to 1:2000. The primary advantage of this method is the clean separation of free steroid from antibody-bound steroid under the mildest conditions possible. This method of separation also has certain disadvantages. For example, each batch of second antibody must be carefully titrated against the steroid antibody to insure that the precipitation reaction is occurring within the "zone of equivalence" and that precipitation is maximal.[36] This may be done by incubating varying dilutions of second antibody in tubes containing first antibody plus radioactive hormone and observing the dilution(s) which result in maximum precipitation. This method is also relatively slow in comparison to other separation techniques. In addition, if one attempts to measure steroid hormones in unextracted serum, immunological components in the serum may interfere with the second antibody reaction. Despite these disadvantages we prefer the second antibody method of separation due to the mild conditions required to effect a clean separation and the high reproducibility of the method. It has also been possible to insolubilize the second antibody by covalently linking it to sepharose which allows the addition of large excesses of anti-γ-globulin and results in shortening the reaction time of the second antibody precipitation step to as little as 30 minutes.[37] This later procedure has all the

[34] D. Mayes, S. Furuyama, D. C. Kem, and C. A. Nugent, *J. Clin. Endocrinol. Metab.* **30,** 682 (1970).

[35] B. Desbuquois and G. D. Aurbach, *J. Clin. Endocrinol. Metab.* **33,** 732 (1971).

[36] A. R. Midgley, Jr., R. W. Rebar, and G. D. Niswender, *Acta Endocrinol.* (*Copenhagen*), *Suppl.* **142,** 247 (1969).

[37] G. D. Niswender, unpublished data (1973).

advantages of both the double antibody and the charcoal separation methods.

Preparation of the Sample. It might be feasible to quantify steroid hormones in unextracted serum in the presence of high affinity antibodies, as have been obtained frequently with steroid antigen. In practice, however, this has not been possible, probably due to the presence of proteins which bind steroid and interfere with the binding of steroid to the antibody. Therefore, it has been necessary to extract the steroid from the sample most commonly by the use of organic solvents. Not only does this procedure eliminate protein but with careful choice of solvent system can also rid the sample of other substances which have the potential to interfere with the assay. For example, petroleum ether extraction of a sample for progesterone not only gives an extract that is free of binding protein but also is relatively free of other steroids. Tritiated steroid hormones are usually used to monitor procedural losses during extraction and whenever chromatographic purification procedures have been necessary. The mass of steroid estimated in the radioimmunoassay can be corrected for procedural losses by determining recovery of the tritiated steroid. Theoretically, this recovery estimate should be made after solubilization of the extract in the aqueous assay buffer since most steroids have a limited solubility in water. In fact, aliquots from the same aqueous solution should be pipetted for radioimmunoassay and for recovery determinations at the same time. For steroid hormones which are present in minute quantities in biological samples (i.e., estrogens and aldosterone) it is necessary that the mass of radioactivity added for recoveries be kept at a minimum and that the final sample estimate be corrected for this mass. For recovery estimates to be meaningful the tritiated steroid should be in equilibrium with the endogenous steroid prior to extraction. A procedure we have found to be very useful in the separation of organic solvents from aqueous samples is to freeze the aqueous phase in a dry-ice acetone bath which simplifies collection of the organic phase by simple decantation.

Numerous thin-layer column and paper chromatographic procedures have been described for the chromatographic purification of steroids. It is beyond the intended scope of this chapter to describe or discuss each of these methods since they vary for different steroids and the methods used for the same steroids differ between laboratories.

Detailed Radioimmunoassay Procedure. The procedure described below is one which we have used successfully to develop radioimmunoassays for several steroid hormones. Steroid hormone standards are initially dissolved in methanol to attain a concentration of 200 μg/ml and further diluted (100- to 100,000-fold) with 0.01 *M* phosphate-buffered saline

(PBS) pH 7.0 containing 0.1% gelatin. The gelatin is included to prevent the binding of steroid hormones to glass and plastic surfaces. All antisera are initially diluted 1:400 with 0.05 *M* EDTA-PBS and then, for reasons discussed previously, further diluted with 1:400 nonimmune serum in 0.05 *M* EDTA-PBS to a concentration which binds 30 to 50% of the radioactive antigen in the absence of nonradioactive antigen.

Triplicate standard curves are included in each assay at 11 dose levels ranging from 1 pg (or 10 pg) to 1 ng (or 10 ng). In the radioimmunoassay, 500 μl of combined standard solution (or sample) plus buffer are incubated with 200 μl of diluted antibody and 100 μl (approximately 40,000 cpm of steroid–TME–^{125}I or approximately 6,000 cpm of [^{3}H]steroid) of the radioactive antigen for 24 hours at 4°. Second antibody, diluted in 0.05 *M* EDTA-PBS, is added (200 μl) to all tubes and incubation is continued for an additional 4 to 24 hours at 4°. At the end of the second incubation, 3 ml cold PBS is added to each tube to "prewash" the precipitate and the tubes are centrifuged at 3000 rpm for 30 minutes. The supernatant is removed by decantation and the precipitates are counted in an automatic gamma counter.

We use two computer programs[38] to facilitate the handling of large numbers of radioimmunoassays. The first program protocols the assay and lists the reagents and samples to be added to each assay tube. Since the computer designed the assay, each tube is merely counted to 10,000 counts (±1% counting error) or to 3 minutes. A second program performs a logit transformation on the data obtained from the standard curves and computes the mean level of steroid along with the 95% confidence limit for each sample. Statistical checks for linearity of the standard curve and parallelism between samples and standard are also computed.

After the production of an antiserum and proof of its specificity in a radioimmunoassay system there are a number of parameters which are necessary to describe the performance of the assay system. Precision may be defined as the extent to which a given set of measurements of the same sample agree with the mean.[30] Precision is usually expressed as a coefficient of variation following repeated estimations of the hormone concentration of the same sample in the assay (within assay variation) and repeated estimates in different assays (between assay variation). Precision in any radioimmunoassay varies at different levels along the standard inhibition curve.[30]

Minor technical details which may not be considered critical under ordinary laboratory conditions play a significant role in achieving preci-

[38] W. G. Duddleson, A. R. Midgley, Jr., and G. D. Niswender, *Comput. Biomed. Res.* **5**, 205 (1972).

sion in radioimmunoassay. Improper choice of micropipetting equipment may increase error enormously.[30] We have found Hamilton syringes with Chaney adapters and the Micromedics automatic pipetting station to be the most reliable for precisely delivering microliter quantities of reagents. Other factors important in minimizing variation in the estimation of a hormone are: contamination free micropipetting equipment, mixing of assay tubes after each addition of reagent, keeping assay conditions uniform for a given assay, prewashing the precipitates with cold buffer before centrifugation, and a precise technique for decanting the supernatant.

The accuracy of a radioimmunoassay system is the degree to which the assay estimate agrees with the actual quantity of steroid hormone in the sample and can be evaluated by the addition of various physiological concentrations of steroid hormone to different sera which contain a wide range of levels of endogenous steroid hormone and sex hormone binding globulin (SHBG). Good agreement between the amount of steroid hormone added and that recovered indicates that the radioimmunoassay accurately measures the steroid hormone added. In addition, agreement also suggests that other factors in the serum sample did not influence the radioimmunoassay.

Blanks. A major requirement in the radioimmunoassay of steroid hormones has been the identification and solution of problems relating to "blanks." Blanks are the result of interference with the binding of radioactive steroid to antibody by nonspecific or contaminating substances. Blanks may originate from any of the reagents or apparatus used for extraction, purification, or quantification of a steroid.

Although commonly practiced, the correction of radioimmunoassay values for blanks is not satisfactory. It is very likely that blank values, due to extraction and purification, vary and cannot be estimated in each sample. In addition, unless the blank produces inhibition curves in the radioimmunoassay which are parallel to those obtained with the standard they cannot legitimately be subtracted. Therefore, the only meaningful solution to blank problems is to eliminate them or reduce them to insignificant levels.

The following precautionary measures may be helpful to eliminate or reduce blanks.

1. Use of freshly opened, glass-distilled solvents, although any carefully purified solvent should be adequate. We have been very successful using glass-redistilled solvents from Burdick-Jackson Laboratories, Muskegan, Michigan, and others have been equally successful using nanograde solvents from Mallinckrodt.
2. Use of disposable glassware wherever possible.

3. Thorough cleansing of nondisposable glassware (i.e., chromic acid wash, heating glassware to 500°, etc.).

4. Use of separate glassware for samples with high levels of steroid hormones versus those with low levels.

5. Use of a redistilled water wash of solvent extract to eliminate protein contamination.

6. Prewash thin-layer chromatography plates in the solvent used to elute the sample. For example, we wash silica gel-coated glass fiber sheets (Gelman Instrument Co., Ann Arbor, Michigan) by submerging them for 20 minutes in a chromatography tank filled with glass-redistilled chloroform. This procedure is usually not possible with ethanol or methanol since both of these solvents tend to elute the binding agent from thin-layer plates.

7. Extraction and/or purification of the samples in an area that is free of steroid contamination.

8. Very careful washing of all micropipetting equipment used in setting up the radioimmunoassay.

Since most blank problems arise during chromatographic purification of the sample, elimination of these procedures by the use of highly specific antisera helps to reduce blanks.

Summary. On the preceding pages we have attempted to summarize the current information relating to the development and routine application of radioimmunoassays for steroid hormones. However, we feel the following points merit re-emphasis.

(1) The specificity of antibodies to steroid hormones is dependent on the site through which the steroid is conjugated to protein. It is possible to design an antibody against a steroid hormone to perform a specific analytical task. For example, it is possible to develop an antiserum that will measure several related steroids following their chromatographic purification with equal precision and sensitivity. On the other hand, it is also possible to reliably quantify an individual steroid without chromatographic purification of the sample using specific antibodies.

(2) It is possible to perform radioimmunoassays of steroid hormones using radioiodinated steroid–tyrosine methyl ester derivatives. The use of these compounds has several advantages including a more sensitive and precise assay system, simpler and less expensive counting procedures, and simpler procedures for determining recoveries following extraction and/or chromatographic purification of the sample.

(3) The major problem in most steroid radioimmunoassays is the identification and solution of blanks. Blanks usually arise during extraction and purification of the steroid, and since they probably vary from

sample to sample and since the inhibition curves obtained with varying quantities of "blanks" are usually not parallel to those obtained with the standard, they cannot legitimately be subtracted from the final radioimmunoassay estimates. Therefore, the only meaningful solution of blank problems is to eliminate or reduce them to insignificant levels. This may be easily done by eliminating those purification steps which most often result in blanks (i.e., chromatography).

[3] Use of Plasma-Binding Proteins for Steroid Hormone Assays

By CHARLES A. STROTT

I. Introduction

So-called protein-binding methods for steroid analysis were first introduced almost 10 years ago.[1] The reader should refer to other sources for more detailed reviews and for references to specific assays.[2-5] The protein-binding methods employ essentially three different classes of proteins, i.e., naturally occurring plasma proteins, immunoglobulins (Section II, Chapter 2), and specific tissue proteins (Section II, Chapter 4). This chapter will deal in a general sense with methodology relating to the use of the naturally occurring plasma proteins, viz., CBG (corticosteroid binding globulin), SSBG (sex steroid binding globulin), and PBP (progesterone binding protein). In addition, a specific method for a single steroid (17-hydroxyprogesterone) is included for illustrative purposes.

It would be quite impossible to elaborate on all the steroid assays available based on the protein-binding principle. Thus, the aim of this chapter is to help orient the person who is unfamiliar with these methods but who nevertheless is interested in developing a steroid assay. An attempt has been made to point out some of the pitfalls and potential problems not infrequently encountered. Hopefully the prospective assayist

[1] B. E. P. Murphy, W. Engelberg, and C. J. Pattee, *J. Clin. Endocrinol. Metab.* **23**, 293 (1963).

[2] R. L. Hayes, F. A. Goswitz, and B. E. P. Murphy, eds., Radioisotopes in Medicine: *In Vitro* Studies, Sect. 1, U.S. At. Energy Comm., Oak Ridge, Tennessee, 1968.

[3] E. Diczfalusy, *Acta Endocrinol. (Copenhagen), Suppl.* **147** (1970).

[4] B. E. P. Murphy, *Recent Progr. Horm. Res.* **25**, 563 (1969).

[5] W. D. Odell and W. H. Daughaday, eds., "Principles of Competitive Protein-binding Assays." Lippincoett, Philadelphia, Pennsylvania, 1971.

can obtain a "feel" for the subject and be able to use this chapter as a general guide in developing a protein-binding method for steroid analysis.

Steroid-Protein-Binding Assay Resume

With few exceptions steroid assays based on the protein-binding principle are performed in a similar fashion, i.e., variations on a theme. In general, the various assay steps involved are: (1) extraction of samples with an organic solvent after a radioactive tracer is added to monitor procedural losses; (2) purification of the steroid extract by some chromatographic procedure; (3) incubation of the purified samples and steroid standards in an aqueous protein solution at 0–4° until equilibrium is reached; (4) separation of bound and unbound steroid moieties by adsorption to particulate material; (5) measurement of the radioactivity in an aliquot of the supernatant (the bound fraction); and (6) development of a standard curve and determination of the unknown mass after correction for procedural losses.

II. General Principle

The general principle of protein-binding assays for steroid hormones is the same as that for the radioimmunoassay of peptide hormones first developed by Yalow and Berson for assaying plasma insulin,[6] and for the protein-binding assay of serum thyroxine developed by Ekins.[7] The principle is as follows: Labeled steroid is bound to a specific protein, forming a protein-label complex; unlabeled steroid is introduced into the system, resulting in displacement of some of the label from protein, i.e., replacement of unlabeled for labeled steroid on protein. [Note: The label can be added in first followed by the unlabeled steroid (displacement assay) or the labeled and unlabeled steroids can be added in simultaneously (competitive assay).] After equilibrium has been achieved, which is time- and temperature-dependent, the two steroid fractions (i.e., steroid bound to protein and steroid not bound to protein) are separated by some physical means (e.g., gel filtration or adsorption to particulate material). The radioactivity in the bound and/or the free fraction is determined and a standard curve plotted. The quantity of labeled steroid bound to protein at equilibrium will be inversely proportionate to the amount of

[6] R. S. Yalow and S. A. Berson, *Nature (London)* **184,** 1648 (1959).
[7] R. P. Ekins, *Clin. Chim. Acta* **5,** 453 (1960).

TABLE I
21-CARBON STEROIDS

11β,17α,21-Trihydroxy-4-pregnene-3,20-dione (cortisol)
11β,21-Dihydroxy-4-pregnene-3,20-dione (corticosterone)
17α,21-Dihydroxy-4-pregnene-3,20-dione (deoxycortisol)
21-Hydroxy-4-pregnene-3,20-dione (deoxycorticosterone)
17α-Hydroxy-4-pregnene-3,20-dione (17-hydroxyprogesterone)
4-Pregnene-3,20-dione (progesterone)

unlabeled steroid present. Maximum binding of label occurs in the absence of competing unlabeled steroid, and, conversely, minimum binding of label occurs in the presence of maximal amounts of competing unlabeled steroid. Standard curves are generally plotted as percent of label bound versus mass of unlabeled steroid standard. Standards and unknowns are processed identically; and following development of the standard curve, the steroid mass contained in an unknown sample is determined from the standard curve.

In considering protein-binding methods for steroid analysis, three factors are of primary importance: (1) the choice of a binding solution; (2) the preparation of the sample to be analyzed; (3) the separation of the bound and the unbound steroid fractions.

III. Steroid-Binding Proteins and Solutions

A. CBG

CBG is an α-globulin which binds the 21-carbon steroids listed in Table I with high affinity ($K \simeq 10^8 M^{-1}$ at 4°); it is common to all vertebrate species examined to date.[8] Human plasma is probably the most common source of CBG used in assays, although animal plasma (e.g., dog) can also be used. It is advantageous to obtain blood from someone who has been taking an estrogen preparation (e.g., ethinyl estradiol, 50–100 μg/day) for at least 3–4 weeks which increases the CBG concentration.[8] It is also helpful to administer a potent glucocorticoid (e.g., dexamethasone, 2 mg/day in divided doses) for 2 days prior to drawing blood in order to suppress endogenous cortisol levels. After separating the plasma from the red and white blood cells, it can be mixed with glycerol (1:1)

[8] U. S. Seal and R. P. Doe, *in* "Steroid Dynamics" (G. Pincus, T. Nakao, and J. F. Tait, eds.), p. 63. Academic Press, New York, 1966.

TABLE II
19-Carbon Steroids

17β-Hydroxy-4-androsten-3-one (testosterone)
17β-Hydroxy-5α-androstan-3-one (5α-dihydrotestosterone)
5α-Androstane-3α,17β-diol
5α-Androstane-3β,17β-diol
5-Androstene-3β,17β-diol
4-Androstene-3β,17β-diol

and stored in small aliquots (0.5–1.0 ml) in a freezer compartment (—15°). Stored in this fashion, the CBG is stable for years.

B. SSBG

SSBG is a β-globulin which binds the 19-carbon steroids listed in Table II with high affinity ($K \simeq 10^9\ M^{-1}$ at 4°); this protein also binds estradiol-17β with high affinity. SSBG has not been as extensively studied as CBG in animals other than the human being and thus its distribution in vertebrates is not fully known. For the present, human plasma is the primary source of SSBG. It is suggested that plasma be obtained and stored as described for CBG; if the source of plasma is an essentially normal woman, endogenous androgen levels should not pose a problem.

C. PBP

PBP, a globulin found in pregnant guinea pigs, is very active in binding progesterone ($K \simeq 10^{10}\ M^{-1}$).[9] PBP has been used in an assay for progesterone, but experience with it is quite limited.[10] The specificity of PBP for progesterone, although not complete, is improved over CBG; this should make it advantageous for routine progesterone determinations. If this protein is to be used, it may be necessary to reduce endogenous progesterone levels by administering a steroid biosynthetic inhibitor (e.g., aminoglutethimide, CIBA Corporation) prior to sacrificing the animal.[11]

D. Protein-Binding Solutions

The preparation of the protein solution for binding assays is an uncomplicated matter. An aliquot of the stored, crude plasma is diluted

[9] R. M. Burton, G. B. Harding, N. Rust, and U. Westphal, *Steroids* **17,** 1 (1971).
[10] C. A. Nugent, Y. Miyakoshi, D. M. Mayes, and D. C. Kem, *Proc. Int. Congr. Horm. Steroids, 3rd, 1970* p. 255 (1971).
[11] S. R. Glasser, R. C. Northcutt, F. Chytil, and C. A. Strott, *Endocrinology* **90,** 1363 (1972).

with a buffer at physiological pH; phosphate buffer, 0.05–0.15 M, is commonly used; however, other buffers such as borate and barbitol have also been used; unbuffered water has been used.

The degree of plasma dilution will depend on the sensitivity required: For more sensitive assays, a higher dilution will need to be made. It should be noted that while increasing the plasma dilution may improve sensitivity, it also narrows the useful range of the standard curve. The diluted plasma solution is generally prepared fresh prior to each assay; however, it can be refrigerated and reused over a few days. This is seldom necessary since the plasma supply is essentially inexhaustible.

The volume of the protein solution used in the assay usually ranges from 0.1 to 1.0 ml depending to some degree on the sensitivity of the assay (another way to increase sensitivity is to reduce the volume of the incubation medium).

IV. Preparation of the Sample to Be Analyzed

Individual investigators must make a decision concerning the degree of purification needed in the particular assay they are trying to develop. This decision will be based on the amount of competing steroids present and the assay convenience. The amount of assay convenience, of course, is dependent on the total number of steps involved in the assay: the more steps, the less convenience, and vice versa.

Assay specificity is primarily determined according to two factors: (1) steroid specificity of the binding protein; (2) degree of steroid purification. The plasma globulins that bind steroids do so only in a quasi-specific fashion. It is known that a large number and variety of steroids are not bound to any significant degree by CBG and SSBG. Tables I and II, however, list six steroids each that are bound by CBG and SSBG, respectively (these tables are not necessarily inclusive). Thus, to obtain a "high" degree of specificity, samples will have to be adequately purified. This purification is generally achieved by subjecting plasma and tissue extracts to some form of chromatography. The form of chromatography chosen will depend, for the most part, on the adequacy of the steroid isolation achieved by the particular method used, time requirements, and blank values. It is beyond the scope of this chapter to discuss in depth the various chromatography methods used in purifying steroids. A few comments, however, are in order in regards to some techniques which are frequently used.

Because of ease and rapidity, thin-layer chromatography (TLC) has been extensively used in purifying steroids for assay by protein-binding methods. Thin-layer chromatography has been particularly useful in assays for glucocorticoids, mineralocorticoids, androgens, and progestins. A variety of precoated thin-layer chromatograms can be purchased. Paper chromatography, on the other hand, is slower and more cumbersome and, thus, has not been used to as great an extent in protein-binding assay procedures. A recently developed procedure which is gaining in popularity is column chromatography on Sephadex LH-20.[12] These columns are easy to build and operate; they can be reused for a long period of time. Sephadex LH-20 has been used successfully for purifying all classes of steroid hormones, including estrogens. Another form of column chromatography worth mentioning is the Celite column used in estrogen assays (Section II, Chapter 4); this column is also useful in androgen and progestin assays.

If a "high" degree of specificity is not required, for instance, if one is trying to estimate the amount of a particular steroid which is present in a large concentration relative to competing steroids, then a chromatographic step may not be necessary. Two such examples are as follows. (1) Plasma cortisol in human beings is normally present in such a large concentration that it has been assayed following extraction without further purification.[13] (2) Plasma progesterone in pregnant women is also present in such a large concentration that it too has been assayed following extraction without further purification.[14]

V. Separation of the Protein-Bound and Unbound Steroid Fractions

This separation can be done satisfactorily utilizing several different techniques; one technique however—adsorption to particulate material—is most commonly used because of simplicity and convenience. This technique involves the use of either a silicate or charcoal. These agents are efficient in removing the *free* steroids from solution and not the protein-bound steroids.

On selecting an adsorbent for use in an assay, an optimal amount to remove tracer from solution will need to be determined: This will be volume-dependent, i.e., the greater the volume the more adsorbent required. The adsorbent should be evaluated for its ability to adsorb the steroid tracer being used in the absence of protein; also, the influence

[12] B. R. Carr, G. Mikhail, and G. L. Flickinger, *J. Clin. Endocrinol. Metab.* **33**, 358 (1971).

[13] C. A. Nugent and O. Mayes, *J. Clin. Endocrinol. Metab.* **26**, 1116 (1966).

[14] E. D. B. Johansson, *Acta Endocrinol. (Copenhagen)* **61**, 607 (1969).

of the presence of a large amount of nonradioactive steroid should be checked. In general, greater than 90% of the tracer should be removed from solution. The medium will consist of either water alone or a buffered solution (several buffers can be checked).

Several siliceous materials have been employed, e.g., Fuller's earth, Lloyd's reagent, and Florisil. Of these, Florisil is the most convenient to use; it can be dispensed easily and requires no centrifugation. A dispenser of appropriate size (to deliver an amount by weight) can be made of plastic or metal. The accuracy of the Florisil delivery is not terribly critical and a deviation of ±5–10% is acceptable. Florisil added to an assay tube quickly settles to the bottom unless the tube is shaken. Thus, following mixing it quickly settles out, leaving a clear supernatant which can be pipetted without prior centrifugation. Batches of Florisil can differ in their binding characteristics; thus, each batch with a different lot number will have to be evaluated. The Florisil can be used as obtained from the supplier; but occasionally it is necessary to wash it and remove the fines. A simple water wash may be all that is necessary, but additional washes with acidic and alcoholic solutions can also be performed if the situation demands it. The Florisil should be kept dry, but it does not require activation at a high temperature.

Dextran-coated charcoal is another commonly used adsorbent. The kind of charcoal used, e.g., Norit A and Darco G, is probably not critical. Charcoal can generally be used as obtained from the supplier; however, acid washing has been found in some cases to improve binding. Several different preparations of dextran have been used to coat the charcoal, and there is no way to predetermine which preparation will give the best result; only trial and error can determine this for any particular steroid assay. Unlike Florisil, which can be dispensed in dry form, charcoal must be used as a suspension. This makes the use of charcoal slightly less convenient than Florisil but it offers no serious problems. Dextran–charcoal solutions have been used with the charcoal concentration ranging from 0.25–0.5% and the dextran ranging from 0.025–20.0%; most assays use approximately 0.5% of each. Because the charcoal is in suspension (which tends to settle), it must be mixed thoroughly before pipetting. The use of a viscous solution (20% dextran), which will not mix in the assay until shaken, has been used to improve convenience.[15] Once charcoal has been added and mixed adequately to allow removal of the free steroid fraction, it must be centrifuged before the supernatant containing the bound steroid fraction can be removed.

[15] R. L. Rosenfield, W. R. Eberlein, and A. M. Bongiovanni, *J. Clin. Endocrinol. Metab.* **29**, 854 (1969).

VI. Methodological Considerations

A. Glassware

The problem of how glassware should be handled need not be a difficult one, but it is important that it be considered, the reason being that improperly handled glassware can create serious problems for the assayist in terms of poor precision and high blank values. All reusable glassware should be soaked in acid (generally a combination of chromate and sulfuric acid) followed by exhaustive water rinsings. Additional rinsings in alcohol and/or ether, while usually not, may be necessary. An alternative is to place glassware in a self-cleaning oven.

Disposable glassware, such as pipettes and assay tubes, will probably not need to be acid washed, but alcohol and ether rinsings may be required; again, all disposable glassware can be placed in a self-cleaning oven. It is important when considering such problems as high blanks, poor precision, and inconsistent binding that glassware be examined.

B. Steroid Standards

(1) Crystalline steroids can be purchased in pure form from numerous suppliers. It is advantageous, however, to evaluate the purity because an occasional bad batch may crop up, also mislabeling is not rare. There are several ways to examine the quality of the steroid: Melting point determination is easily performed (should essentially agree with that quoted by the supplier as well as the literature); if either gas–liquid chromatography or mass spectrometry is available, these are useful ways to examine purity. Once the assayist is satisfied that a steroid is pure, an appropriate standard solution can be made using absolute ethanol. A relatively concentrated stock solution (0.5–1.0 mg/ml) can be prepared, and further working solutions made from this. The final dilute solution is such that small volumes (5–50 μl) can be *reproducibly* pipetted using micropipetting techniques. It is important when pipetting the standard solution that the solvent be placed directly into the bottom of the tube and not be allowed to run down the side. The dilute standard alcohol solution when stored in the refrigerator is reasonably stable for several months at least. If it is not utilized very rapidly, however, it should probably be made fresh once a year. Sometimes standards are made up in dilute aqueous solutions to prevent having to dry down the standard (occasionally drying down the standard can lead to poor reproducibility). In this case the standard which is stored in the refrigerator may deteriorate within a few weeks; thus, this kind of standard solution will need to be made fresh more frequently.

(2) The radioactive steroid, as obtained from the supplier, is generally in a highly pure form. This fact notwithstanding a check on purity should be made (also mislabeling has been known to occur). There are several ways to examine the purity of the radioactive steroid: (a) Check to see if the tracer co-chromatographs with a known pure nonradioactive standard (e.g., thin-layer, paper, or column chromatography); (b) mix the tritiated tracer with some ^{14}C tracer, chromatograph them, and determine $^{3}H/^{14}C$ ratios of the eluted fractions (should be constant); (c) mix the tritiated tracer with pure nonradioactive standard and determine the specific activity following recrystallization (dpm/mg) (should remain constant). Radioactive steroids are probably best stored in benzene at a temperature just above freezing to prevent radiochemical damage (benzene adsorbs radiation better). Stored in this fashion the radioactive steroids are stable for several months. Aliquots can be removed and further diluted using absolute ethanol which should be stored at —15 to —20°. These latter solutions will deteriorate more rapidly and, thus, will need to be checked periodically.

C. Internal Standard

A tracer amount of radioactive steroid (less than 5% of the amount of radioactivity to be used in the assay) should be added to each sample prior to extraction. The recovery of this tracer through the various steps leading up to the assay is used to correct for procedural losses. This is important, for the use of chromatographic methods (some more so than others) can result in inconsistent recoveries. Tracer recoveries are determined following the final elution just prior to the assay. This is generally done by transferring known aliquots of the final eluate into counting vials (to determine tracer recovery) and assay tubes (to determine steroid mass).

D. Solvent Extraction

The choice of a solvent used to extract samples will depend on such considerations as efficiency, convenience, and "cleanliness." The efficiency can be quickly checked using a radioactive steroid as an internal standard. Convenience has to do with how rapidly a solvent will evaporate and whether it has to be treated in some special fashion. "Cleanliness" refers to the degree that a nonspecific residue might contribute to the method blank. Three solvents have proved most useful for sample extraction in steroid assays, viz., diethyl ether (ether), dichlormethane, and petroleum ether. These solvents are volatile and evaporate quickly. Ether should be used from freshly opened 1-lb cans (peroxides form after a can has been opened which can be troublesome); this solvent is very clean, does not require additional purification, and efficiently extracts most steroids except the more polar ones such as cortisol. Dichlormethane efficiently extracts most steroids including cortisol, but generally requires some additional purification. The degree that dichlormethane will need to be repurified will have to be determined by individual labs. Often, simply washing the solvent with distilled water is sufficient, but occasionally it may have to be either redistilled or passed through a silica gel column. Petroleum ether is used primarily for extracting progesterone and is relatively inefficient in extracting many other steroids; it is thus useful for differentially extracting progesterone which can then be directly assayed without further purification.

E. Solvent Evaporation

Evaporation of solvents used in steroid analysis has led to much confusion and uncertainty in the literature. Different gases have been suggested such as air, nitrogen, and helium with the former two being most commonly used. Whether one uses nitrogen or air will need to be determined by individual assayists, particularly since many established investigators cannot agree which is preferable. It may be important to do two things to the gas being used: dry it and filter it. Unfiltered gas containing water vapor has been accused from time to time of causing both poor recoveries and high blanks.

VII. Evaluation of a Protein-Binding Assay

A. Method and Plasma Blank

The blank is undoubtedly the most serious problem the assayist will encounter. Theoretically, if one starts with zero steroid in an extraction tube, one should end up with zero steroid in the assay. Any recording above zero is referred to as the blank. There are really two blanks: the method blank and the plasma or sample blank.

(1) The method blank is determined by processing water or the "contents" of an empty extraction tube through the entire procedure. There are, of course, innumerable things that can contribute to a method blank, some of which are related to the following: glassware, caps and stoppers, solvents, gas used in evaporating solvents, the apparatus used in evaporating solvents (i.e., manifold, tubing, etc.), chromatography material,

dry chemicals (e.g., anhydrous Na_2SO_4), or chemical solutions, actually any material that can come in contact with the sample. Thus, when an unacceptable blank is encountered, a careful systematic search for the contributing factor(s) will have to be made. Solvents may need to be repurified (e.g., redistilled, specially washed, or passed through a column). Thin-layer chromatography plates may need to be prewashed in methanol and/or ether prior to use; TLC tanks may need to be specially washed. It has been reported that the assay "area" may become "contaminated." In dealing with a blank problem, *all* steps in the assay procedure will need to be carefully considered, as well as the person performing the assay and the environment in which the assay is performed. It is unfortunate, but true nevertheless, that blank problems can come and go for no apparent reason.

(2) The plasma blank is determined by processing "steroid-free" plasma (from the same specie as the plasma to be assayed) through the entire procedure. It is quite possible that the method and plasma blank values will differ; in most instances the method blank will be lower than the plasma blank, and in most instances the plasma blank will not be volume dependent. What actually causes the plasma blank is uncertain and probably is the result of multiple factors. Assaying the same steroid: The plasma blank can vary from specie to specie, as well as between plasmas of the same specie. This is a complex matter and is poorly understood; the problem is not aided by the realization that what is considered a "steroid-free" plasma may not in actuality be steroid-free. Be that as it may, the plasma blank must be determined. The best source of steroid-free blood is from an animal whose adrenals and gonads have been totally removed; hypophysectomized animals can be used. While plasma which has had the steroids "stripped" by agents such as charcoal can be used, the former source is preferable.

B. Accuracy

Accuracy refers to the extent to which the measured amount of a steroid corresponds to the actual amount present. This is determined by adding a known amount of steroid to plasma and then estimating its recovery. Protein-binding assays have been found to be generally accurate with recoveries ranging for the most part from 90 to 110%.

C. Precision

The precision of a method is determined by assaying the same sample in replicate. The most common statistic used to determine precision has been the coefficient of variation. Since the precision may vary from one part of a standard curve to another, it is advisable to examine reproducibility at different levels of the curve. The between-assay precision should also be examined; this can be simply done by assaying a plasma pool over a period of time. An intra-assay coefficient of variation of less than 15% is acceptable; an interassay coefficient of variation of less than 30% is acceptable.

D. Specificity

With the use of protein-binding assays, specificity is essentially determined by the relative affinity of the protein being used for all steroids and by the extent of purification of the steroid sample. Absolute specificity is probably not possible, and the assayist must decide what degree of specificity is required; this decision, in turn, is greatly influenced by the balance between reliability and practicability. When examining the problem of specificity, it is helpful to consider the concentration of possible competing steroids and the relative affinity of the binding protein for them. For instance, if a competing steroid binds very weakly to protein but is present in a large concentration, it may pose a problem and thus need to be removed from the sample. On the other hand, if a competing steroid binds strongly to protein but is present in a small concentration, it may not pose a problem and thus may not have to be removed from the sample. There are several ways to examine specificity: (1) assaying a sample at various stages of an extensive purification; (2) adding large amounts (10–100-fold excess) of possible competing steroids to a blank plasma sample and assaying; (3) compare results obtained in a protein-binding assay with a totally independent technique (e.g., double isotope dilution method or gas–liquid chromatography) when available.

E. Sensitivity

Sensitivity can be thought of in two ways: (1) it refers to the sensitivity of the standard curve, i.e., the smallest amount of measured steroid mass that is significantly different from zero; and (2) it can be defined in terms of the blank value. The question of blank subtraction is still unsettled: Some assayists do and some do not. This question is largely settled by considering the value of the blank itself and the range of the standard curve. For example, if a plasma blank is 0.1 ng and the standard curve covers the range of 0–10 ng, a blank subtraction is probably not necessary, particularly if accuracy studies indicate essentially a 100%

recovery from plasma at the low end of the curve. If, however, the plasma blank is 0.1 ng and the standard curve range is 0–1 ng, a blank subtraction may be required especially if accuracy studies indicate an overestimation at the low end of the curve. Most assayists arbitrarily delineate a point on the standard curve, which is "well" away from the blank value, as the lower end of the useful (or working) range of the standard curve; and, if at this point no serious overestimation seems to be occurring, a blank subtraction is deemed unnecessary.

The sensitivity of an assay is dependent on the slope of the standard curve which, in turn, is dependent on the binding-protein dilution. The sensitivity of a protein-binding method is to an extent adjustable and the assayist will want to take advantage of this. For instance, since the plasma testosterone concentration is normally relatively low in females, a more sensitive curve will be required; however, cortisol and corticosterone are normally present in a large concentration, and thus, a less sensitive assay will be adequate. In addition, the more sensitive an assay, the more serious the blank problems that may be encountered.

VIII. Assay of Plasma 17-Hydroxyprogesterone (17-OHP)

A. Materials

1. [7α-^{3}H]-17-*Hydroxyprogesterone* (New England Nuclear Corp., SA 30 μCi/μg) is purified by thin-layer chromatography. After chromatography a trace amount of the label is added to recrystallized, authentic 17-OHP with a trace amount of [^{14}C]-17-OHP and chromatographed before and after acetylation. The ^{3}H/^{14}C ratio should not be altered by this procedure. A stock solution of 1 μCi of [^{3}H]-17-OHP/20 μl of absolute ethanol is prepared (stable for several months stored at −16°).

2. 17-*Hydroxyprogesterone* (K and K Laboratories, Inc.), recrystallized from ethanol three times and once from benzene and acetone, has a melting point of 217–222°. A stock solution of 1 mg/ml is prepared in absolute ethanol. From this stock a working solution of 1 ng/20 μl is prepared (stable for several months stored in the refrigerator).

3. *Precoated thin-layer plates* (silica gel, F 254, E. Merck A.G., Darmstadt, Germany) are washed by ascending chromatography twice with methanol and once with ether before use and stored in air-tight containers.

4. *Solvents* are analytical reagent grade and are not purified further. Absolute ether (Mallinckrodt Chemical Works) is used from freshly opened 1-lb cans.

5. *Corticosteroid-binding globulin (CBG):* Plasma is obtained by plasmaphoresis from a woman without endocrine disease who is given dexamethasone after receiving an estrogen preparation for 4 weeks. The plasma is diluted with glycerol (1:1) and stored in 1 ml aliquots at —16° (stable for years). Prior to use an aliquot of the stock [^{3}H]-17-OHP standard is diluted (1:1000) with 0.05 *M* phosphate buffer (pH 7.4) to a concentration of about 9500 dpm/0.1 ml. This solution is used to make a 240-fold dilution of plasma. The CBG solution is prepared freshly prior to each assay.

6. *Florisil* (Floridin Co.), 60/100 mesh, is used as obtained from the supplier.

B. Method

1. *Extraction.* 0.25 ml of 1 *N* NaOH and 900 dpm (0.01 ng) of [^{3}H]-17-OHP are added to 2.5 ml of heparinized plasma (serum can also be used) in a 60-ml separatory funnel. After mixing, the sample is extracted twice with 25 ml of ether. The ether extracts are combined, washed with 2.5 ml of 1% acetic acid and 2.5 ml of distilled water, dried over Na_2SO_4, and evaporated in a 40° water bath under air.

2. *Chromatography.* Using methanol:ether (1:1) the residues are spotted on thin-layer plates which are developed in an unsaturated ether:benzene (2:1) system. The dye, isatin (K and K Laboratories, Inc.), is used as a marker (co-chromatographs with 17-OHP). After development an appropriate area of each sample lane corresponding to the dye marker is scraped off using a razor blade (total area approximately 4 cm^2). The loosened silica is sucked into a disposable pipette which has been tightly plugged with a small wad of glass wool (previously washed with methanol and ether). The samples are eluted with 1.5 ml of 0.5% methanol in ether into 13-ml glass-stoppered, conical, centrifuge tubes. The solvent is evaporated under air, and the residues acetylated by adding 3 drops each of pyridine and acetic anhydride, vortexing briefly, and let stand overnight in the dark. After acetylation the tubes are washed down with methanol and the solvents evaporated under air. The residues are chromatographed in ether:benzene (2:1) as described above. The plates are dried *in vacuo* at 50° for 15–30 minutes, and the 17-OHP area eluted with 1 ml of 0.5% methanol in ether. (Note: Acetylation is to remove contaminates, for 17-OHP cannot be acetylated under the conditions used.) Three-hundred microliters of the eluate is transferred to counting vials (to determine tracer recovery). Six-hundred microliters are added to 3.5-ml round-bottom glass tubes (i.d. 8 mm) and the solvent evaporated under air (to determine mass recovery).

3. *Protein-Binding Assay.* The entire assay is performed in a walk-in cold room maintained at 4°. Zero, 0.25, 0.5, 1.0, 1.5, and 2.0 ng of the standard 17-OHP solution are evaporated in assay tubes. Two-hundred microliters of the CBG solution containing 19,000 dpm of [^{3}H]-17-OHP are added to each tube. The tubes are placed vertically in a horizontal shaker by wedging them individually in rows between sponge padding and incubated for 1 hour. At the end of the incubation period, 3 mg of Florisil are added to each tube using a dispenser. The tubes are shaken for an additional 30 minutes and 100 μl of the supernatant transferred to counting vials to which 10 ml of scintillation fluid are added [4% v:v liquifluor and 10% v:v Biogel (Beckman Instruments) in toluene]. The samples are counted in a liquid scintillation spectrophotometer.

The percent [^{3}H]-17-OHP bound to protein is plotted against nanograms of authentic 17-OHP added and a standard curve constructed. The mass in the unknown samples is determined from the curve and corrected for procedural losses. The reliability of this assay has been demonstrated.[16]

IX. Concluding Remarks

The use of naturally occurring plasma proteins in steroid binding assays is, in principle, extremely simple. The potential assayist needs to be aware of the fact that in reality these assays are far from simple matters. When steroid–protein-binding assays function well, they are quite convenient for the most part and allow large numbers of samples to be processed relatively quickly. The truth of the matter is they do not consistently function well; and it must be admitted that many times poor quality assays occur for no apparent reason. In order to attain a reasonable degree of consistency with a reliable assay, the assayist must also acquire consistent habits in performing the assay. The person who is always doing something different will find it difficult to develop a reliable assay. If procedural changes are to be made, they should be done so systematically; thus, a change in the results can be related to a specific methodological alteration. Very careful attention to detail is required and success necessitates consistency.

It is important that a quality control be built into the assay procedure. For instance, each assay run should contain a plasma blank and a plasma pool. When the assayist knows the plasma blank is low (the expected level) and that the pool is reproducible, full confidence can be placed in the results. It is also advisable to assay standards and unknowns in replicate.

[16] C. A. Strott and M. B. Lipsett, *J. Clin. Endocrinol. Metab.* **28**, 1426 (1968).

[4] Use of Receptor Proteins for Steroid-Hormone Assays

By S. Korenman

Study of the molecular mechanism of steroid hormone action resulted in the discovery of intracellular receptors in target tissues, first for estrogens and subsequently for androgens, cortisol, aldosterone, and progesterone. While most of the effort in the field was and continues to be directed toward understanding how the hormone–receptor interaction exerts control over cellular processes, it occurred to us that the characteristics of the soluble estrogen receptor of uterus were ideal for the measurement of estrogen concentrations and for study of the potency of putative estrogens and estrogen inhibitors.[1-4] The principal useful properties of the cytoplasmic estrogen receptor are specificity for target organ and biological activity and a high association constant $K = 10^{10}\ M^{-1}$ at 0°.

Materials and Methods

Preparation of Uterine Cytosol

Adequate receptor preparations have been prepared from immature rat, bovine, ovine, equine rabbit, and human uterus as well as 6-day pregnant rabbit uterus. Currently, immature rabbit uterus (Pel Freeze) is employed. Cytosol is prepared by homogenizing uteri at 4° in three volumes of a buffer consisting of 0.01 *M* Tris-HCl, pH 8.0, 0.001 *M* EDTA, 0.25 *M* sucrose, in a Waring Blender using four 30-second pulses at 2-minute intervals. After centrifugation at about 1000 *g* for 15 minutes, cytosol is prepared by centrifugation at 105,000 *g* for 1 hour or its equivalent. Others have reported use of a lower speed supernatant. The material thus prepared is stable in liquid N_2 for at least 3 months.

Assay Procedure

For estradiol (E_2) assay, aliquots of cytosol (usually 20 μl) are incubated with about 10,000 cpm of tritiated E_2 ($E_2{}^3H$) and standards or unknowns. The volume is taken up to 0.5 ml with a 0.01 *M* Tris buffer,

[1] S. G. Korenman, *J. Clin. Endocrinol. Metab.* **28,** 127 (1968).

[2] S. G. Korenman, L. E. Perrin, and T. P. McCallum, *J. Clin. Endocrinol. Metab.* **29,** 879 (1969).

[3] S. G. Korenman, *Steroids* **13,** 163, (1969).

[4] S. G. Korenman, *Endocrinology* **87,** 1119 (1970).

pH 8.0, 1 mM EDTA. After incubation at 4° overnight or at 23° for 1 hour, 0.5 ml of a suspension of 0.01 M Tris, pH 8.0, containing 0.5% activated charcoal (Norit A) and 0.5% Mann D-grade is added. The tubes are incubated in an ice bath for 15 minutes and centrifuged at 4° for 15 minutes. Neither the amount of charcoal or of dextran, nor the time of incubation, nor the time of centrifugation is critical. The supernatant is decanted into a counting vial and 10 ml of a counting solution consisting of 1 liter of toluene to which 50 ml of Liquifluor (Packard Instruments) and 250 ml of Biosolve (Beckman Instruments) has been added. The standard curve is always run in triplicate and the unknown in duplicate. Analysis of variance of standard curves gave an index of precision λ of 0.06 for E_2.[2]

Estrone (E_1) may be measured by a similar technique using tritiated E_1 as radioligand.[5] A similar procedure may be developed for the measurement of any potent estrogen or competitive inhibitor because binding to the receptor is necessary for estrogenicity. The ability of a substance to compete with $E_2{}^3H$ for receptor sites has been employed as a measure of its biological potency either as an estrogen or as a competitive inhibitor by calculation of the ratio of its association constant (RAC) to that for E_2.[4]

Measurement of the RAC of a Putative Estrogen or Competitive Anti-Estrogen

Increasing amounts of unradioactive E_2 or of test substance (H) are incubated with $E_2{}^3H$ as indicated above. The percent bound $E_2{}^3H$ is plotted against the log of mass added. The RAC may be calculated as follows[4]:

$$\mathrm{RAC} = \frac{K_{\mathrm{H}}}{K_{\mathrm{E}_2}} = \frac{\mathrm{R(RA)}}{\mathrm{R} + 1 - \mathrm{(RA)}}$$

where R is the free/bound $E_2{}^3H$ in the absence of unradioactive hormone.

$$\mathrm{RA} = \frac{\text{moles } \mathrm{E}_2}{\text{moles H}} \text{ required to reduce binding to 50\% of } 1/\mathrm{R}$$

Note that RAC values are specific for a species, an organ, and a temperature.

Although binding experiments of this type may be useful in clarifying the nature of putative estrogens or competitive inhibitors and in judging biological potency, the myriad influences on the effective concentration

[5] D. Tulchinsky and S. G. Korenman, *J. Clin. Endocrinol. Metab.* **31**, 76 (1970).

over time of substances given to the whole animal limit the value of this technique in estimating *in vivo* potency which is highly variable in itself.[3]

Estrogen Assay in Blood

The principal technical problems are separation of the major circulating estrogens and elimination of materials which nonspecifically inhibit the estrogen–receptor interaction without introducing new contaminants into the system. It is no mean trick and each component is important.

Preparation of Blood for Assay

Either serum or lightly heparinized plasma are used. After addition of about 1000 cpm of $E_2{}^3H$ and 1000 cpm $E_1{}^3H$, extraction is carried out twice with 2½ volumes of diethyl ether obtained from a freshly opened can. The combined extracts are dried at 30–40° in a stream of N_2 for column chromatography.

The Celite Column

Celite (acid washed, Johns Manville) is washed in 6 *N* HCl, water to pH 5, methanol and ether, dried, and stored in a muffle furnace at 540°C to avoid contamination on storage. Aliquots are thoroughly mixed with ½ w/v of spectroquality ethylene glycol and packed into 5-ml disposable pipets to a height of 5 cm. The dried extract is taken up in spectroquality isooctane, 2 × 1 ml, applied to the column of celite, and elution is carried out as noted below (Table) and with the assistance of 1 lb/in.^{2} N_2 pressure.

The elution of estrogens from a column of ethylene glycol (stationary phase) and Celite (support).

Elution of estriol is associated with so much ethylene glycol that an assay was not possible using this system.

The column eluates are dried under N_2 taken up in assay buffer consisting of 80% .01 *M* Tris, pH 8.0, 1 m*M* EDTA 20% ethylene glycol,

TABLE

Eluant	Volume (ml)	Estrogen recovered
Isooctane	8	None
15% Ethyl acetate–isooctane	4	Estrone
30% Ethyl acetate–isooctane	4	Estradiol

an aliquot taken for counting, and the remainder of the sample split for assay in duplicate against a standard curve prepared in the same buffer. Assay is carried out at 23° for 1 hour or 4° overnight. E_1 is assayed in an identical manner except that $E_1{}^3H$ is used for the radioligand and incubation is always carried out in the cold.

The concentration of estradiol during most of pregnancy is so high that a small aliquot of serum (10–20 μl) may be assayed for E_2 without prior extraction or chromatography.[6]

Technical Notes

1. For preparing clean Celite, both washing and maintenance in a state hostile to adsorption of atmospheric organic compounds is necessary.
2. Fresh ether must be used to avoid the effects of peroxides.
3. Transfer of samples must be done carefully with rinsing employing good volumes of isooctane because substantial losses will otherwise occur.
4. Separation of E_1 and E_2 should be complete.
5. Poor recoveries may be due to inadequate mixing of ethylene glycol with the Celite. We use a household fork and mix thoroughly for 15 minutes.
6. Each solvent used should be tested for blank in the assay system.
7. Elution of ethylene glycol from Celite columns will result in severe inhibition of antibody binding so that use of these celite columns for radioimmunoassay may be difficult.
8. Either ethylene glycol, protein, or salts is necessary to prevent estrogen adsorption to glass surfaces in the presence of buffer.
9. Each assay should be run with water blanks to ensure that contaminants in celite and solvents are not present.

[6] D. Tulchinsky and S. G. Korenman, *J. Clin. Invest.* **50**, 1490 (1971).

[4a] Assay of Cellular Steroid Receptors Using Steroid Antibodies

By EVANGELINA CASTAÑEDA and SHUTSUNG LIAO

The cellular receptors for steroid hormones are generally characterized by their high affinity toward specific groups of active steroids. Their identification and quantitation are often complicated by other tissue and blood proteins that bind the same steroids firmly but nonspecifically. In addition, steroid receptors are known to exist in multiple forms (including aggregates) by intra- or intermolecular interaction with cellular macromolecules or particulates. One of the most useful techniques for distinguishing specific steroid–receptor complexes from nonspecific steroid binders has been to identify the 8 S form in gradient media with low ionic strength. The 8 S form is an unstable entity, however, and very often does not indicate the total receptor content.[1-4]

The technique described below eliminates these difficulties. The method is based on our observation that many commercially available steroid antibodies are effective in removing [^{3}H]steroids bound to nonreceptor proteins, but not [^{3}H]steroids tightly bound to the hydrophobic pockets[4-6] of their own receptors. In gradient media containing 0.4 *M* KCl, the [^{3}H]steroid–antibody complexes sediment at about 8 S and, therefore, can be separated from all [^{3}H]steroid–receptor complexes that are dissociated from the large forms or particulates in the high-ionic-strength media and sediment as a 3–5 S entity. If insolubilized antibodies are used in the assay system, no laborious manipulation or costly setup is needed since the free [^{3}H]steroids and [^{3}H]steroid dislodged from the nonspecific binders and bound to the antibodies can be removed by brief centrifugation in clinical centrifuges. The [^{3}H]steroid–receptor content can then be estimated by measurement of the radioactivity remaining in the solutions. Numerous small samples can be assayed within 3–4 hours. The methods have been employed for the assay of steroid receptors in the target tissues (human and in experimental animals) of male and female steroid hormones, and of other steroid hormone receptors as well.

Samples of [^{3}H]Steroid–Receptor Complex

If the [^{3}H]steroid–receptor complex is in a soluble form, it can be used without further processing. If the complexes are bound to nuclei or cellular particulates (or intact tissues), they can be extracted for 30 minutes at 0° with 2 volumes of 0.6 *M* KCl (final concentration 0.4 *M* KCl) containing 2 m*M* EDTA and 50 m*M* Tris-HCl buffer, pH 7.5. The

[1] E. V. Jensen and E. R. DeSombre, *Annu. Rev. Biochem.* **41,** 203 (1972).

[2] G. C. Chamness and W. L. McGuire, *Biochemistry* **11,** 2466 (1972).

[3] G. M. Stancel, K. M. Leung, and J. Gorski, *Biochemistry* **12,** 2130 (1973).

[4] S. Liao, *Int. Rev. Cytol.* (in press).

[5] S. Liao, T. Liang, S. Fang, E. Castañeda, and T.-C. Shao, *J. Biol. Chem.* **248,** 6154 (1973).

[6] P. A. Bell and A. Munck, *Biochem. J.* **136,** 97 (1973).

insoluble particulates can be removed by centrifuging of the mixture at 12,000 *g* for 20 minutes, but normally a brief centrifugation in a clinical centrifuge is sufficient. For quantitative measurement, reextraction of the particulates is advisable. For the measurement of the receptor in the samples not previously exposed to [^{3}H]steroid, the sample solutions can be incubated with 1 pmole/ml of [^{3}H]steroid (specific radioactivity preferably above 20 Ci/mmole). Since the rate of the association of steroids to receptor molecules is a slow process at 0°, the incubation time should be at least 30 minutes.[7] In some cases, maximum binding is seen only at 1 hour or more. It is also important to expose samples to [^{3}H]steroid as soon as they are ready, since receptors in the crude tissue extract are unstable unless they are protected by steroids. If whole tissue samples are to be analyzed, it is convenient to homogenize them in 2 volumes of the 0.6 *M* KCl (final concentration: 0.4 *M*) solution described above which also includes the required [^{3}H]steroid. The homogenates are then allowed to stand at 0° for 1 hour, centrifuged, and the extract used for analysis.

Steroid Antibodies

The commercially available antibodies being used for the steroid assay have been found to be satisfactory. Most of these antibodies (sheep or rabbit) are made against steroids covalently linked to albumins through oxygen functions at positions 3 and 17 or at 20.[8] Antibodies with low steroid specificity are as effective as those that are highly specific. It is important to assure that the capacity of the antibody used is sufficiently large for quantitative removal of all the [^{3}H]steroid added to the assay tubes, since variations among different batches of antibodies are very large. For the receptor assay described, the use of antibodies in large excess is not harmful.

Insoluble Steroid Antibodies

a. Precipitation of Antibody from Serum. To 1 ml of serum (about 2 mg protein), 180 mg of Na_2SO_4 are added and the mixture is stirred

[7] See other sections of this volume for the most ideal conditions for the formation of individual [^{3}H]steroid–receptor complex.

[8] Some of the useful steroid antibodies are those made against cortisone-21-hemisuccinyl-RSA, estradiol-17β-succinyl-BSA, DHT- or testosterone-3-(*o*-carboxymethyl)-oxime-BSA, progesterone-20-(*o*-carboxymethyl)-oxime-BSA. They are available from Endocrine Science or from Research Plus. See S. Lieberman, B. F. Erlanger, S. M. Beiser, and F. Agate, Jr., *Recent Progr. Horm. Res.* **15**, 165 (1959).

at 25° until all of the salt is dissolved. The tube is kept at the same temperature for 1 hour and is then centrifuged at room temperature in a clinical centrifuge for 10 minutes at 3000 rpm. The antibody precipitate is washed twice with 2.5 ml of 18% (w/v) Na_2SO_4. The final sediment is suspended in 1 ml of 0.1 *M* $NaHCO_3$ containing 0.5 *M* NaCl. The suspension can be stored at —20° or kept at 4° if it is to be used on the same day.

b. Coupling of Antibody to Solid Phase. Various activated polymers can be used.[9-11] The following procedure has been found to be satisfactory for coupling to cyanogen bromide (CNBr)-activated Sepharose. Two-hundred milligrams of CNBr–Sepharose (Pharmacia) are swollen and washed in a glass filter with 1 m*M* HCl (40 ml) for 15 to 30 minutes at room temperature. The washed CNBr–Sepharose is suspended in 4 ml of 0.1 *M* $NaHCO_3$ containing 0.5 *M* NaCl and mixed with 1 ml of the steroid antibody prepared as described in (a). The contents are mixed gently by rotation of the tube end-over-end for 2 hours at room temperature or overnight at 4°. During the mixing, fragmentation of the gel beads should be avoided. The mixture is centrifuged in a clinical centrifuge. The gel sediment is reacted with 1 *M* ethanolamine (pH 8) for 1 hour at 0° to eliminate any remaining active groups. The gel is washed 3 times with 0.1 *M* acetate buffer containing 1 *M* NaCl at pH 4, and 3 times with 0.1 *M* borate buffer containing 1 *M* NaCl at pH 8. The washed gel is suspended in 2 ml of 20 m*M* Tris-HCl buffer containing 1.5 m*M* EDTA at pH 7.5, and stored at —20°. About 90% of the antibody added can be coupled by this procedure.

Assay Method I (with Gradient Centrifugation)

All manipulations are carried out at 0–2°. The sample solution containing [^{3}H]steroid–receptor complexes (0.05–0.2 ml) is mixed with a solution (0.05–0.2 ml) of antibody that has a capacity to bind all the [^{3}H]steroid present. The final concentration of KCl is made to 0.4 *M* with the addition of 2 *M* KCl. The mixture is allowed to stand for 2 hours. A portion of the sample (0.1 or 0.2 ml) is then layered on the top of sucrose (5–20%) gradient medium containing 20 m*M* Tris-HCl buffer, 0.4 *M* KCl, and 1.5 mM EDTA, pH 7.5. The tube is centrifuged at 54,000 rpm for 18 hours in a Beckman-Spinco L2-65B ultracentrifuge using a SW-56 rotor. After centrifugation, the fractions can be collected

[9] L. Wide, *Karolinska Symp.* **1,** 207 (1969).
[10] K. J. Catt, *Karolinska Symp.* **1,** 222 (1969).
[11] M. Wilchek, V. Bocchini, M. Becker, and D. Givol, *Biochemistry* **10,** 2828 (1971).

either from the top or bottom of the tube and the radioactivity of each fraction measured by a liquid scintillation spectrometer.[5]

Assay Method II (with Solid Phase Antibody)

Incubation of the sample is carried out in the same way as in method I, except that antibody coupled to a solid phase (such as Sepharose) is used. After standing for 2 hours, the tube is centrifuged in a clinical centrifuge for 10 minutes and an aliquot of the supernate is taken for the measurement of radioactivity.

Data Analysis

A typical result for method I is shown in Fig. 1. In this assay, a cytosol preparation (4.4 mg protein) previously exposed to [^{3}H]dihydrotestosterone (DHT) was mixed with a steroid antibody (0.1 ml of 20 μg protein/ml prepared against testosterone-3-(*o*-carboxymehyl)-oxime-bovine serum albumin in rabbit. In the absence of the antibody,

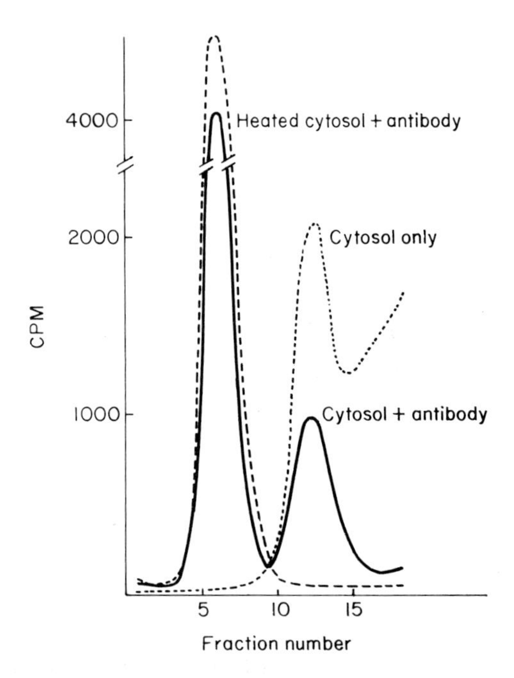

FIG. 1. Assay of the receptor for DHT in the cytosol of rat ventral prostate. Gradient fractions were collected and numbered from the bottom of the centrifuge.

[^{3}H]DHT was bound to the nonreceptor protein as well as to the receptor protein and sedimented as a peak in the 3–4 S area. The addition of the antibody eliminated the nonspecific and low affinity binding as well as the free [^{3}H]DHT present. As a result, a new radioactive peak in the vicinity of 8 S was formed. The amount of the antibody was sufficient in this experiment since, if the sample was heated before the assay at 60° or above for 10 minutes to destroy the DHT–receptor, essentially all the radioactivity was found in the 8 S region. The receptor content is determined from the radioactivity present in the 3–4 S region in the sample that contains the steroid antibody. When an insolubilized antibody was used, the result was identical.

In Fig. 2, a DHT–receptor preparation that contained no nonspecific binder was mixed with the steroid antibody. The antibody effectively removed the free [^{3}H]DHT remaining within the sample, but did not interfere with [^{3}H]DHT bound to the androgen receptor.

For both methods, the control tubes can be set up by omitting the receptor samples or by using samples heated at 70° for 10 minutes before the addition of the radioactive steroids. Samples heated together

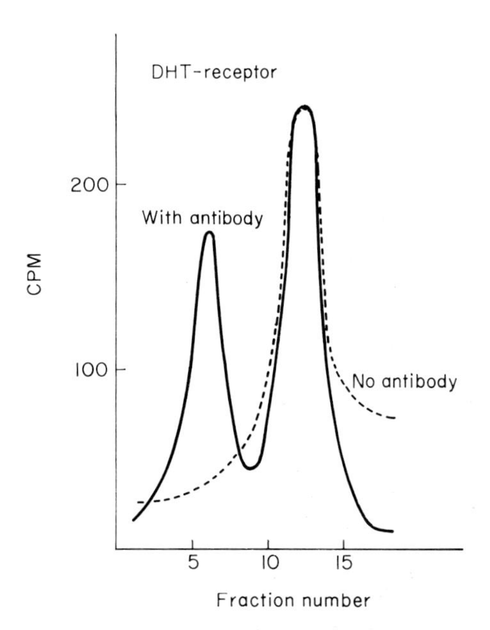

Fig. 2. Sedimentation patterns of the [^{3}H]DHT–receptor complex of rat ventral prostate in the presence and absence of the antibody for testosterone. Fractions were numbered from the bottom of the centrifuge tube.

with the radioactive steroids very often give high background counts. The sensitivity of the assay method II is about 0.5 fmole per assay tube.

Acknowledgment

This work was supported by Grants AM-09461 and HD-0711 and a research fellowship TW-01959 from the National Institutes of Health.

[5] Electron-Capture Techniques for Steroid Analysis

By MARVIN A. KIRSCHNER

Gas-liquid chromatography using electron-capture (EC) detection offered the first "rapid" approach for measurement of steroid hormones at nanogram and picogram levels. Although some of these GLC-EC methods have been superseded by radioimmunoassay techniques, they remain as useful and independent methods of hormone measurement that can be applied to biological samples of diverse origin without interference from extraneous proteins or salts that may be present in the sample.

Principle

The principles of gas-liquid chromatography have been amply considered in previous discussions.[1,2] Gas-liquid chromatography offers the opportunity to both purify and quantify a sample in a single operation by combining high efficiency chromatography with a sensitive detection system. One of the most sensitive GLC detection systems developed to date is electron-capture spectrometry,[3] utilizing the ability of certain compounds to absorb electrons. When such a compound enters a low voltage field, its ability to absorb or "capture" electrons results in disruption of the existing current. This negative deflection is amplified and results in a very sensitive detection system.

Unfortunately, most steroidal compounds have little inherent electron-capturing properties. It is thus necessary to prepare a derivative of the steroid which will enhance its ability to capture electrons, thereby

[1] H. A. Szymanski, "Biomedical Applications of Gas Chromatography." Pp. 1–39. Plenum, New York, 1964.

[2] E. C. Horning and K. B. Eik-Nes, "Gas Phase Chromatography of Steroids." Pp. 13–14 and 47–54. Springer-Verlag, Berlin and New York, 1968.

[3] R. A. Landowne and S. R. Lipsky, *Anal. Chem.* **35**, 532 (1963).

enabling the steroid derivative to register a strong signal in the EC detector.

Acyl Derivatives for EC

Polyhalogenated acyl derivatives have been most widely used in preparing steroids for EC detection. The properties of several commonly used acyl derivatives are compared in Table I.

The chloroacetates are listed for historical interest, since these were the first derivatives prepared for EC detection. The most widely used EC derivative is the heptafluorobutyrate. This derivative enhances EC properties approximately fivefold vs. chloroacetates and is easily prepared from volatile reagents which can be removed under nitrogen after the derivative is prepared. Newer derivatives such as hexadecafluorononanoates (HFN) and eicosafluoroundecanoates (EFU) produce even greater molar responses in EC spectrometry, but have longer retention times on polar and nonpolar phases and present problems of reagent residues which need purification. Since other hydroxyl-containing compounds present in the biological sample will also react to form these derivatives, it is usually necessary to perform preliminary "clean-up" of the sample before and/or after the electron-capturing derivative is prepared. Thin-layer chromatography (TLC) has been most widely used for these clean-up steps.

TABLE I
PROPERTIES OF ELECTRON-CAPTURING DERIVATIVES OF TESTOSTERONE

	Relative molar area responses	Relative retention time		Ease of removing excess reagents
		SE-30[a]	XE-60[b]	
Testosterone (no derivative)	—	1.10	2.50	—
Chloroacetate	0.22	2.86	2.50	Need solvent partition
Heptafluorobutyrate (HFB)	1.00	1.00	1.00	Blow off under N_2
Perfluorooctanoate (PFO)	1.82	1.61	1.12	Blow off under N_2
Hexadecafluorononanoate (HFN)	2.30	2.80	4.30	Need solvent partition
Eicosafluoroundecanoate (EFU)	2.55	3.46	5.10	Need solvent partition

[a] 3.0% SE-30 on Gas-chrom Q, 80–100 mesh 220°. HFB-Testo = 1.00 = 6.0 minutes.

[b] 2.0% XE-60 on Diatoport S, 60–80 mesh, 200°, HFB-Testo = 1.00 = 4.2 minutes.

Our laboratory has utilized electron-capture GLC for determining nanogram quantities of testosterone and androstenedione in biological fluids[4,5] and for providing a sensitive method to accurately determine estrone and estradiol excretion in human urine and for measuring urinary estrogen production rates.[6,7] The use of EC detection has been applied to the determination of aldosterone[8] and dehydroepiandrosterone,[9] but these will not be discussed further.

A. Determination of Testosterone Only

1. A volume of biological fluid containing at least 2 ng of testosterone is required for accurate quantification. The sample is diluted to 10 ml with distilled water, and approximately 5000 cpm [^{3}H]testosterone are added to monitor procedural losses. With each set of 24–30 samples, a standard solution containing 30 ng testosterone/10 ml and 2–10 ml water blanks are also processed.

2. One-half (0.5) ml of 1 *N* NaOH are added along with tracer, and the plasma sample is shaken to disperse the tracer and NaOH.

3. The sample is extracted twice with 25 ml ether.

4. The ether extracts are combined and washed twice with $\frac{1}{10}$ volume distilled water and allowed to dry overnight in a fume hood.

5. The sample is then transferred from a beaker to a stoppered 13-ml centrifuge tube with ether and dried under nitrogen.

6. The HFN derivative is made by adding 0.1 ml of 1% hexadecafluoronanoyl chloride reagent in benzene (Penninsular Chem Research, Gainesville, Florida). After stirring, the tube is placed diagonally in a sand bath at 56° for $\frac{1}{2}$ hour so that only the bottom of the tube is heated.

7. The reagents are dried under nitrogen.

8. Two (2) ml of ethyl acetate are added, and the excess reagents are removed by partitioning twice between ethyl acetate and 1 ml water. The water layer in each case is drawn off by a fine capillary pipette attached to vacuum.

9. The HFN derivative is dried under nitrogen, taken up with benzene, and developed by thin-layer chromatography, in the system: ben-

[4] M. A. Kirschner and G. D. Coffman, *J. Clin. Endocrinol. Metab.* **28**, 1347 (1968).

[5] M. A. Kirschner and J. P. Taylor, *Anal. Biochem.* **30**, 339(1969).

[6] D. W. R. Knorr, M. A. Kirschner, and J. P. Taylor, *J. Clin. Endocrinol. Metab.* **31**, 409 (1970).

[7] M. A. Kirschner and J. P. Taylor, *J. Clin. Endocrinol. Metab.* **35**, 513 (1972).

[8] G. L. Nicolis and J. L. Gabrilove, *J. Clin. Endocrinol. Metab.* **29**, 1519 (1969).

[9] F. H. De Jong and H. J. Van der Molen, *J. Endocrinol.* **53**, 461 (1972).

zene:ethyl acetate (90:10), along with testosterone-HFB, used as a mark. (Testo HFB has identical retention time as Testo HFN in this system.)

10. Following TLC, the markers are identified under ultraviolet light, and the sample is scraped off the plate into an inverted Pasteur pipet, packed with siliconized glass wool. The sample is eluted with 3 ml of 10% methanol in benzene and dried under nitrogen.

11. Fifty (50) to 100 μl of a solution containing 0.2 ng/ml of testosterone EFU as internal standard are added. Ten microliters are taken for counting, to correct for procedural losses and 10 μl are injected into the GLC system. The GLC systems used are either an F & M Model 400 containing tritium foil detector operating at 225° with a pulse interval of 150 μsec and a pulse width of 0.75 μsec, or a Packard Model 7400 using ^{63}Ni detector operating at 2.5 V. In both systems the electrometer gain is established such that full-scale deflection is 3×10^{-9} A. Siliconized glass columns containing 2–3% stationary phases such as SE-30, XE-60, or OV-225 on inert supports of 60–100 mesh (prepared by Supelco Inc., Bellefonte, Pa.) are used. From the GLC tracings obtained, a standard curve is established with each set of determinations as follows.

One-half (0.5) ng of standard testosterone made into its HFN derivative is mixed in a syringe with 2 ng of testosterone-EFU, or other suitable internal standard. The area responses of both compounds in GLC are then measured and expressed as a ratio. Similar ratios are obtained relating area response of 1.0 and 2.0 ng testosterone-HFN to that of 2 ng of internal standard. The ratios are plotted on the ordinate vs. nanograms testosterone on the abscissa. From the chromatogram of a given plasma sample or water blank, the area of testosterone-HFN and the internal standard are calculated and their ratios obtained. By referring this ratio to the standard curve, the amount of testosterone present in the final 10-μl aliquot is obtained. Since an equal aliquot is counted,

$$\text{Nanograms of testosterone in starting sample} = \text{ng of testosterone in final aliquot} \times \frac{\text{cpm } ^3\text{H-testo added to sample}}{\text{cpm } ^3\text{H in final aliquot}}$$

A representative GLC tracing obtained using this method is shown in Fig. 1.

B. Determination of Both Testosterone and Androstenedione in the Same Sample

1. [^{3}H]Testosterone and [^{3}H]androstenedione are added to the biological sample estimated to contain 3–5 ng of each substance.

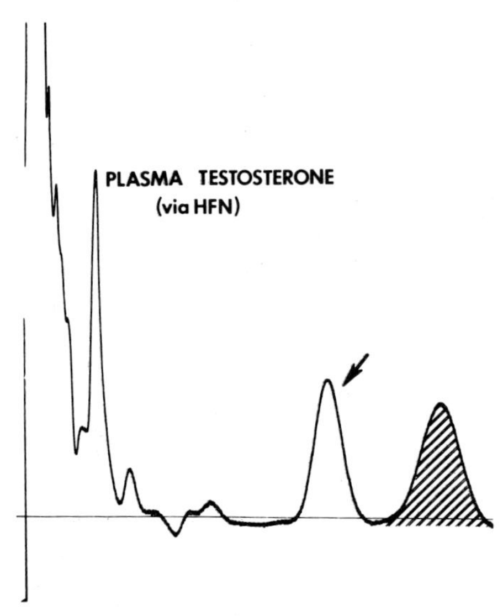

FIG. 1. GLC tracing obtained from 4 ml extract of male plasma via HFN method. The shaded area represents the internal standard. Column temp. 225°; column consisted of 3% OV-225 on Diatoport S, 60–100 mesh.

2. After addition of 0.5 ml of 1 *N* NaOH, extraction, washing, and transferring steps are the same as above.

3. Separation of androstenedione and testosterone is accomplished by TLC in the system: benzene:ethyl acetate (65:35).

4. The testosterone and androstenedione fractions are separately scraped off the plate, eluted with 4 ml ethyl acetate, and dried.

5. The testosterone fraction is converted to its heptafluorobutyrate (HFB) by addition of 0.1 ml of a solution containing 4 ml hexane, 80 μl heptafluorobutyric anhydride (Penninsular Chem Research, Gainesville, Florida), and 40 μl tetrahydrofuran. After stirring, the tube is placed in a sand bath for 30 minutes, as above.

6. Following incubation, the excess reagents are dried under nitrogen, and the HFB is taken up in benzene and developed in a second thin-layer chromatographic system: benzene:ethyl acetate (90:10) along with testosterone-HFN, as marker. (The partition step to get rid of excess reagent is not required with HFB.)

7. Following TLC No. 2, the sample is eluted with 10% methanol in benzene, dried, and GLC is performed. Sample GLC tracings obtained by this method are presented in Fig. 2.

8. The androstenedione fraction is eluted with 4 ml ethyl acetate,

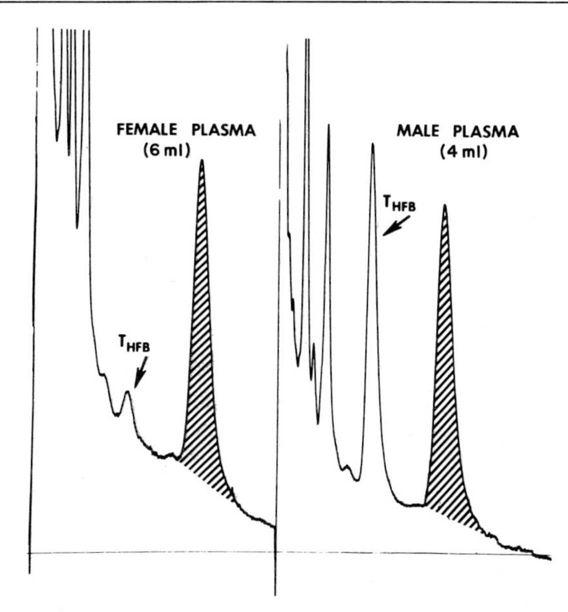

FIG. 2. GLC tracings obtained via "HFB" method. Column temp. 210°; column consisted of 3% OV-225 on Diatoport S, 60–100 mesh.

dried, and converted to testosterone by addition of borohydride as follows:

Four-tenths (0.4) ml ethanol is added to the dried sample, followed by 20 μl of the freshly made borohydride solution (20 mg KBH_4 dissolved in 1.0 ml water). The mixture is stirred and left to stand 20 seconds *by the clock*. The reaction is then terminated by adding 3 drops of glacial acetic acid and immediately stirring. Water is then added (1.6 ml), and the newly formed testosterone is extracted once with 8 ml methylene chloride.

9. The sample is then dried and converted to the HFB and treated thereafter as testosterone, with repeat TLC, HFB, formation, and GLC.

C. Determination of Testosterone in Urine

1. Since testosterone is excreted in human urine primarily as a glucuronide conjugate, hydrolysis of this conjugate precedes the analysis as follows:

One-twentieth to one-tenth of a 24-hour urine is adjusted to pH 5.0 by the addition of $\frac{1}{10}$ volume of 1 *M* acetate buffer and is incubated with 500 Fishman units of Ketodase/ml × 3 days at 37°.

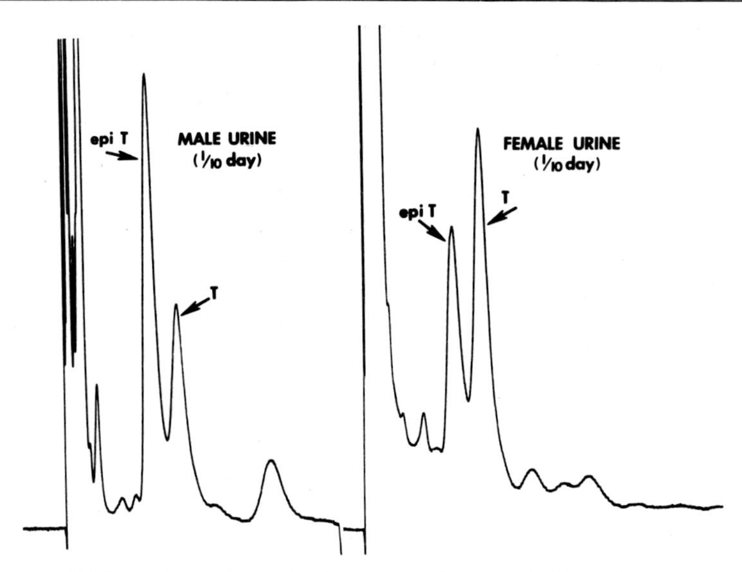

FIG. 3. GLC tracings obtained from normal male (left) and normal female urines (right), processed by "HFB" method.

2. [^{3}H]Testosterone tracer is added and the incubated urine is extracted three times with equal volumes of ether. The ether extract is washed twice with $\frac{1}{10}$ volume of 1 N NaOH, followed by a $\frac{1}{10}$ volume water wash. The ether is dried in a fume hood, transferred to a centrifuge tube, and the procedure B above is followed. A representative GLC-EC tracing is shown in Fig. 3.

D. Determination of Estrone and Estradiol in Low-Level Urines

Electron-capture GLC has been successfully applied to quantify estrone (E_1) and estradiol (E_2) excretion in urines of normal subjects, postmenopausal women, and endocrine-ablated patients.[6] This method has also been applied to estimate the mass of urinary E_1 and E_2 as part of urinary production rate determinations.[7] In this method, E_1 and E_2 are isolated, purified, and converted into estradiol, 3-methyl ether, 17-hexadecafluoronanoate which is then quantified by EC-GLC.

Method in Detail

1. An aliquot of urine estimated to contain at least 4 ng of E_1 and E_2 is chosen. Generally $\frac{1}{10}$ of a day's urine is used, except for children or endocrine-ablated patients, where $\frac{1}{3}$ of a 24-hour collection is used.

The urine is adjusted to pH 5.0 with 1 *M* acetate buffer and incubated for 72 hours with 500 Fishman units/ml of Ketodase. Acid hydrolysis has uniformly resulted in lower values for urinary E_1 and E_2 (presumably due to destruction) and is thus not used for low-level urines.[7]

The incubated urine is extracted three times with equal volumes of ether, and the ether extract is partitioned after the method of Brown,[10] using combinations of NaOH–bicarbonate. The washed extracts are taken to dryness in a fume hood overnight, transferred to centrifuge tubes, and spotted on thin-layer plates.

2. Thin-layer chromatography No. 1—the extracts are developed in the system: benzene:ethyl acetate, 63:35. The E_1- and E_2-containing zones are separately scraped from the plate and eluted with 4 ml ethyl acetate.

3. Methylation—the microprocedure of Bush[11] is used as follows: One-tenth (0.1) ml ethanol is added to the dried sample followed by 2.0 ml of a mixture containing 25 ml 2% boric acid, 0.5 ml dimethyl sulfate, and sufficient 20% NaOH to adjust pH to 11.0 to 11.5. The tubes are shaken by hand for 1 minute, followed by 20-minute incubation at 37°. An additional 0.05 ml of dimethyl sulfate is added along with 0.1 ml of 20% NaOH and the sample is shaken and incubated an additional 10–15 minutes. On completion of the incubation, 0.5 ml 25% NaOH is added and the methyl ethers are extracted once with 9 ml of ethyl acetate, washed with 2 ml 12% acetic acid and then 2 ml water. The entire methylation and extraction procedure is carried out in a single stoppered centrifuge tube. The lower layer is drawn off by a fine capillary pipet attached to house vacuum. The ethyl acetate is then dried.

4. Borohydride reduction of estrone methyl ether (E_1-Me). The E_1-Me is dissolved in 0.1 ml ethanol, then 0.2 ml of borohydride reagent (80 mg KBH_4 in 10 ml 60% ethanol) is added. The samples are stirred and after 4 minutes the reaction is terminated by addition of 1 ml 12% acetic acid. After addition of 1 ml water, the product (E_2-Me) is extracted once with 9 ml ethyl acetate and the ethyl acetate layer is washed with 2 ml water. Centrifugation is usually required to separate the phases at this point. The samples are then dried.

5. Thin-layer chromatography No. 2. Four samples are spotted on a 20-cm plate and developed in the system: benzene:ethyl acetate (80:20). Samples are eluted as above with 4 ml ethyl acetate and dried.

6. HFN formation. The dried samples are dissolved in 0.1 ml of 1% HFN-acid chloride in benzene as previously described. The tubes are

[10] J. B. Brown, *in* "Estrogen Assays in Clinical Medicine" (C. A. Paulson, ed.), p. 346. Univ. of Washington Press, Seattle, 1966.

[11] I. E. Bush, "Chromatography of Steroids," p. 367. Pergamon, Oxford, 1961.

stirred and placed diagonally in a sand bath at 56° for 30 minutes. Excess reagents are blown off under nitrogen. The samples are dissolved in 2 ml ethyl acetate and washed with an equal volume of distilled water to remove reagent residues. Mechanical stirring at this point seems more effective than shaking by hand. The washed samples are then dried under nitrogen.

7. Thin-layer chromatography No. 3. Five samples per plate are spotted and developed in the system: hexane:ether:acetic acid (100:10:0.5). The R_f for E_2-Me, 17-HFN in this system is 0.55. The samples are eluted with 4 ml of benzene:methanol mixture (10:1) and dried for GLC.

8. Gas-liquid chromatography. The sample is dissolved in 50 μl benzene containing 0.1 ng/μl of epitestosterone-HFN or testosterone-HFN used as internal standard. Ten (10) μl are taken for counting at the same time that 10 μl or less is introduced into the GLC system. Area responses of standard amounts of E_2 made into the E_2-Me, 17-HFN derivatives and internal standard are calculated as above. Representative GLC tracings representing a standard, urinary E_1 fraction, E_2 fraction, and water blank are shown in Fig. 4.

In many cases a minimal deflection at the proper R_t for E_2-ME, 17-HFN is noted in the GLC tracing of the accompanying water blanks. This small peak (Fig. 4D) is quantified, corrected for recovery, and this "water blank" value is subtracted from the quantity found in the starting sample. The value assigned to the water blank represents a "maximal

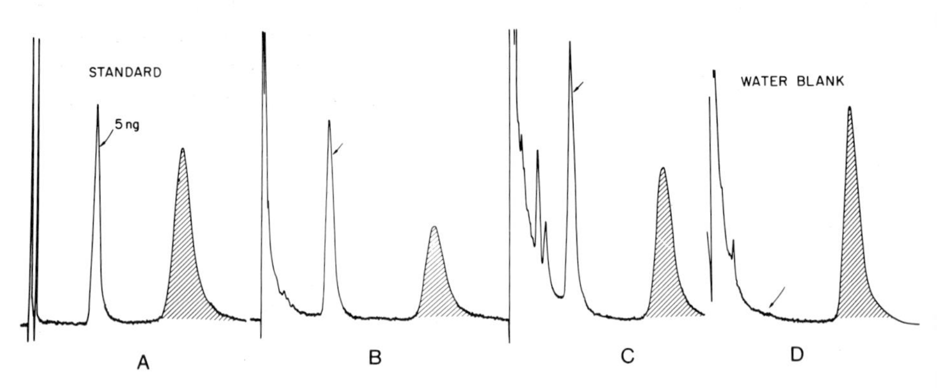

FIG. 4. GLC tracings of urinary estrogens in 35-year-old man. One-tenth of a 24-hour urine collection was used. (A) Five nanograms of estradiol made into its 3-Me, 17-HFN along with internal standard (shaded). (B) Estrone fraction. (C) Estradiol fraction, (D) Water blank showing minimal deflection corresponding to E_2-3Me, 17-HFN. (Reproduced by permission from *Journal of Clinical Endocrinology & Metabolism.*)

TABLE II
LIMITS AND PRECISION ESTIMATES OF ELECTRON-CAPTURE GLC METHODS

	Minimal sample size (ng/sample)	Coefficient of variation	
		(ng/sample)	(percent)
Testosterone (via HFN)	2	<7	30
		7–15	13
		>15	4
Testosterone (via HFB)	3	<8	28
		8–15	11
		>15	3
Urinary estrone	4	<20	18
		>20	8
Urinary estradiol	4	<20	25
		>25	6

estimate." In general, the quantity of E_1 or E_2 in the starting sample is many times that of the water blank, making this correction unnecessary.

Comments

The above procedures for determining testosterone, androstenedione, and urinary estrogens have been shown to be specific for the substance being measured, accurate, and contain adequate precision at lower levels of measurement (Table II). Although some of the newer radioimmunoassays are indeed more sensitive (require smaller samples), the use of the EC-GLC approach serves as an independent method of hormone analysis at lower levels.

[6] Methods for Monitoring *in Vivo* Steroid Hormone Production and Interconversion Rates*

By C. WAYNE BARDIN, LESLIE P. BULLOCK, and DONALD PROUGH

The production rate of a steroid is defined as the mass of that hormone which enters the blood per unit time from all sources. For an individual

* Supported in part by PHS Grant No. HD05276 and Contract No. NIH-NICHD-72-2730.

steroid, this is equal to that secreted from endocrine glands plus that formed in peripheral tissue from prehormones. Prehormones may be defined as substances which are converted peripherally to more potent compounds which then contribute significantly to an overall biological effect.[1] The peripheral conversion can be in any tissue including target organs. Androgens and estrogens are notable examples of groups of steroids which may interconvert to alter their biologically active blood pools. For example, several relatively nonactive androgens may be converted in a variety of tissues to testosterone and dihydrotestosterone which may be active locally at the site of production or which may reenter the blood and contribute to the overall total androgen pool. Under certain conditions, androgens such as testosterone and androstenedione may be prehormones for the blood estrogens, estrone, and estradiol, which interconvert with one another. In the discussion which follows, techniques are presented for studying the *in vivo* interconversion and production rates of steroids in blood and in tissue. Although testosterone and androstenedione will be used as typical examples, it should be emphasized that the methods are general and may be applied to any group of interconverting compounds.

Theory and Calculations

The first technique for studying androgen production and interconversion in blood utilized the specific activity of appropriate urinary metabolites after simultaneous administration of the ^{3}H- and ^{14}C-compounds under study. The validity of this model depends on several assumptions, one of which is that the urinary metabolites are uniquely derived from the specific plasma compartments.[2] That the procedures using urinary metabolites would not work for androgens was first suggested by Korenman and Lipsett,[3,4] who demonstrated that testosterone is synthesized in a tissue compartment from plasma dehydroepiandrosterone and androstenedione and then conjugated without reentry into the blood as free steroid. Thus, testosterone glucuronide is not uniquely derived from the plasma testosterone compartment. These findings, coupled with the observation that the testosterone production rates of women are higher when calculated from the specific activity of urinary testosterone glucuronide

[1] D. T. Baird, R. Horton, C. Longcope, and J. F. Tait, *Perspect. Biol. Med.* **11**, 384 (1968).
[2] R. L. Vande Wiele, P. C. MacDonald, E. Gurpide, and S. Lieberman, *Recent Progr. Horm. Res.* **19**, 275 (1963).
[3] S. G. Korenman and M. B. Lipsett, *J. Clin. Invest.* **43**, 2125 (1964).
[4] S. G. Korenman and M. B. Lipsett, *Steroids* **5**, 509 (1965).

than when estimated from the specific activity of plasma testosterone,[5] indicated that the urinary isotope dilution technique overestimates both blood production and blood interconversion rates.

Tait and Horton[6] described a steady-state model for androgen interconversion which utilized two anatomical (blood and tissue) and two chemical (testosterone and androstenedione) compartments (Fig. 1). Although this is the simplest model which can be described for the production and interconversion of testosterone and androstenedione, it is readily apparent that as the precursors of plasma testosterone such as dehydroepiandrosterone are included, the model will become increasingly complex. The interconversion of the two androgens in the same and between anatomical compartments are defined by "ρ values." Although the tissue compartment of this system has not been fully characterized under steady-state conditions, the blood production rates and the overall interconversions [ρ] (ρ values) can be determined from the specific activity of androgens in the blood compartment following introduction of radioactive androgens into blood. The production of steroids into the blood compartment can be estimated by the product of the metabolic clearance rate (MCR) and the plasma steroid concentration. The MCR is defined as the volume blood or plasma irreversibly cleared of steroid per unit time.[6]

The symbols and calculations for the several parameters of two interconverting steroids such as testosterone and androstenedione are those of Tait and Horton.[6] The superscripts indicate steroid, the subscripts the compartment (Fig. 1). The symbol z refers to the isotope in testosterone and x to the isotopes in androstenedione. The formulas are as follows:

$$MCR^{T} = Rz^{T}/z^{T} \quad \text{and} \quad MCR^{A} = Rx^{A}/x^{A}$$

where MCR^{T} is the metabolic clearance rate of testosterone, Rz^{T} is the disintegrations per minute of testosterone infused per unit time, and z^{T} is the concentration of isotope in testosterone in disintegrations per minute per unit volume of blood. The androstenedione expressions are analogous.

$$P_{B}{}^{T} = MCR^{T} \times i^{T} \quad \text{and} \quad P_{B}{}^{A} = MCR^{A} \times i^{A}$$

where $P_{B}{}^{T}$ is the blood testosterone production rate and i^{T} is the blood testosterone concentration.

$$C_{BB}{}^{AT} = x^{T}/x^{A}$$

[5] J. F. Tait and R. Horton, *Steroids* **4**, 365 (1964).

[6] J. F. Tait and R. Horton, *in* "Steroid Dynamics" (G. Pincus, T. Nakao, and J. F. Tait, eds.), p. 393. Academic Press, New York, 1966.

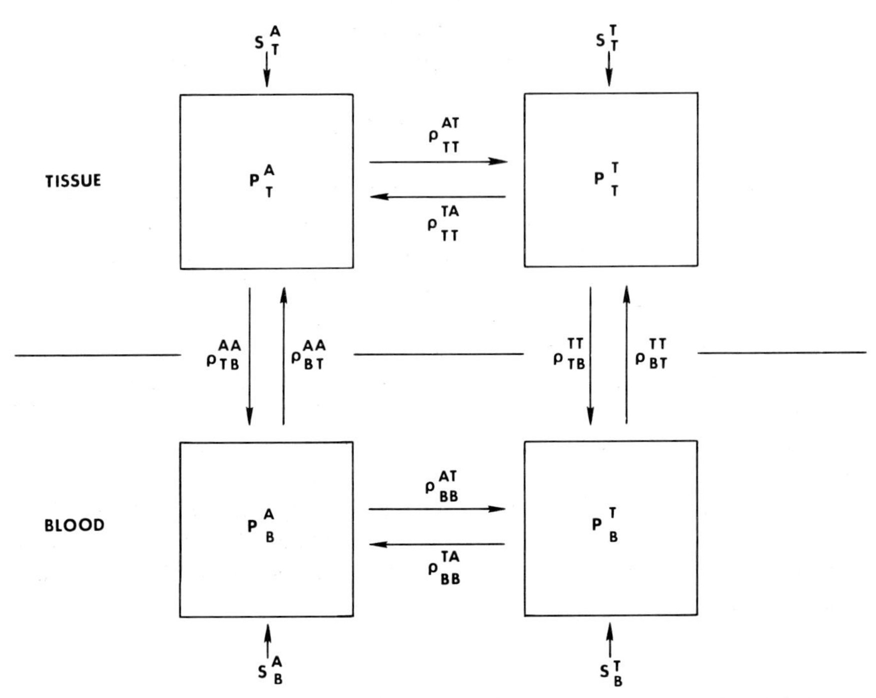

FIG. 1. Model for secretion and interconversion of testosterone and androstenedione. P_B^T equals production of testosterone in blood; P_T^T equals production of testosterone in tissue; S_B^T equals secretion of testosterone in blood; the symbols for androstenedione are analogous. ρ is defined as the fraction of one compartment, pool or production rate that is converted to another compartment, pool or production rate.[7,8] When written according to Tait and Horton[6] the superscripts represent the steroid or chemical compartments involved, e.g., TA = testosterone to androstenedione; and the subscripts are the anatomical compartments involved, e.g., TB = from tissue to blood, BB = in the blood. With the use of both superscripts and subscripts, the entire process is described. For example, ρ_{BB}^{AT} equals the fraction of androstenedione converted to testosterone in the blood compartment. The "ρ values" refer to interconversions between specific compartments and do not involve pathways through intermediate compartments. The $[\rho]$ values are those obtained *in vivo* irrespective of the number of participating pathways and compartments.

where C_{BB}^{AT} is the conversion ratio of androstenedione to testosterone in blood.

$$[\rho]_{BB}^{AT} = (MCR^T/MCR^A) \times x^T/x^A$$

[7] E. Gurpide, P. C. MacDonald, R. L. Vande Wiele, and S. Lieberman, *J. Clin. Endocrinol. Metab.* **23**, 346 (1963).

[8] E. Gurpide, J. Mann, and S. Lieberman, *J. Clin. Endocrinol. Metab.* **23**, 1155 (1963).

where $[\rho]_{BB}{}^{AT}$ is the fraction of the blood androstenedione pool converted to the testosterone pool. This expression is also termed a "transfer factor."

$$([\rho]_{BB}{}^{AT} \times P_B{}^A)/P_B{}^T = (x^T/x^A) \times (i^A/i^T)$$

where the expression is the fraction of the blood testosterone production that is derived from the blood androstenedione production. A similar expression can be derived from androstenedione production from testosterone. The correction factor, $1 - [\rho]_{BB}{}^{AT} \times [\rho]_{BB}{}^{TA}$, for the amount of precursor originating from product is ignored, since the product of the two $[\rho]$ values is approximately 0.03.

The statistical analysis of conversion ratios ($C_{BB}{}^{AT}$) and transfer factors ($[\rho]_{BB}{}^{AT}$) are very similar to the statistical analysis of $^3H/^{14}C$ ratios. It is easier to base the statistical analysis of a ratio on its natural log rhythm. The variance of the natural log rhythm of the ratio (R) can be calculated from the Taylor series expansion argument and its standard error is SE (ln R) = square root of the variance (ln R). This number can be used to calculate the 95% confidence interval for R. For the details of this type of analysis, see Hembree *et al.*[9]

Procedure for Blood Studies

The technique for the simultaneous determination of testosterone and androstenedione clearances and interconversions using two isotopes is presented.[10] If, however, one wishes to study the conversion of a single steroid such as testosterone to multiple blood products (as in Fig. 2), then only tritiated precursor should be infused. The procedure described has been used for studies in man and other large animals such as donkey, sheep, and dog. For small animals the modifications described below for tissue may be used.

All subjects are studied under basal conditions, if possible. Approximately 10–15 μCi of [^{3}H]androstenedione and 0.25–0.5 μCi of [^{14}C]testosterone are injected intravenously in 20 ml of 5% ethanol in saline as a priming dose, and starting 30 minutes later, three times these quantities are given as a constant infusion for 120 minutes in 200 ml of 5% ethanol in saline. With the infusion of androstenedione and testosterone at a disintegration per minute ratio of 30:1, sufficient tritium and carbon counts are present to determine the metabolic clearance of both steroids and

[9] W. C. Hembree, C. W. Bardin, and M. B. Lipsett, *J. Clin. Invest.* **48**, 1809 (1969).

[10] C. W. Bardin and M. B. Lipsett, *J. Clin. Invest.* **46**, 891 (1967).

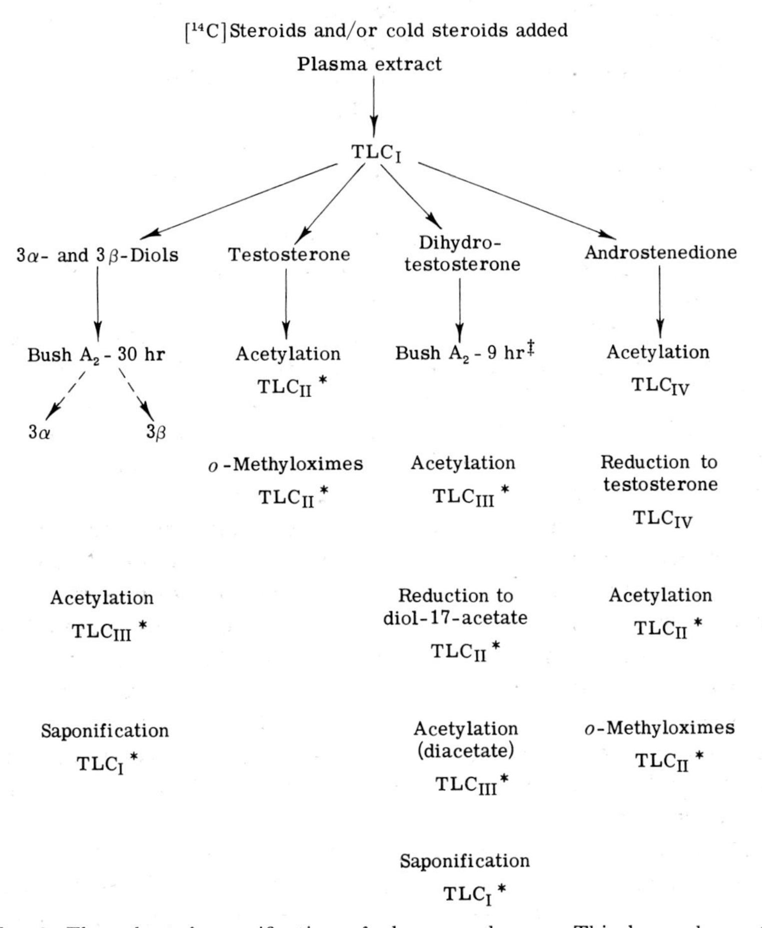

FIG. 2. Flow chart for purification of plasma androgens. Thin-layer chromatography systems used: I, chloroform:methanol (98:2); II, benzene:ethyl acetate (8:2); III, benzene:ethyl acetate (9:1); IV, benzene:ethyl acetate (6:4). Paper chromatography [Bush A$_2$, ligroine:methanol:water (10:7:3)]. (*) Fraction counted for a specific activity determination. (‡) This paper chromatography is necessary in testosterone and androstenedione infusions to separate androsterone from dihydrotestosterone, but in dihydrotestosterone infusions the separation by the two following steps are sufficient and the paper step is omitted. Adapted from Mahoudeau *et al.*[10]

the conversion ratio of androstenedione to testosterone. If the conversion ratio of testosterone to androstenedione is to be measured, then the amount of [14C]testosterone infused should be doubled.

The radioactive steroids are infused with the Bowman Infusion Pump equipped with a siliconized latex pump tube and Teflon tubing. The latex pump tubing is replaced after every fifth infusion. The tubing effluent is monitored at frequent intervals, generally three to five times during the infusion to determine whether actual rate of infusion of labeled steroids agrees closely with the calculated rate in each instance. The metabolic clearance rate is then calculated from the steroid infusion rate and the concentration of isotopic steroid per volume of plasma as noted above.

Isotopic testosterone and androstenedione concentrations are determined by reverse isotope dilution on 40 ml of heparinized blood obtained at 15–20-minute intervals over the last hour of infusion. The plasma is separated and 200 μg of each testosterone and androstenedione are added. The plasma samples are extracted twice with two volumes of ether after the addition of 1 ml of 1.0 *N* NaOH per 20 ml of plasma. The extracts are washed with water, dried, and purified by chromatography and derivative formation as outlined in Fig. 2. Acetylation and suponification were accomplished as previously described.[10] *o*-Methyloximes were made according to Fales and Luukkainen.[11] Steroid reductions were performed with freshly prepared 2% potassium borohydride in water. Twenty seconds were required to convert androstenedione to testosterone and 60 seconds to reduce dihydrotestosterone-17β-acetate to 3β-hydroxy-5α-androstane-17γ-acetate, in over 90% yield.[12] After chromatography of each derivative, the specific activity of the plasma steroid was determined. The plasma concentration of isotopes in the individual steroids was calculated from the recovery of added unlabeled steroid when two isotopes were infused simultaneously[10] and from the recovery of ^{14}C when only a single [^{3}H]steroid was infused.[12]

Whenever possible, measurements made on plasma samples are used for calculation of production and interconversion rates. These determinations from plasma can be corrected by the percent red cell volume (hematocrit) to obtain corresponding values for blood. In several instances, however, steroids may bind to red cells and in such cases an additional correction factor is required.[13] This difficulty is entirely obviated if whole blood rather than plasma is used for the steroid extraction. Blood extraction is essential when studying androgen kinetics in sheep since red cells can metabolize testosterone and androstenedione even in the brief period required to remove them from plasma.[14]

[11] H. M. Fales and T. Luukkainen, *Anal. Chem.* **37**, 955 (1965).

[12] J. A. Mahoudeau, C. W. Bardin, and M. B. Lipsett, *J. Clin. Invest.* **50**, 1338 (1971).

[13] D. T. Baird, R. Horton, C. Longcope, and J. F. Tait, *Recent Progr. Horm. Res.* **25**, 611 (1969).

[14] R. Lippert, M. Borger, and C. W. Bardin, *Steroids and Lipid Res.* (in press).

Procedure for Tissue

The techniques for studying the steady-state conversion of product to precursor in tissues are not as well developed as those for blood. Nonetheless, a procedure has been developed for infusing steroids into rats for 6 to 8 hours. Using this technique, tissue concentrations of precursors such as testosterone can be related to the levels of various products such as dihydrotestosterone and the androstanediols (5α-androstane-3α,17β-diol and 5α-androstane-3β,17β-diol).

A rat is lightly anesthetized with ether and a 19-gauge, 12-in. Teflon intercath inserted into the external jugular vein. The animal is then placed into a rat holder (Plas-Labs) and the intercath is connected to a syringe containing 5 ml of [^{3}H]testosterone (10 μCi) in 5% ethanol (75 to 750 ng/ml). One-fifth (0.4) ml of the infusate is given from the syringe as a priming dose and the remainder is infused at a rate of 0.58 ml/hour. Blood samples (0.25 ml) are collected from a tail vein hourly during the last 4 hours of infusion. The animals are killed with d-tubocurarie and the tissues to be studied are rapidly removed and placed on ice. From

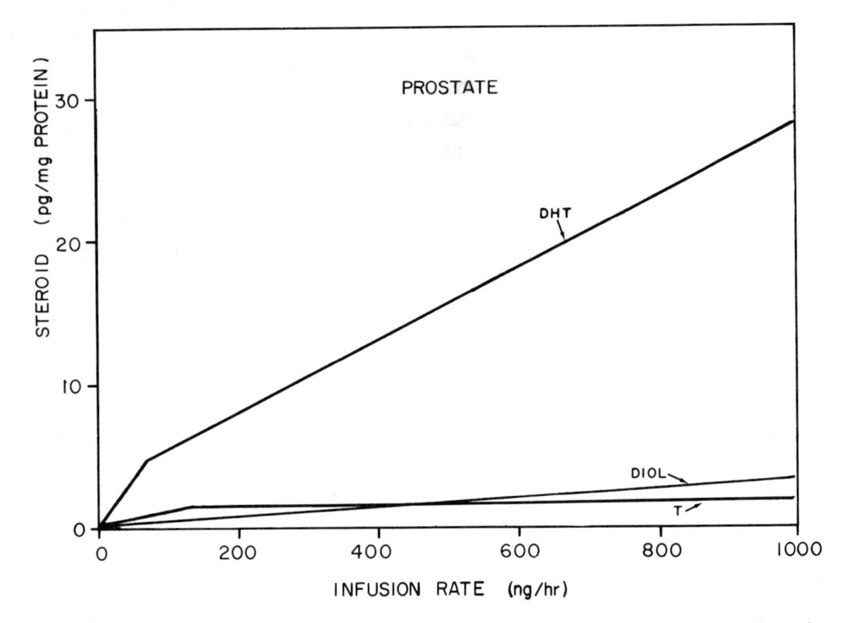

FIG. 3. Androgen levels in prostate of a castrate rat as a function of various testosterone infusion rates. T, testosterone; DHT, dihydrotestosterone; DIOL, 5-α androstanediols. The relative concentrations of these steroids were similar in preputial glands but in muscle all three (T, DHT, and DIOL) were similar to T and DIOL shown above.

75 to 1000 mg of each tissue are minced, placed in 1 to 5 ml of saline along with [^{14}C]steroids for recovery. Samples are homogenized and extracted immediately with methylene chloride. The extracts are washed twice with water and dried. Precursor testosterone and various products are isolated from blood and tissues are illustrated in Fig. 2. In most studies, the blood concentration of testosterone metabolites reaches a steady state within 2 to 4 hours. The tissue levels of testosterone and several of its metabolites that can be expected in prostate at various infusion rates of testosterone are shown in Fig. 3.

[7] A Tracer Superfusion Method to Measure Rates of Entry, Exit, Metabolism, and Synthesis of Steroids in Cells

By ERLIO GURPIDE

An incubation method suited for the study of steroid dynamics in cell suspensions or tissue slices has been developed recently.[1,2] The method is characterized by a continuous flow of medium containing two tracers between which isotope exchange can be observed. If these two compounds are called A and B, isotopic data obtained from perfusate and tissue when a steady state is reached serve to estimate the following parameters:

(a) rates of entry of A and B into the cells
(b) rates in intracellular conversion of A to B or B to A and total rates of irreversible metabolism of these compounds to other products
(c) rates of exit of A and B from the cells
(d) rates of synthesis of A and B in the cells

In this description of the tracer superfusion method, the derivation of formulas which relate rates to experimental data is followed by comments on tests necessary to validate the calculations and investigate the existence of intracellular compartmentalization.

Derivation of Formulas

Rates of Movement and Exchange of Material in Two Pools Embedded in a System of Multiple Pools (Fig. 1). The formulas applicable to

[1] E. Gurpide and M. Welch, *J. Biol. Chem.* **244**, 5159 (1969).
[2] L. Tseng, A. Stolee, and E. Gurpide, *Endocrinology* **90**, 390 (1972).

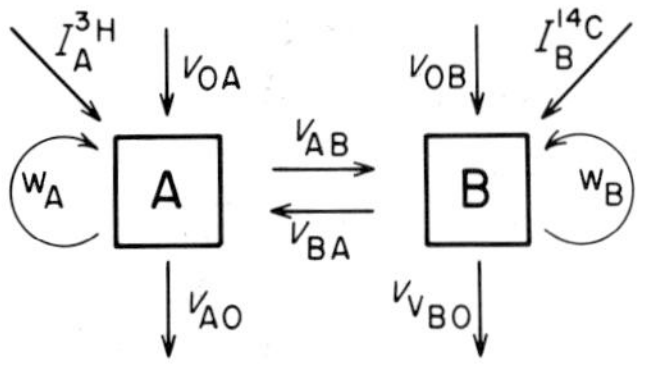

FIG. 1. Model representing 2 pools embedded in a multicompartmental system. The v's denote rates (nmoles/hour), the w's denote rates of recycle of material around each of the 2 pools without the intermediacy of the other, and $I_A{}^{3H}$ and $I_B{}^{14C}$ are the rates of infusion (dpm/hour) of tracers ^{3}H-A and ^{14}C-B, respectively.

the superfusion system are derived with the aid of the model shown in Fig. 1. In this model, A and B denote the intracellular pools of compounds A and B. The rates v_{AB} and v_{BA} represent the rates of exchange of material between the two pools (expressed in nmole/hour); v_{OA} and v_{OB} are the rates at which the compounds are entering the pools from other sources, for the first time; v_{AO} and v_{BO} denote the rates of irreversible removal of material from each of these pools; w_A corresponds to the rate at which compound A leaves the pool and returns to it without the intermediate formation of compound B, and w_B is the rate of recycle of the material in pool B by paths not involving pool A.

A most important feature of this model is the implicit inclusion of an undetermined number of pools exchanging with A and B. These other pools may represent other spaces of distribution of compounds A and B or other compounds which are reversibly converted to A or B. Thus, the conversion of A to B indicated in Fig. 1 is not necessarily direct but may occur through other intermediate pools. Similarly, the recycling indicated by the rates w_A and w_B may involve the reversible conversion of A and B to other compounds. At the steady state, the rates of entry and exit of material into and out of a pool are identical, as expressed by the following equations:

$$v_{OA} + v_{BA} + w_A = v_{AO} + v_{AB} + w_A$$

and

$$v_{OB} + v_{AB} + w_B = v_{BO} + v_{BA} + w_B$$

This particular choice of definitions of rates facilitates their calculation from isotopic data.[3]

If a tracer of compound A is infused into pool A at a constant rate (dpm/hour) and the infusion is continued for an adequate period of time, constant values of the specific activities (dpm/nmole) of compounds A and B can be expected. If a tracer of compound B is simultaneously infused, another set of steady-state specific activities is obtained. If, for instance, ^{3}H-A and ^{14}C-B are infused simultaneously, four different specific activities $a_A{}^{^3H}$, $a_B{}^{^3H}$, $a_A{}^{^{14}C}$, and $a_B{}^{^{14}C}$ are experimentally obtained at the isotopic steady state. The symbol a denotes a specific activity at the steady state, the superscript indicates the isotope and the subscript the compound. All the v's in Fig. 1 can be estimated from these specific activities and the rates of infusion of the tracers. The corresponding formulas are derived from equations expressing the conditions of isotopic steady state in each of the pools, as follows.

Consider the isotope balance with respect to ^{3}H in pool A. At the steady state, the rates at which ^{3}H-A enters and leaves the pool are equal, i.e.,

$$I_A{}^{^3H} + v_{BA}a_B{}^{^3H} + w_A a_A{}^{^3H} = (v_{AB} + v_{AO} + w_A)a_A{}^{^3H}$$

In this expression, $I_A{}^{^3H}$ denotes the rate of infusion of ^{3}H-A, the product $v_{BA}a_B{}^{^3H}$ corresponds to the rate at which ^{3}H coming from pool B enters pool A, and $w_A a_A{}^{^3H}$ corresponds to the rate at which labeled compound A is returning to the pool without passing through pool B. The rate at which ^{3}H-A leaves the pool is given by the sum of all the rates of exit multiplied by the specific activity of A with respect to ^{3}H. The term $w_A a_A{}^{^3H}$ appears in both sides of the equation and can be eliminated, i.e.,

$$I_A{}^{^3H} + v_{BA}a_B{}^{^3H} = (v_{AB} + v_{AO})a_A{}^{^3H} \tag{1}$$

The balance of the rates of entry and exit of ^{14}C into and out of pool A at the steady state leads to the following:

$$v_{BA}a_B{}^{^{14}C} = (v_{AB} + v_{AO})a_A{}^{^{14}C} \tag{2}$$

From Eqs. (1) and (2), the values of the unknowns v_{BA} and $(v_{AB} + v_{AO})$ can be calculated from the experimental data using for instance the method of determinants.

The equations describing the steady-state conditions with respect to ^{3}H and ^{14}C in pool B are the following:

$$v_{AB}a_A{}^{^3H} = (v_{BA} + v_{BO})a_B{}^{^3H} \tag{3}$$

$$I_B{}^{^{14}C} + v_{AB}a_A{}^{^{14}C} = (v_{BA} + v_{BO})a_B{}^{^{14}C} \tag{4}$$

Solution of the system of Eqs. (3) and (4) yields values for v_{AB} and $(v_{BA} + v_{BO})$ in terms of the specific activities and the infusion rate of ^{14}C-B.

It follows from these results that all the v's in Fig. 1 can be estimated from the experimental data. The corresponding formulas are summarized in Table I. Note that the w's are not obtained from these data.

TABLE I
LIST OF FORMULAS APPLICABLE TO THE MODEL IN FIG. 1[a]

Parameter	Formula
Rates	$(D = a_A^{^3H} a_B^{^{14}C} - a_B^{^3H} a_A^{^{14}C})$
	$v_{OA} = I_A^{^3H}(a_B^{^{14}C} - a_A^{^{14}C})/D$
	$v_{OB} = I_B^{^{14}C}(a_A^{^3H} - a_B^{^3H})/D$
	$v_{AB} = I_B^{^{14}C} a_B^{^3H}/D$
	$v_{BA} = I_A^{^3H} a_A^{^{14}C}/D$
	$v_{AO} = (I_A^{^3H} a_B^{^{14}C} - I_B^{^{14}C} a_B^{^3H})/D$
	$v_{BO} = (I_B^{^{14}C} a_A^{^3H} - I_A^{^3H} a_A^{^{14}C})/D$
Transfer or conversion factors	$\rho_{AB} = v_{AB}/(v_{AB} + v_{AO}) = \left(\frac{^3H}{^{14}C}\right)_B \div \left(\frac{^3H}{^{14}C}\right)_{infused}$
	$\rho_{BA} = v_{BA}/(v_{BA} + v_{BO}) = \left(\frac{^3H}{^{14}C}\right)_{infused} \div \left(\frac{^3H}{^{14}C}\right)_A$
Contribution factors	$\Delta_{AB} = v_{AB}/(v_{AB} + v_{OB}) = a_B^{^3H}/a_A^{^3H}$
	$\Delta_{BA} = v_{BA}/(v_{BA} + v_{OA}) = a_A^{^{14}C}/a_B^{^{14}C}$
Production rates	$PR_A = v_{OA} + \rho_{BA} v_{OB} = I_A^{^3H}/a_A^{^3H}$
	$PR_B = v_{OB} + \rho_{AB} v_{OA} = I_B^{^{14}C}/a_B^{^{14}C}$

[a] Symbols: $I_A^{^aH}$, infusion rate (dpm/hour) of ^{3}H-A into pool A; $a_A^{^3H}$, specific activity (dpm/nmole) with respect to ^{3}H of the compound in pool A, at the steady state; $(^3H/^{14}C)_A$, isotope ratio in pool A, at the steady state; $(^3H/^{14}C)$ infused, ratio (dpm/dpm) of infusion rates of tracers. A similar notation is used for pool B and the ^{14}C isotope. The v's (nmoles/hour) are rates as indicated in Fig. 1. The basic formulas shown in this table were obtained by solving the system of Eqs. (1) to (4).

Table I also includes other parameters which can be calculated from the isotopic data and are found useful in defining the system. One of these parameters is the transfer or conversion factor (ρ) of material from one pool to another. The definition of the transfer factors from A to B (ρ_{AB}) in terms of rates and the resulting expression in terms of isotopic data are given as

$$\rho_{AB} = \frac{v_{AB}}{v_{AB} + v_{AO}} = \frac{I_B^{^{14}C}\, a_B^{^3H}}{I_A^{^3H}\, a_B^{^{14}C}} = \frac{(^3H/^{14}C)_B}{(^3H/^{14}C)_{infused}} \tag{5}$$

It follows from these formulas that ρ_{AB} indicates the fraction of new material coming into pool A that is transferred to pool B.[3] Note that a transfer factor can be estimated simply from isotopic ratios in the infusion mixture and in one of the pools without need for measuring specific activities. Also note that the isotope ratio in B is established by the extent

[3] J. Mann and E. Gurpide, *J. Clin. Endocrinol. Metab.* **26**, 1346 (1966).

of conversion of A to B; further metabolism of B does not alter this ratio.

Another parameter listed in Table I is the contribution factor of pool A to pool B (Δ_{AB}). This factor is defined by the following ratio of rates and the corresponding experimental expression:

$$\Delta_{AB} = \frac{v_{AB}}{v_{AB} + v_{OB}} = \frac{a_B{}^{^3H}}{a_A{}^{^3H}}$$

The contribution factor Δ_{AB} indicates the fraction of B made up by material received from pool A; $(1 - \Delta_{AB})$ represents the contribution of other sources to the formation of pool B.

The last parameter included in Table I is the production rate (PR) of a compound in a pool, i.e., the rate at which it appears *de novo* in the pool. It follows from the previous definitions of rates and transfer factors that

$$PR_A = v_{OA} + \rho_{BA}v_{OB} = I_A{}^{^3H}/a_A{}^{^3H}$$

The experimental expression to estimate production rates from isotopic data is intuitively evident. For instance, the steady-state specific activity of compound A with respect to ^3H should equal the ratio of the rate at which the labeled and unlabeled compound appear *de novo* in the pool ($a_A{}^{^3H} = I_A{}^{^3H}/PR_A$).

The model in Fig. 1 has been extensively used in the design and interpretation of *in vivo* experiments conducted to study the quantitative aspects of steroid metabolism. Pools A and B have been chosen to represent two compounds in systemic circulation[4] or the same hormone in fetal and maternal circulations.[5]

Model for Tracer Superfusion Experiments (Fig. 2). When a cell suspension or tissue slices are superfused with a solution containing ^3H-A and ^{14}C-B, only a fraction of each of the superfused tracers enters the cells. If these fractions (α_A and α_B) were known, then the rates corresponding to the dynamics of compounds A and B in the cells could be studied using the formulas derived on the basis of the model in Fig. 1 since

$$I_A{}^{^3H} = \alpha_A \phi c_A{}^{^3H}$$

and

$$I_B{}^{^{14}C} = \alpha_B \phi c_B{}^{^{14}C}$$

[4] D. Baird, R. Horton, C. Loncope, and J. F. Tait, *Perspect. Biol. Med.* **11,** 384 (1968).

[5] E. Gurpide, J. Tseng, L. Escarcena, M. Fahning, C. Gibson, and P. Fehr, *Amer. J. Obstet. Gynecol.* **113,** 21 (1972).

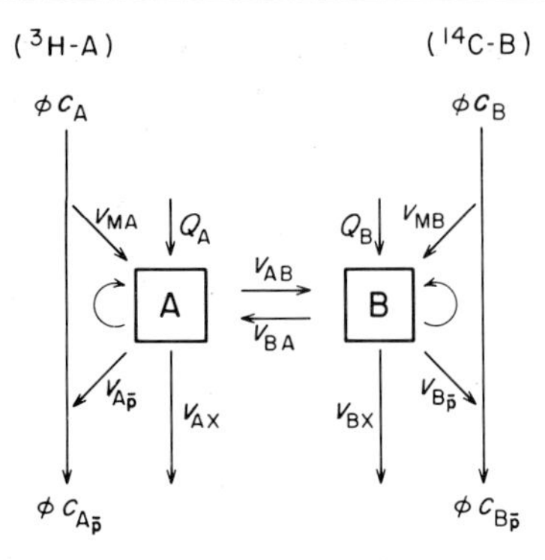

FIG. 2. Model representing 2 compounds, A and B, in tissue superfused with a mixture of ^{3}H-A and ^{14}C-B tracers, at a constant rate, without recycle. Symbols: ϕ, flow rate (ml/hour); c_A and c_B, concentrations of A and B in the inflowing medium (dpm/ml); v_{MA} and v_{MB}, rates of entry of the compounds into cells (nmoles/hour); v_{AB} and v_{BA}, rates of intracellular conversion between A and B; Q_A and Q_B, rates of synthesis of A and B from other intracellular precursors; $v_{A\bar{p}}$ and $v_{B\bar{p}}$, rates of exit of A and B from the tissue to the medium; v_{AX} and v_{BX}, rates of irreversible metabolism of A and B to all other products; $c_{A\bar{p}}$ and $c_{B\bar{p}}$, concentration of A and B in perfusate (outflowing medium).

where ϕ represents the flow rate of the medium (ml/hour) and $c_A{}^{^3H}$, $c_B{}^{^{14}C}$ denote the concentrations of ^{3}H-A and ^{14}C-B in the perfusion medium (dpm/ml).

Assuming an identical behavior of the tracer in all the cells, the rates calculated from the specific activity of A and B isolated from the tissue at the steady state correspond to the total rates of entry, interconversion, synthesis, and metabolism of these compounds in the tissue sample superfused.

Fraction of a Superfused Tracer That Enters the Cells (α). If α_A is the fraction of superfused ^{3}H-A entering the tissue, $1 - \alpha_A$ is the fraction which bypasses the cells. According to this definition,

$$1 - \alpha_A = \frac{c_{A\bar{p}}{}^{^3H} \text{ (bypassing tissue)}}{c_A{}^{^3H}} \tag{6}$$

The total concentration of ^{3}H-A in the perfusate, which is denoted by the symbol $c_{A\bar{p}}{}^{^3H}$, is formed by tracer that bypassed the tissue and ^{3}H-A which after entering the cells is released back into the medium:

$$c_{A\bar{p}}{}^{^3H} = c_{A\bar{p}}{}^{^3H} \text{ (bypassing tissue)} + c_{A\bar{p}}{}^{^3H} \text{ (from tissue)}$$

The estimation of α then requires the experimental distinction between the concentrations in the perfusate of ^{3}H-A that either bypassed or entered and left the cells. This discrimination can be accomplished by intracellular labeling of compound A with another isotope. Thus, if ^{3}H-A and ^{14}C-B are superfused as a mixture and B is converted to A by enzymatic intracellular reactions, compound A in the tissue will be labeled with both ^{3}H and ^{14}C. At the isotopic steady state, the ^{3}H/^{14}C ratio in intracellular compound A can be expected to be identical to the isotope ratio in compound A released from the cells into the medium. Therefore, by measuring the concentration and ^{14}C-A in the perfusate ($c_{A_{\bar{p}}}{}^{^{14}C}$), the concentration in the perfusate of ^{3}H-A derived from the tissue can be estimated. Thus, if

$$\frac{c_{A_{\bar{p}}}{}^{^{3}H}\ (\text{from tissue})}{c_{A_{\bar{p}}}{}^{^{14}C}} = \left(\frac{^{3}H}{^{14}C}\right)_{AT}$$

then

$$c_{A_{\bar{p}}}{}^{^{3}H}\ (\text{from tissue}) = (^{3}H/^{14}C)_{AT}\, c_{A}{}^{^{14}C} \tag{7}$$

where $(^{3}H/^{14}C)_{AT}$ denotes the isotopic ratio in compound A isolated from tissue at the steady state.

The formula to calculate α from the experimental data can be obtained from Eqs. (6) and (7):

$$\alpha_A = 1 - \frac{c_{A_{\bar{p}}}{}^{^{3}H} - (^{3}H/^{14}C)_{AT}\, c_{A_{\bar{p}}}{}^{^{14}C}}{c_{A}{}^{^{3}H}} \tag{8}$$

If the conversion of B to A is reversible, as indicated in Fig. 2, the symmetry of the model makes apparent that

$$\alpha_B = 1 - \frac{c_{A_{\bar{p}}}{}^{^{14}C} - (^{14}C/^{3}H)_{BT}\, c_{B_{\bar{p}}}{}^{^{3}H}}{c_{B}{}^{^{14}C}}$$

Fraction of a Superfused Tracer Returning from the Cells to the Medium (β). If β_A is defined as the ratio of the concentration in the perfusate of ^{3}H-A released by the cells and the concentration of the tracer in the perfusion medium, i.e.,

$$\beta_A = \frac{c_{A_{\bar{p}}}{}^{^{3}H}\ (\text{from tissue})}{c_{A}{}^{^{3}H}},$$

it follows from Eq. (7) that

$$\beta_A = \frac{(^{3}H/^{14}C)_{AT}\, c_{A_{\bar{p}}}{}^{^{14}C}}{c_{A}{}^{^{3}H}} \tag{9}$$

Fraction of Superfused Tracer Appearing in the Perfusate as a Metabolite (γ). The fraction of superfused ^{3}H-A appearing in the perfusate as a tritiated metabolite L is

$$\gamma_{AL} = \frac{c_{L_{\bar{p}}}{}^{^{3}H}}{c_{A}{}^{^{3}H}}$$

Conversion Factors (ρ). According to the definition of ρ_{AB} given by Eq. (5) and the expressions to estimate the entry of the tracer into the cells [Eq. (8)], it follows that

$$\rho_{AB} = \frac{I_B{}^{^{14}C}}{I_A{}^{^{3}H}} \frac{a_{BT}{}^{^{3}H}}{a_{BT}{}^{^{14}C}} = \frac{(^3H/^{14}C)_{BT}}{(\alpha_A/\alpha_B)(^3H/^{14}C)_{superfused}}$$

Similarly,

$$\rho_{BA} = \frac{I_A{}^{^{3}H}}{I_B{}^{^{14}C}} \frac{a_{AT}{}^{^{14}C}}{a_{AT}{}^{^{3}H}} = \frac{(\alpha_A/\alpha_B)(^3H/^{14}C)_{superfused}}{(^3H/^{14}C)_{AT}}$$

Rates in the Superfusion Model (Fig. 2). Comparison of the models in Figs. 1 and 2 indicates that the rate of entry of material into pool A, indicated by v_{OA} in Fig. 1, appears in Fig. 2 resolved into two rates, i.e.,

$$v_{OA} = v_{MA} + Q_A$$

One of these rates (v_{MA}) corresponds to the entry of superfused compound A into the cells and is given by the expression

$$v_{MA} = \alpha_A \phi c_A$$

where c_A is the concentration of A in the perfusion medium. The other rate (Q_A) represents the rate of intracellular formation of A from sources other than compound B. Since the total rate of entry into the pool (v_{OA}) can be calculated from the specific activity of A and B in the tissue and the concentrations of the labeled compounds in the perfusion medium and the perfusate,

$$v_{OA} = \phi \alpha_B c_B{}^{^{14}C} \frac{a_{AT}{}^{^{3}H}}{a_{AT}{}^{^{3}H} a_{BT}{}^{^{14}C} - a_{BT}{}^{^{3}H} a_{AT}{}^{^{14}C}}$$

then Q_A can be estimated by subtracting v_{MA} from v_{OA}.

A list of formulas for the calculation of rates from isotopic data obtained during superfusion experiments is given in Table II.

The rate of removal of material from pool A represented by the rate v_{AO} in Fig. 1 appears resolved into two different rates in Fig. 2: One ($v_{A\bar{p}}$) corresponds to the rate of release of A back into the medium, the other (v_{AX}) corresponds to the rate of irreversible metabolism of A to other products without the intermediary of B. The estimation of the rate $v_{A\bar{p}}$ is based on the intuitively apparent relationship.

$$\frac{v_{A\bar{p}}}{PR_A} = \frac{\beta_A}{\alpha_A} \tag{10}$$

which indicates that the ratio of the rate at which compound A is released by the tissue into the medium and the rate at which it appears

TABLE II
LIST OF FORMULAS APPLICABLE TO THE MODEL IN FIG. 2[a]

Parameter	Formula
Fraction of superfused tracer entering cells	$\alpha_A = \frac{v_{MA}}{\phi c_A} = 1 - \frac{c_{A_{\bar{p}}}{}^{3H} - (^3H/^{14}C)_{AT} c_{A_{\bar{p}}}{}^{14C}}{c_A{}^{3H}}$
	$\alpha_B = \frac{v_{MB}}{\phi c_B} = 1 - \frac{c_{B_{\bar{p}}}{}^{14C} - (^{14}C/^3H)_{BT} c_{B_{\bar{p}}}{}^{3H}}{c_B{}^{14C}}$
Fraction of superfused tracer released by tissue into medium	$\beta_A = (^3H/^{14}C)_{AT} c_{A_{\bar{p}}}{}^{14C}/c_A{}^{3H}$
	$\beta_B = (^{14}C/^3H)_{BT} c_{B_{\bar{p}}}{}^{3H}/c_B{}^{14C}$
Conversion factors	$\rho_{AB} = \left(\frac{^3H}{^{14}C}\right)_{BT} \div \frac{\alpha_A}{\alpha_B}\left(\frac{^3H}{^{14}C}\right)_{\text{superfused}}$
	$\rho_{BA} = \frac{\alpha_A}{\alpha_B}\left(\frac{^3H}{^{14}C}\right)_{\text{superfused}} \div \left(\frac{^3H}{^{14}C}\right)_{AT}$
Rates	$(D = a_{AT}{}^{3H} a_{BT}{}^{14C} - a_{BT}{}^{3H} a_{AT}{}^{14C})$
	$v_{MA} + Q_A = \phi \alpha_A c_A{}^{3H}(a_{BT}{}^{14C} - a_{AT}{}^{14C})/D$
	$v_{MA} = \phi \alpha_A c_A$
	$v_{MB} + Q_B = \phi \alpha_B c_B{}^{14C}(a_{AT}{}^{3H} - a_{BT}{}^{3H})/D$
	$v_{MB} = \phi \alpha_B c_B$
	$v_{AB} = \phi \alpha_B c_B{}^{14C} a_{BT}{}^{3H}/D$
	$v_{BA} = \phi \alpha_A c_A{}^{3H} a_{AT}{}^{14C}/D$
	$v_{A_{\bar{p}}} = v_{AX} = \phi(\alpha_A c_A{}^{3H} a_{BT}{}^{14C} - \alpha_B c_B{}^{14C} a_B{}^{3H})/D$
	$v_{A_{\bar{p}}} = (\beta_A/\alpha_A) PR_A$
	$PR_A = \phi \alpha_A c_A{}^{3H}/a_{AT}{}^{3H}$
	$v_{B_{\bar{p}}} + v_{BX} = \phi(\alpha_B c_B{}^{14C} a_{AT}{}^{3H} - \alpha_A c_A{}^{3H} a_{AT}{}^{14C})/D$
	$v_{B_{\bar{p}}} = (\beta_B/\alpha_B) PR_B$
	$PR_B = \phi \alpha_B c_B{}^{14C}/a_{BT}{}^{14C}$
Intracellular clearances	$IC_A = (PR_A/W) \div c_{AT} = (\phi \alpha_A c_A{}^{3H}/W) \div c_{AT}{}^{3H}$
	$IC_B = (PR_B/W) \div c_{BT} = (\phi \alpha_B c_B{}^{14C}/W) \div c_{BT}{}^{14C}$

[a] Symbols: ϕ, flow rate (ml/hour) of superfusion medium; W, weight (g) of tissue superfused; $c_A{}^{3H}$, concentration (dpm/ml) of ^{3}H-A in the inflowing medium; $c_{AT}{}^{3H}$, concentration (dpm/g) of ^{3}H-A in tissue; $c_{A_{\bar{p}}}{}^{3H}$, concentration (dpm/ml) of ^{3}H-A in perfusate (outflowing); c_A and $c_{A_{\bar{p}}}$, concentrations (nmoles/ml) of compound A in inflowing medium and perfusate, respectively; c_{AT}, concentration (nmoles/g) of compound A in tissue; $a_{AT}{}^{3H}$, specific activity (dpm/nmole) with respect to ^{3}H of compound A in tissue; $(^3H/^{14}C)_{AT}$, isotope ratio (dpm/dpm) in compound A in tissue; $(^3H/^{14}C)$ superfused, isotope ratio in the inflowing medium.

de novo intracellulary equals the fraction of tracer A entering the tissue that is released back into the medium.

Then, from Eqs. (9) and (10),

$$v_{A_{\bar{p}}} = \frac{\beta_A}{\alpha_A} PR_A = \frac{\beta_A}{\alpha_A} \frac{\phi \alpha_A c_A{}^{3H}}{a_{AT}{}^{3H}} = \phi \frac{c_{A_{\bar{p}}}{}^{14C}}{a_{AT}{}^{14C}}$$

The rate of irreversible removal of A from the cells ($v_{AO'}$ Fig. 1) can be estimated from the experimental data,

$$v_{AO} = v_{A_{\bar{p}}} + v_{AX} = \phi \frac{\alpha_A c_A{}^{^3H} a_{BT}{}^{^{14}C} - \alpha_B c_B{}^{^{14}C} a_{BT}{}^{^3H}}{a_{AT}{}^{^3H} a_{BT}{}^{^{14}C} - a_{BT}{}^{^3H} a_{AT}{}^{^{14}C}}$$

Consequently, v_{AX} can be calculated from the same data ($v_{AX} = v_{AO} - v_{A_{\bar{p}}}$).

Intracellular Clearance (IC). Another important parameter which is included in Table II is the intracellular clearance, defined as the ratio of the production rate of the intracellular compound to the amount of compound in the tissue or, equivalently, as the ratio of the production rate per unit weight of tissue divided by the intracellular concentration of the compound, i.e.,

$$IC_A = \frac{PR_A}{W \times c_{AT}} = \frac{\alpha_A \phi c_A{}^{^3H}}{W \times c_{AT}{}^{^3H}},$$

where W is the weight of the superfused tissue. This parameter has the dimensions of a rate constant (hr^{-1}). Other rate constants are obtained by dividing rates corresponding to various processes of conversion or transfer of a compound by the amount of the compound in the pool, e.g.,

$$k_{AB} = \frac{v_{AB}}{W \times c_{AT}}$$

A Special Case: Nonsteroidogenic Tissue. When nonendocrine tissues or cell suspensions are used, the sole source of intracellular steroid at the steady state is the superfusion medium. Under these circumstances, the specific activity of compounds A and B in the tissue can be calculated knowing the amount of intracellular radioactivity in these compounds and the specific activity of the superfused tracers. Thus, if all intracellular A derives from superfused compounds A and B, it follows that at the steady state

$$c_{AT} = \frac{c_{AT}{}^{^3H}}{a_A{}^{^3H}} + \frac{c_{AT}{}^{^{14}C}}{a_B{}^{^{14}C}}$$

Concentrations of radioactivity are much more easily measured than specific activities, using isotope dilution methods in which carriers of the compounds are added to the tissue homogenate or subcellular preparations.

Tests for Compartmentalization. Various experimental approaches are available for the investigation of intracellular compartmentalization. The most obvious one is based on the superfusion of two metabolically related tracers, ^{3}H-A and ^{14}C-B, and comparison of isotope ratios of the compounds isolated from subcellular fractions.

Compartmentalization can be investigated by comparing the specific activity of an intracellular compound determined by direct analysis of the tissue with the specific activity calculated indirectly from isotopic data. The indirect calculation of a specific activity is based on the following formula:

$$a_{\mathrm{AT}}{}^{^3\mathrm{H}} = \frac{c_{\mathrm{A}_{\bar{\mathrm{p}}}}{}^{^3\mathrm{H}} - (1 - \alpha_{\mathrm{A}})c_{\mathrm{A}}{}^{^3\mathrm{H}}}{c_{\mathrm{A}_{\bar{\mathrm{p}}}} - (1 - \alpha_{\mathrm{A}})c_{\mathrm{A}}}$$

This expression indicates that the specific activity of the intracellular compound A is expected to equal the ratio of the rate at which unlabeled and labeled compound A are released from the tissue into the medium. Discrepancy of this calculated value with the experimentally determined specific activity could serve as evidence for the existence of nonmetabolic or passive pools of the compound in the tissue, as has been postulated for cholesterol. Furthermore, the nonexchangeable fraction of the total amount of the compound in the cell would equal the quotient of these two specific activities.

A powerful tool to obtain evidence for intracellular compartmentalization of a compound is a perfusion of two of its precursors labeled with different isotopes, followed by a comparison of the isotopic ratio of the compound in tissue and in perfusate at the steady state. Thus, if two precursors of X are labeled with different isotopes and perfused as a mixture, e.g., ^{3}H-A and ^{14}C-B, and the isotope ratio in X in perfusate and tissue is found to be significantly different at the steady state, it can be concluded that more than one intracellular pool of X is present.[6]

Although interpretation of results obtained by the tracer superfusion method must take into account compartmentalization, it should be noted that the calculation of rates of entry into cells is based on the condition expressed by Eq. (7), regardless of the intracellular distribution of the compound.

Experimental Considerations

Superfusion can be carried out in a simple apparatus in which the medium containing the tracers is placed in a syringe driven by a mechanical pump which forces the solution through a capillary glass tube into a chamber fitted with a porous glass filter. The top of the chamber may consist of a perforated Teflon plug connected to a capillary tubing which

[6] J. Tseng and E. Gurpide, *J. Biol. Chem.* **248**, 5634 (1973).

carries the overflow to tubes immersed in a chilling solution. The chamber and the tubing connecting the syringe with the chamber are immersed in a constant temperature bath.

When the chamber has been filled with the superfusion medium, the tissue slices or the cell suspension are placed in the chamber. Flow rates ranging from 8–20 ml/minute are adequate and do not disturb the tissue or most cell preparations. In fact, even nuclei can be perfused in this manner.

Since the fraction of tracer entering the tissue depends on the amount and geometry of the sample, identical chambers and amounts of material are used in parallel runs in which the influence of various factors on the dynamics of the uptake and metabolism of the steroids is investigated.

Verification of the Achievement of Steady State. Analysis of various fractions of perfusate collected during the superfusion reveal the time necessary to achieve constant levels of the labeled compounds. Sampling of the tissue at different times of perfusion and measurement of the intracellular concentration of the labeled compounds serves a similar purpose. Furthermore, biochemical tests (e.g., oxygen consumption) provide information about the viability of the system during the perfusion. Survival of dispersed cells during the experimental period can be tested by the dye uptake procedure.

Extracellular Enzymatic Activity. Since the calculations are based on the assumption that the interconversions between compounds A and B occur in the cells, the presence of extracellular enzymatic activities would create artifacts. In order to test for the possibility that the enzymes leak out from the cells during the perfusion, released to the medium by broken, dead, or even intact cells, the tissue may be superfused with medium without tracers. The presence of enzymatic activity in the perfusate might then be detected by using labeled substrates.

Removal of Medium from the Tissue before Analysis. When the superfusion is stopped, it is important to wash the tissue or cell suspension quickly and thoroughly before analysis, in order to remove contaminating superfusion medium. A fast washing under suction on a Millipore filter is usually adequate. Analysis of fractions of the filtrate obtained at various times during washing provides information about the completeness of the removal of medium. For instance, the $^{3}H/^{14}C$ ratio in compound A in the filtrate is expected to decrease with the number of washes and become constant and equal to the final ratio in A isolated from the tissue. The identity of the final isotope ratio in the compound washed from the tissue and the isotope ratio in the compound isolated from the washed tissue need be established to validate the condition indicated by Eq. (7).

Enzymatic reactions involving compounds A and B may proceed during the washing period. Therefore, the use of a cold buffer or enzymatic inhibitors during this step is recommended. In order to evaluate the extent of isotope exchange between A and B occurring during the 1–2 minutes of washing, kinetic calculations based on a system consisting of only two pools can be performed using the rate constant determined from the superfusion experiments, as described elsewhere.[7]

Comments on Parameters Measured Using Tracer Superfusion Methods

Rates of Entry. Unidirectional flow rates, in and out of cells, can be measured in tracer superfusion experiments. Therefore, permeability studies including determination of kinetic constants and action of inhibitors can be performed using this experimental design.

A constant value of α at various concentrations of the compounds in the perfusion medium may indicate that the entry occurs by passive permeation. If such were the case, the rate of entry of the compound would be proportional to its concentration in the medium and to the area available for entry, or the weight of the tissue if the geometry is maintained constant. In this case,

$$v_{\mathrm{MA}} = kc_{\mathrm{A}}W$$

and

$$\alpha_{\mathrm{A}} = \frac{v_{\mathrm{MA}}}{\phi c_{\mathrm{A}}} = \frac{kW}{\phi}$$

On the other hand, the same result could be expected if the K_m for a facilitated transport system were much lower than the concentration of the compound in the perfusion medium. According to the kinetic expression characteristic of carrier mechanisms,[8]

$$\alpha_{\mathrm{A}} = \frac{v_{\mathrm{MA}}}{\phi c_{\mathrm{A}}} = \frac{V_{\max}}{\phi(K_m + c_{\mathrm{A}})}$$

Rates of Intracellular Reactions. It is possible with this system to measure rates of metabolic reactions as they occur at normal intracellular concentrations and distribution of enzymes, cofactors, and substrates. Since the intracellular concentration of substrates can be modified and measured, kinetic constants applicable to a reaction as it occurs in the cell can be determined.

[7] L. Tseng, A. Stolee, and E. Gurpide, *Endocrinology* **90,** 405 (1972).

[8] W. D. Stein, "The Movement of Molecules across Cell Membranes." Academic Press, New York, 1967.

Other parameters can be used with less accuracy to evaluate the extent of these reactions. The conversion factor [ρ, Eq. (5)] is one of these parameters. The ratio of intracellular concentrations of labeled compounds is another. It follows from the formulas listed in Table II that

$$\frac{c_{AT}^{^3H}}{c_{BT}^{^3H}} = \frac{IC_B}{\rho_{AB} IC_A}$$

as can be verified by substituting IC_A, IC_B, and ρ_{AB} by their corresponding expressions in terms of isotopic data. This ratio may be quite different from the ratio obtained after a prolonged batchwise incubation of the tissue with a labeled precursor.

Rates of Exit. Measurements of the release of a compound by the tissue into the medium are relevant to studies of the processes of secretion of the steroids by endocrine glands or of elimination of hormones from target tissue. The various factors which might regulate the rate of exit of a compound, e.g., rates of formation and metabolism, binding to proteins, can be evaluated by means of tracer superfusion experiments.

If the transfer of the steroids across the cell membrane is considered to occur by simple diffusion with similar permeability coefficient in both directions, it is possible to estimate the fractions of the intracellular compound which is in a diffusible form (d) according to the following self-evident equation:

$$\frac{\beta_A}{\alpha_A} = \frac{d_A c_{AT}^{^3H}}{c_A^{^3H}}$$

or

$$d_A = \frac{\beta_A}{\alpha_A} \frac{c_A^{^3H}}{c_{AT}^{^3H}}$$

Intracellular Receptors. Concentrations of hormonal receptors can be determined from measurements of intracellular levels of receptor-bound labeled hormone. Prolonged superfusion of the target tissue with an excess of labeled compound results in a complete exchange of the intracellular hormone. Therefore, the amount of hormone saturating receptors is estimated from values of specifically bound radioactivity knowing the specific activity of superfused hormone or precursors.[9]

[9] L. Tseng and E. Gurpide, *Amer. J. Obstet. Gynecol.* **114,** 995 (1972).

Section II

Serum-Binding Proteins for Steroid and Thyroid Hormones

[8] Characterization of Steroid-Binding Glycoproteins: Methodological Comments

By ULRICH WESTPHAL, ROBERT M. BURTON, and GEORGE B. HARDING

The principal procedures used in our laboratory for the purification and binding analysis of the steroid-binding serum proteins have been previously described in "Methods in Enzymology."[1] For this reason, the following sections will be limited to a number of unpublished observations made over the years, all relevant to preparative and analytical studies in the field of steroid–protein interactions.

Charcoal Treatment of Serum or Plasma

A method for removal of endogenous steroid from serum by gel filtration at 45° has been described previously.[1] Complete elimination of steroid, e.g., cortisol or progesterone, from the high-affinity binders such as CBG[2] can be achieved. However, the technique is not practical in processing large volumes, or in the routine analysis of many samples, especially when the original protein concentration has to be reconstituted from the dilute eluates. Heyns *et al.*[3] have described a simpler method by which the endogenous cortisol is adsorbed to powdered charcoal (50 mg Norit A/ml plasma) during 30 minutes at room temperature; the charcoal is then removed by centrifugation and filtration. The resulting plasma is essentially free of corticosteroid.

The charcoal technique has been used in many laboratories to eliminate endogenous steroids from serum or plasma. The investigators assume that essentially all steroid is adsorbed to the charcoal, usually without verifying this assumption. The success of removal depends mainly on the dissociation rate of the steroid–protein complex under the conditions of the charcoal adsorption; this rate increases with rising temperature.

The conditions used by Heyns *et al.*[3] are suitable for the stated objective, i.e., to remove corticosteroids from CBG in human plasma. The complex of progesterone with human CBG, which at 37° has a three-times

[1] U. Westphal, Vol. 15, p. 761 (1969).

[2] Abbreviations used: CBG, corticosteroid-binding globulin or transcortin; AAG, α_1-acid glycoprotein; HSA, human serum albumin; PBG, progesterone-binding globulin; SDS, sodium dodecyl sulfate.

[3] W. Heyns, H. Van Baelen, and P. DeMoor, *Clin. Chim. Acta* **18**, 361 (1967).

TABLE I
CHARCOAL TREATMENT OF SERUM FOR REMOVAL OF ENDOGENOUS PROGESTERONE[a]

Serum of	Treatment[b]	Progesterone content after treatment (ng/ml)	% Removal
A. Pregnant woman	None	50	—
	1 hour at 22°	0.01	100
	1 hour at 37°	0.03	100
	1 hour at 45°	0.0	100
B. Pregnant guinea pig	None	200	—
	1 hour at 37°	160	20
	1 hour at 45°	118	41
	2 hours at 40°	98	51
	4 hours at 40°	77	62
	6 hours at 40°	34	83

[a] Unpublished data with C. V. Rao.
[b] The sera were gently shaken with 50 mg Norit A/ml.

higher association constant than the cortisol complex,[4] can be dissociated under similar conditions: Charcoal treatment for 1 hour at 22° or at higher temperatures frees the serum completely of the endogenous progesterone (Table I,A).

However, the same or more vigorous treatment accomplishes only a partial removal of progesterone from the serum of the pregnant guinea pig as may be seen in Table I,B. Only 20% of the endogenous progesterone is removed after charcoal incubation for 1 hour at 37°, and even after 6 hours at 40° as much as 17% of the original progesterone is still present in the bound form. The explanation for this slow adsorption to charcoal is seen in a low dissociation rate associated with the relatively high binding affinity of the progesterone–PBG complex.[5,6] It should be noted that the incubations at the elevated temperatures applied (Table I) do not affect the binding affinity of CBG or PBG irreversibly, as verified in additional experiments.

In view of these results, the charcoal method cannot be assumed to completely remove endogenous steroid from serum or plasma in all cases.

[4] U. Westphal, "Steroid-Protein Interactions," p. 214. Springer-Verlag, Berlin and New York, 1971.
[5] U. Westphal, "Steroid-Protein Interactions," p. 311. Springer-Verlag, Berlin and New York, 1971.
[6] E. Milgrom, P. Allouch, M. Atger, and E. E. Baulieu, *J. Biol. Chem.* **248,** 1106 (1973).

The quantity of steroid that can be adsorbed to the charcoal under the conditions employed depends on the affinity-related dissociation rate. In an unknown system, the extent of steroid removal has to be determined by analysis.

Adsorption of Steroids to Various Materials

In any work with very low steroid concentrations, possible adsorption to the containing surfaces becomes important. Loss of progesterone and desoxycorticosterone by adsorption to the lusteroid (nitrocellulose) tubes used in ultracentrifugation has been described previously.[7] The steroid initially forms a monomolecular layer on the nitrocellulose surface and then diffuses into the plastic in a pseudo first-order reaction. Passage of progesterone, testosterone, estradiol, and other steroids into silicone rubber has also been observed.[8]

In view of the increased experimentation with very dilute aqueous steroid solutions in many laboratories, we have tested the adsorption of steroids to several rubber and plastic materials that are frequently used in such studies. Tubing of the following materials was purchased: rubber (yellow gum); silicone rubber; polyethylene; Plexiglas (polyacrylate and methacrylate); tygon (polyvinylchloride); and Teflon (polytetrafluoroethylene). One-inch sections of the tubing were thoroughly washed with hot water and soap, rinsed with hot water, and with distilled water, and finally with absolute methanol. Two each of the dry pieces were then gently shaken at 22° in a 25-ml Erlenmeyer flask with 10 ml solution of radiosteroid in 50 m*M* phosphate buffer of pH 7.4. Aliquots were counted after given time intervals as indicated in Table II.

The table indicates that the adsorption of the steroids to the various polymers shows considerable differences. As to be expected, the least hydrophilic one, progesterone, is adsorbed most strongly, with silicone rubber and tygon leading. Teflon seems to be least attractive for progesterone. With increasing numbers of hydroxy groups in the steroid molecule, the adsorption to the (predominantly hydrophobic) surfaces decreases; virtually no attachment was detected for cortisol and aldosterone although a slight adsorption of the latter to silicone rubber was observed. In general, the relative adsorption of the steroids to the polymers tested was comparable to their hydrophobic bonding to HSA[9] and AAG.[10]

[7] U. Westphal, *J. Lab. Clin. Med.* **51,** 473 (1958).

[8] P. J. Dziuk and B. Cook, *Endocrinology* **78,** 208 (1966).

[9] U. Westphal, "Steroid-Protein Interactions," p. 55. Springer-Verlag, Berlin and New York, 1971.

[10] U. Westphal, "Steroid-Protein Interactions," p. 395. Springer-Verlag, Berlin and New York, 1971.

TABLE II
ADSORPTION OF STEROIDS TO RUBBER AND PLASTICS: TWO 1-IN. SECTIONS OF TUBING, 1/4 IN. O.D., 1/8 IN. I.D., WERE EXPOSED AT ROOM TEMPERATURE WITH GENTLE SHAKING TO 10 ML RADIOSTEROID SOLUTION IN 50 mM PHOSPHATE BUFFER, pH 7.4; VALUES SHOW PERCENT OF TOTAL ADSORBED DURING HOURS INDICATED

Steroid	Hours	Rubber	Silicone rubber	Polyethylene	Plexiglas	Tygon	Teflon
Progesterone	1	68	89	10	7	90	7
(2.0 ng/ml)	4	88	95	17	13	95	7
	24	92	97	35	15	96	8
Desoxycorticosterone	1	11	48	3	3	47	0.8
(16 ng/ml)	4	20	74	2	4	73	0.2
	24	58	87	3	5	86	0.6
Corticosterone	1	4	3	0.7	1	0	0
(2.3 ng/ml)	5	3	4	0	2	6	2
	24	7	9	0	0	15	0
Cortisol (2.3 ng/ml)	1–24	0	0	0	0	0	0
Testosterone	1	3	33	0.6	0	21	0.8
(1.3 ng/ml)	5	14	77	0.4	0.6	54	0
	24	40	91	2	0	75	0
Aldosterone	1	0.6	2	0	0	0.8	0
(8.4 ng/ml)	4	0	1	0	0	2	0
	24	0.8	4	0	0.6	2	1

The results indicate the importance of determining total recoveries in experiments where radiolabeled or unlabeled steroids are exposed to surfaces, especially for extended periods of time. The outside solutions of equilibrium dialysis systems in plastic chambers are especially pronc to such error by steroid adsorption since their steroid concentrations are usually low, and binding proteins are absent. Another consideration is the possible contamination of container material with radioactive steroids that may be subsequently transferred to other experiments. We observed[11] and measured directly a very strong adsorption of radiolabeled progesterone to tygon tubing that connected a chromatographic column to the fraction collector. Obviously, the radiosteroid could be taken up by binding protein in subsequent experiments.

[11] S. D. Stroupe, unpublished observations.

TABLE III
INFLUENCE OF DELIPIDATION ON APPARENT ASSOCIATION CONSTANTS (nK) OF PROGESTERONE COMPLEXES WITH HSA PREPARATIONS—EQUILIBRIUM DIALYSIS IN 50 mM PHOSPHATE, pH 7.4, 4°

HSA preparation	I ($M^{-1} \times 10^{-5}$)	II ($M^{-1} \times 10^{-5}$)	III ($M^{-1} \times 10^{-5}$)
Untreated	1.7	2.6	1.9
Delipidated	13.0	18.9	20.9
Relipidated	0.3	0.5	—

Elimination of Binding Inhibitors

The association of steroids with serum albumin is inhibited by fatty acids and other lipids.[12] Removal of lipids from commercial crystalline albumin preparations of highest available grades results in greatly enhanced binding affinity; increases of the association constant up to 10-fold have been observed repeatedly. The apparent enhancement of the affinity constant can be reversed by readdition of the extracted lipid. Table III gives examples of these results.

To remove traces of lipid contamination from the glassware used in the determination of binding parameters, the vessels are treated in our laboratory with concentrated nitric acid for at least 15 hours and rinsed with double distilled water until free of acid. Thorough rinsing in all washing operations is essential. Traces of laboratory detergent remaining in the glassware can decrease the binding affinity as Table IV shows. Other experiments in which an average detergent concentration of 13 μg/ml was used gave 44% binding inhibition.

Reversible binding inhibition by lipid has also been observed for the interaction between progesterone and AAG.[13] As in the case of HSA, the protein has to be thoroughly delipidated to obtain maximal binding affinity. Readdition of the lipidic material, extracted from the glycoprotein in the delipidation procedure, results in very low apparent association constants. In addition to lipids, certain heavy metal ions inhibit steroid binding to AAG.[14] This inhibition is also reversible. We are using double-distilled deionized water in all binding studies and perform, whenever pertinent, the equilibrium dialysis in the presence of mM EDTA.[15]

[12] U. Westphal, "Steroid-Protein Interactions," p. 106. Springer-Verlag, Berlin and New York, 1971.

[13] U. Westphal, "Steroid-Protein Interactions," p. 382. Springer-Verlag, Berlin and New York, 1971.

[14] J. Kerkay and U. Westphal, *Arch. Biochem. Biophys.* **129,** 480 (1969).

[15] U. Westphal, "Steroid-Protein Interactions," p. 392. Springer-Verlag, Berlin and New York, 1971.

TABLE IV
INHIBITION OF PROGESTERONE-HSA INTERACTION BY LABORATORY DETERGENT[a]

Detergent concentration (μg/ml)	Bound/unbound progesterone	% Inhibition
0	30.7	—
0.67	31.1	0
6.7	22.2	28
66.7	4.2	87

[a] Equilibrium dialysis for 48 hours in 50 mM phosphate buffer, pH 7.4, 4°. Inside solution: 2 × 4 ml (duplicate bags) delipidated HSA, 0.98 mg/ml; outside solution: 16 ml [^{3}H]progesterone in buffer, 0.01 μg/ml. The detergent, containing sodium *n*-alkyl benzene sulfonates, nonylphenol polyethoxy ethanols, and condensed sodium phosphates, was added to the outside solution; the detergent concentration was calculated for even distribution through the total volume.

TABLE V
INFLUENCE OF SKIN SECRETIONS ON PROGESTERONE BINDING TO α_1-ACID GLYCOPROTEIN AND HUMAN SERUM ALBUMIN[a]—EQUILIBRIUM DIALYSIS IN 50 mM PHOSPHATE, pH 7.4, 4°

	nK Values obtained			
Protein	With gloves ($M^{-1} \times 10^{-5}$)	Without gloves ($M^{-1} \times 10^{-5}$)	% Inhibition	Investigator
AAG-1	13.2	5.0	62	OE
AAG-2	13.7	5.4	61	OE
AAG-3	14.5	5.4	63	OE
AAG-3	15.3	5.8	62	TK
AAG-4	17.8	5.8	67	TK
HSA-1	4.9	4.3	11	GBH
HSA-2	4.3	3.2	26	GBH
HSA-3	2.5	1.8	28	GBH

[a] Unpublished results with O. Edelen and T. Kute.

It has been our experience over the years that the reproducibility of the association constants of steroid–protein complexes of moderate affinity, such as those of AAG, is difficult to control. It was surprising to find that one cause of the erratic and greatly reduced affinity values is the contact of the cellulose dialysis bags with human skin, i.e., thoroughly washed and rinsed hands, during handling and tying the bags. Covering the hands with surgical gloves eliminated this source of error and enhanced the apparent binding affinity of the AAG complexes almost three times (Table V). It would be of interest to determine which compo-

TABLE VI
INFLUENCE OF SKIN SECRETIONS ON STEROID BINDING TO CBG AND PBG[a]: EQUILIBRIUM DIALYSIS IN 50 mM PHOSPHATE, pH 7.4, 4°—THE RESULTS ARE GIVEN AS BOUND/UNBOUND STEROID

Binding protein	Steroid	Bound/unbound steroid obtained[b]	
		With gloves	Without gloves
CBG, purified by hydroxylapatite chromatography, 7 μg/ml	Cortisol	11.5	10.9
PBG, purified by hydroxylapatite chromatography, 40 μg/ml	Progesterone	48.6	47.2
PBG, purified by SP-Sephadex chromatography, 5 μg/ml	Progesterone	45.7	40.9
		51.7	40.0
		40.5	35.2

[a] Unpublished results with K. Acree.
[b] The steroid: protein ratios were different in the various experiments, but identical in the comparison with and without gloves.

nent(s) of the human skin secretions are responsible for the drastic effects seen in Table V.

In comparison to these observations with AAG and HSA, steroid binding to the specific high-affinity binders of the serum seems less influenced by the skin secretions. Table VI shows that the binding values for the progesterone–PBG complex, and that for the cortisol–CBG complex, are only slightly reduced when the dialysis bags are handled without gloves.

Anomalous Mobility of Glycoproteins in Dodecyl Sulfate–Polyacrylamide Electrophoresis

It has been recognized for several years that glycoproteins behave abnormally in gel filtration when their carbohydrate content exceeds about 5%.[16] When the Stokes' radii are related to those obtained with carbohydrate-free proteins, apparent molecular weights of up to twice the actual values have been observed.[17-20] No simple relationship has been

[16] P. Andrews, *Methods Biochem. Anal.* **18,** 1 (1970).
[17] P. Andrews, *Biochem. J.* **96,** 595 (1965).
[18] P. Andrews, *Protides Biol. Fluids* **14,** 573 (1967).
[19] D. N. Ward and M. S. Arnott, *Anal. Biochem.* **12,** 296 (1965).
[20] G. J. Chader, N. Rust, R. M. Burton, and U. Westphal, *J. Biol. Chem.* **247,** 6581 (1972).

found between carbohydrate content and the deviation from the standard elution volumes which would permit correction factors to be applied.

There are certain similarities between the degree of retardation during filtration over a gel column and the mobility in SDS gel electrophoresis. They are based on the expanded structures of the glycoproteins, presumably resulting from greater hydration of the carbohydrate moieties, in comparison with polypeptide chains. In addition to size, the net negative charge determines the electrophoretic mobility in the gel. Bretscher[21] and Segrest *et al.*[22] observed that the mobility of glycoproteins relative to polypeptide proteins increased with increasing acrylamide concentration and attributed this to the relatively smaller influence of charge at the higher gel concentrations; at the lower porosity, the effect of size (Stokes' radius) becomes predominant. In accordance with earlier reports[23] indicating the polypeptide moiety being responsible for SDS binding, Segrest *et al.*[22] found less SDS bound to glycoproteins than to polypeptide proteins.[24] It appears then that the erroneously high molecular sizes observed for glycoproteins in SDS gel electrophoresis are the result of both size and charge effects.

The general procedures for measurement of molecular weights by electrophoresis on SDS–acrylamide gel have been described.[25] Table VII gives examples of high apparent molecular weights for several glycoproteins, determined by this technique. In accordance with reported observations,[21,22] the discrepancy between the apparent and actual values becomes smaller with increasing acrylamide concentration. As in the case of gel filtration, no quantitative relationship to the carbohydrate content is evident.

It follows from Table VII that the molecular weights assessed for glycoproteins by determination of the Stokes' radii by SDS gel electrophoresis cannot be considered correct values.

Isoelectric Focusing of Steroid-Binding Glycoproteins

Isoelectric focusing[26] of a radiosteroid–protein complex has been commonly used for the determination of the isoelectric point of the binding protein. The location of the radiolabel at the completion of the run has

[21] M. S. Bretscher, *Nature (London), New Biol.* **231**, 229 (1971).

[22] J. P. Segrest, R. L. Jackson, E. P. Andrews, and V. T. Marchesi, *Biochem. Biophys. Res. Commun.* **44**, 390 (1971).

[23] R. Pitt-Rivers and F. S. A. Impiombato, *Biochem. J.* **109**, 825 (1968).

[24] J. A. Reynolds and C. Tanford, *Proc. Nat. Acad. Sci. U.S.* **66**, 1002 (1970).

[25] K. Weber, J. R. Pringle, and M. Osborn, Vol. 26, p. 3 (1972).

[26] O. Vesterberg, Vol. 22, p. 389 (1971).

TABLE VII

ERRONEOUS MOLECULAR WEIGHTS OF GLYCOPROTEINS DETERMINED BY SDS–POLYACRYLAMIDE ELECTROPHORESIS—THE MOBILITIES OF STANDARD CARBOHYDRATE-FREE PROTEINS ARE USED FOR CALIBRATION

			Molecular weight		
Glycoprotein	Carbohydrate (%)	Polyacrylamide (%)	Apparent, from SDS–gel electrophoresis	Actual, from ultracentrifugation or other physical methods	Reference
Immunoglobulin, μ-type heavy chain	12	7.5	100,000	65,000	a
Immunoglobulin, α-type heavy chain	10–15	7.5	70,000	40–50,000	a
Ribonuclease, pig, higher	35	12.5	24,000	21,000	b
mol. wt. species	35	5	41,000	21,000	b
Ribonuclease, pig, lower	20	12.5	20,000	17,000	b
mol. wt. species	20	5	31,000	17,000	b
α_1-Acid glycoprotein, human	41.9	8	48,500	37,000	c
Ovomucoid	23	8	37,000	27,000	c
CBG, rabbit	29.2	8	48,000[d]	35,000	c
Luteinizing hormone, sheep	14.3	10	34,400	29,100	e
Vitamin B_{12}-binding protein	33	5	121,000	56,000	f
PBG, guinea pig, different	62	8	94,000	69,100	g
mol. wt. species	|	8	119,000	77,000	g
	to	8	119,000	80,800	g
	↓	5	150,000	99,800	g
	72	5	200,000	117,300	g

[a] D. Schubert, *J. Mol. Biol.* **51,** 287 (1970).

[b] J. P. Segrest, R. L. Jackson, E. P. Andrews, and V. T. Marchesi, *Biochem. Biophys. Res. Commun.* **44,** 390 (1971).

[c] G. J. Chader, N. Rust, R. M. Burton, and U. Westphal, *J. Biol. Chem.* **247,** 6581 (1972).

[d] Under reducing conditions a double band is obtained, corresponding to apparent molecular weights of 51,000 and 46,000.

[e] D. Gospodarowicz, *Endocrinology* **90,** 1101 (1972).

[f] R. H. Allen and P. W. Majerus, *J. Biol. Chem.* **247,** 7702 (1972).

[g] R. M. Burton, G. B. Harding, W. Aboul-Hosn, D. T. MacLaughlin, and U. Westphal, *Biochemistry* (in press).

been taken to indicate the isoelectric point of the protein, assuming that the complex remains stable during the electrofocusing. Experience in our laboratory has shown that erroneous results may be obtained if this assumption is not verified. Particularly for the highly acidic glycoproteins such as PBG,[26a] the isoelectric point of the steroid binder must be validated by direct determination of the binding activity. An analogous conclusion was reached by Van Baelen *et al.*[27] in studies on CBG.

The following procedure has been found useful in our laboratory for determination of the isoelectric point of PBG of the pregnant guinea pig.[27a] A 0–46% sucrose gradient containing 0.8% Ampholine pH 3–10 was pumped from a gradient mixer into a 110-ml LKB column[28] and a pH gradient formed for 24 hours at 2 or 10° ("prefocusing"). An initial power was maintained at or below 3 W; when possible the potential was increased to 960 V for the final 4–6 hours of focusing. The use of previously prepared pH gradients has been described by Van Baelen *et al.*[27]; we have found that PBG focusing can be completed in such gradients within 10 hours. Fractions of 2.0 ml were collected at 1-minute intervals at 4° and pH values measured at 25°. Approximately 400 μg desalted lyophilized protein containing 0.4 μg or 0.12 μCi bound tritiated progesterone were dissolved in the pH 4.6 fraction. The prefocused ampholyte fractions were reapplied to the column and a potential of 960 V was maintained for 10 hours at 2 or 10°. The contents of the column were then eluted at a constant rate of 1.0 or 2.0 ml/minute, and 1.0-ml fractions were collected. Radioactivity was determined in 0.2-ml aliquots of the fractions from pH 0.9 to 10. The fractions were immediately neutralized with 0.2 *M* tribasic sodium phosphate. For determination of progesterone-binding activity, every two adjacent fractions were pooled for a multiple equilibrium dialysis[1] at 4° for 60 hours. Tritiated progesterone (15 ng, 1.61 μCi) and three times the total inside volume of 50 m*M* sodium phosphate buffer, pH 7.4, were added outside. Streptomycin sulfate and potassium penicillin G were included.[1] The final equilibrium pH values in different experiments ranged from 7.0 to 7.4. The result is expressed as bound [^{3}H]progesterone/ml inside solution.

The results of an electrofocusing experiment performed at 10° for 10 hours are presented in Fig. 1. About half the total radioactivity in the column remained near pH 4.6, i.e., at the point of application; the other half was found around pH 3.5. This pattern was identical to that seen in a similar focusing experiment carried out for 48 hours. Multiple equilibrium dialysis at neutral pH showed, however, that neither of these radioactive zones contained any progesterone-binding activity. Rather

[26a] R. M. Burton, G. B. Harding, W. Aboul-Hosn, D. T. MacLaughlin, and U. Westphal, *Biochemistry* (in press).

[27] H. Van Baelen, M. Beck, and P. DeMoor, *J. Biol. Chem.* **247**, 2699 (1972).

[27a] G. B. Harding, R. M. Burton, S. D. Stroupe, and U. Westphal, *Life Sciences* (in press).

[28] LKB Producter AB · S-16125, Bromma 1, Sweden.

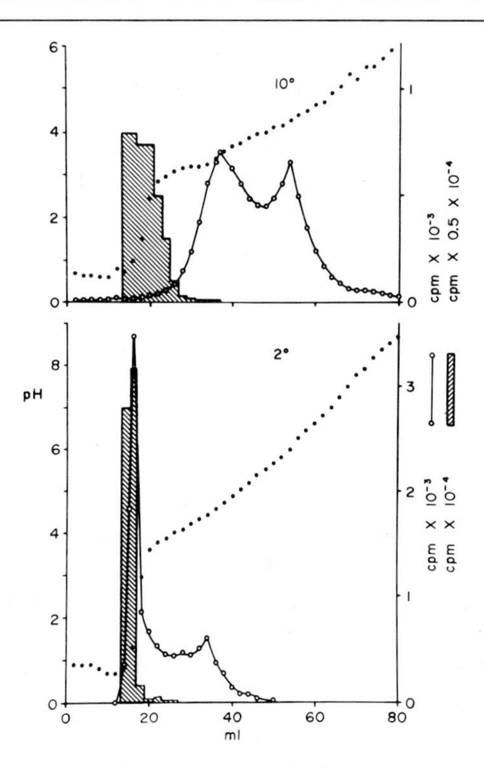

FIG. 1. Isoelectric focusing of PGB. An enriched preparation was obtained from pregnant guinea pig serum by hydroxylapatite and Sephadex G-200 chromatography and was saturated with tritiated progesterone. Dotted line, pH; open circles, radioactivity in 0.2 ml of eluate fractions; shaded area, progesterone-binding activity, determined by equilibrium dialysis and expressed as bound [^{3}H]progesterone per milliliter.

the binding affinity was associated with the steep portion of the gradient in the pH range of approximately 1 to 3.

This result was confirmed in an experiment performed at 2° (Fig. 1). At this temperature, much less dissociation occurred during the electrophoretic migration so that more than half of the progesterone radioactivity was detected in the bottom fractions of pH 0.9 to approximately 3, close to the binding activity measured by equilibrium dialysis. In order to rule out the possibility of precipitation of PBG into the acidic region, the entire experiment was repeated with reversed polarity. It was found that PBG focused as before at the pH region of about 1 to 3, this time at the top of the column.

It was evident that the isoelectric point of PBG is quite low, certainly below the range of the pH 3–10 or 3–5 ampholytes. The true isoelectric point was more closely estimated by a modified procedure in which a gradient could be established with a reasonable slope from pH 2.4 to 5.0.[28a] This was accomplished by adding 60 mg each of aspartic and glutamic acid to the 0.8% Ampholine solution used in the 110-ml LKB column; the phosphoric acid at the anode was replaced by 0.5 *M* citric acid. In electrofocusing experiments with this extended pH gradient, the active progesterone-binding globulin was focused at pH 2.8.

Without an independent determination of progesterone-binding activity, incorrect pI values would have been deduced from the localization of the radiolabeled steroid, due to dissociation of the steroid–protein complex in the acidic milieu. This explanation was substantiated[27a] by direct measurement of the pH effect on the binding activity of PBG. The PBG–progesterone complex, in equilibrium dialysis at 4° for 48 hours, dissociated completely at pH values below 4.0. In contrast to other serum proteins with high affinity for steroids, acidic conditions as low as pH 2 did not destroy the steroid-binding ability of PBG: More than 80% of binding activity was regained even after long exposure to pH 2.

Determination of Sedimentation Coefficients at Low Protein Concentrations

Because of extensive hydration and of nonspherical shapes of glycoproteins, analysis of sedimentation rates at very low concentrations may be necessary for accurate extrapolation of $S_{20,w}$ values to zero concentration. For many workers the sensitive but expensive ultraviolet scanner[29] is not available. Richards and Schachman[30] have described the use of Rayleigh optics at usual protein concentrations in obtaining S values of greater precision than those estimated with schlieren methods. These optics have been used in our laboratory to obtain sedimentation velocity data at protein concentrations as low as 0.03%.

We employed a Spinco Model E ultracentrifuge with optics focused for optimal interference patterns utilizing the Spinco manuals[31–33] for general instructions regarding centrifuge operation and measurements with

[28a] This technique was applied before Ampholine carrier ampholytes below pH 3 became available.

[29] H. K. Schachman, *Biochemistry* **2**, 887 (1963).

[30] E. G. Richards and H. K. Schachman, *J. Phys. Chem.* **63**, 1578 (1959).

[31] K. E. Van Holde, "Fractions," No. 1, p. 1. Beckman Instruments, Inc., Palo Alto, California, 1967.

[32] C. H. Chervenka, "Manual of Methods for the Analytical Ultracentrifuge," p. 10. Beckman Instruments, Inc., Palo Alto, California, 1969.

[33] M. Zeller, "Beckman Model E Analytical Ultracentrifuge-Instruction Manual." Beckman Instruments, Inc., Palo Alto, California, 1964.

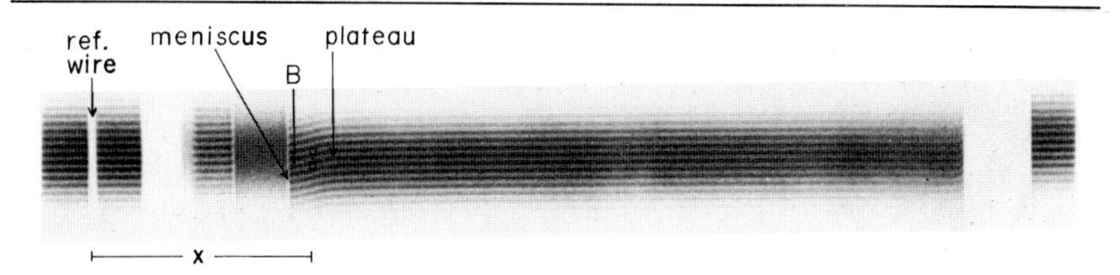

FIG. 2. Interference diagram for sedimentation velocity run at 47,488 rpm on a 0.03% solution of PBG in 0.1 *M* NaCl, 5°. Sedimentation is to the right.

interference optics. A filled Epon center, double-sector cell with a 12-mm light path was equipped with sapphire windows. The patterns obtained with quartz windows at speeds above 10,000 rpm are not clear enough for accurate measurements.[31] A photographic polarizing filter placed atop the filter holder greatly improved the quality of the fringe pattern obtained.[34] Pictures were taken on Kodak Royal X Pan film and developed with a Kodak D-19 developer. One chamber of the double-sector cell was filled with 0.2 ml of protein solution and the other chamber with the dialysate. Concentrations ranged from 0.03 to 0.5%. A standard sedimentation velocity run was performed.[35]

Films were aligned on a microscope comparator by reference to the air fringes of the reference holes. A typical fringe pattern from a run at 47,700 rpm with PBG at 0.03% is seen in Fig. 2. The best measure of the height of the central white fringe near the meniscus where the protein concentration is approximately zero (baseline, B) was obtained by averaging the heights of three white and two black fringes. The height of the fringe in the plateau region of the cell was determined in the same way.

The plateau height plus the baseline height of the fringe divided by two is equal to the half-height, corresponding in radial position to the maximum of a dc/dx schlieren curve; at this half-height, the distance x of the fringe from the reference wire image was measured as indicated by the dot. For greater accuracy, the x values of the two adjacent white fringes were also measured (squares) at their half-heights; these half-heights were calculated as the half-height of the central fringe plus and minus the fringe displacement, characteristic of the optical system. The resulting x values were averaged. These measurements were repeated for each picture, and the radial distance r in centimeters for each exposure calculated using the known magnification factor and the r value of the

[34] C. H. Chervenka, personal communication.
[35] H. K. Schachman, Vol. 4, p. 32 (1957).

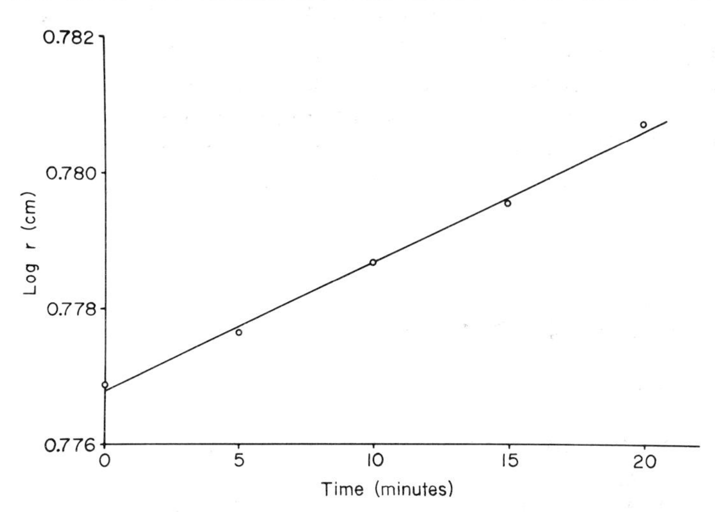

Fig. 3. Plot of sedimentation velocity data from experiment described in Fig. 2.

reference wire. The resulting log r vs. time plot, shown in Fig. 3, was used to calculate the sedimentation coefficients.[35] It is apparent that a fair degree of precision is possible in this procedure.

Acknowledgment

These studies were supported by a grant from the National Institute of Arthritis and Metabolic Diseases (AM-06369) and a research career award (U.W.) from the Division of General Medical Sciences (GM-K6-14,138) of the United States Public Health Service.

[9] Isolation and Purification of Corticosteroid-Binding Globulin from Human Plasma by Affinity Chromatography

By William Rosner and H. Leon Bradlow

Introduction

Corticosteroid-binding globulin (CBG) has been isolated in several laboratories by conventional methods.[1-3] These methods are, unfortu-

[1] U. S. Seal and R. P. Doe, *J. Biol. Chem.* **237**, 3136 (1962).

[2] W. R. Slaunwhite, Jr., S. Schneider, F. C. Wissler, and A. A. Sandberg, *Biochemistry* **5**, 3527 (1966).

[3] T. G. Muldoon and U. Westphal, *J. Biol. Chem.* **242**, 5636 (1967).

nately, time consuming and impractical if one desires to work with large volumes. The application of affinity chromatography significantly decreases the time required for the isolation and allows one to work with large volumes of starting material (plasma) and hence to isolate large amounts of CBG in an ordinary laboratory. The procedure to be described is a modification of the one we have previously published.[4]

Method

Preparation of the Affinity Column

Reagents

1. Sepharose 4B (Pharmacia Fine Chemicals)
2. Cyanogen bromide
3. 3,3′-Diaminodipropylamine (Eastman Chemicals)
4. Cortisol hemisuccinate (Steraloids Inc. or prepare from cortisol and succinic anhydride in pyridine)
5. *N*,*N*′-Dicyclohexylcarbodiimide (Aldrich Chemicals)

The object of this portion of the procedure is first to couple the diamino compound (reagent 3) to Sepharose and then to form an amide between the cortisol hemisuccinate and the free amino groups on the Sepharose. Preparation of amino-Sepharose is based on previously published methods.[5] Dilute 200 mmoles of reagent 3 with water to a volume of 100 ml, adjust the pH to 10 with concentrated HCl, and cool to 4°. Transfer 100 ml of Sepharose in a total water–slurry volume of 200 ml to a beaker, adjust to pH 10 with 8 *N* NaOH, and bring the temperature to 20° with ice chips. Place the beaker in a well-ventilated hood on a magnetic stirring motor and insert into it a thermometer and a pH electrode. At this time prepare 25 g of cyanogen bromide (1 bottle) by placing it in a plastic bag in a hood and crushing it with a hammer or rolling pin. Add the crushed cyanogen bromide to the Sepharose, and as the pH falls, add 8 *N* NaOH to the reaction mixture at a rate sufficient to maintain the pH at 10. In addition, ice chips will have to be added to keep the reaction at 20°. After 15–20 minutes, although the pH is still falling, approximately 300 ml of ice chips are added to stop the reaction. Pour the Sepharose–ice mixture onto a Büchner funnel fitted with a coarse fitted disk and wash under suction with about 500 ml of ice cold water

[4] W. Rosner and H. L. Bradlow, *J. Clin. Endocrinol. Metab.* **33**, 193 (1971).
[5] P. Cuatrecasas, *J. Biol. Chem.* **245**, 3059 (1970).

that has been previously adjusted to pH 10. Disconnect the suction and *immediately* add the previously prepared solution of reagent 3 to the activated Sepharose. Transfer the mixture to a beaker and stir at 4° overnight (about 18 hours). Wash the amino-Sepharose with at least 2 liters of water and store it at pH 4–5 with a bacterial retardant. This procedure routinely yields a gel which contains 10 μmoles of free amino groups/ml of packed gel. The yield can be checked by observing the color developed with 2,4,6-trinitrobenzene sulfonate.[5] The gel is now ready for coupling with cortisol hemisuccinate. Transfer 100 ml of packed amino-Sepharose that has been washed with 5 volumes of dioxane to a beaker, allow the gel to settle, and remove as much of the supernatant dioxane as possible. In separate containers dissolve 10 mmoles of cortisol hemisuccinate and an equimolar amount of dicyclohexylcarbodiimide in minimal volume of dioxane. Add each reagent to the amino-Sepharose slurry and stir for 3 hours at room temperature. The total volume of the reaction mixture should not exceed 250 ml. Pour the reaction mixture into a glass column, wash with 8–9 liters of dioxane to remove the unreacted cortisol hemisuccinate, then wash with 500 ml of water and store the gel at pH 4–5 with a bacterial retardant. The procedure should yield 1–2 μmoles of cortisol hemisuccinate/ml of gel. The actual yield can be checked by using tritiated cortisol hemisuccinate and counting the gel, but this is not imperative since the capacity of the gel for CBG should be ascertained before use, and it is this capacity which is the important operational variable.

Utilization of the Affinity Column

Preliminary Data to Be Obtained. We have used outdated plasma anticoagulated with acid–citrate–dextrose (ACD) as our sole source of CBG. The concentration of CBG in such plasma has varied from 15–27 mg CBG/liter. Surprisingly we have found reasonable amounts of CBG in plasma that has been stored at room temperature for months and was malodorous on arrival in the laboratory. Hemolysis does not interfere with isolation.

Prepare ACD plasma for affinity chromatography by first adding solid $CaCl_2$ to a final concentration of 20 m*M* and removing the fibrin clot which forms. After this is complete, add 230 g/liter of solid ammonium sulfate, stir for 30 minutes at 4°, and harvest the CBG, which is soluble in this concentration of ammonium sulfate, by centrifugation at 16,000 *g* for 30 minutes. Remove any endogenous steroid by adding norite A, 50 g/liter, and shaking at room temperature for 30 minutes. Remove the charcoal by centrifugation and adjust the CBG solution to pH 5.50 with 3 N HCl.[5a] The ratio of ammonium sulfate supernate (hereafter referred to as serum) to cortisol–Sepharose is determined by preliminary capacity studies in which varying quantities of serum are added to 0.5 ml of freshly washed cortisol–Sepharose. Specifically 25, 50, and 75 ml of serum are added to 0.5 ml cortisol–Sepharose in round-bottom flasks and stirred for 1.5 hours at 4°. Remove the supernatant serum from each flask, readjust the pH to pH 7.0, and assay it for CBG activity.[6] The ratio of serum-to-gel to be used is that which gives 50–70% absorption of the CBG which is present. This will vary not only with a given gel but with the concentration of CBG in the serum being used. That is, serum with a low concentration of CBG will require more gel for a given volume of serum. If a large gel excess is used, i.e., all the CBG in the sample is taken up, then it will be difficult or impossible to recover the CBG from the gel. Typically 100 ml of gel is used for 5 liters of serum. This may vary from 2 to 12 liters/100 ml of gel.

Absorption and Elution of CBG from Cortisol–Sepharose

Add 5 liters of serum (prepared as above) to 100 ml of freshly washed (2 liters of 80% methanol followed by 500 ml of water) cortisol–Sepharose and stir at 4° for 1.5 hours. Transfer the serum–cortisol–Sepharose slurry to a Büchner funnel fitted with a coarse fritted disk and remove most of the plasma by suction. Then wash the gel under suction with ice cold sodium phosphate buffer, pH 7.0, 0.05 M, 0.2 M in NaCl, until the wash appears grossly free of serum (about 1.5 liters of buffer are required. Transfer the gel to a glass column 4–6 cm in diameter and continue to wash with phosphosaline buffer at 4° until the optical density (280 nm) of the effluent (collected on a fraction collector at 4°) falls to 0.2 to below. Then wash the gel with 100 ml of cold sodium phosphate buffer, pH 7.0, 0.001 M, at which time the optical density in the effluent will have fallen to 0. At this point add approximately 100 ml of sodium phosphate buffer, pH 7.0, 0.001 M, containing 200 μg/ml tritiated cortisol (spe-

[5a] The pH adjustment is extremely important. At pH's greater than 5.50 the succinate ester of cortisol is partially hydrolyzed by plasma after 1–2 hours. This creates noncovalently bound cortisol and renders the affinity gel useless. There is presumably an esterase in plasma which is not active at this pH and which had been inactivated in the outdated plasma (minimum of 1–2 months old) which we used in our original experiments.[4] This fact probably accounts for the inability of Le Galliard *et al.*[5b] to obtain stable affinity columns as they appear to have used reasonably fresh plasma in their attempts to purify CBG by affinity chromatography.

[5b] F. LeGalliard, A. Racadot, N. Racadot-Leroy, and M. Dautrevaux, *Biochimie* **56,** 99 (1974).

[6] G. A. Trapp and C. D. West, *J. Lab. Clin. Med.* **73,** 861 (1969).

cific activity, 30 cpm/μg), and continue to elute until radioactivity appears in the effluent or, alternatively, until cortisol is detected at 241 nm. Remove the column from the cold room and add one "cortisol volume" (that volume it took for the cortisol to appear in the previous step) of sodium phosphate buffer (24°, pH 6.5, 0.05 M, 0.1 M NaCl) containing 200 μg/ml cortisol. Wait 1 hour to allow the entire column to come to room temperature and elute with an additional 200 ml of the pH 6.5 phosphate buffer containing cortisol. More than 95% of the CBG will be found in the first 200 ml eluted. Actually almost all the CBG is in the first half of this pool. This can be ascertained simply by collecting 8–10 ml fractions and examining 10 μl aliquots against an antibody to purified CBG on Ouchterlony plates. If this antibody is not available (as it generally is not) it is not worthwhile to perform CBG assays on each fraction as this is more time consuming than accepting the greater degree of impurity involved in taking the entire 200 ml. At 4° concentrate the protein to a volume of 30 ml by thin channel pressure ultrafiltration in a model TCF 10 Amicon cell using a PM 10 membrane (Amicon Corp.), followed by further concentration to 10 ml in a model 52 cell. Finally, dialyze the concentrated protein solution overnight against 1 liter of sodium phosphate buffer, pH 7.0, 0.001 M. It is convenient at this time to subject a small aliquot of the concentrated protein to paper electrophoresis at pH 8.6. Since the duration of electrophoresis is 15–18 hours, the composition of the pool will be known the following morning. The major contaminant will be albumin with varying but lesser amounts of γ-globulin. The CBG at this stage constitutes 20–50% of the total protein and is easily visualized as an inter-α-globulin on the paper strip. The entire procedure of eluting the gel takes one working day, starting from decantation of the serum and ending with putting the concentrated eluate in the dialysis tank and setting up the paper electrophoresis.

Hydroxylapatite Chromatography

The concentrated protein should contain approximately 75 optical density units (280 nm) and can now be applied to a 2.5 × 2.5-cm column of hydroxylapatite (Bio-Rad laboratories) which has previously been washed and freed of fines with sodium phosphate buffer, pH 7.0, 0.001 M. The CBG elutes at the front and can be detected by monitoring the effluent at 280 nm. After the CBG has been eluted, the contaminating proteins (mostly albumin) can be eluted with potassium phosphate buffer, pH 7.0, 0.5 M. The CBG is now lyophilized and its purity and activity examined by disc electrophoresis and equilibrium dialysis, respectively.[4] In our experience the initial hydroxalapatite chromatography yields a product which is 85–100% pure. If the product is not pure, a single rechromatography on the same hydroxylapatite system will remove the offending protein. Do not regenerate the hydroxylapatite as we find that its reuse leads to poor results.

Discussion

The method we have described lends itself to isolation of CBG from large quantities of plasma in a reasonable period of time. A similar approach has been used by Trapp *et al.*[7] who observed, as we have, several places in the procedure which are sources of variability. As the concentration of CBG relative to other protein falls, there is an apparent decrease of the capacity of the cortisol–Sepharose. This is presumably due to effective competition, by a massive molar excess of albumin, for sites on the column. Depending on the composition of the starting material, the fraction of CBG adsorbed to the affinity column may vary from 50–70% of that present. Of the CBG which is initially adsorbed to the affinity column, 31–100% is recovered as the pure protein. This figure includes losses sustained in the desorption procedure and the hydroxylapatite chromatography. The average yield from plasma in six separate experiments was 39.5% and the average recovery of that CBG which initially bound to the column was 60.5%.

Increasing the scale of the procedure by a factor of 5 has been accomplished with no difficulty. We have processed 31 liters of serum at a time without significant variation from the procedure as outlined.

[7] G. A. Trapp, U. S. Seal, and R. P. Doe, *Steroids* **18,** 421 (1971).

[10] Isolation of Human Testosterone-Estradiol-Binding Globulin

By WILLIAM ROSNER and REX SMITH

Introduction

The method to be presented for the isolation of testosterone-estradiol-binding globulin (TeBG) is not only previously unpublished, but is still evolving in this laboratory. We feel justified in departing from custom by presenting new data in "Methods in Enzymology" because previously published methods contain major deficiencies resulting in poor yields of impure and inactive products.[1-6]

Since the concentration of TeBG in serum is small, an adequate source of starting material is the initial obstacle in purification. Assuming a molecular weight of about 100,000,[1-3,7] and a single binding site per molecule, then it can be calculated from the binding capacity[8,9] that the concentration of TeBG in male serum is about 3.2 mg/liter. If the correct molecular weight is 50,000[6] then the concentration would be 6.4 mg/liter. Values in women are double those in men. Cohn fraction IV precipitate,[10] which contains active TeBG,[6] is normally discarded by blood centers and is an ideal source of TeBG.

The procedure to be described depends on affinity chromatography[11] for the major purification and the presence of Ca^{2+} at all stages for stabilization of TeBG activity. The affinity column to be described is novel in that the side chain, azodianiline (ADA) (Fig. 1), is a bifunctional reagent permitting the facile attachment of ligands and the chemical separation of the ligand from the gel by cleavage of the azo bond with dithionite.

Method

Preparation of the Affinity Column

Preparation of Azodianiline Sepharose

REAGENTS

1. Sepharose 4B (Pharmacia Fine Chemicals)
2. Epichlorhydrin (Aldrich Chemical Co.)
3. Azodianiline (K and K Laboratories), ADA
4. Dimethylformamide, DMF

[1] J. L. Guériguian and W. H. Pearlman, *J. Biol. Chem.* **243**, 5226 (1968).
[2] H. Van Baelen, W. Heyns, E. Schonne, and P. De Moor, *Ann. Endocrinol.* **29**, 153 (1968).
[3] W. Rosner, W. G. Kelly, and N. P. Christy, *Biochemistry* **8**, 3100 (1969).
[4] S. H. Burstein, *Steroids* **14**, 263 (1969).
[5] M. A. Pizarro, *Ann. Endocrinol.* **30**, 206 (1969).
[6] C. Mercier-Bodard, A. Alfsen, and E. E. Baulieu, *Acta Endocrinol.* (*Copenhagen*), *Suppl.* **147**, 204 (1970).
[7] P. L. Corvol, A. Chrambach, D. Rodbard, and C. W. Bardin, *J. Biol. Chem.* **246**, 3435 (1971).
[8] W. Rosner, *J. Clin. Endocrinol. Metab.* **34**, 983 (1972).
[9] W. Heyns and P. De Moor, *Steroids* **18**, 709 (1971).
[10] P. Kistler and Hs. Nitschmann, *Vox Sang.* **7**, 414 (1962).
[11] P. Cuatrecasas, *J. Biol. Chem.* **245**, 3059 (1970).

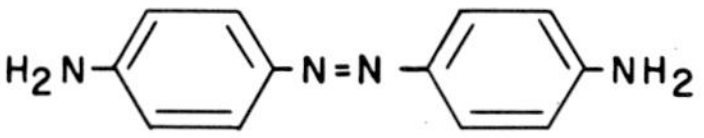

$H_2N-C_6H_4-N=N-C_6H_4-NH_2$

FIG. 1. Azodianiline.

The object of this portion of the procedure is to couple ADA to Sepharose. Epichlorhydrin[12] is used as the activating agent in preference to cyanogen bromide[11] since we found that cyanogen bromide-coupled ADA is not sufficiently stable for use in the isolation of TeBG.

Wash 100 ml of Sepharose 4B with 1 liter of deionized water on a funnel fitted with a coarse fritted disk under suction. Transfer the moist cake of Sepharose 4B to a round-bottom flask, add 100 ml of 1 *N* NaOH, followed by 10 ml epichlorhydrin, and mix fairly vigorously with a magnetic stirrer for 2 hours at room temperature. Transfer the gel back to the funnel and wash with 2–3 liters of water. Remove most of the water, transfer the gel to a beaker, add 5.3 g (25 mmoles) ADA in 200 ml 50% (v/v) DMF, and stir gently for 2 hours. Wash the coupled gel under suction with 3 liters 80% DMF followed by 3 liters water. The coupled gel is burnt orange in color and contains about 7 μmoles of free amino groups/ml of gel.

Preparation of Androstanediol-3-hemisuccinate

Dissolve 12 g succinic anhydride in a minimal volume of pyridine, add 10 g 3β-hydroxy-5α-androstan-17-one (epiandrosterone), and let stand at room temperature overnight (about 18 hours). Acidify the reaction mixture to pH 3 with concentrated HCl and pour over crushed ice. Wash the resulting precipitate with ice cold acidified water (pH 3), then acidified 20% methanol, and dry *in vacuo*. The yield of epiandrosterone-3-succinate is about 85%. Dissolve 5 g of the synthesized epiandrosterone-3-hemisuccinate in 200 ml absolute ethanol at 4° and slowly add 4 g of sodium borohydride maintaining the temperature at 0–4° in an ice bath. Allow the reaction mixture to stand for 2–3 hours at 0–4°, acidify to pH 3 with concentrated acetic acid, and reduce the volume on a flash evaporator until the product just starts to precipitate (volume at this point is 20–30 ml). Add an equal volume of water, adjusted to pH 3 with acetic acid; wash the precipitate which forms with ice cold acidified water and then with ice cold ether. Dry the precipitate, which is a white powder, *in vacuo*. The yield is 4 g of androstanediol-3-hemisuccinate.

[12] J. Porath and N. Fornstedt, *J. Chromatogr.* **51**, 479 (1970).

Preparation of Androstanediol–Sepharose

Add 1.4 g (3.6 mmoles) of androstanediol hemisuccinate, dissolved in 50 ml of 50% DMF, to 50 ml of ADA–Sepharose previously suspended in 50 ml of 50% DMF, followed by the addition over 5 minutes of 0.7 g (3.7 mmoles) of 1-(3-dimethylaminopropyl)-3-ethyl carbodiimide (Aldrich). Stir for 2 hours at room temperature and wash with 10 liters of 80% DMF and then 5 liter of water just prior to use.

Isolation

General Procedures

Measure protein concentrations by the microbiuret method of Itzhaki and Gill[13] using bovine serum albumin as a standard. When this method is used with calcium-containing buffers, a precipitate forms but the optical density of the supernatant at 310 nm obeys Beer's law up to a protein concentration of 0.45 mg/ml.

Measure the activity of TeBG by the method of Rosner.[8] This method depends on saturation of the TeBG binding sites with dihydrotestosterone (DHT) and precipitation of TeBG-bound steroid with ammonium sulfate. Results are expressed as μg DHT bound. Dilute concentrations of protein will not precipitate, and, therefore, equilibrium dialysis and analysis of the data with modified Scatchard plots[14] is the technique used as an alternative when appropriate. Further, the bound DHT from any of the several Sephadex chromatographies used in the isolation agrees well with the values obtained by the first two methods.

Conduct all procedures at 4° unless otherwise specified; this includes all chromatographic steps and centrifugation.

Purification

REAGENTS

1. Cohn fraction IV precipitate,[10] lyophilized
2. Ammonium sulfate
3. Charcoal, alkaline norite A
4. 17β-Hydroxy-5α-androstan-3-one (dihydrostestosterone, DHT)
5. 1,2-^{3}H-DHT, specific activity 44 Ci/mmole
6. Sephadex G-50 (medium) and G-150
7. $Na_2S_2O_4$, dithionite (Fisher Chemicals)
8. Buffers—pH to be adjusted at the temperature of use

[13] R. F. Itzhaki and D. M. Gill, *Anal. Biochem.* **9,** 401 (1964).

[14] A. A. Sandberg, H. Rosenthal, S. L. Schneider, and W. R. Slaunwhite, Jr., *in* "Steroid Dynamics" (G. Pincus, T. Nakao, and J. F. Tait, eds.), p. 1. Academic Press, New York, N.Y., 1966.

a. Tris-HCl, pH 7.5, 0.05 *M*, containing 0.05 *M* $CaCl_2$
b. Tris-HCl, pH 7.0, 0.3 *M*, containing 0.15 *M* $CaCl_2$

Ammonium Sulfate Precipitation. Dissolve 280 g Cohn fraction IV in buffer 8a so that the final volume is 8 liters. Add 300 g/liter solid ammonium sulfate, stir for 30 minutes, and recover the precipitate by centrifugation at 16,000 *g* for 30 minutes. Wash the precipitate once with 50% ammonium sulfate and make it up to 4 liters in buffer 8a.[15] Remove any endogenous steroid by adding norite A, 50 g/liter, and shaking at room temperature for 30 minutes.[17] Remove the charcoal by centrifugation as above.

Affinity Chromatography. Add the steroid-free solution of protein to 50 ml of androstanediol–Sepharose and stir in a 5-liter round-bottom flask for 2 hours.[18] Transfer the plasma–Sepharose mixture to a coarse sintered glass funnel, remove the plasma by suction, and wash on the funnel with 50–100 ml of buffer 8a. Transfer the gel to a 5 × 20-cm glass column, wash with buffer 8a, containing 1 *M* NaCl, until the optical density (280 nm) falls to 0, and then with 100 ml of buffer 8b.[20] The gel is now ready

[15] Of several buffers examined, we find that Tris has a definite stabilizing effect on TeBG[16]; this is the major reason for using Tris buffers throughout the purification. Other agents which stabilize the protein are Ca^{2+}, glycerol, and dihydrotestosterone, and these are introduced where appropriate in the purification procedure.[16]

[16] W. Rosner, S. Toppel, and R. N. Smith, *Biochim. Biophys. Acta.* In press (1974).

[17] W. H. Heyns, H. Van Baelen, and P. De Moor, *Clin. Chim. Acta* **18,** 361 (1967).

[18] The capacity of the gel can be examined as described by Rosner and Bradlow in Chapter 9, this volume, by using small amounts of gel with appropriately smaller volumes of Cohn fraction IV. There are, however, two important differences. First, the capacity study for TeBG should be done with the buffers and times described in this chapter; second, one should be certain that substances which might interfere with the assay procedure are not present in the material assayed. Leudens *et al.*[19] have shown that material which competes for binding sites on specific proteins may come off of the affinity columns during chromatography, making it appear as if binding protein adsorbed to the column when in truth it has not. We routinely assay the supernatant plasma from capacity studies in a way which takes possible interference into account. Specifically, we assay the supernatant as such, and again after passage through a small column of Sephadex G-50 to remove possible excess unbound interfering small molecules. Under the above conditions, we detect no interfering substances. The gel described in this section is not as efficient in binding TeBG as the gel described for the purification of corticosteroid-binding globulin (Chapter 9, this volume). Hence, we choose to use a large protein to gel ratio and accept the relatively small fraction of TeBG bound by the affinity column (see Table).

[19] J. H. Leudens, J. R. de Vries, and D. D. Fanestil, *J. Biol. Chem.* **247,** 7533 (1972).

[20] The increase in concentration of Tris and Ca^{2+} is in preparation for the dithionite cleavage of the ADA. With lower concentrations of Tris there is a fall in pH. Dithionite is known to inactivate TeBG,[21] but we have found that high Ca^{2+} concentrations markedly slow the rate of inactivation.[16]

[21] W. H. Pearlman, I. F. F. Fong, and J. H. Tou, *J. Biol. Chem.* **244,** 1373 (1969).

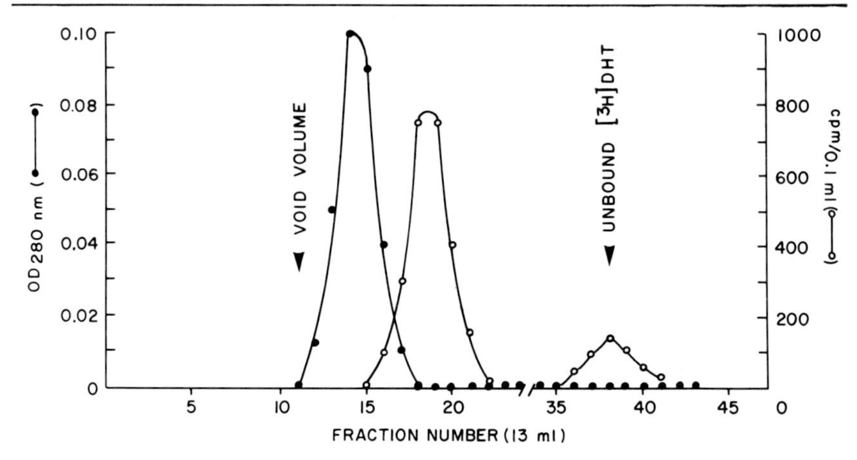

FIG. 2. Sephadex G-150 chromatography of the eluate from affinity chromatography. The column is 2.6 × 92 cm and fractions are collected at 15 ml/hour.

for cleavage. The reagent for cleavage consists of 4 mg [^{3}H]DHT (1000 cpm/μg) and 3.5 g dithionite (20 mmoles) in 200 ml of buffer 8b.[22] Transfer the gravity-drained gel to a beaker, add 200 ml of the cleavage mixture, and stir until the gel has lost most of its color, about 30–45 minutes. Transfer the gel to a coarse sintered glass funnel and recover the cleavage mixture by suction. Wash the gel with 200 ml of buffer 8b, add the wash to the recovered cleavage mixture, and make up to 1 liter with the same buffer.[24] Concentrate the diluted cleavage mixture to 2–3 ml by pressure ultrafiltration through PM 10 membranes (Amicon Corp.). This can be done in 2–3 hours by the sequential use of Amicon cells, numbers TCF-10 and 8MC. Measure accurately the volume of the concentrated sample, count 25–30 μl, and raise the specific activity of the DHT by adding the sample to about 2×10^6 cpm [^{3}H]DHT, previously dried in a small tube.[25] Chromatograph the sample on a 2.5 × 30-cm column of Sephadex G-50 (packed in and eluted with buffer 8a) to remove the excess DHT, dithionite, and [17β-hydroxyandrostan-3-yl succinamide]-*p*-aminoben-

[22] It is important that the cleavage mixture stands for *3 hours* prior to use. This assures solution of DHT but more importantly yields the proper concentration of the active species of the dithionite. Dithionite deteriorates in soluton[23] and since it is a reagent capable of denaturing TeBG,[21] standard conditions must be adhered to. The DHT serves both to stabilize the protein and to act as a marker in the remainder of the purification.

[23] M. Wayman and W. J. Lem, *Can. J. Chem.* **48**, 782 (1970).

[24] The dilution of the dithionite by a factor of 5 retards its harmful effects on TeBG. Further, we have found that higher concentrations of dithionite clog the ultrafiltration membranes.

[25] Precise knowledge of the adjusted specific activity is important since all calculations of binding capacity are based on the measurement of tritium activity.

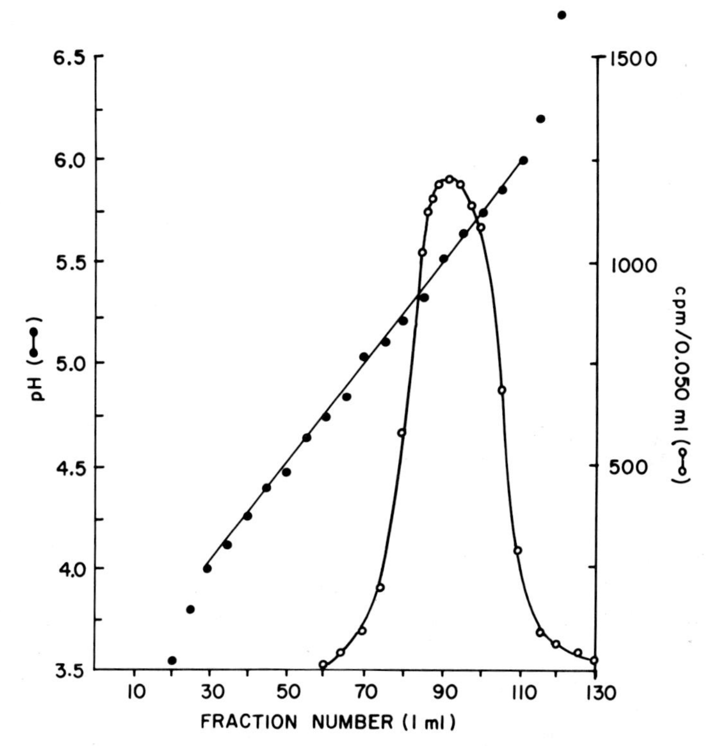

FIG. 3. Isoelectric focusing using ampholytes in the range from pH 4–6. In addition to the standard sucrose gradient used in isoelectric focusing, this column is made 10% in glycerol. Fractions 17–21 from Sephadex G-150 (Fig. 2) were pooled, concentrated, and applied at the center of the gradient in a volume of 1 ml.

zene (the dithionite cleavage product). The protein can be detected in the void volume at 280 nm along with 10–12% of the radioactivity. Concentrate the protein to a volume of 5 ml in preparation for the next chromatography. The sequence from affinity chromatography through the Sephadex G-50 column takes place during the course of one working day. *The protein is not left in the dithionite solution overnight.*

Chromatography on Sephadex G-150. Enrich the concentrated G-50 eluate with an additional 2×10^6 cpm of [^{3}H]DHT[25] and chromatograph on a 2.6×92-cm column of Sephadex G-150 packed and eluted with buffer 8a at a rate of approximately 15 ml/hour. Pool the tubes containing bound radioactivity (see Fig. 2) and again concentrate by pressure ultrafiltration to a volume of 1.0 ml in preparation for isoelectric focusing.

Isoelectric Focusing. Conduct isoelectric focusing experiments in an LKB model 8101 column (110 ml capacity) according to the instructions supplied by the manufacturer. Focusing is achieved from pH 4–6, and the final ampholyte concentration is 1%. Form the sucrose gradient (0–40%) with the LKB gradient maker with the addition of 10% glycerol to both chambers.[15] When approximately one-half of the gradient has been formed add the sample (1 ml) mixed with an equal volume of heavy solution by layering it on top of the portion of the gradient already formed. Add the remaining 50% of the gradient. Cool the column at 4° and set the initial voltage at 500. The amperage is 2–3 mA and care should be taken not to exceed 1.5 W. Approximately 18 hours later increase the voltage to 800 V and terminate the column in another 5–7 hours. Drain the column with a peristaltic pump collecting 1-ml fractions and measure the pH at 4° and radioactivity in 50-μl aliquots. Pool the fractions containing [^{3}H]DHT (there is too little optical density to detect) (Fig. 3), concentrate to a volume of 5 ml by ultrafiltration, and

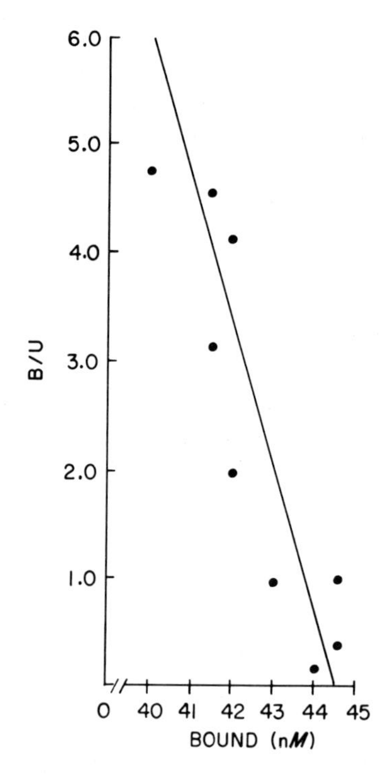

FIG. 4. Equilibrium dialysis of the protein isolated from isoelectric focusing, fractions 85–100. Dialysis is in a Tris-calcium buffer and takes place over 18 hours at 4°. There is 2.8 μg of protein in 0.5 ml inside the dialysis bag and 6.0 ml of buffer outside. The line is fitted by the method of least squares and reveals an association constant of $1.1 \times 10^9\ M^{-1}$ and a capacity of 2.3 μg DHT/mg protein.

remove the ampholytes and sucrose on a 2.5 × 30-cm column of Sephadex G-50 in buffer 8a. The protein eluted from this column is used for protein determination (biuret) after concentration, equilibrium dialysis, and polyacrylamide gel electrophoresis.

Analysis of the Isolated Protein

Equilibrium Dialysis

Conduct equilibrium dialysis in screw-top tubes at 4° using 6 ml of buffer 8a outside the dialysis sac and 0.5 ml of the protein solution from the last G-50 column inside the dialysis sac. The protein should be diluted so that its concentration is about 5.6 μg/ml (biuret), and an additional 10,000 cpm of [^{3}H]DHT should be added to the outside solution as there are only 400 cpm on the 2.8 μg of protein in the dialysis sac. Figure 4 illustrates the results of the equilibrium dialysis. The association constant for DHT calculated from this plot is 1.1×10^9 M^{-1}. The range of bound steroid values is small (40–45.5 nM), because the isolated protein is fully bound. More adequate plots would be obtained if the protein were stripped of steroid, but we have not yet examined the stability of the isolated protein in the absence of DHT.

Polyacrylamide Gel Electrophoresis (PAGE)

Conduct PAGE in duplicate according to Davis[26] but let electrophoresis take place at 4°.[7] Concentrate the sample from the last G-50 column so that 25 μg of protein (biuret) can be applied in 50–100 μl. Stain one gel and slice the other for determination of radioactivity (25 μg contains about 4000 cpm and no additional [^{3}H]DHT need be added). The results of PAGE are shown in Fig. 5. The [^{3}H]DHT migrated with the central, heaviest staining protein band.

Discussion

The isolation of active TeBG has eluded investigators for several years. Although we have not presented the supporting data in this chapter, a major reason for this is the specific choice of buffer systems and the type of isolation procedure used. Calcium is an extremely important stabilizing agent, and any isolation technique that removes calcium during the course of purification will have a deleterious effect.[16] Further,

[26] B. J. Davis, *Ann. N.Y. Acad. Sci.* **121**, 321 (1964).

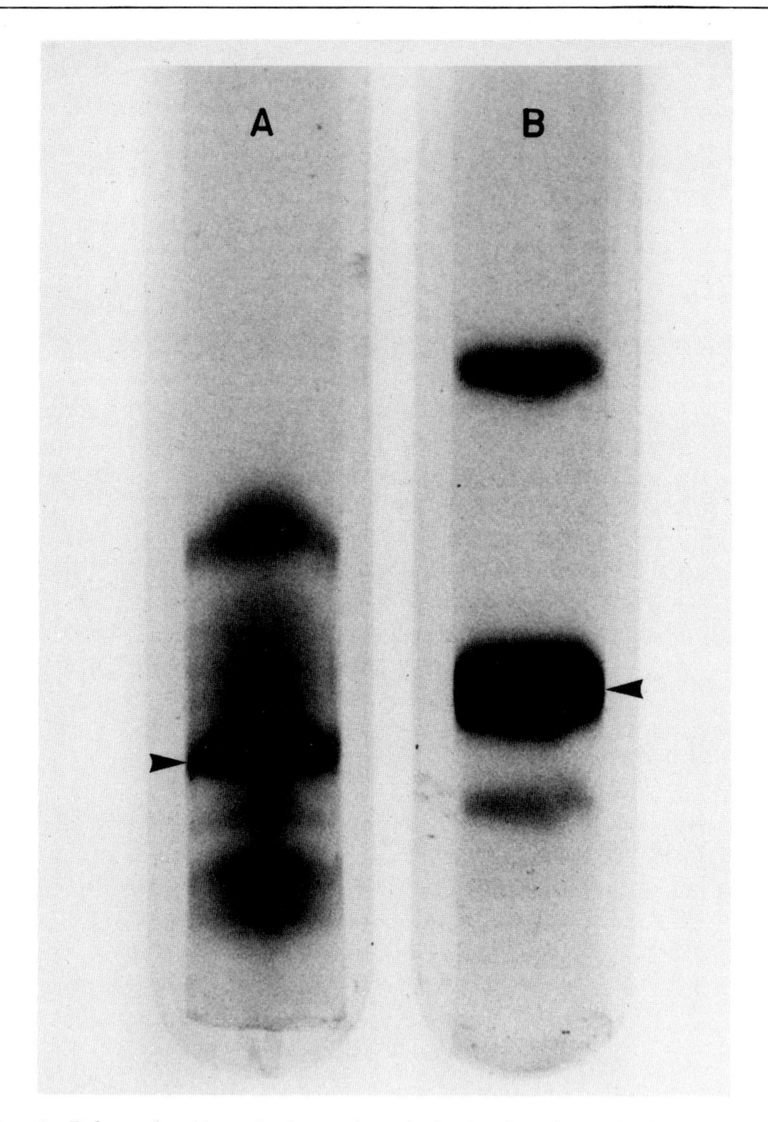

FIG. 5. Polyacrylamide gel electrophoresis in 7.2% polyacrylamide with 2.6% cross-linkage and an operative pH of 9.5. Tris-glycine buffer, pH 8.3 at 4°, is used in both anode and cathode chambers. The anode and direction of migration are toward the top. (A) Eluate from affinity column. (B) Final preparation after isoelectric focusing. Not shown are duplicates of the gels which were assayed for [^{3}H]DHT. The radioactivity migrated with an R_f of 0.35 (horizontal arrows), the same R_f which we find in whole serum. The apparent difference in the rate of migration of the TeBG on the two gels is corrected when R_f values are used.

TABLE
SUMMARY OF ISOLATION OF TeBG

Procedure	Protein applied	Activity applied[a]	Protein recovered	Activity recovered	Specific activity[b]	Purification from previous step	Cumulative purification	Cumulative recovery
Cohn fraction IV	280 g	82 μg	—	—	0.3	—	—	—
Ammonium sulfate	280 g	82 μg	160 g	82 μg	0.5	1.7	1.7	100%
Androstanediol sepharose	160 g	82 μg	30 mg	10.6 μg[c]	353	700	1200	60%[c]
Sephadex G-150	30 mg	10.6 μg	6.5 mg	10.5 μg	1615	4.6	5470	59%
Isoelectric focusing	6.5 mg	10.5 μg	3.0 mg	7.0 μg	2333	1.4	7670	40%

[a] Activity is expressed as μg DHT bound. One microgram of binding activity is equivalent to 0.35 mg TeBG (see text).
[b] The specific activity is defined as: activity[a]/total protein (g). The theoretical specific activity at purity is 2900 μg/g.
[c] The gel bound 17.6 μg, of which 10.6 μg were eluted and 64.4 μg recovered in the supernatant. Recoveries subsequent to this step are based on the 17.6 μg of activity bound to the gel.

TeBG is more stable both in Tris and imidazole buffers than in phosphate buffers,[16] yet phosphate buffers were used either exclusively or at some stage in previous attempts at purification.[1-6]

We also found that adsorption of TeBG to ion-exchange resins causes major losses of activity when recoveries are examined using quantitative techniques.[16] If one chooses an ion-exchange system that allows free passage of TeBG then no activity is lost but almost no purification is achieved, i.e., contaminating proteins are also not adsorbed. This eliminates the one commonly used kind of chromatography that has both high capacity and the potential for good separations. Our solution to the problem uses affinity chromatography for the major purification, a method allowing the use of large amounts of starting material and yielding reasonable quantities of active TeBG.

The final preparation of TeBG (Table and Fig. 5) is not completely pure, and we cannot be certain it is completely active. The evidence, however, indicates that the isolated protein is most probably fully active. The state of purity, as estimated from the specific activity, (see Table) indicates 20% contamination. This figure is in good agreement with the results of PAGE (Fig. 5) which show that 80–85% of the stainable protein migrates as TeBG. This agreement would be unlikely if there are serious errors in our estimates of the specific activity.

We have stored the protein at 4° in Tris-calcium containing 10% glycerol for three weeks without detectable loss of activity, and this stability should allow final purification on a narrower pH gradient as a final step.

[11] Purification of Progesterone-Binding Plasma Protein (PBP) from Pregnant Guinea Pigs

By Edwin Milgrom, Pierre Allouch, and Michel Atger

During pregnancy a progesterone-binding protein (PBP) appears in guinea pig plasma.[1,2] It may be distinguished from corticosteroid-binding globulin (CBG) by many techniques among which are sucrose gradient ultracentrifugation, Sephadex G-200 chromatography, and resistance to heating. The purification of this protein is easier than that of other steroid-binding proteins since its concentration is 10–100 times higher (15 μM at the end of the pregnancy).

[1] E. Milgrom, M. Atger, and E.-E. Baulieu, *Nature (London)* **228**, 1205 (1970).

[2] R. M. Burton, G. B. Harding, N. Rust, and U. Westphal, *Steroids* **17**, 1 (1971).

Purification[3]

Reagents

1. [^{3}H]Progesterone (specific activity 20–50 Ci/mmole) purified by alumina column chromatography[4]
2. Ammonium sulfate ultra-pure (Mann)
3. Sephadex G-200 and DEAE Sephadex A-50 (Pharmacia)
4. Buffer Tris 0.01 *M* HCl pH 8.6 containing 0.11 or 0.13 *M* KCl (Tris-KCl buffer)

Procedure (*Table I*). Forty- to sixty-day pregnant Hartley guinea pigs are used. When the actual age of pregnancy is unknown it is possible to obtain an approximate value by measuring the size and weight of the fetuses.[5] After ether anesthesia, blood is collected on heparin by cutting the blood vessels of the neck. After centrifugation 1200 $g \times 15$ minutes the plasma is kept frozen at —20° (no decrease in binding is observed after 3 months).

Seventy-two (72) ml of plasma are incubated with 0.1 μM [^{3}H]progesterone for 15 minutes at 37°.

TABLE I
PBP PURIFICATION FROM PREGNANT GUINEA PIG PLASMA

	Bound [^{3}H]progesterone (dpm $\times 10^6$)	Protein (mg)	% Recovery of bound [^{3}H]progesterone	Specific activity[a]	Purification
Plasma	67.26	4,460	100	1.5	1
Heating	67.02	4,200	100	1.6	1
Ammonium sulfate precipitation	41.76	1,320	62	3.2	2
Sephadex G-200 chromatography	36.40	136	54	26.8	18
DEAE Sephadex chromatography	14.94	18	22	83.0	55
2nd Sephadex G-200 chromatography	14.05	13.5	21	104.1	69

[a] Specific activity = [^{3}H]progesterone dpm/mg protein $\times 10^4$. Seventy-two (72) ml of plasma were used. Protein concentration was measured by using optical density at 280 nm ($E^{1\text{ cm}}_{1\text{ mg/ml}} = 1$).

[3] E. Milgrom, P. Allouch, M. Atger, and E.-E. Baulieu, *J. Biol. Chem.* **248,** 1106 (1973).

[4] T. K. Lakshmanan and S. Lieberman, *Arch. Biochem.* **53,** 258 (1954).

[5] R. L. Draper, *Anat. Rec.* **18,** 369 (1920).

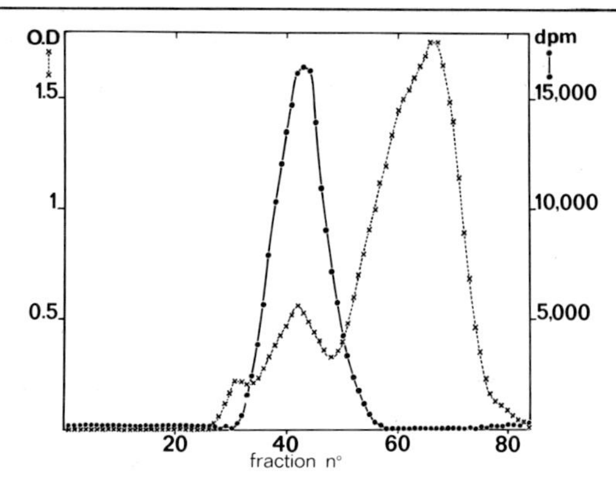

FIG. 1. Purification of PBP, first Sephadex G-200 chromatography. Seventy-two (72) ml of pregnant guinea pig plasma was treated as described in purification steps 1 and 2. Half of the partially purified material was applied to a Sephadex G-200 column equilibrated with Tris 0.11 *M* KCl buffer (diameter, 5 cm; gel height, 37 cm, 8-ml fractions; 4 fractions per hour; 0°). In each fraction absorption at 280 nm was measured and an aliquot (50 μl) was used to measure radioactivity.

STEP 1. The plasma is then heated 30 minutes at 60° (CBG is thus eliminated). After a 35,000 $g \times 30$ minutes centrifugation the supernatant is saved for further steps all performed at 0–4°.

STEP 2. The plasma is diluted with 1 volume of deionized water and solid ammonium sulfate added while stirring to obtain 50% saturation (pH being kept at 7 with HCl if necessary). The mixture is kept 4–16 hours at 0° and the precipitate eliminated by centrifuging 30 minutes at 35,000 *g*. Ammonium sulfate is added to the supernatant in order to obtain 80% saturation. The precipitate obtained as previously described is dissolved in Tris 0.11 *M* KCl buffer (0.16 ml/1 ml of plasma initially used for purification) and dialyzed 48 hours against 2 liters of the same buffer which is changed three times.

STEP 3. The dialyzate is then chromatographed on a Sephadex G-200 column in two runs (Fig. 1). The fractions containing the peak of bound radioactive progesterone from both chromatographies are pooled, concentrated to 0.03 ml/1 ml of purified plasma, and dialyzed for 24 hours against 1 liter of Tris 0.11 *M* KCl.

STEP 4. The partially purified protein is chromatographed on a column of DEAE Sephadex A-50 (Fig. 2). The column is washed with Tris 0.11 *M* KCl buffer and PBP eluted by Tris 0.13 *M* KCl. The bulk of the proteins may be eluted by 1 *M* KCl.

STEP 5. The appropriate fractions are pooled and concentrated and rechromatographed on a Sephadex G-200 column in Tris 0.11 *M* KCl at 0°. The central part of the radioactive peak is kept (Fig. 3).

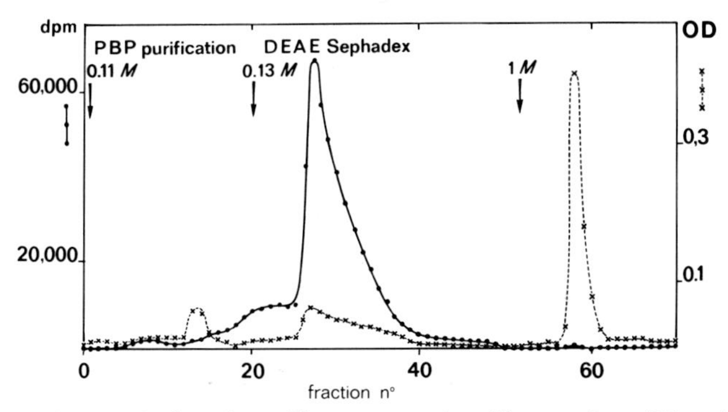

FIG. 2. DEAE Sephadex A-50 Chromatography. Twenty-five (25) ml pregnant guinea pig plasma was treated as described in steps 1 to 3 of the purification procedure. The 3.5 ml of the partially purified PBP solution was equilibrated by dialysis with the Tris 0.11 *M* KCl buffer and was layered on an A-50 DEAE Sephadex column (22 × 1.5 cm) and equilibrated with the same buffer. Elution was with Tris 0.11 *M* KCl buffer (160 ml), Tris 0.13 *M* KCl buffer (250 ml), and Tris 1 *M* KCl buffer (250 ml). The chromatography was performed at 0° with a flow of 12 ml/hour and 8-ml fractions. Absorbance at 280 nm and radioactivity was measured in each fraction.

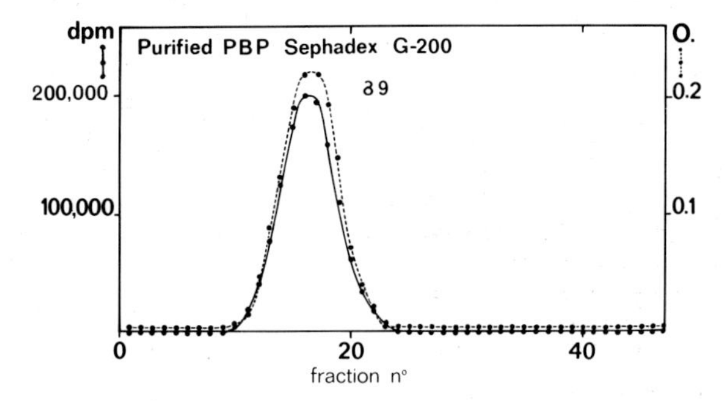

FIG. 3. Second Sephadex G-200 chromatography. Purified PBP (9 mg in 2.5 ml Tris 0.11 *M* KCl buffer) was chromatographed at 0°, through a Sephadex G-200 column (height, 32 cm; diameter, 1.5 cm) equilibrated with the same buffer. Flow rate, 10 ml/hour; 2-ml fractions.

Homogeneity of the Purified Protein[3]

The purified material shows superposition of bound [^{3}H]progesterone and optical absorption at 280 nm when examined by Sephadex G-200 chromatography and sucrose gradient ultracentrifugation.

A single sharp peak is observed during sedimentation velocity measurements in the analytical ultracentrifuge. Polyacrylamide gel electrophoresis of the native protein is unsuccessful in various buffer systems PBP (as detected by radioactivity), being spread across the first one-third of the gel. This is probably due to aggregation or nonstacking. No other protein is detected.

Gel electrophoresis in presence of SDS[6] shows one single protein band.

TABLE II
AMINO ACID AND CARBOHYDRATE COMPOSITION OF PBP[a]

	g/100 g Polypeptide	g/100 g Glycoprotein	Residue/mole glycoprotein
Lysine	7.1 ± 1	3.6	19.1
Histidine	4.2 ± 0.5	2.2	11.0
Arginine	4.8 ± 0.4	2.5	11.1
Aspartic acid	11.3 ± 0.7	5.8	33.7
Threonine	7.0 ± 0.4	3.6	23.4
Serine	6.1 ± 0.1	3.1	22.8
Glutamic acid	10.8 ± 0.1	5.5	28.9
Proline	4.1 ± 0.9	2.1	14.1
Glycine	3.1 ± 0.4	1.6	16.5
Alanine	3.6 ± 0.1	1.8	15.6
½ Cystine	0.5 ± 0.4	0.3	1.9
Valine	5.3 ± 0.4	2.7	17.8
Methionine	2.1 ± 0.2	1.1	6.0
Isoleucine	4.3 ± 0.5	2.2	13.0
Leucine	12.1 ± 0.3	6.2	36.6
Tyrosine	4.3 ± 0.2	2.2	9.4
Phenylalanine	8.6 ± 0.9	4.4	20.6
Tryptophane	2.1	1.1	4.1
Hexoses	—	18.8	81.3
Hexosamines	—	18.5	80.0
Fucose	—	0.4	2.1
Sialic acid	—	11.0	33.3

[a] Amino acid composition of three different preparations of PBP were determined with a Technicon amino acid analyser. The values were corrected for losses during hydrolysis by the use of a standard solution of amino acids submitted to the same procedure. Average value ± SEM are given. Tryptophan content was measured by the spectrophotometric technique of Edelhoch.[7] Carbohydrate content values are the mean of two different determinations. The techniques of Rimington,[8] Blix,[9] Svennerholm,[10] and Dische and Shettles[11] were used for the determination of hexoses, hexosamines, sialic acid, and fucose, respectively.

[6] R. Weber and M. Osborn, *J. Biol. Chem.* **244,** 4406 (1969).

[7] M. Edelhoch, *Biochemistry* **6,** 1948 (1967).

[8] C. Rimington, *Biochem. J.* **34,** 931 (1940).

[9] G. Blix, *Acta Chem. Scand.* **2,** 467 (1948).

[10] L. Svennerholm, *Biochim. Biophys. Acta* **24,** 604 (1957).

[11] Z. Dische and L. B. Shettles, *J. Biol. Chem.* **175,** 595 (1948).

Characteristics of the Purified PBP[3]

Physicochemical Characteristics

The protein is stable (as judged by binding properties) between pH 5 and 9 and at all the ionic strengths tested (0.01–2.00).

Its molecular weight (equilibrium sedimentation, polyacrylamide electrophoresis in sodium dodecylsulfate) is 77,500, whereas the sedimentation coefficient ($s^{\circ}_{20,w}$) is 4.5.

The Stokes radius of PBP (Sephadex G 200 chromatography) is 47 Å, the fractional ratio (f/f_0 calculated from Stokes radius and molecular weight[12]) 1.685, its pHi (isoelectrofusing) 3.6.

The extinction coefficient of PBP ($E^{1\ cm}_{1\ mg/ml}$) is 0.49.

The chemical composition of this glycoprotein is given in Table II.

TABLE III
STEROID SPECIFICITY OF PBP[a]

Unlabeled steroid	Competing efficiency with [³H]progesterone
Progesterone	100
5α-Pregnane-3,20-dione	97
20α-Hydroxypregn-4-en-3-one	90
21-Hydroxypregn-4-ene-3,20-dione	86
17β-Hydroxyandrostan-3-one	69
Testosterone	40
3β-Hydroxypregn-5-en-20-one (pregnenolone)	33
5β-Pregnane-3,20-dione	32
20β-Hydroxypregn-4-en-3-one	14
17α-Hydroxypregn-4-ene-3,20-dione	11
Cortisol	0
Corticosterone	0
Estradiol	0

[a] Pregnant guinea pig plasma was diluted 80-fold with Tris-NaCl buffer and incubated 3 hours at 0° with 10 nM [³H]progesterone and in some experiments 1 μM unlabeled steroid. Two-fifths (0.2) ml of the incubate was centrifuged 18 hours at 45,000 rpm in a SW 50.1 rotor. Bound radioactivity (B) was calculated. Competing efficiency of various steroids was compared to that of unlabeled progesterone by calculating

$$\frac{\text{(B) with [}^3\text{H]progesterone alone} - \text{(B) with [}^3\text{H]progesterone and competing steroid}}{\text{(B) with [}^3\text{H]progesterone alone} - \text{(B) with [}^3\text{H]progesterone and unlabeled progesterone}} \times 100$$

[12] L. M. Siegel and K. J. Monty, *Biochim. Biophys. Acta* **112**, 346 (1966).

Steroid Binding

PBP binds 1 mole of progesterone per mole of protein and has an affinity for progesterone of $9.10^8\ M^{-1}$ (K_A at 4°) and for testosterone of $1.6\ 10^7\ M^{-1}$.

The relative binding affinities of various steroids for PBP are given in Table III.

Distribution of PBP

PBP is found during pregnancy only in the maternal plasma and cannot be detected in the fetus or in the umbilical vein or arteries.

PBP is absent in nonpregnant animals and cannot be induced by injections of estradiol, progesterone, or a combination of both.

[12] Methods for Measuring the Thyroxine-Binding Proteins and Free Thyroid Hormone Concentration in Serum[1]

By RALPH R. CAVALIERI and SIDNEY H. INGBAR

Background

The two thyroid hormones which exist in the circulation, L-thyroxine (T_4) and 3,5,3′-triiodo-L-thyronine (T_3), are almost entirely bound to specific plasma proteins. In man, T_4 is bound by an inter-α-globulin, thyroxine-binding globulin (TBG), by a pre-albumin (TBPA), identical to retinol-binding protein, and by serum albumin. T_3 is bound by TBG and by albumin, but not to any significant extent by TBPA. The generally accepted concept that the protein-bound fraction of the hormones is physiologically inert, while the unbound fraction is the active form, makes measurement of the latter important in physiological and clinical studies. Since the concentration of unoccupied sites on the binding proteins is a major determinant of the proportion of total circulating hormone present in the unbound form, measurement of the binding capacities of the thyroid hormone-binding proteins also assumes importance.

[1] Supported in part by Research Grant No. AM 16497 from the National Institute of Arthritis and Metabolic and Digestive Diseases, National Institute of Health, Bethesda, Maryland.

The relationship between the free and bound forms of the hormone can be considered as a dynamic equilibrium

$$T + P_i \rightleftharpoons TP_i$$

and

$$K_i = \frac{[TP_i]}{[T][P_i]}$$

where T is the free hormone, TP_i is the occupied binding sites, P_i is the unoccupied binding sites, K_i is the association constant for any class of binding site, i, and is molar concentration. The approximate association constants, in liters per mole, for TBG are: $K_{T_4} = 2\text{–}4 \times 10^{10}$ and $K_{T_3} = 1\text{–}2 \times 10^9$; for TBPA, $K_{T_4} = 1\text{–}20 \times 10^7$; and for albumin, $K_{T_4} = 6\text{–}16 \times 10^5$ and $K_{T_3} = 2\text{–}3 \times 10^5$.

The reader is referred to detailed reviews[1a-3] for more complete discussion of the theoretical and physiological aspects of hormone binding.

The proportion of unbound hormone can be measured directly either by equilibrium dialysis or by ultrafiltration, but the former is more widely used. The concentrations of the major saturable binding proteins, TBG and TBPA, are generally measured in terms of their maximal binding capacity. However, the concentration of the protein can be measured directly by electrophoretic techniques in the case of TBPA, and by immunochemical methods in the case of both TBPA and TBG.

Measurement of the Proportion of Free Thyroxine in Serum

General Considerations. In principle, the method involves the addition of a tracer quantity of radiolabeled T_4 to the test serum and separation of the unbound from the protein-bound fraction. Two general methodological approaches have been employed to achieve this separation: equilibrium dialysis and ultrafiltration. In either case, one major drawback is encountered. All preparations of radioiodine-labeled T_4 are contaminated in varying degrees with labeled iodide and in lesser amounts with organic iodinated constituents [such as 3,5,3′-triiodothyronine (T_3) and 3,3′,5′-triiodothyronine (reverse T_3)].[4,5] All such contaminants are less

[1a] J. H. Oppenheimer and M. I. Surks, *in* "The Thyroid" (S. C. Werner and S. H. Ingbar, eds.), 3rd ed., pp. 52–65. Harper, New York, 1971.

[2] K. A. Woeber, *in* "The Thyroid" (S. C. Werner and S. H. Ingbar, eds.), 3rd ed., pp. 256–266. Harper, New York, 1971.

[3] J. Robbins, *in* "Methods in Investigative and Diagnostic Endocrinology" (S. A. Berson, ed.), Vol. 1, pp. 241–254. Amer. Elsevier, New York, 1972.

[4] G. C. Schussler and J. E. Plager, *J. Clin. Endocrinol. Metab.* **27**, 242 (1967).

[5] E. M. Volpert, M. Martinez, and J. H. Oppenheimer, *J. Clin. Endocrinol. Metab.* **27**, 421 (1967).

firmly bound to serum proteins than is T_4. The result is that, unless steps are taken to circumvent the difficulties that these contaminants pose, the apparent percent of free T_4 is spuriously high. In the case of the equilibrium dialysis method, two steps are taken to avoid this problem. First, the preparation of labeled T_4 is subjected to preliminary purification, by either chromatography or preparative dialysis, in order to remove a large proportion of the less firmly bound labeled contaminants. Second, following equilibrium dialysis, the labeled T_4 is separated from labeled non-T_4 components in the diffusate by one or another procedure. The published methods involving equilibrium dialysis differ mainly in the approach used in this latter procedure.[6-8]

Procedure

Preliminary Purification of Labeled T_4. Fifty to one-hundred microcuries of radiodinated (^{125}I) T_4, specific activity 35 to 70 μCi/μg (Abbott Laboratories, North Chicago, Illinois), are added to 2 to 4 ml human serum albumin (HSA), 3 gm/100 ml, in 0.15 *M* sodium phosphate buffer, pH 7.4, or in 0.9% sodium chloride. The labeled T_4–HSA mixture is placed inside a sac made from dialysis tubing (Visking casing, 23 mm in diam., Union Carbide Corp., Chicago, Illinois). The sac is sealed and completely immersed in a vessel containing at least 2 liters distilled water to which is added 40 mg methimazole in order to minimize oxidative degradation of T_4. Dialysis is performed at 4° for 16–20 hours, following which the contents of the sac are removed and stored at 4° until used.

Equilibrium Dialysis Procedure. Fifty microliters of the dialyzed labeled T_4–HSA solution are added to 1.5 ml of the serum to be tested (approximately 0.8 μCi labeled T_4/ml serum). Sacs are made from dialysis membrane tubing, ¼ in. wide, which is prepared by soaking the membrane in distilled water at 4° for 1 hour. Exactly 1 ml of the labeled T_4–serum mixture is placed in the sac, which is then sealed and completely immersed in 5.0 ml 0.15 *M* sodium phosphate buffer, pH 7.4, within a 50-ml plastic (Lusteroid) round-bottomed centrifuge tube. The tube is stoppered and incubated in a shaking water bath at 37° for 20 hours (sufficient time to achieve equilibrium in the system). Following dialysis, 1 ml of the outside buffer (diffusate) is added to 1 ml outdated blood bank plasma. The proteins in this mixture are precipitated with

[6] J. H. Oppenheimer, R. Squef, M. I. Surks, and H. Haver, *J. Clin. Invest.* **42,** 1769 (1963).

[7] S. H. Ingbar, L. E. Braverman, N. A. Dawber, and G. Y. Lee, *J. Clin. Invest.* **44,** 1976 (1965).

[8] K. Sterling and M. A. Brenner, *J. Clin. Invest.* **45,** 153 (1966).

8 ml cold 20% trichloroacetic acid (TCA) containing carrier iodide (0.2 mg KI/ml), the precipitate is washed twice with 8 ml each of 10% TCA, and the radioactivity of the precipitate is determined (sample A). A counting standard is prepared by added 100 μl buffer of the original labeled T_4-test serum to a mixture of 1 ml outdated plasma and 0.9 ml buffer. The proteins in the standard are precipitated by TCA as described above, and the radioactivity in the residue is determined (sample B).

Calculation of Results

$$\text{Percent free } T_4 = \frac{\text{net counting rate in A} \times 100}{\text{net counting rate in B} \times 10}$$

In the serum of normal adults, with na abnormalities in thyroxine binding, the percent free T_4 by this method ranges from 0.020 to 0.040, averaging 0.030%. The particular values found are influenced by small methodological variations. Therefore, when needed, the normal range should be established in each laboratory. The absolute level of free T_4 is calculated as the product of the percent free T_4 times the total concentration of T_4 in the serum, determined separately by any one of several methods, preferably one based on the competitive protein-binding principle.[9]

The proportion of free T_3 in serum can be measured by methods entirely comparable to those described above for measuring the proportion of free T_4. In normal serum, values average about 10 times those for free T_4, approximately 0.30%.[7] The absolute concentration of free T_3 can then be calculated as the product of the proportion of free T_3 and the total concentration of T_3 (see chapter by Surks and Oppenheimer.)

Determination of the Maximal Thyroxine-Binding Capacity of Serum TBG and TBPA by Paper Electrophoresis

General Considerations. The method involves the addition of labeled T_4 and sufficient amounts of unlabeled T_4 to saturate the binding sites on the specific proteins. Following separation of the proteins by electrophoresis, the proportions of labeled T_4 associated with the specific binding components are determined and used together with the known concentration of added plus endogenous T_4 in the sample to calculate the maximal binding capacity of each protein. Various electrophoretic support media have been employed; cellulose acetate and agar films adsorb less labeled

[9] L. E. Braverman, A. G. Vagenakis, A. E. Foster, and S. H. Ingbar, *J. Clin. Endocrinol. Metab.* **32**, 497 (1971).

T_4 than does filter paper, but the use of the latter is technically simpler and is quite satisfactory, provided certain precautions are taken. In the determination of TBG capacity using filter paper electrophoresis the reverse-flow method of Robbins is recommended.[10] In this technique, the hydrodynamic flow of buffer nearly counterbalances the electrophoretic movement of proteins toward the anode, with the result that albumin and pre-albumin migrate anodally from the origin and TBG moves cathodally. This avoids the spuriously high values for TBG capacity observed in the conventional paper electrophoretic method caused by trailing of albumin and pre-albumin over the path traversed by TBG. In the determination of TBPA capacity the reverse-flow technique is not required, owing to the rapid anodal migration of this protein. Indeed, the separation of TBPA from albumin is less satisfactory in the reverse-flow than in the conventional system. Factors of buffer composition and pH appear to influence the values obtained for TBPA capacity to a greater extent than they do the TBG capacity. Barbital inhibits binding of T_4 to TBPA.[11] Glycine-acetate buffer yields considerably higher values than Tris-maleate buffer, both at pH 8.6. Oppenheimer *et al.*[12] have compared the results obtained with glycine-acetate buffer at pH 8.6 and 9.0, employing paper electrophoresis in a Durrum-Spinco apparatus. They found that the TBPA capacity was somewhat greater at the higher pH and there was also less variation using different lots of filter paper.

Although, in the methods to be described, a Tris-maleate buffer is recommended for TBG capacity and a glycine-acetate buffer for TBPA capacity determination, the latter buffer is quite satisfactory for both proteins. Furthermore, the quantities of stable T_4 in stock solutions and the dilutions employed can be varied to provide from a single weighing, concentrations of added T_4 which are appropriate for each of the proteins.

TBG Capacity. The method to be described is based on the reverse flow technique of Robbins, modified by Elzinga *et al.*,[13] for use in the standard Durrum-type electrophoretic apparatus (Spinco, Model R, Series D). Tris-maleate buffer, 0.1 *M*, pH 8.6, is prepared as follows: 24.2 g Tris (hydroxymethyl) aminomethane (Sigma Chemical Co., St. Louis, Missouri), 23.2 g maleic acid, and 173 ml 2 *N* NaOH are added to sufficient distilled water to make 2 liters. The final pH is adjusted to 8.6 with NaOH or HCl. The anodal buffer chamber of the electropho-

[10] J. Robbins, *Arch. Biochem. Biophys.* **63**, 461 (1956).

[11] S. H. Ingbar, *J. Clin. Invest.* **42**, 143 (1963).

[12] J. H. Oppenheimer, M. Martinez, and G. Bernstein, *J. Lab. Clin. Med.* **67**, 500 (1966).

[13] K. E. Elzinga, E. A. Carr, and W. H. Beierwaltes, *Amer. J. Clin. Pathol.* **36**, 125 (1961).

retic cell is filled with 525 ml and the cathodal side with 475 ml of buffer. Strips of filter paper (Schleicher and Schuell No. 2043, 3.0 × 30.6 cm) wet with buffer are placed in the cell, which holds a maximum of eight strips. Heavy filter paper sections (S and S No. 470, 31.8 × 5.1 cm) are used as wicks connecting the ends of the strips with the buffer reservoirs. The cell is closed for a preliminary period (2 hours) of equilibration to saturate the inside atmosphere. Preparation of serum samples with labeled and stable T_4: ^{125}I-labeled T_4 is diluted with 1% human serum albumin (HSA) to a concentration of approximately 120 μCi/ml, and 7 mg synthetic L-thyroxine, or its equivalent as the sodium salt (Sigma Chemicals), is dissolved in 2 *N* NaOH and made up to 100 ml with 1% HSA, yielding a solution containing 3.5 μg T_4/50 μl. To 3 ml of the serum to be tested is added 50 μl of the labeled T_4 solution to yield a concentration of approximately 2 μCi/ml serum (solution A). To exactly 2 ml of solution A is added 50 μl of the stable T_4, yielding a concentration of added stable T_4 of 1.71 μg/ml (solution B). One milliliter each of solutions A and B are mixed to form solution C, containing 0.85 μg/ml of added stable T_4. Two samples each of solutions B and C are employed for the determination of TBG capacity. Electrophoresis: Following the preliminary equilibration, the Durrum-type cell is opened and a 20-μl sample of each solution is applied to the filter paper strips at a point 8 cm from the anodal end of each strip. (Immediately prior to the application of the sample, excess buffer is removed from the point of origin by gentle blotting with the edge of a piece of dry filter paper wick, which procedure minimizes unwanted diffusion of serum.) The cell is sealed and a constant current of 10 mA is applied for 20 hours at 25 ± 2°. Following electrophoresis, the paper strips are dried quickly at 110–120°. The amount of radioactivity in each region of the strip is determined either by scanning, or by radioautography followed by well-counting of the individual regions. The TBG area is identified as the radioactive zone closest to the cathodal end.

TBPA Capacity. Serum is labeled with ^{125}I T_4 (2 μCi/ml) and enriched with stable T_4 to a level of approximately 6 μg/ml. (Twelve milligrams synthetic L-T_4 are dissolved in diluted NaOH and 1% HSA to a total volume of 50 ml. To 2 ml of serum is added 50 μl of the stable T_4 solution, yielding a final concentration of added stable T_4 of 5.85 μg/ml.) For electrophoresis, glycine-acetate buffer, pH 9.0, is employed (7.8 g glacial acetic acid, 15.0 g glycine, 6.8 g NaOH pellets, and sufficient water to make 1 liter, adjusting to final pH with 50% NaOH or acetic acid). A Durrum-Type cell may be used, with equal volumes of buffer in each reservoir, since the reverse-flow technique is not desirable for determination of TBPA capacity. The procedures for applying samples, conditions of electrophoresis, drying strips, and counting of radioactive zones are the same as described above.

Calculation of Maximal Binding Capacity. The total T_4 concentration in the enriched and labeled sera is computed as the sum of the endogenous T_4, determined separately, and the added T_4. The latter includes the T_4 added as labeled T_4, which is calculated from the stated specific activity of ^{125}I T_4, supplied by the manufacturer. (This is negligible usually, except at the lowest level of added stable T_4.) The maximum binding capacity is calculated as the product of the fraction of total radioactive T_4 associated with the protein times the total T_4 concentration present in the sample tested. The normal values in adults for TBG capacity range from 0.18 to 0.25 μg T_4/ml serum, averaging 0.22 μg/ml. Normal values for TBPA capacity range from 2.0 to 3.4 μg T_4/ml serum, averaging 2.7 μg/ml.

Direct Measurement of Binding Proteins

TBPA. The concentration of TBPA in serum of normal individuals is approximately 30 mg/100 ml, which is sufficient to permit direct measurement of this protein. The method of Surks and Oppenheimer[14] involves electrophoretic separation of proteins in serum in starch gel, according to the methods of Smithies,[15] followed by staining with aniline blue-black. The concentration of the fastest migrating component (PA-1), determined by densitometry of the stained gel, correlates directly with the maximum T_4-binding capacity of TBPA, measured by conventional paper electrophoresis.

TBG. This protein is present in human serum in a concentration of approximately 1 mg/100 ml, too low to permit measurement of the protein itself. The development of a specific radioimmunoassay by Levy, Marshall, and Valayo[16] has enabled these workers to detect less than 1 ng TBG in serum.

[14] M. I. Surks and J. H. Oppenheimer, *J. Clin. Endocrinol. Metab.* **24,** 794 (1964).
[15] O. Smithies, *Biochem. J.* **71,** 585 (1959).
[16] R. P. Levy, J. S. Marshall, and N. L. Velayo, *J. Clin. Endocrinol. Metab.* **32,** 372 (1971).

Section III

Cytoplasmic Receptors for Steroid Hormones

Methods in Enzymology Vol. 36
Hormone Action part. A.
Steroid Hormones Ed. by B W O'Malley
J G Hardman, Academic Press 1975

[13] Autoradiographic Techniques for Localizing Steroid Hormones

By WALTER E. STUMPF and MADHABANANDA SAR

I. Introduction

Autoradiography is one of the most sensitive techniques for the localization of hormones or other substances in tissue structures or compartments. This is attributable to the use of radioisotopic labels and the photographic exposure-related amplification. The technique is so sensitive that the number of tagged molecules occupying an individual cell or even a subcellular organelle can be determined, without disrupting the topographic relationships of the tissue components.

Information about the tissue localization of substances, such as hormones, is of interest regarding the interpretation and understanding of their mechanisms of action and the identification of the sites of transport, storage, action, metabolism, and excretion. Light and electron microscopic autoradiography has in the past made important contributions to the understanding of the mechanism of biological regulation and can be expected increasingly to do so in the future. In general, slices or surfaces of intact specimens containing the label are used and exposed to radiation sensitive material, and subsequently, the localized radiation effect on the photographic film and the histological structure can be viewed simultaneously. Thus, information can be gained through autoradiography which may not be obtainable otherwise, provided the technical approach used did not betray the authenticity of the information gleaned from the object.

In autoradiography two major objectives exist: (1) to obtain useful localization and (2) to interpret the meaning of the localization. In the case of steroids and similar small molecular weight substances which are not covalently bound to macromolecules, loss and redistribution easily occur during tissue preparation and photographic emulsion application.

If this is not considered adequately in the selection of the technical steps, autoradiography may produce artifacts rather than meaningful results. For the localization of steroids many autoradiographic techniques have been applied and reported in the literature. Most of them, however, provided unsatisfactory results. Some of the technique-related pitfalls and the conflicting data obtained and published in the literature have been studied by Stumpf and Roth[1] and reviewed by Stumpf.[2-4]

Special prerequisites have been established for the autoradiography of diffusible substances: (a) to exclude all solvents and fluids for the histological treatment of the tissue, or (b) to use a fixative, which when penetrating into the tissue does not translocate the labeled molecules and tissue constituents, but immobilizes the hormones at their original sites by specific linkage between tissue components, fixative, and hormone, or by trapping of the hormone to such a degree as to prevent loss and translocation during tissue treatment subsequent to fixation. Furthermore, (c) a recommended new technique should be tested beforehand for its utility with compounds known to be diffusible and localized. These theoretical and practical prerequisites need to be observed for the design and use of autoradiographic techniques and the judgment of their validity—in order to render them suitable for the study of diffusible substances. In cases where it is not known whether all, a portion, or none of the labeled material may be translocated during one or several of the technical steps used, an adequate control must be employed which follows the above prerequisites. Apparently, a proper immobilizing fixative for steroids has not been found and no generally applicable procedure can be recommended which employs liquid or vapor fixation. Encouraging progress has been made by V. Mizuhira[5] who utilizes digitonin precipitation for steroids with a 3β-OH group for the study of the synthetic pathways using tritiated pregnenolone. While it is doubtful that a general "fixative" for steroids can be found which immobilizes without translocation and leaching, more hope is justified toward finding a specific fixative for a certain group of compounds which would satisfy the immobilization imperative.[4]

In autoradiography, general prerequisites, in addition to the above-mentioned special prerequisites for diffusible compounds, must be considered in order to be able to obtain meaningful results.

[1] W. E. Stumpf and L. J. Roth, *J. Histochem. Cytochem.* **14**, 274 (1966).
[2] W. E. Stumpf, *J. Histochem. Cytochem.* **18**, 21 (1970).
[3] W. E. Stumpf, *Acta Endocrinol. (Copenhagen), Suppl.* **153**, 205 (1971).
[4] W. E. Stumpf, *Amer. Zool.* **11**, 725 (1971).
[5] V. Mizuhira, *Proc. Int. Congr. Histochem. Cytochem., 4th, Jap. Soc. Histochem. Cytochem., Kyoto, Japan* p. 35 (1972).

(1) Depending on the resolution desired, a proper choice of radioactive isotope must be made. Highest resolution on a cellular and subcellular level is obtained with ^{3}H, followed by ^{125}I. Track forming harder β-emitters, such as ^{14}C, ^{35}S, or ^{32}P, give only poor cellular and no subcellular resolution in most of the biological tissues.

(2) The radioactive label must be in such a position within the tagged molecule that loss and easy exchange of the label, without metabolism, does not occur.

(3) It is desirable that every single molecule be labeled and the specific activity be high, i.e., in the range of curies per millimoles in order to be able to use physiological doses and to reduce the autoradiographic exposure time.

(4) The labeled compound must be radiochemically pure.

(5) The dose of the compound administered must be commensurate with the biological effects studied.

(6) Identification of the chemical nature of the radioactivity in the tissue at the time of preparation of the autoradiogram is necessary for the interpretation of the findings. Tissue extraction with radiochemical assay and autoradiographic competition with unlabeled antagonists can be used to establish specificity.

II. Autoradiographic Techniques

Autoradiographic techniques may be divided according to the resolution obtainable into (A) macroautoradiography, (B) light microscopic autoradiography, and (C) electron microscopic autoradiography. The techniques suitable for the localization of steroid hormones and other diffusible noncovalently bound substances will be listed and discussed accordingly.

A. Macroautoradiography

Macroautoradiographic techniques are all so-called apposition techniques in which the surface of an object, or slices or thin sections of it, are brought in temporary close contact with photographic film for exposure, with subsequent separation for photographic processing and viewing. Slices or sections of whole rodents, such as rats or mice, or of individual organs of animals, such as rabbits, cats, or monkeys, may be studied in this manner. After the administration of the radioactively labeled compound, the biological material in general is frozen in order to preserve the *in vivo* state. While the frozen surface, slice, or section has been used by some investigators, it appears preferable to freeze-dry the sections and the surface of the slice, perhaps for 100 μm, prior to the assembly of object and film, in order to reduce the occurrence of pressure artifacts and to assure more even contact. Among the various macroautoradiographic techniques reported in the literature, three are cited here as the ones which have been most perfected and utilized.

(1) Section-Scotch tape technique[6,7]
(2) Sawbench technique[8]
(3) Free-hand slice technique[9,10]

All these techniques may be used for the study of hormone and drug localization. They provide information on the general distribution and can be used for scanning without optical aid. Since no cellular and subcellular resolution is obtained, the techniques are of limited application. Techniques 1 and 2 depend on the availability of special equipment, that is a heavy large-stage microtome mounted in a freezer or cold room or a sawbench operated under ambient conditions. Macroautoradiographic techniques will not be described in detail here, and the investigator is referred to the above-cited papers.

B. Light Microscopic Autoradiography

Autoradiographic techniques which can be used for the study of steroid hormone localization with optical cellular and subcellular resolution have become available only during the last decade.[4] Among the many autoradiographic techniques proposed in the literature only the following fulfill the outlined prerequisites and are therefore recommended.

(1) Dry-autoradiography with unembedded freeze-dried frozen sections, dry-mounted on emulsion coated slides[3,11-13]
(2) Thaw-mounting of frozen sections on emulsion coated slides[1-3]
(3) Touch- or smear-mounting on emulsion coated slides[3]

[6] S. Ullberg, *Acta Radiol., Suppl.* **118,** 1 (1954).
[7] S. Ullberg, *Proc. U.N. Int. Conf. Peaceful Uses At. Energy, 2nd, 1958* Vol. 24, p. 248 (1958).
[8] F. Kalberer, *Advan. Tracer Methodol.* **3,** 139 (1966).
[9] J. Schoolar, C. F. Barlow, and L. J. Roth, *Proc. Soc. Exp. Biol. Med.* **91,** 347 (1956).
[10] A. V. Lorenzo, C. F. Barlow, and L. J. Roth, *Amer. J. Physiol.* **212,** 1277 (1967).
[11] W. E. Stumpf and L. J. Roth, *Stain Technol.* **39,** 219 (1964).
[12] W. E. Stumpf, *in* "Radioisotopes in Medicine: *In Vitro* Studies" (R. L. Hayes, F. A. Goswitz, and B. E. P. Murphy, eds.), p. 633. U.S. At. Energy Comm., Oak Ridge, Tennessee, 1968.
[13] W. E. Stumpf, *in* "Introduction to Quantitative Cytochemistry II" (G. L. Wied and G. F. Bahr, eds.), Vol. 2, p. 507. Academic Press, New York, 1970.

In addition to these techniques, which will be detailed below, other autoradiographic techniques shall be mentioned, which have been applied to the localization of steroids with varying and limited degrees of success. For instance, Attramadal[14,15] studied the localization of [^{3}H]estradiol in the rat, using epoxy embedding after freeze-drying or conventional liquid fixation and liquid dehydration of tissue blocks. During the various steps of this wet tissue preparation, differing amounts of radioactivity were reported to be lost.[14,15] This argues for the possibility of translocation and redistribution of the labeled compound within the tissue. These factors may account for some of the variability within Attramadal's studies of estradiol localization[14-16] and also for the difference between their results and ours obtained with the dry-autoradiographic technique.[17-20]

Tuohimaa[21] criticized the technique of Appleton,[22] and the freeze-drying and paraffin-embedding technique of Hammarström *et al.*[23]: "Severe chemographic artifacts in the radioautograms hampered the results when using the method of Appleton," while the "method of Hammarström *et al.* resulted in extensive diffusion of radioactivity."[21] Therefore, Tuohimaa[24] developed a "Dry-Fixation Method" in which cryostat sections are melted on a glass slide, dried, and subsequently fixed in osmium tetroxide and paraformaldehyde vapor, prior to dipping into liquid emulsion. As judged from Tuohimaa's published autoradiograms, considerable diffusion may occur with this approach. While some of his results on [^{3}H]estradiol and [^{3}H]testosterone localization agree with ours, others are in disagreement.

1. Dry-Autoradiography

The dry-mount autoradiographic technique consists of the following steps (Fig. 1): (a) Simultaneous tissue freezing and mounting on tissue

[14] A. Attramadal, *Proc. Int. Congr. Endocrinol., 2nd, 1964, Excerpta Medica* Int. Congr. Ser. No. 83, p. 612 (1965).

[15] A. Attramadal, Universitetsforlaget, Oslo, 1970.

[16] A. Attramadal, *Z. Zellforsch. Mikrosk. Anat.* **104,** 572 (1970).

[17] W. E. Stumpf, *Science* **162,** 1001 (1968).

[18] W. E. Stumpf, *Science* **163,** 958 (1969).

[19] W. E. Stumpf, *Amer. J. Anat.* **129,** 207 (1970).

[20] W. E. Stumpf, *Z. Zellforsch. Mikrosk. Anat.* **92,** 23 (1968).

[21] P. J. Tuohimaa, Ph.D. Thesis, University of Turku, Finland (1970).

[22] T. C. Appleton, *J. Roy. Microsc. Soc.* [3] **83,** 277 (1964).

[23] L. Hammarström, L. E. Appelgren, and S. Ullberg, *Exp. Lab. Res.* **37,** 608 (1965).

[24] P. Tuohimaa, *Histochemie* **23,** 349 (1970).

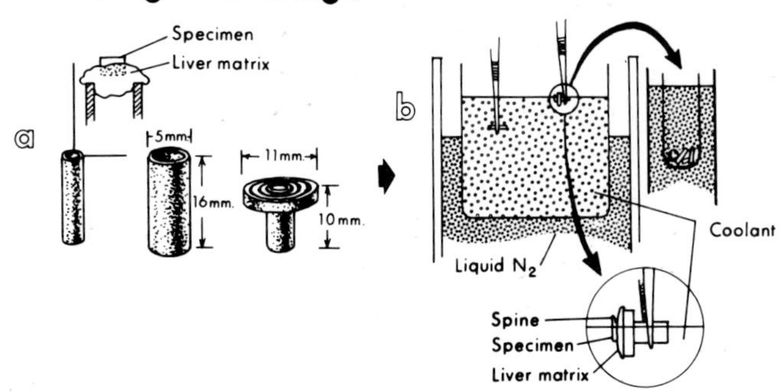

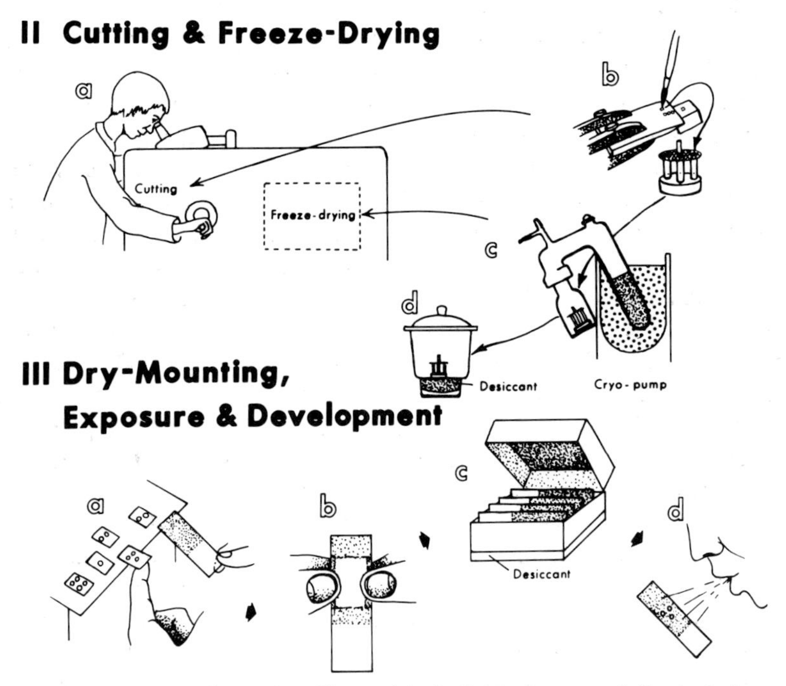

FIG. 1. Dry-autoradiography, illustrating individual steps of the technique: Ia, mounting of the specimen with different types of tissue holders; Ib, freezing of a small and large (circle) specimen; IIa and b, cutting of frozen sections with IIc, freeze-drying in Cryo-pump; IId, storage of freeze-dried sections, and IIIa and b, dry-mounting of the freeze-dried sections on emulsion-coated slides, IIIc, exposure; IIId, brief moistening of section area at the end of exposure.

holder, (b) cryostat sectioning of thin sections, (c) freeze-drying of sections, (d) dry-mounting of freeze-dried sections on emulsion precoated slides, and (e) photographic processing and staining of the slides.

a. Simultaneous Tissue Freezing and Mounting on Tissue Holder. Radioactively labeled steroids are obtainable from commercial sources in ethanol–benzene solution. Prior to the experiment, the material is evaporated to dryness in a stream of dry N_2 gas and dissolved in 5 or 10% ethanol saline. The labeled steroid is injected intravenously or subcutaneously and animals may be killed by decapitation at different time intervals afterwards according to the design of individual experiments. Tissue samples of 1–10 mm^3 are excised and mounted on a tissue holder (Fig. 1, Ia), the size of which is selected according to the size of the tissue piece. For small pieces, a hollow rod may be used and for samples of brain tissue a solid brass mount of larger size may be used (Fig. 1, Ia). A small layer of minced liver is placed on the top of the tissue holder and serves as an adhesive for the base of the excised tissue. The tissue holder with the tissue attached is plunged into liquefied propane at about —180°, thus providing simultaneous quenching and mounting and eliminating the possibilities of diffusion that may accompany separate quenching and mounting. Natural grade propane is in a gaseous state and is liquefied prior to the experiment by condensing it in a 250-cm^3 round-bottom flask immersed in a Dewar containing liquid nitrogen. Liquefied propane is then transferred to a beaker which is immersed in liquid nitrogen. During freezing of the tissues, propane is kept at about —180° in a liquid state. While small tissue blocks may be frozen by sudden immersion, larger blocks of tissue, above about 2 mm^3, require a two-step freezing in order to prevent cracking of the tissue. In this case, freezing needs to be controlled by observation. Initially, the tissue mount is immersed only partially (rapid step), leaving one-third or less of the tissue above the level of liquid propane (slow step). The tissue above propane level freezes slower and the advancement of the interphase between frozen and unfrozen tissue can be followed. At the nonsubmerged small part of the tissue and liver support an outgrowth of tissue, "terminal spine," may become visible (Fig. 1, Ib). Since this may distort a marginal portion of the tissue, the important part of the tissue should be immersed first in order to obtain optimal preservation. The temperature of the liquefied propane should be controlled during the experiment in order to assure rapid freezing and to minimize ice crystal disruption. The tissue mount is kept in the liquid propane for a few seconds and is then rapidly transferred to a plastic test tube (Fig. 1, Ib) which afterwards is stored in a liquid nitrogen tissue storage container. The tissue can be kept in liquid nitrogen for any length of time without further damage.

b. Cryostat Frozen Sectioning. The test tube containing the tissue mounts is removed from the liquid nitrogen storage container and transferred in liquid nitrogen to the microtome cryostat. Using a precooled long forceps, the tissue mount is removed from the test tube within the cryostat and inserted into the tissue holder of the microtome. During this process, as well as during cutting and frozen tissue handling, care must be taken not to expose the tissue to body heat or ambient air. Tissue temperatures should always be kept below —30°. The Wide-Range Cryostat (Harris Manufacturing Co., North Billerica, Mass.) has proved most suitable for thin frozen sectioning and safe handling of specimen. It has the additional advantage that freeze-drying or other histochemical procedures, which require low temperatures, can be performed within the large cryostat chamber (Fig. 1, IIa), without need for transfer of tissue. The Wide-Range Cryostat is equipped with a microtome having a cutting range from 1 to 40 μm in 1-μm increments. The cryostat temperature is variable and can be maintained between —30 to —50°. This low temperature has the advantage of facilitating thin sectioning[25] as well as safer tissue handling compared to microtome–cryostats which furnish —30° as the lowest possible cabinet temperature. The lower temperature reduces the danger of ice crystal disruption due to body heat or convection of warm air. The Wide-Range Cryostat is also equipped with a dissecting microscope and cold light fiber optic illumination (Fig. 2). Sections are cut under observation with a dissecting microscope, at a thickness between 1 to 4 μm, according to desired resolution and stainability. Observation with a dissecting microscope facilitates successful cutting of thin sections; it helps to orient the tissue block, to guide the sections with a brush during cutting and transfer, and to judge section quality. The sections are carried with the bristles of a fine brush from the knife and transferred, up to 20 or 30 sections, to a Polyvial (Fig. 1, IIb), situated in a carrier in the vicinity of the knife. The carrier may contain several vials. It is useful to clean the vials prior to cutting with a jet of clean air in order to prevent contamination of sections with dust particles. The vial is covered with a fine punched out wire mesh (#40, Small Parts, Inc., 6901 NE Third Avenue, Miami, Florida) which helps dislodging the sections from the bristles of the brush and preventing loss of sections during freeze-drying and breaking of the vacuum. For freeze-drying, the carrier with the Polyvials containing frozen sections is transferred by a long precooled forceps to the precooled sample chamber of the Cryo-pump (Thermovac Industries Corp., Copiague, L.I., N.Y.) within the cryostat (Fig. 1, IIc).

[25] W. E. Stumpf and L. J. Roth, *Nature (London)* **205,** 712 (1965).

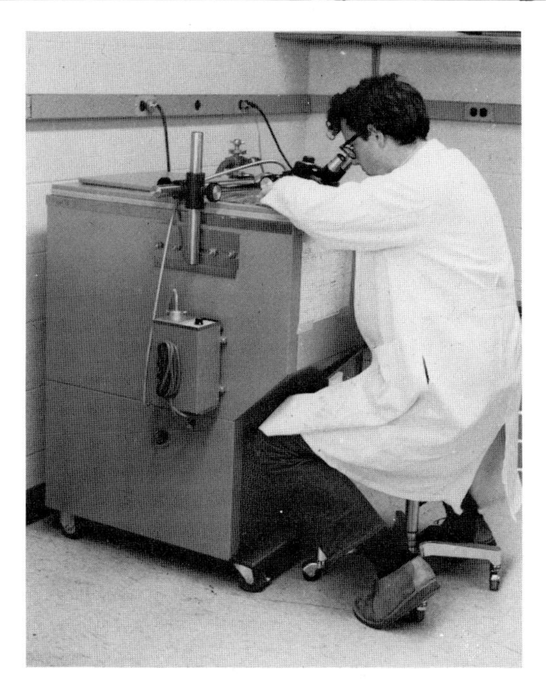

FIG. 2. The Wide-Range Cryostat (Harris Manuf. Co.) for low temperature tissue preparation, equipped with fiber-optics cold light and a dissecting microscope which can easily be moved into place for cutting of thin frozen sections.

c. Freeze-Drying. Although any good vacuum system which provides a vacuum of about 10^{-4} to $^{-5}$ torr can be used, the compact, clean, and nonmechanical Cryo-pump is preferred. The sample chamber and cryosorption chamber of the pump are assembled within the cryostat with the help of a vacuum produced by a mechanical forepump. The latter facilitates self-seating of the O-ring joint and activation of the molecular sieve of the Cryo-pump. After preevacuation for 10 to 15 minutes, the forepump is disconnected and the cryosorption chamber of the assembled Cryo-pump is inserted into a Dewar filled with liquid nitrogen. The liquid nitrogen Dewar with the Cryo-pump is kept inside the cryostat (Fig. 1, IIc), thus, the cryostat provides the cooling temperature for the specimen chamber of the Cryo-pump. If a suitable cryostat, as described above, is not available, freeze-drying of sections can be done outside of the cryostat. In this case, two Dewar flasks are prepared, one filled with liquid nitrogen and the other filled with dry ice–alcohol slush, as a coolant for the molecular sieve and specimen chamber, respectively. After assembly and preevacuation within the cryostat, the Cryo-pump is disconnected from the forepump and transferred to the two outside Dewars, immersing simultaneously into the coolants the two chambers of the Cryo-pump. A vacuum between 10^{-5} to 10^{-6} torr is achieved. Freeze-drying is complete after 12 to 24 hours, depending on the specimen temperature.[26] One hour before breaking the vacuum with dry nitrogen gas, the sample chamber is removed from the cryostat or coolant and allowed to equilibriate to room temperature, while the cryosorption chamber remains in liquid nitrogen. The vials containing the freeze-dried sections are removed and stored in a desiccator at room temperature. The Cryo-pump is removed from the liquid nitrogen and the molecular sieve will desorb the trapped water and reactivate for reuse at room temperature or at 80° in an oven for 2 hours.

d. Dry-Mounting of Freeze-Dried Sections. The liquid emulsion coating of slides is carried out in an absolutely lightproof darkroom in the presence of a safelight with Kodak filter Wratten Series No. 2 or, preferably, in full darkness. Kodak NTB 3 or NTB 2 bulk emulsion may be used. The emulsion, shipped in 4-oz. plastic bottles, is placed in a lightproof metal container, which is in part filled with water, and heated in a water bath to 40–45° for liquification. After liquification within 1 to 2 hours, a portion of the emulsion is gently poured into a beaker which then is covered and again kept in a second metal container in the water bath for another 30 minutes to 1 hour in order to permit air bubbles to disappear. Presence of air bubbles in the emulsion would cause uneven coating of the slides. The emulsion may be used undiluted or diluted with distilled water. Microscope slides with one end frosted for labeling may be cleaned with absolute ethanol and then placed into a plastic slide grip (Lipshaw Manufacturing Co., Detroit, Mich.) which holds five slides. Several slide grips are placed on a rack and are preheated at 40° in an incubator before dipping.[12] Individual slide grips are dipped into liquid emulsion, which is kept in the water bath, by gentle down and up movement. Excess emulsion is allowed to drip off and the slide grip is transferred and hung on a rack for air drying at room temperature. The drying of the slides takes at least 2 to 3 hours. No desiccants or hot air need to be used since rapid drying may cause reticulation of the emulsion. After drying the slides are removed from the grips, without touching the emulsion, and placed into a modified Clay-Adams black desiccator box, where slides should be kept under Drierite for at least 24 hours before use. They may be stored up to 2 months under refrigeration. Before use of the coated slides a sample is developed and the number of reduced silver grains is counted with a 100 × oil objective under the microscope,

[26] W. E. Stumpf and L. J. Roth, *J. Histochem. Cytochem.* **15**, 243 (1967).

using an eye-piece reticule. The silver grain background should not exceed 4 silver grains per 1000 μm^2. Dry-mounting is accomplished with Teflon supports which are used to press the sections onto desiccated emulsion-coated slides. Teflon may be obtained in 12-in. sq. sheets of a preferred thickness of $\frac{1}{16}$ in. from Crane Parking Co., Morton Grove, Ill. Teflon pieces of about 2×1 cm size are cut, cleaned in chloroform, and kept under dust-free conditions. Mounting of the freeze-dried sections should be done at a relative humidity between 20 to 45%. In geographical areas with high summer humidity, a humidity controlled room or box at ambient temperature may be required. If the humidity is too high, both emulsion and section may pick up moisture which may cause diffusion of the label and autolysis of the tissue. If the humidity is too low, section handling is impaired and electrostatic discharge may cause artifacts. Sections are transferred from the Polyvial to the Teflon with a fine forceps and, if necessary, positioned under a dissecting microscope. Folds in sections should be avoided, or such sections or the folded parts of them should be removed. Several pieces of Teflon may be prepared in this manner and then pushed over the rim so that the forefinger can be placed under it. Under safe light, an emulsion-coated slide is taken from the desiccator box and placed over the Teflon with sections on it (Fig. 1, IIIa). Teflon and slide are pressed together between forefinger and thumb, using perpendicular pressure without sliding. After release of pressure, the Teflon falls off and the sections adhere to the emulsion. The mounted slides are stored in sealed modified Clay-Adams desiccator boxes for exposure at $-15°$ (Fig. 1, IIIc). The length of exposure is influenced by many factors, such as, specific activity of the radioactively labeled compound, the dose, the mode of administration, the time interval between application and tissue preparation, the thickness of the section, the specific concentration in tissues, and the sensitivity of the emulsion. For this reason, the length of exposure time cannot be assessed theoretically. Therefore, autoradiograms need to be developed at different time intervals. Also, different exposure times may provide specific information.

e. Photographic Processing and Staining. For development, the slide box is removed from the freezer and allowed to warm to room temperature in order to avoid precipitation of moisture on sections when opened. A slide is removed and the area of the section is breathed on once or twice in order to improve adherence of the section to the emulsion. If moisture is not applied, loss of sections may occur during photographic processing and staining. The slide is developed in Kodak D19 developer for 1 minute at 20°, briefly dipped in running tap water, fixed in Kodak fixer for about 5 minutes, rinsed in tap water for 5 to 10 minutes, and subsequently stained in methylgreen-pyronin for 30 seconds to 1 minute.

The excess dye is removed from the emulsion and sections by washing in tap water. During the washing, emulsion on the back of the slide can easily be wiped off. The temperature of all the fluids should be similar in order to avoid reticulation of the emulsion. The slide is air-dried and mounted with Permount and a coverslip. Methylgreen-pyronin has been found to be a useful and simple single-step stain for autoradiograms. It permits easy recognition of cytoplasm and cell nuclei without obscuring or fading of silver grains and only minimal background coloration of the gelatin of the emulsion. The recipe for the preparation of the dye has been described.[13]

2. Thaw-Mounting of Frozen Sections on Emulsion-Coated Slides

Freezing of tissues and frozen sectioning is performed as described in the dry-autoradiography procedure above. For frozen sectioning, the cryostat is moved into a darkroom. Sections are cut under fiber optics illumination. Several sections may be kept on the knife. The red light (Kodak Wratten No. 2) is then turned on. An emulsion-coated slide is removed from the black desiccator box and placed over the sections on the knife. Avoiding sliding movement, the slide is brought into brief contact with the upper surface of the knife so that the sections adhere to the emulsion by melting from the heat of the slide. This step requires practice and the development of some skill in order to be able to obtain flat and minimally distorted sections. If the emulsion-coated slide is precooled, the thaw-mounting may be facilitated by slight and brief pressure of the slide. Other modes of section thaw-mounting may be employed by placing a section onto a precooled slide within the cryostat with subsequent warming of the bottom of the slide with the finger. While several possibilities of thaw-mounting seem feasible, mounting without melting appears not to be feasible and may also lead to loss of sections during photographic processing. After mounting of the section the slide is transferred to the desiccator black box and kept at room temperature. The slide-mounted section may also be kept in the cryostat so that refreezing of the section occurs. Whether or not this variation has advantage or disadvantage is questionable and needs to be investigated. The section-mounted slides are exposed at $-15°$. Photographic processing and staining are done as described above in the dry-mount procedure.

3. Smear- and Touch-Mounting on Emulsion Precoated Slides

Autoradiographic recording of hormone localization may be utilized in combination with biochemical experiments, using tissue homogenates or cellular and subcellular fractions, as well as in combination with tissue culture experiments, using monolayers of cells or cell suspensions.

a. Smear-Mounting. In the darkroom under safe light (Kodak Wratten No. 2), photographic emulsion-coated slides are removed from the desiccator box and a drop of the sample is placed on the emulsion without touching it. The drop is then spread with a piece of Teflon by placing the Teflon flat over the droplet so that the material spreads out over the emulsion without scratching. Instead of placing the droplet on the emulsion-coated slide, it may be applied to the Teflon and the slide then laid over the Teflon. The Teflon is removed, the slide air-dried for a few minutes and then transfered to the black desiccator box for exposure. If Teflon is not available, a glass rod may be used for spreading. Exposure, photographic processing, and staining may be done as in the dry-mount technique.

b. Touch-Mounting. This procedure is almost identical with the smear-mounting, except that the specimen is picked up by brief appositioning of the emulsion-coated slide to the surface of the specimen, e.g., bone marrow, tissue culture suspension, or layer, in order to pick up material that can be released from the surface.

4. Limitations and Possible Artifacts

Although the described techniques fulfill prerequisites for the authentic tissue localization of steroid hormones and other noncovalently bound hormones and drugs, certain limitations and possible artifacts need to be considered for each individual technique. Less than optimal results may be attributable to such factors as inadequate freezing of tissue blocks and handling of frozen and freeze-dried sections, section-mounting with folds and shifting movement, and high humidity during mounting. Freezing, if not done properly, that is, slow freezing or large block size, may lead to ice-crystal disruption of the tissue. Ice-crystal growth may also occur as recrystallization during handling of frozen specimen, sectioning, and freeze-drying, if the temperature of the specimen is permitted to rise above a critical point which may be, for most of the tissues, at about —25° for the optical level and at about —100° for the ultrastructural level. The freeze-drying system needs to be effective, that is, a vacuum better than 10^{-4} torr must be maintained, and the freeze-drying must not be terminated before the ice is removed from the specimen. Premature termination may lead to melting of residual ice, causing tissue shrinkage and diffusion of compounds. During dry-mounting, the use of sections with folds in them, or shifting of the Teflon and slide, may cause mechanical artifacts. While the occurrance of such artifacts can be avoided or minimized, their presence is, in general, easily recognized by their localization in the same section and variability between different sections. Also, the appearance and extent of pressure artifacts can be recognized in control sections without radioactivity. Control sections may be placed on the Teflon support along with the radioactive sample and mounted simultaneously.[12] Since the tissue is not denatured by fixatives, many of the enzymes are viable. If freeze-dried sections are exposed to high humidity, not only diffusion of the radioactive label but also autolysis of tissue constituents may take place. Since the dry-autoradiography procedure is based on the exclusion of fluid and moisture, using dried sections and desiccated emulsion, chemical interaction between tissue components and photographic emulsion is excluded or minimized. Therefore, so-called positive or negative chemography remains a domain of the "wet" techniques.

In the thaw-mount and smear- and touch-mount autoradiographic procedures, wet interaction between tissue and emulsion, albeit short, imposes the need to control for positive and negative chemographic artifacts. Melting may not only cause chemical interaction between section and emulsion, it may also cause varying degrees of diffusion. The extent of the latter may be dependent on the type of compound and the section thickness. The thaw-mount procedure, if controlled, is time saving since the vacuum freeze-drying is excluded and mounting of sections simplified. The dry-mount autoradiographic technique is a necessary control for thaw-mount autoradiography. Currently both techniques are used in our laboratory to advantage, with the thaw-mounting being used preferably for screening and surveying. It appears that in the dry-autoradiography the cellular and subcellular resolution as well as the tissue preservation and tinctorial differentiation are superior compared to the thaw-mount technique. In the case of smear- and touch-mount autoradiography, the use of whole cells may obscure resolution and cause superimposition pitfalls in the interpretation of the autoradiograms.[2,12]

C. Electron Microscopic Autoradiography

No electron microscopic technique exists which can be recommended for the autoradiographic study of steroid hormone localization. This is so, despite several efforts made in different laboratories during the last decade. There is a need to have such a technique available, so that the physiomorphology for steroid hormone synthesis can be better studied and understood and the subcellular sites of steroid hormone action elucidated. The ultrastructural correlate is needed to confirm and supplement the now available biochemical data on intracellular steroid transport and binding.

Ideally, for ultrastructural steroid localization an autoradiographic technique should follow the principles outlined[2,4] with minimal tissue treatment and avoidance of fluids and embedding media. Histocryotechniques for the light microscopic level were developed and applied in our laboratories[25,26] and provided encouragement for electron microscopists.[27] The feasibility of routinely cutting thin frozen sections at temperatures below —70°[24] and to freeze-dry sections at very low temperatures[26] has been demonstrated. Although further progress in the development of tools and techniques has been made, thus far none of the laboratories, including ours, has been able to provide ultra-thin sections of satisfactory quality. Ice-crystal disruption, lack of contrast, vulnerability of the unembedded, and unfixed freeze-dried ultra-thin sections are all major obstacles which are added to the difficulties of handling the cut frozen sections. It seems that the progress made in recent years does not match the expectations. Modifications, perhaps, with the use of cryoprotective agents will have to be attempted. Also, further development of equipment will be necessary.

In the meantime, adaptation of conventional fixation and embedding techniques needs to be pursued more effectively. This can be advocated, since steroids are bound in target tissues, and proper tissue treatment may not detach them from their binding sites, that is, if suitable conditions of pH, temperature, ionic concentration, and probably others can be maintained. Also, approaches similar to the one recommended by V. Mizuhira[5] in which digitonin precipitation of 3β-OH-steroids seems to facilitate the localization of at least this one type of steroids should be considered and further established.

In recent years, morphological studies of transcription have been facilitated by a technique developed by Miller and Bakken[50] utilizing spreading of nuclear contents. This technique is now modified and applied in our laboratory as well as other laboratories and promises to provide most valuable insights at the molecular level into effects of steroids and other compounds on RNA, DNA, and protein synthesis. Steroid hormone localization may be accomplished simultaneously in the same preparation along with the identification of specific active genes and ultrastructural mapping of chromosomes.

III. Applications

The authenticity of the results obtained by the techniques described here have been validated by control experiments with diffusible substances known to be localized.[1] The utility, especially of the dry-mount

[27] A. K. Christensen, *in* "Autoradiography of Diffusible Substances" (L. J. Roth and W. E. Stumpf, eds.), p. 349. Academic Press, New York, 1969.

autoradiography, has been demonstrated in the localization of [^{3}H]estradiol,[19,20,28–30] [^{3}H]testosterone,[31–33] [^{3}H]corticosterone,[4,34,35] [^{3}H]progesterone,[36,37] [^{3}H]ecdyson,[38] [^{3}H]digoxin,[39] and other hormones and diffusible compounds such as [^{125}I]thyroxine and [^{125}I]triiodothyronin,[40] [^{3}H]acetazolamide,[39] [^{3}H]TSH-releasing hormone,[41] [^{3}H]proline,[41] [^{3}H]inulin,[42] [^{3}H]mannitol,[1] [^{3}H]sorbitol,[12] [^{3}H]urobilinogen,[43] as well as [^{125}I]LH and FSH (unpublished). Thus, the utility of the techniques developed in our laboratories and their superiority has been demonstrated. A comparison of results has been reviewed earlier.[2,4,12] The results obtained with our techniques have been, almost as a rule, at variance with those stemming from other approaches and published in the literature.

Figures 3 to 8 are examples of the dry-autoradiographic technique. Figures 3 and 4 compare the localization of [^{3}H]estradiol in rat uterus of a castrated untreated (Fig. 3) and castrated progesterone-treated (Fig. 4) mature animal demonstrating shift of binding sites. Figure 5 shows the distribution of progesterone in the guinea pig oviduct. The selective concentration of androgen in nuclei of pituitary gonadotrophs is demonstrated in Fig. 6. Corticosterone[4,34,35] and cortisol[4,34] have first been localized in our laboratory, as in the case with androgen in the pituitary[44] and brain,[45] progesterone in the oviduct, and the topographic distribution of estrogen in the brain,[19] spinal cord,[46] and pituitary.[20] Figure 7 shows

[28] W. E. Stumpf, *Endocrinology* **83**, 777 (1968).
[29] W. E. Stumpf, *Endocrinology* **85**, 31 (1969).
[30] C. Tachi, S. Tachi, and H. R. Lindner, *J. Reprod. Fert.* **31**, 59 (1972).
[31] M. Sar, S. Liao, and W. E. Stumpf, *Endocrinology* **86**, 1008 (1970).
[32] M. Sar and W. E. Stumpf, *Fed. Proc., Fed. Amer. Soc. Exp. Biol.* **30**, 363 (1971).
[33] W. E. Stumpf and M. Sar, *Proc. Int. Congr. Horm. Steroids, 3rd, 1970,* p. 503 (1971).
[34] W. E. Stumpf, *Fed. Proc., Fed. Amer. Soc. Exp. Biol.* **30**, 309 (1971).
[35] W. E. Stumpf and M. Sar, *Progr. Brain Res.* **39**, 53 (1973).
[36] W. E. Stumpf and M. Sar, *J. Steroid Biochem.* **4**, 1 (1973).
[37] M. Sar and W. E. Stumpf, *Endocrinology* **94**, 1116 (1974).
[38] J. A. Thomson, D. C. Rogers, M. M. Gunson, and D. H. Horn, *Cytobios* **2**, 79 (1971).
[39] W. E. Stumpf and L. J. Roth, *in* "Autoradiography of Diffusible Substances" (L. J. Roth and W. E. Stumpf, eds.), p. 69. Academic Press, New York, 1969.
[40] W. E. Stumpf and M. Sar, *55th Meet. Endocrine Soc.* **92** (suppl), A. 273 (1973).
[41] W. E. Stumpf and M. Sar, *Fed. Proc., Fed. Amer. Soc. Exp. Biol.* **32**, 221 (1973).
[42] D. A. Brown, W. E. Stumpf, and L. J. Roth, *J. Cell Sci.* **4**, 265 (1969).
[43] W. E. Stumpf and R. Lester, *Lab. Invest.* **15**, 1156 (1966).
[44] M. Sar and W. E. Stumpf, *Science* **179**, 389 (1973).
[45] M. Sar and W. E. Stumpf, *Endocrinology* **92**, 251 (1973).
[46] D. A. Keefer, W. E. Stumpf, and M. Sar, *Proc. Soc. Exp. Biol. Med.* **143**, 414 (1973).

the distribution of [^{3}H]corticosterone in an area of the hippocampus, and Fig. 8 demonstrates the localization of radioactivity in anterior pituitary cells after the injection of tritium-labeled thyrotropin releasing hormone.

In a number of our autoradiographic studies, the results not only confirm but extend the observations stemming from biochemical experiments and these may not have been obtainable otherwise. It is the advantage and potential of histochemical approaches to inform about individual cells and cell types within the complex texture of organs. The thaw-mount autoradiographic technique—combined with dry-mount controls—has been applied to the study of the localization of steroids in neonatal brain, providing for the first time direct evidence about the topographic distribution of estrogen neurons in neonatal rats.[47]

The smear-mount autoradiographic technique is useful in immediate connection with biochemical techniques. For instance, the distribution of [^{3}H]estradiol in homogenate of rat uteri[12] and of [^{3}H]androgen in prostatic nuclei[31] after incubation of minced ventral prostate with [^{3}H]testosterone could be demonstrated.

From the autoradiograms the number of "receptors" or of molecules occupying certain cellular or subcellular tissue compartments at a given time after the administration of the radioactively labeled material may be calculated. This can be done only under strict observation of certain prerequisites. The injected material must be radiochemically pure and the silver grains must be known to represent the compound in question. In addition, the *silver grain yield* needs to be assessed, that is, the *specific sensitivity,* expressed as the number of disintegrations per silver grain, linked to the specific conditions of the autoradiographic procedure. In assessing the specific sensitivity several factors must be considered, such as type of radioisotope, specific activity, section thickness, density of cellular and subcellular components,[48] embedding material, conditions and lengths of exposure, and photographic processing.

Quantitative autoradiography for the determination of the number of binding sites in target cells has not been utilized for steroid hormones. It can be expected that the autoradiographic techniques outlined in this chapter can be applied.

IV. Conclusions

Several techniques have been recommended for the autoradiographic localization of hormones and used with varying degrees of success. In

[47] P. J. Sheridan, M. Sar, and W. E. Stumpf, *55th Meet. Endocrine Soc.* **92** (suppl), A. 67, (1973).

[48] W. Maurer, and E. Primbsch, *Exp. Cell Res.* **33,** 8 (1964).

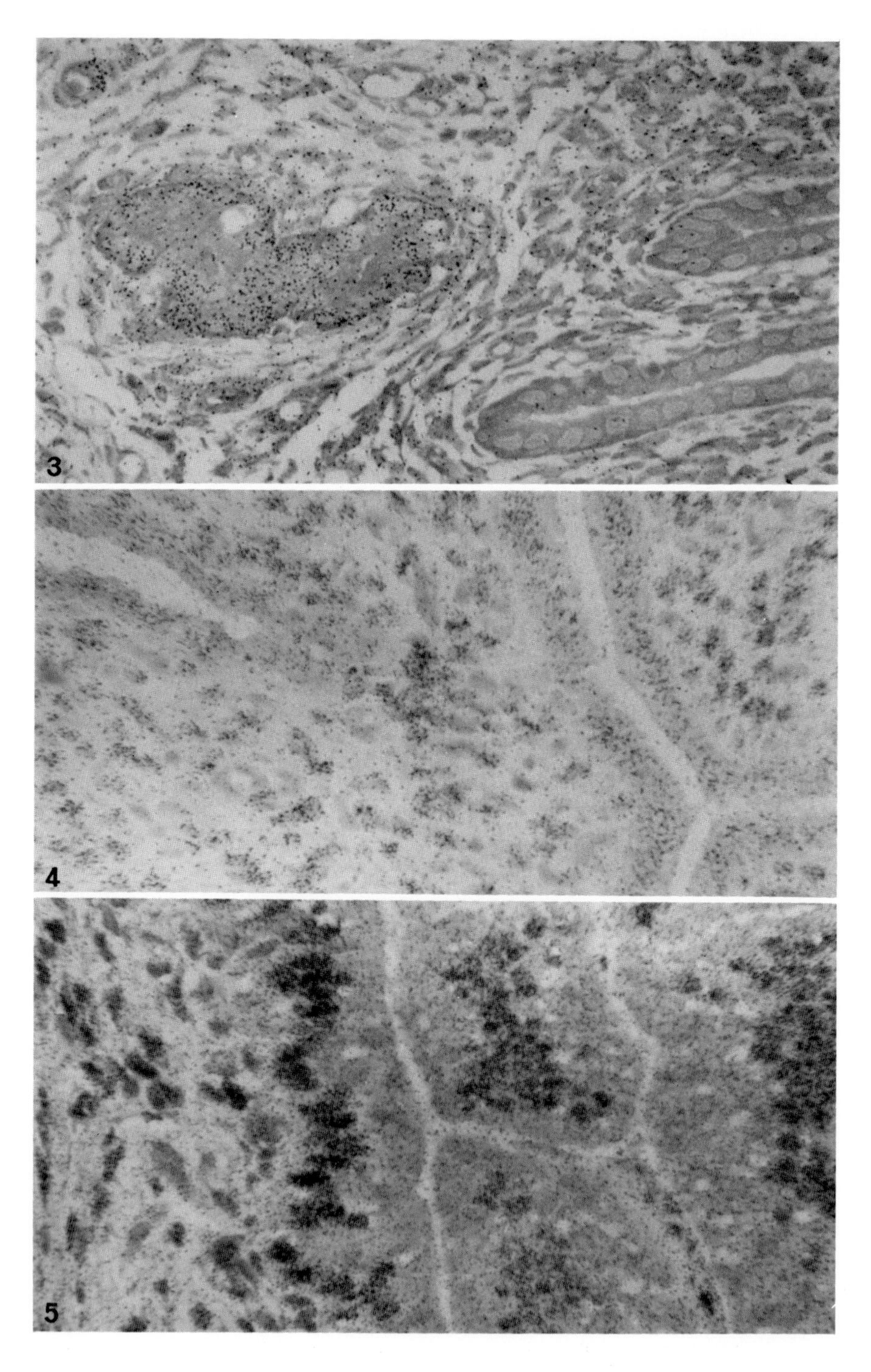

FIGS. 3–5.

general, the choice of a technique will depend on the resolution desired and the compound used, that is, the degree of tissue binding and the potential diffusibility. All the macroautoradiographic apposition procedures satisfy the theoretical demands for the localization of diffusible compounds and, if properly used, will give useful information. Because of their limited histological resolution, not only ^{3}H but also harder β-emitters such as ^{14}C, ^{35}S, and ^{32}P can be used to advantage. Autoradiography has mostly been used for the study of *target tissues* and *sites of deposition* or metabolism. However, the potential to study and identify *sites of hormone production* has barely been recognized and efforts in this direction are sparse. From the successful work of Applegren[49] who used radioactively labeled pregnenolone and progesterone with the section-Scotch tape technique, the feasibility to utilize autoradiography for the study of *sites of hormone synthesis* is demonstrated. For the light microscopic localization of steroid hormones and other diffusible compounds, dry-autoradiography appears to be superior from the viewpoint of preservation of the molecules at their original sites in the tissue as well as the demonstrated resolution obtained. This judgement is derived from our own comparative studies with five different techniques as well as from the evidence taken from autoradiograms published in the literature. In the dry-autoradiographic technique, the observation of optimal conditions for the sectioning of 1–4 μm thin sections, the maintenance of low temperatures, proper freeze-drying, and the exclusion of high humidity after the freeze-drying are all essential. The thaw-mounting of frozen sections, unfixed, which also was first tested and used in our laboratory, has the advantage that freeze-drying prior to section mounting is circumvented, but has the dis-

[49] L. E. Applegren, *Acta Physiol. Scand., Suppl.* **301,** 1 (1967).

FIGS. 3–5. Dry-mount autoradiography of [³H]estradiol (Figs. 3 and 4) and [³H]progesterone (Fig. 5), stained with methylgreen-pyronine.

Figs. 3 and 4. Show rat uterus with a different distribution of [³H]estradiol, spec. act. 95 C/m*M* 1 hour after injection of 0.2 μg (Fig. 3) or 1.0 μg (Fig. 4) per 100 g body weight. Exposure time 110 days (Fig. 3) or 42 days (Fig. 4), 2 μm, ×500. After pretreatment with progesterone, the luminal epithelium is free of radioactivity while the glandular epithelium shows increased concentration. Without pretreatment, the luminal epithelium of an ovariectomized rat concentrates [³H]estradiol (Fig. 4).

FIG. 5. Guinea pig oviduct, 15 minutes after the injection of 1.0 μg/100 g body weight of [³H]progesterone, spec. act. 110 C/m*M*, showing nuclear concentration of progestin in the luminal epithelium, muscle, and connective tissue cells. Note the unlabeled cell nuclei in the luminal epithelium, which are visible close to the lumen as compared with the more basally located labeled nuclei. Exposure time 170 days, 4 μm, ×500.

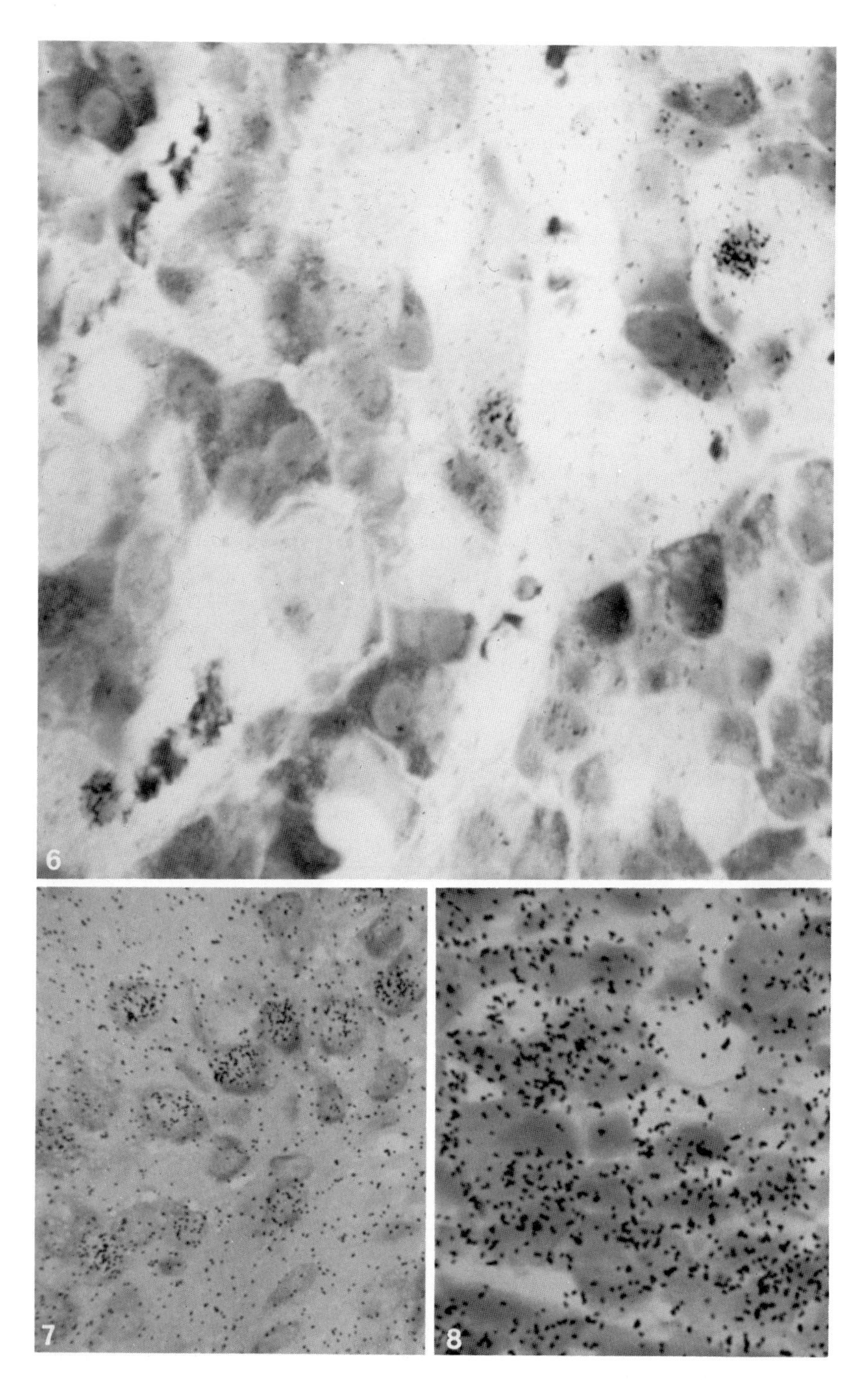

FIGS. 6–8.

advantage that thawing, albeit "brief," is introduced. The latter entails varying degrees of limited diffusion and possible interaction between tissue and emulsion during the state of wetness. Cryostat sections of 1–4 μm, thinner than commonly used, have been found to keep the extent of diffusion minimal. Due to the melting, ice-crystal artifacts introduced during freezing and cutting are less conspicuous, but controls are required against negative or positive chemography through possible interaction between tissue enzymes or other chemical groups and the gelatin or silver salts of the photographic emulsion. The dry-autoradiographic procedure excludes or minimizes the possibility for such interactions and therefore is recommended as a control for the validation of the thaw-mount and other autoradiographic techniques. In our laboratory, a combination of the dry-mount and thaw-mount techniques proved most effective for the mapping of steroids in vertebrate brains.

From the survey of autoradiographic techniques, their applications at the optical level, and the pictorial evidence provided, it appears that dry-autoradiography is truely a *general* histoautoradiographic technique—a claim that has often been made unjustifiably for other techniques—useful for the localization of bound and unbound substances, drugs and hormones of different kinds, including steroids, polypeptides, and proteins. It can be used as a *control* for applications of autoradiography where less established procedures are used and the possibility of translocation artifacts needs to be considered. Even in the study of polypeptide and protein hormones, which may be retained to a large degree at the original sites during fixation and subsequent wet tissue preparation, the possibility of translocation artifacts and loss of material must be kept in mind and adequate controls applied. Information from the use of autoradiographic techniques relates to structure and topography, which is often deficient in biochemical approaches. A combined use of both tech-

---

FIGS. 6–8. Dry-mount autoradiograms showing the cellular and subcellular distribution of radioactivity after the injection of [^{3}H]testosterone (Fig. 6), [^{3}H]corticosterone (Fig. 7), and [^{3}H]TSH-releasing hormone (Fig. 8).

FIG. 6. Rat pituitary, pars distalis, 4 weeks after orchidectomy, showing nuclear concentration of androgen in gonadotrophs (castration cells) 1 hour after intravenous injection of [^{3}H]testosterone, spec. act. 91 C/mM. Exposure time 154 days. Stained with modification of aldehyde fuchsin and Masson's trichrome, 4 μm, $\times$1280.

FIG. 7. Rat hippocampus, showing nuclear concentration of [^{3}H]corticosterone in neurons, 1 hour after intravenous injection, spec. act. 44 C/mM. Exposure time 95 days. Stained with methylgreen-pyronine, 3 μm, $\times$700.

FIG. 8. Rat pituitary, pars distalis, showing nuclear concentration of radioactivity 1 hour after the injection of 0.9 μg/100 g body weight of [^{3}H]TSH-releasing hormone, spec. act. 40 C/mM. Exposure time 64 days. Stained with methylgreen-pyronine, 4 μm, $\times$1260.

niques is advocated and promises to provide optimal morphophysiological evidence.

In the future, important new information on hormone and drug actions at the molecular level can be expected by the spreading of nuclear contents as introduced by Miller and Bakken.[50] Strong encouragement is given here to utilize and develop this approach which is likely to provide insights superior and supplementary to those currently obtained from the use of biochemical techniques.

Acknowledgments

The work was supported by PHS Grant No. NS09914, PHS Grant No. HD05700, AEC Grant AT(40-1)-4057, and a grant from the Rockefeller Foundation to the Laboratories for Reproductive Biology, University of North Carolina, Chapel Hill, North Carolina.

[50] O. L. Miller, Jr., and A. H. Bakken, *Acta Endocrinol. Suppl.* (*Kbh.*) **168,** 155 (1972).

[14] Receptor Identification by Density Gradient Centrifugation

By David O. Toft and Merry R. Sherman

Introduction

Density gradient centrifugation provides one of the most useful means for detecting and separating subcellular particles and macromolecules. A number of reviews on the various applications of this technique have been published.[1-6] This chapter, like those of Stancel and Gorski[7] and McGuire,[8] focuses on kinetic density gradient centrifugation, a technique

[1] D. H. Moore, *in* "Physical Techniques in Biological Research" (D. H. Moore, ed.) 2nd ed., Vol. 2, Part B, p. 285. Academic Press, New York, 1969.

[2] M. K. Brakke, *Arch. Biochem. Biophys.* **107,** 388 (1964).

[3] M. K. Brakke, *Methods Virol.* **2,** 93 (1967).

[4] V. N. Schumaker, *Advan. Biol. Med. Phys.* **11,** 245 (1967).

[5] J. Vinograd and R. Bruner, *Biopolymers* **4,** 131 (1966).

[6] J. Sykes, *in* "Methods in Microbiology" (J. R. Norris and D. W. Ribbons, eds.), Vol. 5B, p. 55. Academic Press, New York, 1971.

[7] G. M. Stancel and J. Gorski, this volume [15].

[8] W. McGuire, this volume [21].

developed by Martin and Ames for enzyme characterization[9] and applied extensively to steroid hormone receptors since 1966.[10]

The rate of movement of a macromolecule through a homogeneous medium in a centrifugal field is expressed as a sedimentation coefficient (s) in Svedberg units (1 S = 10^{-13} seconds) defined by

$$s = \frac{1}{\omega^2} \frac{d \ln r}{dt} \tag{1}$$

where ω is the angular velocity of the centrifuge rotor in radians per second [$\omega = \pi(\text{rpm})/30$], r is the distance in centimeters of the sedimenting band or boundary from the axis of rotation, and t is the time of centrifugation in seconds.[11] When a thin layer of macromolecules is centrifuged through an appropriate gradient of increasing sucrose concentration, the superimposed density and viscosity gradients cancel the effects of the inhomogeneous radial field. The macromolecular band then moves as a linear rather than exponential function of time and is stabilized by the gradient against convection when the centrifuge is stopped.[9]

Centrifugation through a sucrose gradient has become a basic method for the characterization of steroid-binding macromolecules, since in most systems it permits clear resolution of specific target organ receptors (s = 7–10 S) from many other proteins including the less specific steroid-binding components of serum ($s \simeq 4$ S). In addition, it is a relatively gentle procedure that can be applied to small quantities of unpurified material. A cytoplasmic or nuclear extract labeled with tritiated steroid is layered onto a preformed sucrose gradient in a cylindrical tube and centrifuged in a swinging bucket rotor in a preparative ultracentrifuge. The experiment is terminated before the molecules of interest have reached the bottom of the tube. The gradient is fractionated and analyzed to obtain a distribution of radioactivity as a function of fraction number, corresponding in information content to a single photograph in analytical band ultracentrifugation. From this limited data, the sedimentation coefficient for the steroid-binding components can be evaluated either by computer analysis of the characteristics of the gradient[12] or more easily by comparison with macromolecules of known parameters.[9]

Technical Considerations

Gradient Solutions. Although 5–20% (w/v) sucrose gradients have been routinely used in most receptor studies, in some cases alternative

[9] R. G. Martin and B. N. Ames, *J. Biol. Chem.* **236**, 1372 (1961).
[10] D. Toft and J. Gorski, *Proc. Nat. Acad. Sci. U.S.* **55**, 1574 (1966).
[11] H. K. Schachman, Vol. 4 [2].
[12] C. W. Dingman, *Anal. Biochem.* **49**, 124 (1972).

gradients may be more useful. An expanded range of sedimentation coefficients can be observed within a single gradient of higher density, such as 10–30% sucrose. The resolution of closely sedimenting substances, however, is then diminished. Gradients of 8–35% glycerol (w/v), spanning a comparable density range to 5–20% sucrose,[13] should also be considered, since glycerol is conducive to the stability of various enzymes[14] and receptors for several steroids.[15,16] Another alternative is the addition of 10% glycerol throughout a 5–20% sucrose gradient.[17] It should be noted that linear sedimentation of macromolecules with time is essential for calculation of *s* by the method of Martin and Ames.[9] This linearity should be verified when gradients other than 5–20% sucrose are used.

Salt-induced changes in receptor structure are often studied by density gradient centrifugation. The sample may be exposed to a medium of high ionic strength either before or during centrifugation. Treatment of receptors with salts, followed by centrifugation through gradients of lower ionic strength, may lead to aggregation of the binding components or to reversal of the salt-induced effect during centrifugation. On the other hand, the interpretation of data obtained when salts are included in the gradient is complicated by a progressive decrease in the effectiveness of the salt (as measured by conductivity) with increasing concentration of the dense constituent such as sucrose or glycerol. For example, the conductivity of 0.1 *M* $CaCl_2$ is reduced by 50% in the presence of 30% (w/v) glycerol, and similar effects have been measured in a wide variety of salt solutions.[18]

Receptor stability during centrifugation may be enhanced in gradients of hyperphysiological pH (e.g., 8.4) or in gradients containing chelating agents or sulfhydryl protective reagents such as thioglycerol, mercaptoethanol, or dithiothreitol.[18,19] Errors introduced by these additives, as well as by sucrose or glycerol, in subsequent protein determinations are discussed elsewhere.[20]

Conventional gradient centrifugation techniques are not applicable to receptors that have rapid rates of dissociation of the bound [^{3}H]steroid

[13] "Iscotables, A Handbook of Data for Biological and Physical Scientists," 4th ed. Instrument Specialties Company, Lincoln, Nebraska, 1972.

[14] S. L. Bradbury and W. B. Jakoby, *Proc. Nat. Acad. Sci. U.S.* **69,** 2373 (1972).

[15] I. S. Edelman, *J. Steroid Biochem.* **3,** 167 (1972).

[16] B. P. Schaumburg, *Biochim. Biophys. Acta* **261,** 219 (1972).

[17] P. D. Feil, S. R. Glasser, D. O. Toft, and B. W. O'Malley, *Endocrinology* **91,** 738 (1972).

[18] M. R. Sherman, S. B. P. Atienza, J. R. Shansky, and L. M. Hoffman, *J. Biol. Chem.* (1974) in press.

[19] W. W. Cleland, *Biochemistry* **3,** 480 (1964).

[20] M. R. Sherman, this volume [18].

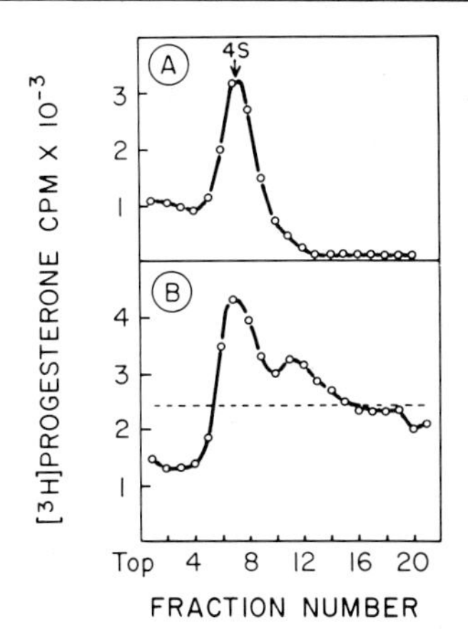

FIG. 1. Use of labeled steroid throughout a density gradient to demonstrate an unstable or rapidly dissociating hormone–receptor complex. The cytoplasmic supernatant fraction (cytosol) was prepared from uteri of castrated rats primed with estrogen for 3 days. Aliquots (0.2 ml) containing 4.8×10^{-9} *M* [^{3}H]progesterone (50 Ci/mmole) and 2.7×10^{-6} *M* unlabeled cortisol were layered onto gradients of 5–20% sucrose in 10 m*M* Tris, 1 m*M* EDTA, 10 m*M* KCl, and 10 m*M* thioglycerol, pH 7.5. Gradient B also contained 3.0×10^{-10} *M* [^{3}H]progesterone, resulting in the initial level of radioactivity indicated by the dashed line. Centrifugation for 16 hours at 3° and 202,000 *g* (average force) revealed an increment of radioactivity in the 7 S region of B, whereas no peak was detectable in A.

or are unstable in the absence of the steroid. The presence of labeled steroid through the gradient may permit characterization of such receptors, as demonstrated in studies of glucocorticoid receptors of rat liver cytosol[21] and in the data of Fig. 1. Rat uterine cytosol was incubated with [^{3}H]progesterone and excess unlabeled cortisol, to compete with [^{3}H]progesterone for less specific steroid-binding sites. Aliquots were centrifuged through parallel 5–20% sucrose gradients containing or lacking [^{3}H]progesterone. In the gradient containing no added steroid (Fig. 1A) only a 4 S peak of bound radioactivity was detected. In the gradient containing [^{3}H]progesterone (Fig. 1B), as in 5–20% sucrose gradients containing no steroid but 10% glycerol throughout,[17] an additional 7 S complex was observed. While these results demonstrate that the 7 S complex is not a glycerol-induced artifact, they are compatible with various interpretations. Both glycerol and progesterone may stabilize the 7 S form of the receptor, or the sole effect of glycerol may be to decrease the intrinsic rate of steroid dissociation from the receptor.

[21] M. Beato and P. Feigelson, *J. Biol. Chem.* **247**, 7890 (1972).

For any particular receptor in a given state of purification the optimum solutions to be used for gradient preparation should be selected on the basis of studies of receptor stability in both the presence and absence of steroid. Apparent rates of dissociation of the steroid from binding sites in crude preparations may reflect progressive loss of active receptor through denaturation or interactions with other constituents, including proteolytic enzymes, superimposed on the intrinsic dissociation rates. Agents that appear to stabilize an impure receptor may have no effect or adverse effects on the same receptor following partial purification, for example, after resolution from most contaminating proteins by gradient centrifugation.

Preparation of Gradients. Continuous gradients of sucrose or glycerol may be prepared from the two limiting solutions by means of inexpensive mixing devices[6,22] (e.g., Buchler[23]) or more elegant programmable equipment (e.g., ISCO). Linear gradients can be formed rapidly and uniformly by the use of two motor-driven syringes which deliver the solutions of high and low density through a mixing chamber into the centrifuge tube. Methods have been devised for the use of syringes that are driven at either constant or variable speed. A reliable apparatus of the latter type which forms three gradients simultaneously is obtainable from Beckman Instruments. When gradient-forming devices are used with polyallomer centrifuge tubes, which are essentially nonwettable and nonsteroid-binding, it is advisable to slant the tubes to reduce droplet formation and disruption of the gradient. This problem may also be avoided by connecting the mixing device to a vertical needle that is stably mounted and positioned to deliver the gradient solution onto the bottom surface of the centrifuge tube. In this case, the least dense solution is delivered first and is displaced upward by solution of increasing density. The needle is then gently removed without disturbing the gradient significantly.

Gradients may also be prepared by repeatedly freezing and thawing a single sucrose solution in the centrifuge tube[24] or by manually layering solutions of decreasing densities into a centrifuge tube and allowing the layers to diffuse for a determined period of time.[2,3] The latter two methods require more time between preparation and use of the gradients and may provide less linear or consistent gradients. Gradients prepared at room temperature by any method must be chilled for at least 1 hour before applying the sample. The form and reproducibility of the resultant gradients may be verified refractometrically, conductimetrically if an

[22] E. A. Peterson and H. A. Sober, *Anal. Chem.* **31**, 857 (1959).

[23] Mention of specific equipment, trade products or a commercial company does not constitute endorsement over similar products or companies not named.

[24] K. L. Baxter-Gabbard, *FEBS Lett.* **20**, 117 (1972).

electrolyte is included, or by optical density, if an absorbing solute is added to one of the limiting solutions in a mixing device.

Sample Preparation and Layering. Centrifugal resolution of a steroid–receptor complex from the incubation medium is sufficiently slow that prolonged incubation of the extract with the labeled steroid before layering may not be necessary. Adequate binding may occur during the early hours of centrifugation. There are several conditions, however, under which preincubation is recommended. (1) If the density of the receptor preparation, including the standard proteins, exceeds that of the least dense gradient solution, the sample must be diluted before layering. The same amount of labeled steroid will generally result in more binding and greater receptor stability if the interaction occurs in the more concentrated state. Note that, within certain limits, the width of the receptor band after centrifugation will not be greatly increased by dilution of the sample.[25] The diluent should correspond in ionic strength and other properties to the sample or the gradient solutions. (2) In experiments with relatively stable receptors, in which the ratio of free-to-bound steroid is high, preincubation permits removal of excess unbound steroid before layering. This is accomplished by the addition of dextran-coated charcoal[26] or by Sephadex G-25 treatment in a small column or batchwise procedure.[27] The restrictions on the ionic composition of a diluent apply equally to the medium in which the charcoal or gel is suspended.

Warming of the sample during layering and centrifugation should be rigorously avoided. Following preliminary experiments with a specified receptor, rotor, and gradient, the centrifugal force can be adjusted to provide a convenient time of centrifugation, since the migration is proportional to the product $(\text{rpm})^2t$.

Fractionation and Recovery. The recovery of steroid in density gradient analysis of a receptor depends on the chemistry of the steroid, the centrifuge tubes, and the tubing used in collecting the gradient. Cellulose nitrate tubes, which are readily punctured for fractionation and are transparent, bind steroids with high affinity and capacity. Tubes of polyallomer (a thermoplastic copolymer of ethylene and propylene) have the opposite properties. They are punctured with difficulty and are only translucent, a disadvantage in sample layering, but generally bind steroids to a negligible extent. Like cellulose nitrate tubes, Tygon tubing (trademark Norton Plastics and Synthetics Division) and silicon rubber tubing, which are both suitably elastic for use in peristaltic pumps, retain enormous quantities of steroids. By contrast, polyethylene and teflon

[25] M. M. Rubin (Sherman) and A. Katchalsky, *Biopolymers* **4**, 579 (1966).
[26] C. A. Nugent and D. B. Mayes, *J. Clin. Endocrinol. Metab.* **26**, 1116 (1966).
[27] W. H. Pearlman and O. Crépy, *J. Biol. Chem.* **242**, 182 (1967).

tubing, which have negligible affinity for steroids, are generally too rigid for use in pumps. Assessment of the recovery of bound and free steroid after centrifugation requires extraction of radioactivity from any tubing through which the gradient was collected and from the centrifuge tube cut into its bottom hemispherical portion (containing rapidly sedimenting receptor forms or aggregates) and upper cylindrical portion(s) of convenient length.

The method of fractionation of the density gradient may affect the recovery of steroid as well as the resolution between binding components. A brief discussion of four of the available techniques follows:

1. The bottom of the centrifuge tube is punctured and the effluent gradient is collected through the tubing of a peristaltic pump or under gravity, which results in a decreasing flow rate. If a peristaltic pump is used, pulsing should be avoided and pump tubing should be chosen to reduce steroid uptake and contamination of subsequent fractions or gradients. In an airtight apparatus, the collection rate may be controlled by the rate at which air is pumped into the space above the gradient. The puncturing device may provide interchangeable holders for centrifuge tubes of various sizes (e.g., Buchler Instruments, Universal Piercing Unit). While satisfactory for cellulose nitrate tubes, this method is applied with some difficulty to tubes of tougher composition such as polyallomer. In a simple puncturing device resembling a hypodermic needle, a core of polyallomer may become lodged in the needle. This core can be removed with minimal disturbance of the gradient by inserting a wire through the length of the needle. The problem is avoided by a piercing device in which the solution aperture is perpendicular to the cutting surface (e.g., Beckman Fraction Recovery System).

2. A rigid tube connected to a pumping device is immersed through the center of the gradient to collect it from the bottom by suction. This technique has the drawback of steroid adsorption to pump tubing and loss of resolution. If the apparatus is airtight, the gradient may be forced through the probe out one channel in the tube cover by air pumped into a second channel near the top of the apparatus (e.g., Beckman Fraction Recovery System, which fits one tube size only; Buchler Densi-Flow). This method permits direct collection of the gradient through the metal probe and a short piece of polyethylene tubing. Any procedure that involves upward flow from the bottom of the centrifuge tube creates negative density and viscosity gradients in the collection probe. The resultant turbulence may be more serious when suction rather than positive pressure is applied.

3. A dense solution is injected through a probe or by puncture into the bottom of the centrifuge tube to raise the gradient through a conical

opening (Beckman) and a photometric flow cell[28] (ISCO), enroute to the fraction collector. Excessive free steroid near the meniscus may trail into the region of slowly sedimenting binding components with any method of collection from the top. This method nevertheless causes minimal disturbance of the gradient and probably provides the best resolution of closely sedimenting components.

4. The gradient is collected from the top through a collection probe that is lowered in synchrony with the descending meniscus (Buchler Instruments Auto Densi-Flow). This method suffers from some loss of resolution, as the collection inlet must be slightly below the meniscus to avoid air bubbles, and the gradient itself passes through the pump.

The method of fractionation for a particular system should be chosen on the basis of the required recovery, resolution, and facility of analysis. In many of our studies, the simple puncturing device (method 1) has proved adequate.

Analysis of Results

Receptor Detection. Since a receptor is generally identified by observing the migration of a reversibly bound [^{3}H]steroid, the coincidence of the radioactivity with a macromolecular–steroid complex must be confirmed. If, for example, the steroid were released during prolonged centrifugation, a broad band of radioactivity would be detected in the region between the true complex and the free steroid location. To verify that the radioactivity is receptor-bound, Sephadex G-25[27] or dextran-coated charcoal[26] in the appropriate buffer can be added to each gradient fraction, removed by low speed centrifugation, and the supernatant fluid analyzed for bound radioactivity. When charcoal is used, the protein content of the gradient fractions may need to be supplemented, since binding components that migrate far from most contaminating proteins are readily adsorbed by charcoal despite the dextran coating.[29] Losses may also result from the addition of charcoal–dextran suspensions in media of lower ionic strength than that of the gradient. An abrupt decrease in salt concentration may cause aggregation and coprecipitation of receptor with the charcoal.

An alternative demonstration of the coincidence of radioactivity with a receptor form involves centrifugation of an unlabeled sample, incubation of each gradient fraction with [^{3}H]steroid, and separation of bound from free steroid by one of the methods cited above. Many receptors

[28] M. K. Brakke, *Anal. Biochem.* **5**, 271 (1963).

[29] P. W. Jungblut, S. Hughes, A. Hughes, and R. K. Wagner, *Acta Endocrinol. (Copenhagen)* **70**, 185 (1972).

appear to be considerably less stable in the absence of the steroid. A careful comparison of the physical-chemical behavior of the residual binding activity with that of the steroid–receptor complex is nevertheless a worthwhile undertaking. In this way, subtle but physiologically meaningful steroid-induced changes may be revealed.

Determination of Sedimentation Coefficients. When a linear 5–20% sucrose gradient is employed, Martin and Ames have shown that the distance migrated by macromolecules of the same partial specific volume ($\bar{v}$) is nearly a linear function of the sedimentation coefficient (s).[9] If the gradients are identical in length and form, s for the steroid-binding components may be calculated by comparison with standards centrifuged in parallel gradients. A more reliable procedure is to mix optically, isotopically, or enzymatically distinguishable standards with the receptor preparation before it is layered onto the gradient. Useful protein standards have values of s that are well established and are close to those of the binding components. Values of s obtained from the literature have generally been corrected for concentration dependence by extrapolation to infinite dilution and refer to standard conditions of 20° in water.[30] Standard proteins should therefore be used at low concentrations and have sedimentation coefficients that are not altered by the temperature, salts, or chelating agents used in receptor studies, except according to the usual corrections for solvent temperature, viscosity, and density.[11,31] Introduction of radioisotopes other than tritium permits simultaneous detection of the standard and the [^{3}H]steroid–receptor complex under double label counting conditions. Convenient methods have been described for derivatization by [^{14}C]formaldehyde[7,32] and iodine isotopes,[33] and a few iodinated proteins are commercially available. Liquid scintillation spectrometry of ^{125}I and ^{131}I has been investigated.[34] Precautions must be taken, however, to assure that derivatized and untreated standard pro-

[30] M. H. Smith, *in* "Handbook of Biochemistry" (H. A. Sober, ed.), Sect. C, p. 10. Chem. Rubber Publ. Co., Cleveland, Ohio, 1968.

[31] The widely used value of 4.6 S for bovine serum albumin (BSA), while consistent with an early study of human serum albumin [J. L. Oncley, G. Scatchard, and A. Brown, *J. Phys. Colloid Chem.* **51**, 184 (1947)], exceeds more recent findings for BSA corrected to 20° in water of 4.3 S [G. L. Miller and R. H. Golder, *Arch. Biochem. Biophys.* **36**, 249 (1952)], 4.41 S [G. I. Loeb and H. A. Scheraga, *J. Phys. Chem.* **60**, 1633 (1956)], and similar values cited in the latter reference. The interactions of serum albumins with many ions and steroids make them poor standards for receptor studies.

[32] R. H. Rice and G. E. Means, *J. Biol. Chem.* **246**, 831 (1971).

[33] Y. Miyachi, J. L. Vaitukaitis, E. Nieschlag, and M. B. Lipsett, *J. Clin. Endocrinol. Metab.* **34**, 23 (1972).

[34] E. D. Bransome, Jr. and S. E. Sharpe, III, *Anal. Biochem.* **49**, 343 (1972).

teins cosediment, and that the labeled group is neither released nor transferred to other proteins by contaminating enzymes in the sample.

When two or more standards are included with a receptor preparation, the unknown s can be estimated from a linear plot of gradient fraction (n) vs. s for the standards. With a single internal standard, the calibration line is drawn from its coordinates of s and n to a value of $s = 0$ for the fraction number corresponding to the midpoint of the initial sample layer (*not* the meniscus). The diluted sample plus standards may occupy about 8% of the centrifuge tube without serious loss of resolution.[25] Failure to correct for a large initial band width, however, introduces a significant error in the calculated s for a component near the meniscus.

The observed sedimentation coefficient of a macromolecule is a function of the solvent density (ρ) and the solute parameters $\bar{v}$, molecular weigh (M), and a friction factor (f), which encompasses molecular asymmetry and solvation,[20]

$$s = \frac{M(1 - \bar{v}\rho)}{Nf} \tag{2}$$

where N is Avogadro's number. A fundamental assumption in evaluating s by numerical analysis of the gradient[12,35] or by comparison with protein standards[9] is that $\bar{v}$ of the receptor molecules is known. This assumption deserves critical scrutiny on several counts. The behavior of some intracellular steroid-binding components on gel electrophoresis and gel filtration is indicative of molecular radii that are larger than expected from the values of s in the same solvents.[36,37] These observations are reminiscent of gel chromatographic and electrophoretic anomalies reported for glycoproteins[38,39] including corticosteroid-binding globulin of serum.[40] Values of $\bar{v}$ for glycoproteins containing 15% carbohydrate are about 0.70 cm^3 g^{-1}, compared with values between 0.72 and 0.74 cm^3 g^{-1} for most simple proteins.[30] It has also been suggested that receptors may contain phosphate,[41] which would likewise depress $\bar{v}$. Since the observed sedimentation rate is proportional to $(1 - \bar{v}\rho)$, a 5% error in the assumed value for $\bar{v}$ causes a 15% error in the calculated s.

[35] C. R. McEwen, *Anal. Biochem.* **20,** 114 (1967).

[36] M. R. Sherman, P. L. Corvol, and B. W. O'Malley, *J. Biol. Chem.* **245,** 6085 (1970).

[37] G. A. Puca, E. Nola, V. Sica, and F. Bresciani, *Biochemistry* **10,** 3769 (1971).

[38] P. Andrews, *Methods Biochem. Anal.* **18,** 1 (1970).

[39] D. Schubert, *J. Mol. Biol.* **51,** 287 (1970).

[40] G. J. Chader, N. Rust, R. M. Burton, and U. Westphal, *J. Biol. Chem.* **247,** 6581 (1972).

[41] R. J. B. King, *Res. Steroids* Vol. **4,** 259 (1970).

For several steroid–receptor systems, there have been wide discrepancies among the values of s reported by different laboratories. This may be due in part to alterations in receptor structure by experimental variables such as pH, ionic strength, and the concentrations of the receptor and of contaminants with which it interacts. For example, polycations such as protamine form precipitates with many target–tissue receptors at low ionic strength.[42] Conversely, increasing concentrations of the polyanion heparin added to rat uterine cytoplasmic extracts progressively decrease the apparent s of the estrogen-binding components from about 10 S to 5 S.[8] In conclusion, sedimentation coefficients measured in impure preparations are useful in distinguishing various forms of a hormone–receptor complex, but should not be construed as definitive parameters of macromolecular structure.[43]

[42] A. W. Steggles and R. J. B. King, *Biochem. J.* **118**, 695 (1970).

[43] We gratefully acknowledge research support to D. O. T. by National Institutes of Health Contract 70-2165 and the Vanderbilt University Center for Population Research and Studies in Reproductive Biology (NIH Grant HD-05797) and to M. R. S. by The Paul Garrett Fund, American Cancer Society Grant PRA-83, and NIH Grant CA-08748.

[15] Analysis of Cytoplasmic and Nuclear Estrogen-Receptor Proteins by Sucrose Density Gradient Centrifugation[1]

By GEORGE M. STANCEL and JACK GORSKI

Sucrose density gradient centrifugation has proved useful for studying in target tissues the estrogen-binding proteins generally believed to be estrogen receptors.[1a] The technique is useful because it provides physical data on the structure of estrogen-receptor proteins, and because it provides a method of physically (1) separating intracellular estrogen-binding proteins from serum binding proteins, (2) separating cytoplasmic and nuclear forms of estrogen receptors, and (3) comparing receptor proteins from different tissues and different species. In principle the technique involves the reversible formation of a complex between estrogen-receptor proteins (RP) and a labeled estrogen (E*) as illustrated in Eq. (1).

$$RP + E^* \leftrightharpoons RP - E^* \qquad (1)$$

[1] This work was supported by grants HD 4828 and HD 00181 from the National Institutes of Health and Ford Foundation Training Grant 700-0333.

[1a] E. V. Jensen and E. R. DeSombre, *Annu. Rev. Biochem.* **41**, 203 (1972).

Once formed, the protein hormone complex dissociates very slowly (the half-life for the dissociation of the complex is >20 hours at 0°),[2] and therefore the sedimentation coefficient of the receptor protein may be estimated from the migration of protein-bound radioactivity relative to the migration of internal markers of known sedimentation coefficient.[3] Since tritiated estrogens are used, we have found it convenient to use ^{14}C-labeled marker proteins.[4]

The technique of sucrose density gradient centrifugation is used in a number of types of studies, therefore this chapter will deal primarily with the preparation of estrogen receptors for centrifugation and with the interpretation of results from such experiments. For added detail on the technique of gradient centrifugation per se, the reader is referred to the original article by Martin and Ames[3] and to several other excellent descriptions of methodology.[5-7]

Sample Preparation

In the uterus and other target tissues the estrogen receptor is originally found in the cytoplasmic fraction.[1,8] After the hormone enters the target tissue cell and forms a complex with the receptor, the receptor–hormone complex migrates to the nuclear fraction.[1,8] Since the cytoplasmic and nuclear forms of the estrogen receptor have different sedimentation properties, the preparation of receptors from both cellular fractions is described in this section.

Preparation of Cytoplasmic Receptor

In Vitro. The procedure described below has been routinely used in our laboratory for studies with estrogen receptors from uteri of immature rats, 20–24 days old.[4] For use with other tissues, the concentration of receptor-binding activity and/or the total protein concentration may be adjusted proportionately. See, for example, Notides, for studies with es-

[2] D. L. Williams and J. Gorski, *Biochem. Biophys. Res. Commun.* **45**, 258 (1971).

[3] R. G. Martin and B. N. Ames, *J. Biol. Chem.* **236**, 1372 (1961).

[4] G. M. Stancel, K. M. T. Leung, and J. Gorski, *Biochemistry* **12**, 2130 (1973).

[5] J. Sykes, *in* "Methods in Microbiology" (J. R. Norris and D. W. Ribbons, eds.), Vol. 5B, p. 55. Academic Press, New York, 1971.

[6] R. Trautman and K. M. Cowan, *in* "Methods in Immunology and Immunochemistry" (C. A. Williams and M. W. Chase, eds.), Vol. 2, p. 81. Academic Press, New York, 1967.

[7] J. C. Gerhardt, Vol. 11, 187 (1967).

[8] J. Gorski, D. Toft, G. Shyamala, D. Smith, and A. Notides, *Recent Progr. Horm. Res.* **24**, 45 (1968).

trogen receptors from pituitary,[9] and Shyamala and Nandi for studies with receptors from breast tissue.[10] Similar results are also obtained with mature, ovariectomized animals.[11]

The animals are decapitated and the uteri are excised, stripped of any adhering fat, and placed in chilled TE buffer (0.01 M Tris, 1.5 mM EDTA, pH 7.4) at 0°. The uteri are then homogenized at 0° in the desired volume of TE buffer using an all-glass homogenizer (Kontes). Care should be taken during homogenization so that the temperature remains at 0–4°. For most experiments we have homogenized 1–5 uteri/ml buffer. This procedure yields cytosol preparations having protein concentrations ranging from 1 to 5 mg/ml of total protein and of the order of 10^{-9} M in estradiol-binding sites. If a precise knowledge of these parameters is required for a particular experiment, measurements must be made on aliquots of the cytosol in question.

The resulting homogenate is then spun at high speed (180,000 g) for 60 minutes with the temperature held at 0–4°. If a surface layer of lipids is noted after centrifugation, the tip of a disposable pipet is placed below the lipid layer and the clear cytosol is drawn up.

An aliquot of a concentrated solution of tritiated estradiol is then added to achieve the desired concentration of labeled hormone. For most experiments, a stock solution of estradiol (10^{-6} M) in ethanol is diluted to a final concentration of $1\text{–}10 \times 10^{-9}$ M. After the addition of hormone, the mixture is incubated at 0° for the appropriate length of time. Incubations of 1–3 hours are sufficient for experiments investigating the structure of the receptor itself, while longer incubations (8–12 hours) are required for quantitative estimates of the concentration of receptor–hormone complex when low concentrations of estrogen are used.[12]

In Vivo. If a particular experiment calls for gradient analysis following the *in vivo* formation of a cytoplasmic hormone–receptor complex, the cytosol is prepared as described following an injection of labeled hormone.[13]

A cytoplasmic receptor–estrogen complex may also be prepared using organ culture, by incubation of uterine tissue with labeled estrogens. If organ culture techniques are employed, however, the TE buffer used for homogenization should contain a 100-fold excess of unlabeled hormone. This prevents the artifactual formation of complex between receptors and

[9] A. C. Notides, *Endocrinology* **87,** 987 (1970).
[10] G. Shyamala and S. Nandi, *Endocrinology* **91,** 861 (1972).
[11] A. W. Steggles and R. J. B. King, *Biochem. J.* **118,** 695 (1970).
[12] D. Ellis and H. Ringold, *in* "The Sex Steroids" (K. McKerns, ed.), p. 73. Appleton, New York, 1971.
[13] D. Toft and J. Gorski, *Proc. Nat. Acad. Sci. U.S.* **55,** 1574 (1966).

labeled hormone which was trapped in the tissue, but not specifically bound to receptor proteins prior to tissue homogenization.[2]

It is normally advisable to prepare the cytoplasmic receptor–estrogen complex by an *in vitro* cell-free incubation, since the complex rapidly migrates to the nuclear fraction *in vivo* or in organ culture, and hence the concentration of the complex formed under these conditions is low.

Preparation of Nuclear Estrogen–Receptor Complex

Uteri are excised from immature female rats and stripped of fat. Whole uteri, or uterine slices, are then incubated at 37° for 1 hour in Eagle's HeLa medium (Difco) containing tritiated estradiol at the desired concentration, usually $1–10 \times 10^{-9}$ *M*. Incubations are performed under an atmosphere of 95% O_2:5% CO_2 in a water bath with constant shaking at ~60 cycles per minute. One uterus is used per 2 ml of media. This treatment leads to the migration of approximately 90% of the bound estrogen into the nuclear fraction,[1] and at the higher level of estrogen (10^{-8} *M*) approximately 90% of total receptor activity moves into the nuclear fraction.

Following the incubation the uteri are washed well with chilled TE buffer and homogenized the same as described in a previous section. The nuclear-myofibrillar pellet is then obtained by centrifugation at 1000 *g* for 10 minutes. The volume of the pellet is estimated, and the pellet washed 3–4 times with 10 volumes of TE buffer per wash. After the washings, enough 0.8 *M* KCl (0.02 *M* Tris, 3 m*M* EDTA, 0.8 *M* KCl, pH 7.4) is added to the pellet to yield a suspension with a final KCl concentration of 0.4 *M*.[14]

The suspension is kept at 0° for 1 hour with intermittent vortexing and then centrifuged at 180,000 *g* for 45 minutes to yield the final nuclear extract. This procedure leads to the extraction of 50–60% of the nuclear estrogen receptor.[15]

A similar procedure is used to prepare nuclear receptor–estrogen complex following an injection of labeled hormone in whole animal experiments.

The procedure described above yields a form of the nuclear estrogen receptor which sediments at about 5 S,[1,15] but another procedure, originally described by Giannopoulos and Gorski,[16] yields a different form

[14] It has been reported that a buffer of similar composition, but pH 8.5 rather than 7.4, yields better extraction. See G. A. Puca and F. Bresciani, *Nature (London)* **218,** 967 (1968).

[15] G. M. Stancel, K. M. T. Leung, and J. Gorski, *Biochemistry* **12,** 2137 (1973).

[16] G. Giannopoulos and J. Gorski, *J. Biol. Chem.* **246,** 2530 (1971).

of the nuclear receptor which sediments slightly faster, at about 6 S. In this procedure, nuclear receptor is formed as described above, either *in vivo* or *in vitro,* and the uterine tissue is then *homogenized directly in 0.4 M KCl* (0.01 *M* Tris, 1.5 m*M* EDTA, 0.4 *M* KCl, pH 8.5). The resulting homogenate is vortexed and left at 0° for 1 hour. The extract obtained by high-speed centrifugation of the suspension contains a form of the receptor which sediments at about 6 S.[16]

As in work with all hormone receptors, several criteria must be met with the samples used for gradient centrifugation[1,8]:

(1) Hormone binding must be specific for estrogens.
(2) Hormone binding must be high affinity and low capacity, and therefore abolished by excess concentrations of unlabeled hormone.
(3) Binding proteins must be present in target tissue cells, as opposed to serum or non-target tissue cells.

Preparation of ^{14}C-Labeled Marker Proteins

^{14}C-Labeled protein markers can be easily prepared by a method suggested by Means and Feeney[17] and later modified.[18] We have used this basic method for preparing labeled ovalbumin and γ-globulin.[4] While any proteins could be used, ovalbumin and γ-globulin are well suited to work with estrogen receptors because of their sedimentation coefficients, stability, and low cost.

The method of labeling involves the methylation of amino groups on the marker protein, as illustrated in Eq. (2) below.

$$\text{—NH}_2 + \text{O=C}^*\text{H}_2 \rightleftharpoons \text{—N=C}^*\text{H}_2 \xrightarrow{\text{NaBH}_4} \text{—NH—}\overset{*}{\text{C}}\text{H}_3 \qquad (2)$$

A 1-ml sample of the marker protein at a concentration of 10 mg/ml is prepared in 0.2 *M* sodium borate buffer, pH 9.0, and the solution is cooled in ice. A 10-μl aliquot of 4 m*M* [^{14}C]formaldehyde (New England Nuclear, 10 Ci/mole) is then added, and the mixture gently swirled. After 1 minute, four 2-μl aliquots of sodium borohydride (0.5 mg/ml prepared in distilled water immediately prior to use) are added sequentially. The entire procedure is repeated 4–5 times. Following the last sequence of additions, an extra 10-μl aliquot of sodium borohydride is added to reduce any remaining formaldehyde. Unreacted components are then removed by dialysis or passage of the reaction mixture over a column of Sephadex

[17] G. E. Means and R. E. Feeney, *Biochemistry* **7**, 2192 (1968).
[18] R. H. Rice and G. E. Means, *J. Biol. Chem.* **246**, 831 (1971).

G-25 equilibrated with a suitable buffer. Labeled ovalbumin and γ-globulin may be stored at −20°. The sedimentation coefficients of the markers are 3.6 S and 7 S for ovalbumin and γ-globulin, respectively.

Sucrose Density Gradient Centrifugation

Preparation of Gradients

Linear 5–20% sucrose gradients are prepared as described by Martin and Ames[3] using polyallomer centrifuge tubes. The polyallomer tubes are preferable to nitrocellulose tubes, since estrogen-receptor proteins adhere to the walls of the latter type.[19] For most studies the total volume of the gradient is 3.8 ml with a Beckman SW 56 Rotor and 5 ml with a Beckman SW 50.1 Rotor. After preparation the gradients are kept at 0–4° for 3–6 hours before use.

The sucrose gradients may be prepared with several different buffers. Samples of cytosol receptor are routinely studied with either a dilute TE buffer or high salt TKE buffer (TE buffer containing 0.4 *M* KCl). At the low ionic strength of cytosol receptor sediments in the 8–9 S region of the gradients,[1,4,12] and the high salt medium in the 4 S region of the gradients.[20–22] However, samples of the nuclear form of the receptor, which sediment in the 5 S region[23,24] are examined in TKE buffer, since the nuclear receptor aggregates at low ionic strength.

Centrifugation

A small amount of ^{14}C-labeled marker protein is added to the samples containing the receptor–hormone complex immediately before the centrifuge run is started. The samples are then layered on the gradients in the cold room using a micropipet. Volumes of 100–200 μl are normally used. Centrifugation is then performed at 0–4°.

The length of centrifugation is variable and depends on the separation required between different protein species and the form of receptor (4 S or 8 S) present. At 300,000 *g*, for example, a 10–12 hour run yields satisfactory migration of the 8 S cytosol receptor, while a 14–18 hour

[19] G. Giannopoulos and J. Gorski, unpublished observations.

[20] S. G. Korenman and B. R. Rao, *Proc. Nat. Acad. Sci. U.S.* **61**, 1028 (1968).

[21] T. Erdos, *Biochem. Biophys. Res. Commun.* **32**, 338 (1968).

[22] H. Rochefort and E.-E. Baulieu, *C. R. Acad. Sci., Ser. D* **267**, 662 (1968).

[23] E. V. Jensen, T. Suzuki, T. Kawashima, W. E. Stumpf, P. W. Jungblut, and E. R. DeSombre, *Proc. Nat. Acad. Sci. U.S.* **59**, 632 (1968).

[24] G. Shyamala and J. Gorski, *J. Biol. Chem.* **244**, 1097 (1969).

run is routinely used for the 5 S nuclear receptor or the 4 S cytosol receptor observed in the high salt media. After the centrifuge run, the rotor is allowed to decelerate without the use of the brake.

Gradient Fractionation

After the centrifuge run the tubes are individually fractionated. The tube is securely held in an appropriate device in a vertical position and the bottom of the tube punctured with a fine gauge needle. There are two ways in which the fractions are usually collected. (1) Fractions containing a suitable number of drops may be collected by gravity flow or (2) a dense solution, e.g., 40–50% sucrose, may be forced under the gradient and the contents of the gradient tube forced out of the top of the tube for collection. If the latter procedure is used, the dense solution may be forced up the centrifuge tube with a peristaltic pump or a motor-driven syringe.

In our laboratory we have found that the use of a motor-driven syringe to pump dense sucrose solutions under the gradient yields the most reproducible results and the sharpest bands of receptor proteins and markers. The entire fractionation process is easily automated for routine use with a drop detecting photocell and a fraction collector equipped for direct collection into scintillation vials. All the equipment for such an automated setup is commercially available (ISCO).

The protein-bound estradiol is then located by scintillation counting. In our laboratory, we have found it convenient to collect 100-μl aliquots from the gradients. The aliquots are then counted, after addition of 3 ml of absolute ethanol and 10 ml of toluene-based scintillation fluid (5 g PPO, 0.3 g POPOP per liter toluene). A typical pattern obtained from a sample of uterine cytoplasmic estrogen receptor is depicted in Fig. 1B.

Interpretation of Results

Calculation of Sedimentation Coefficient

The sedimentation coefficient of the receptor is calculated directly from the known sedimentation coefficient of the marker protein and from the relative migrations of the marker and receptor bands from the meniscus.[3]

$$\frac{S_{\text{unknown}}}{S_{\text{marker}}} = \frac{\text{distance}_{\text{unknown}}}{\text{distance}_{\text{marker}}} \tag{3}$$

The values obtained, however, should be considered as approximate in the absence of knowledge of the hydrodynamic properties of the unknown relative to the marker.

Variability of the Sedimentation Coefficient of Estrogen Receptors

Examination of the literature reveals that numerous values have been reported for the sedimentation coefficient of cytoplasmic estrogen receptors,[25] especially using crude homogenates. These variations are due to a number of factors described below. Less variation is observed for the nuclear form of the receptor.[15] Because of this variation in sedimentation values, meaningful comparisons can only be made under carefully controlled conditions, preferably using receptors from the original cytosol preparations.

Specific Proteins. Specific proteins have been shown to affect the sedimentation of estrogen receptors. Ribonuclease, for example, increases the sedimentation coefficient of the cytoplasmic receptor at low ionic strength, but not at high ionic strength.[26]

Sulfated Mucopolysaccharides. Heparin at low concentrations increases the sedimentation coefficient of the cytoplasmic receptor, while higher concentrations of heparin decrease the sedimentation coefficient.[25] Heparin also affects the sedimentation coefficient of the nuclear form of the receptor.[25]

Calcium Stabilization. Treatment of cytoplasmic estrogen receptor with calcium and EDTA yields a stabilized form of the receptor which does not aggregate and which sediments at 4 S at both high and low ionic strength.[23,27]

Aggregation. The cytoplasmic receptor from rat uterus undergoes a slow aggregation with other soluble uterine proteins to produce more rapidly sedimenting species.[4] This aggregation is time-dependent and appears to take place in homogenates during incubation of the cytosol fraction with estrogens and during the gradient centrifugation.[4] This aggregation also depends on the total protein concentration; aggregation becoming minimal at low protein concentrations.[4] The effect of protein concentration is seen by comparing the two panels in Fig. 1. A number of other tissues, e.g., diaphragm, liver, and spleen, also contain soluble proteins which aggregate with the uterine receptor.[4]

[25] G. C. Chamness and W. L. McGuire, *Biochemistry* **11,** 2466 (1972).

[26] E. V. Jensen, M. Numata, P. I. Brecher, and E. R. DeSombre, *in* "The Biochemistry of Steroid Hormone Action" (R. M. S. Smellie, ed.), p. 133. Academic Press, New York, 1971.

[27] G. A. Puca, E. Nola, V. Sica, and F. Bresciani, *Biochemistry* **11,** 4157 (1972).

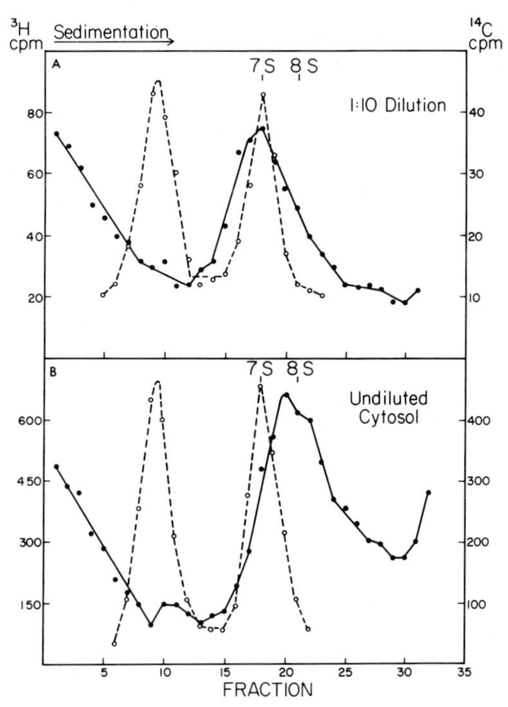

FIG. 1. Effect of dilution on sedimentation coefficient of estrogen receptor in TE gradients. Cytosol from immature rat uterus (5 uteri/ml in TE buffer) was prepared as described in "Preparation of Cytoplasmic Receptor *In Vitro*" and incubated (0°) for 3 hours with 5×10^{-9} *M* tritiated estradiol. One aliquot was used without further treatment (B), and one aliquot was diluted 1:10 with TE buffer before use (A). Centrifugation at 1° was for 12 hours at 50,000 rpm in a Beckman SW 56 Rotor. The dotted line represents the ^{14}C-markers, ovalbumin and γ-globulin, and the solid line the tritium counts. See "Interpretation of Results" for explanation of dilution effect.

In some cases aggregation of uterine receptor proteins may lead to the formation of very high molecular weight species which "pellet" to the bottom of the centrifuge tube. Therefore, if the recovery of protein-bound counts in the peak area is low, it is advisable to determine whether any protein-bound counts are present in the bottom portion of the centrifuge tube.

Denaturing Agents. Low concentrations of urea (less than 4 *M*) completely prevent the aggregation phenomena described above for the cytoplasmic receptor.[15] Furthermore, in the presence of 4 *M* urea, both nuclear and cytoplasmic receptors have identical sedimentation coefficients of approximately 4 S.[15]

Degree of Purification. After cytoplasmic estrogen receptors have been partially purified, the sedimentation coefficients show much less variability.[27]

In most experiments performed in our laboratory, the variability we observe seems to stem mainly from the tendency of cytoplasmic estrogen receptors to aggregate (see above). Therefore, in order to reproducibly measure the sedimentation coefficient of uterine cytoplasmic estrogen receptors we have found it necessary to always use internal ^{14}C-labeled marker proteins, rather than the more commonly used BSA standard in a separate gradient, and to carefully control experimental conditions, especially protein concentration.[4] We strongly recommend that future experiments using this technique adhere to these precautions. Even with these precautions there is some variability in values for different cytosol preparations and, therefore, we further suggest that conclusions concerning structural features of estrogen-receptor proteins be based only on comparison of samples derived from the same original cytosol preparation.

Comparative Studies

Gradient centrifugation has been used to examine estrogen receptors from a number of tissues. In the rat, for example, cytoplasmic receptors from the uterus,[4] pituitary,[9] and breast[10] sediment at approximately 8 S in low ionic strength media. Similar values have also been reported for receptors from calf uteri.[26,27]

Other literature reports, however, are conflicting. Thus, values of 5 S have been reported for cytoplasmic receptors from human uterus[28,29] and monkey uterus,[28,29] while other laboratories have reported values of 8 S for receptors from human uteri.[30] Because numerous factors described in the preceding section can affect the sedimentation of receptors, comparative studies between different tissue and different species should involve mixing experiments as illustrated below (Fig. 2).

Aliquots of A-2* and B-2* containing labeled receptors from the different cytosols may not be run on different gradients. Aliquot A-2* may now be mixed with B-1, and B-2* mixed with A-1, and the resulting mixtures centrifuged. The mixing will then allow one to determine if any

[28] R. H. Wyss, R. Karsznia, W. L. Heinrichs, and W. L. Herrmann, *J. Clin. Endocrinol. Metab.* **28,** 1824 (1968).

[29] R. H. Wyss, W. L. Heinrichs, and W. L. Herrmann, *J. Clin. Endocrinol. Metab.* **28,** 1227 (1968).

[30] A. C. Notides, D. E. Hamilton, and J. H. Rudolph, *Biochim. Biophys. Acta* **271,** 214 (1972).

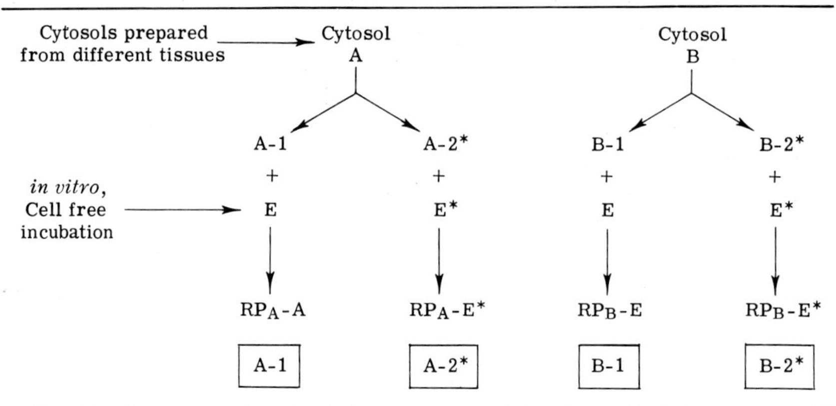

FIG. 2. Mixing experiments. RP, receptor protein; E, unlabeled estrogen; E*, labeled estrogen.

differences observed between receptors in cytosols A (aliquot A-2*) and B (aliquot B-2*) represent structural differences in receptors, or the presence of interfering substances in either tissue preparation.

[16] Extraction Artifacts in the Preparation of Estrogen Receptor

By GARY C. CHAMNESS and WILLIAM L. MCGUIRE

The estrogen receptor found in the cytoplasm of estrogen target tissues assumes a variety of forms on extraction, depending on the conditions of extraction and examination. After binding estrogen in the cytoplasm the receptor can enter the nucleus, either *in vivo* or in cell-free incubation, and this recovered nuclear receptor assumes another variety of forms. Sucrose gradient sedimentation of receptor, introduced in 1966 by Toft and Gorski,[1] has been generally employed to reveal these distinctions. Here we present those procedures currently used in our own laboratory which have consistently produced certain receptor forms.

Cytoplasmic Receptor

Tissues. Either mature uteri from 150–200 g intact Sprague-Dawley rats or immature uteri from Sprague-Dawley rats 21–23 days old are used immediately. Pituitary tumor MtTW5, originally obtained from Dr.

[1] D. Toft and J. Gorski, *Proc. Nat. Acad. Sci. U.S.* **55**, 1574 (1966).

Vincent Hollander, is transplanted serially by subcutaneous implantation into mature intact female Wistar-Furth rats. These rats are ovariectomized at least 2 days prior to removal of 5–10 g tumors, which are frozen in liquid nitrogen and maintained at —56° until use. Although only results from these two tissues will be described in detail, similar procedures are employed for receptor studies in DMBA-induced or transplantable rat mammary tumors[2] and in human breast tumors.[3]

Buffers. TE 7.4 (0.01 *M* Tris-HCl, pH 7.4, with 1.5 m*M* Na_2EDTA) is used throughout. Other solutes are added or EDTA omitted as indicated.

Preparation of Cytosol. All procedures are carried out at 2–4°. Tissues are minced with scissors and homogenized in 2.5 volumes (uterus) or 4 volumes (pituitary tumor) of TE 7.4 in a Duall conical glass-glass homogenizer. The homogenate is centrifuged 40 minutes at 140,000 *g* and the supernatant is recovered with a transfer pipet, leaving behind any floating fat layer. The supernatant is immediately charged with [^{3}H]estradiol (48 or 95 Ci/mmole) at 1×10^{-9} *M* or 4.5×10^{-9} *M* as needed. Radiochemical purity of the estradiol is periodically checked by thin-layer chromatography.[4] After at least 60 minutes the cytosol containing charged, labeled receptor is ready for use, though for some applications free estradiol is removed by passage over Sephadex G-25 or by treatment with dextran-coated charcoal. The charcoal procedure, adapted from Korenman,[5] uses a suspension of 2.5 g/liter of Norit A and 25 mg/liter of dextran in 0.01 *M* Tris-HCl (pH 8.0) with 1.5 m*M* Na_2EDTA. A volume equal to that of the cytosol preparation to be treated is centrifuged 10 minutes at 2000 *g* and the supernatant discarded; the pellet is resuspended directly in the cytosol and, after 15 minutes, centrifuged 10 minutes at 2000 *g*, leaving the supernatant free of unbound estradiol.

Sucrose Gradients. All sucrose solutions contain TE 7.4, and some also include 0.15 *M* or 0.4 *M* KCl. Linear gradients are prepared with an ISCO gradient former in cellulose nitrate tubes for the Beckman SW 56 rotor. (An earlier procedure[6] in which sucrose solutions of decreasing density were layered in the tube and allowed to diffuse overnight to reach a continuous gradient produced equally linear gradients but required more time.) Generally 5–20% sucrose is used for those gradients containing salt and 10–30% sucrose for the others in order to emphasize the respec-

[2] W. L. McGuire, J. A. Julian, and G. C. Chamness, *Endocrinology* **89**, 969 (1971).
[3] W. L. McGuire, *J. Clin. Invest.* **52**, 73 (1973).
[4] R. H. Bishara and I. M. Jakovljevic, *J. Chromatogr.* **41**, 136 (1969).
[5] S. G. Korenman, *J. Clin. Endocrinol. Metab.* **28**, 127 (1968).
[6] G. C. Chamness and W. L. McGuire, *Biochemistry* **11**, 2466 (1972).

tive regions of interest after the standard 15.5-hour centrifugation at 56,000 rpm (308,000 g_{av}). The gradients are chilled after preparation, and the upper 300 μl are gently removed before 200-μl samples are layered. Each sample contains about 1000 cpm of ^{14}C-labeled bovine serum albumin[7] as an internal marker; the ^{14}C-peak provides a check of the quality of each gradient as well as a more precise sedimentation standard than achieved with the usual separate albumin gradient. Neither the presence or absence of 0.4 *M* KCl in the gradient nor any other variable used appears to affect the sedimentation of the ^{14}C-labeled albumin. After centrifugation, 4-drop fractions are collected and suspended in modified Bray's solution (125 g of naphthalene, 7.4 g of 2,5-diphenyloxazole, 0.38 g of 1,4-bis[2-(5-phenyloxazolyl)]benzene, 1 liter of *p*-dioxane) for liquid scintillation counting in a Beckman LS 233; 80–90% of the applied counts are recovered on the gradients. Sedimentation values are determined according to Martin and Ames[8] from the [^{14}C]BSA standard.

Effects of Ionic Strength. When freshly prepared cytoplasmic receptor from either rat uterus (Fig. 1), pituitary tumor,[9] or hormone-dependent mammary tumor[10] is sedimented in TE 7.4 alone, it forms a peak at 8–9 S, independent of the original salt concentration of the cytosol preparation. Erdos[11] first showed that receptor sediments at 4–5 S if the sucrose gradient contains 0.3 or 0.4 *M* KCl; since the transition is rapid, the sample need not be adjusted to the final salt concentration before layering. More recently it has been shown[6,12,13] that extraction and centrifugation at physiological ionic strength (0.15 *M* KCl) produces a distinct receptor form at approximately 6 S (Fig. 1). If the cytosol is originally prepared in TE 7.4 and then brought to 0.15 *M* KCl, the transition from 8 S to 6 S requires many hours at 4°; bringing the sample to 0.4 *M* KCl a few minutes before layering on a gradient containing 0.15 *M* salt essentially eliminates this period.

Effect of Calcium. In calf uterus cytosol, Puca *et al.*[14] have been able to manipulate the addition of 4 m*M* $CaCl_2$, KCl, and ammonium sulfate to produce either stable 8 S or stable 4 S receptors as desired. We find these procedures to have no effect with either rat uterus or rat pituitary tumor cytoplasmic receptor. However, the addition of 4 m*M* $CaCl_2$ and

[7] R. H. Rice and G. E. Means, *J. Biol. Chem.* **246,** 831 (1971).
[8] R. G. Martin and B. N. Ames, *J. Biol. Chem.* **236,** 1372 (1961).
[9] W. L. McGuire, M. DeLaGarza, and G. C. Chamness, *Endocrinol.* **93,** 810 (1973).
[10] W. L. McGuire and J. A. Julian, *Cancer Res.* **31,** 1440 (1971).
[11] T. Erdos, *Biochem. Biophys. Res. Commun.* **32,** 338 (1968).
[12] I. Reti and T. Erdos, *Biochimie* **53,** 435 (1971).
[13] H. Rochefort, *Proc. Int. Congr. Horm. Steroids, 3rd, 1970* p. 23 (1971).
[14] G. A. Puca, E. Nola, V. Sica, and F. Bresciani, *Biochemistry* **10,** 3769 (1971).

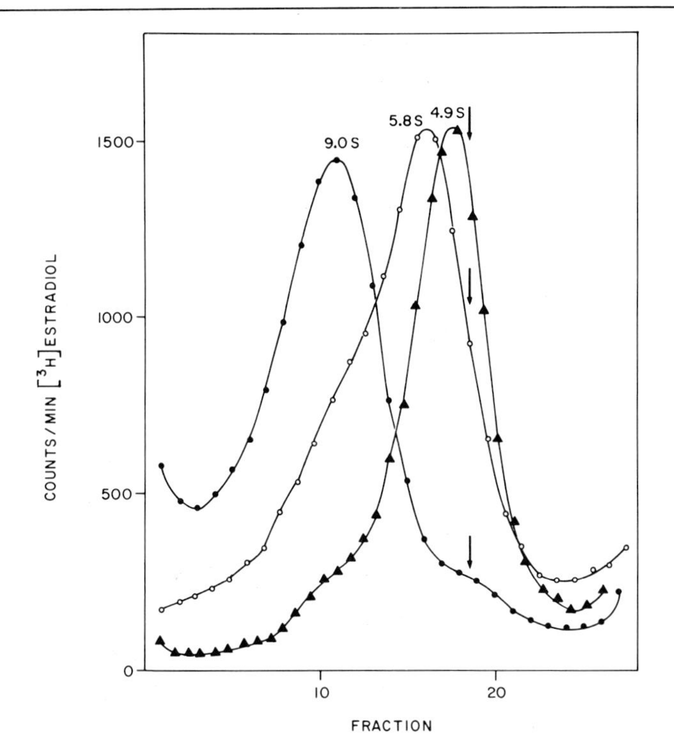

FIG. 1. Sedimentation of mature rat uterus cytoplasmic receptor in 0.4 *M* KCl (▲), 0.15 *M* KCl (○), or TE 7.4 without added salt (●). The 0.15 *M* KCl sample was brought to 0.4 *M* KCl 15 minutes before layering on the gradient (see text). Arrows mark the peaks of the [^{14}C]bovine serum albumin internal standards. (From Chamness and McGuire.[6])

0.4 *M* KCl to rat uterine cytosol in TE 7.4, either 45 minutes or 4.5 hours before centrifugation in TE 7.4 gradients without salt, produces a 6 S peak identical to that formed in physiological salt (Fig. 2).

Temperature Dependent Transformation. Brecher *et al.*[15] first showed that cytoplasmic receptor could be transformed by warming from a form sedimenting at 4 S in 0.4 *M* KCl sucrose gradients to a form sedimenting at "5 S." The actual sedimentation value achieved by this procedure is 5.6–5.8 S, identical to that "6 S" seen above in physiological salt or produced by certain calcium treatments. Figure 3 shows that this transformation occurs when rat uterus cytosol is prepared in T 7.4 (without EDTA), charged 1 hour at 4° with 2×10^{-9} *M* [^{3}H]estradiol, and

[15] P. I. Brecher, M. Numata, E. R. DeSombre, and E. V. Jensen, *Fed. Proc., Fed. Amer. Soc. Exp. Biol.* **26,** 536 (1970).

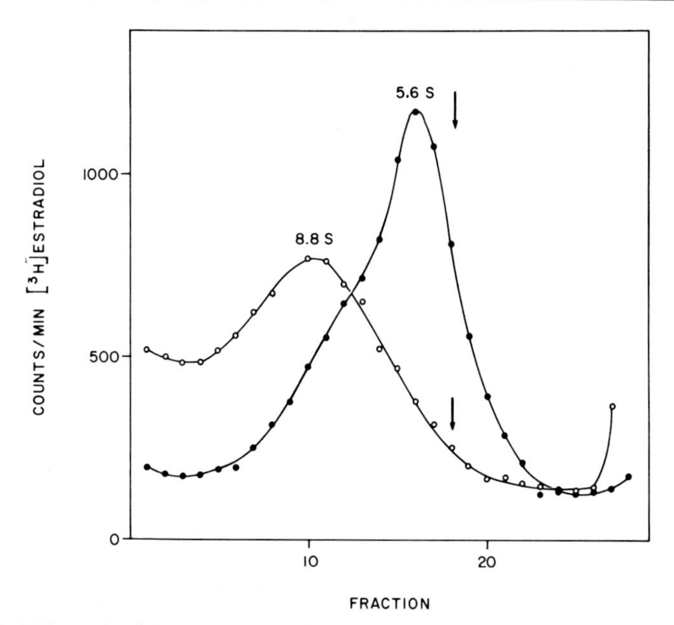

FIG. 2. Effect of calcium ion on sedimentation of mature rat uterus cytoplasmic receptor. Labeled cytosol in TE 7.4 was brought to 0.4 *M* KCl with (●) or without (○) 4 m*M* $CaCl_2$ 45 minutes before layering on TE 7.4 sucrose gradients. Arrows mark the peaks of [^{14}C]BSA. (From Chamness and McGuire.[6])

warmed to 25°. The cytosol is chilled and brought to 0.4 *M* KCl before layering on sucrose gradients containing the same salt concentration. Thirty minutes at 25° is sufficient to fully transform rat uterus receptor, and some estradiol-binding activity is lost in this treatment. Jensen *et al.*[16] and Mohla *et al.*[17] have indicated that this transformation is required for effective nuclear binding and for cell-free stimulation of nuclear RNA polymerase by estrogen-charged receptor. The effect of transformation on cell-free binding of receptor to nuclei will be discussed later.

Aging. A stable 4 S form of the cytoplasmic receptor in pituitary tumor results from aging a TE 7.4 pituitary cytosol at 4° for at least 9 days (Fig. 4). (It is advisable to include 3 m*M* sodium azide to prevent growth of contaminants during aging). This 4 S form is not affected by

[16] E. V. Jensen, M. Numata, P. Brecher, and E. R. DeSombre, *in* "The Biochemistry of Steroid Hormone Action" (R. M. S. Smellie, ed.), p. 133. Academic Press, New York, 1971.

[17] S. Mohla, E. R. DeSombre, and E. V. Jensen, *Biochem. Biophys. Res. Commun.* **46**, 661 (1972).

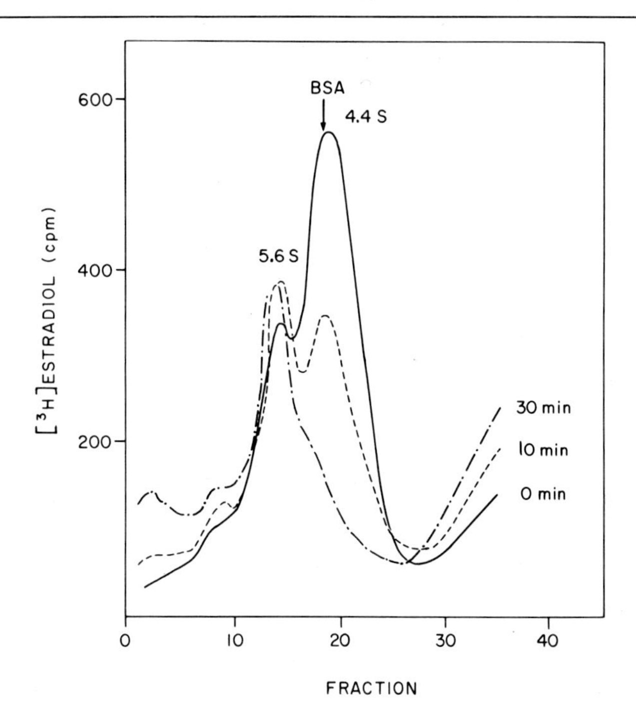

FIG. 3. Transformation of immature rat uterus cytoplasmic receptor. Cytosol in T 7.4 (without EDTA) was charged with 2×10^{-9} *M* [^{3}H]estradiol and incubated at 25° for 0, 10, or 30 minutes before layering on sucrose gradients containing 0.4 *M* KCl.

changes in ionic strength or by treatment with heparin (see below). This conversion is reminiscent of the "stable 4 S" described by others in mature uteri of some strains[18] or resulting from either prolonged KCl treatment or 37° incubation.[19,20]

Polyanion Treatment. Almost any sedimentation value within a wide range can be produced by the appropriate concentration of polyanion.[6] Sodium heparin is added to cytoplasmic receptor preparations at 0.3 to 6.0 mg/ml final concentration just before layering of samples on TE 7.4 sucrose gradients. A single sharp sedimentation peak results in each case, but the sedimentation value decreases from 8–9 S without heparin to 5

[18] A. W. Steggles and R. J. B. King, *Biochem. J.* **118,** 695 (1970).

[19] B. K. Vonderhaar, U. H. Kim, and G. C. Mueller, *Biochim. Biophys. Acta* **208,** 517 (1970).

[20] G. C. Mueller, B. Vonderhaar, U. H. Kim, and M. LeMahieu, *Recent Progr. Horm. Res.* **28,** 1 (1972).

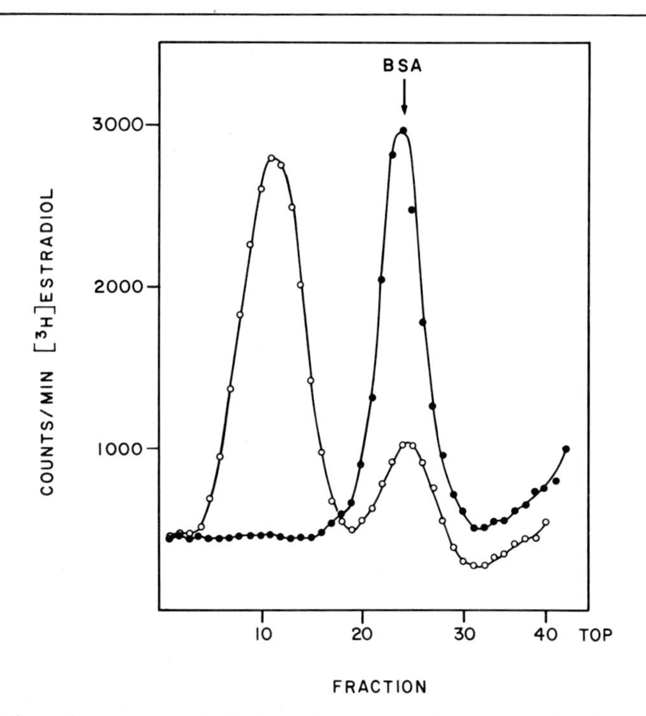

FIG. 4. Effect of aging on rat pituitary tumor cytoplasmic receptor. Cytosols were charged with 1×10^{-9} *M* [^{3}H]estradiol and sedimented immediately (○) or after 9 days at 4° (●) in TE 7.4 sucrose gradients.

S at 6 mg/ml (Fig. 5). The original observation of Shyamala Harris[21] suggests that other polyanions would produce the same effects as heparin. The presence of 0.4 *M* KCl in the gradient abolishes all effect, as would be expected for an ionic interaction. The effect operates on cytoplasmic receptor from rat pituitary and human mammary tumors[22] as well as from rat uterus, also on receptor extracted from nuclei (discussed later) and on cytoplasmic receptor partially purified by precipitation with 30% ammonium sulfate followed by passage over a Bio-Gel A1.5M column; it does not operate on the 4 S cytoplasmic receptor produced by aging. The molecular basis of the polyanion effect has been discussed[6] but no conclusions are possible yet.

Nuclear Receptor

Preparation in Vivo. Mature intact or ovariectomized female rats are given 0.1 μg [^{3}H]estradiol in 0.5 ml saline by intraperitoneal injection.

[21] G. Shyamala Harris, *Nature (London), New Biol.* **231,** 246 (1971).

[22] W. L. McGuire and M. DeLaGarza, *J. Clin. Endocrinol. Metab.* **36,** 548 (1973).

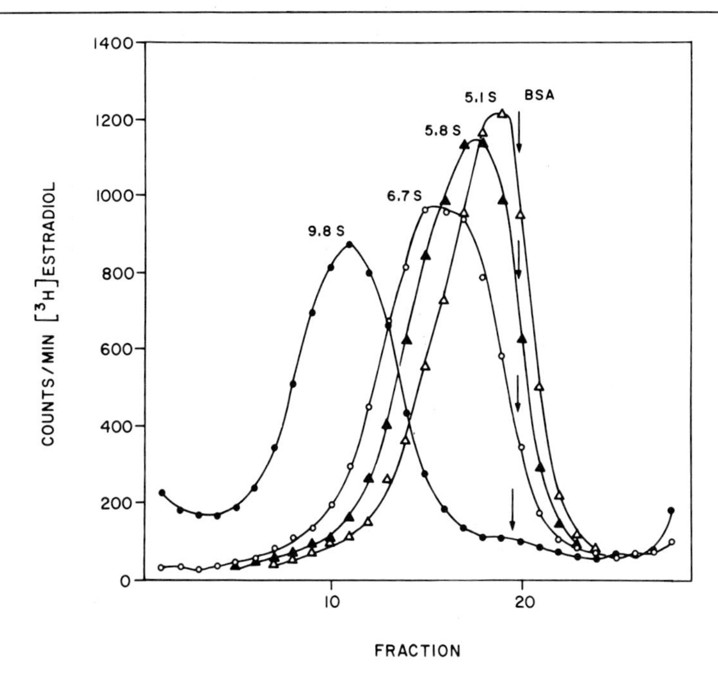

FIG. 5. Effect of polyanion on mature rat uterus cytoplasmic receptor. Immediately before layering of labeled cytosol on TE 7.4 sucrose gradients, sodium heparin was added to 0 mg/ml (●), 0.3 mg/ml (○), 1.5 mg/ml (▲), or 6.0 mg/ml (△). (From Chamness and McGuire.[6])

After 1 hour uteri or tumors are removed, chilled, homogenized in TE 7.4, and centrifuged at 140,000 g as described earlier. The rinsed pellet is rehomogenized in several volumes of TE 7.4 and recentrifuged. This washed pellet is then extracted by gentle homogenization in a small volume of TE buffer, pH 8.5, containing 0.4 M KCl, the homogenization being repeated in the same vessel three to four times during the course of 1 hour. The homogenate is centrifuged at 140,000 g for 20 minutes, after which the supernatant contains at least 50% of the receptor-bound [^{3}H]estradiol previously fixed to the nuclear pellet.

Effect of Ionic Strength. Receptor recovered from nuclei by salt extraction sediments at 4–5 S on sucrose gradients containing 0.4 M KCl. At lower ionic strengths there is no shift to higher sedimentation values which would parallel the behavior of cytoplasmic receptor, although a certain amount of random aggregate is formed in TE 7.4 alone (Fig. 6). Samples for centrifugation may be desalted by passage over a small column of Sephadex G-25 before layering, but omission of this step has no effect on the results.

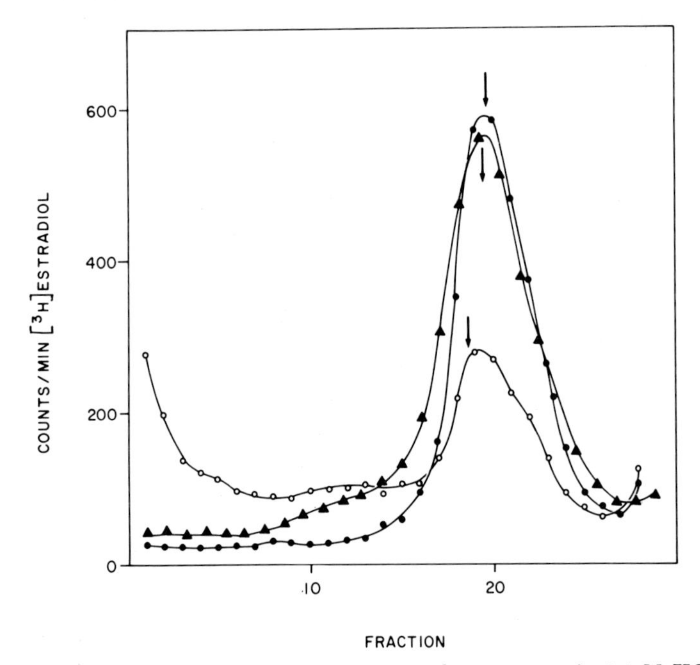

FIG. 6. Sedimentation of mature rat uterus nuclear receptor in 0.4 M KCl (●), 0.15 M KCl (▲), or TE 7.4 without added salt (○). Arrows mark the peaks of [^{14}C]BSA. (From Chamness and McGuire.[6])

Polyanion Treatment. Addition of 0.3 mg/ml sodium heparin to the nuclear extract prior to centrifugation in TE 7.4 sucrose gradients produces a sedimentation peak at 8 S and eliminates the random aggregation, while higher concentrations of heparin progressively reduce the sedimentation rate back toward 4 S (Fig. 7). When samples are desalted as above before addition of heparin, the resulting sedimentation values are identical at each heparin concentration to those found with cytoplasmic receptor.

Preparation by Cell-Free Incubation. From mature female rats ovariectomized two to three days previously, washed nuclear pellets are prepared as described above, using TE 7.4 with 0.15 M KCl for homogenization and washing. Cytoplasmic receptor is also prepared as described using TE 7.4 with 0.15 M KCl. The receptor is charged for 1 hour with either 1 or 4.5×10^{-9} M [^{3}H]estradiol and treated with dextran-coated charcoal. The washed nuclear pellet is homogenized directly in the charged cytosol and incubated at 25°. Experiments show that maximum binding occurs within 15 minutes under these conditions, and this is the

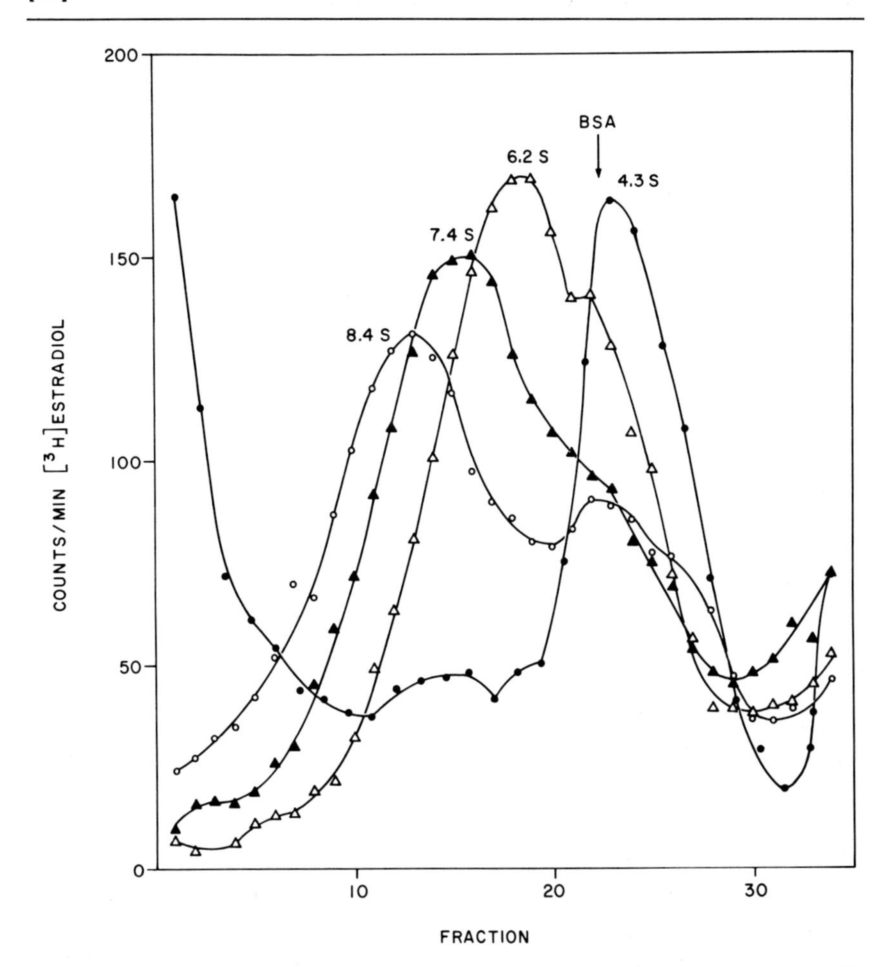

FIG. 7. Effect of polyanion on mature rat uterus nuclear receptor. Samples in TE pH 8.5 with 0.4 M KCl were treated as described in Fig. 5. (From Chamness and McGuire.[6])

period normally used. After incubation, 8 volumes of chilled buffer are added and the suspension is agitated vigorously before centrifugation for 20 minutes at 140,000 g. Further washing removes little if any [^{3}H]estradiol. The pellet is extracted as above, freeing approximately 25% of the bound charged receptor.

For some experiments, purified nuclei are prepared by homogenizing in at least 6 volumes 2.2 M sucrose containing 1 mM $MgCl_2$ and 0.1 mM Na_2EDTA using short bursts in the Polytron. After filtration through two layers of cheesecloth, the homogenate is centrifuged at 96,000

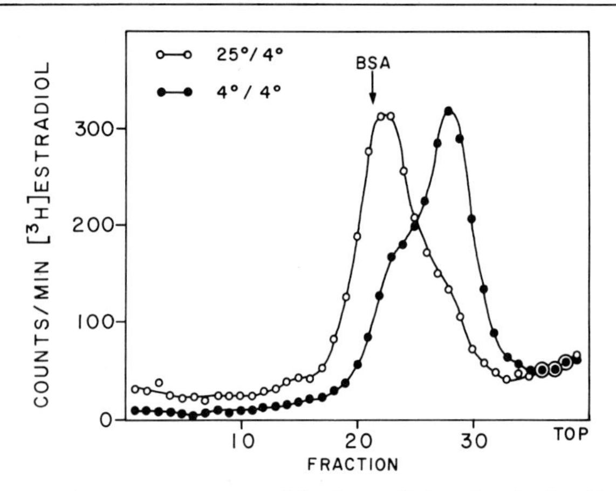

FIG. 8. Receptor from rat uterus nuclei after cell-free incubation in T 7.4 (without EDTA) at 4° for 60 minutes with untransformed rat uterus cytoplasmic receptor (●) or receptor transformed at 25° for 30 minutes (○). (From Chamness *et al.*[23])

g_{av} for 90 minutes. The pellet is gently resuspended in TE 7.4 with 0.15 *M* KCl and centrifuged at 140,000 *g* for 30 minutes. After this washing they are resuspended and incubated with cytoplasmic receptor and extracted as above.

The cell-free nuclear receptor shows the same 4–5 S sedimentation peak and the same response to ionic strength and to heparin as that formed *in vivo;* the appearance of a second peak at 3 S in these preparations will be discussed below.

Effect of Transformation on Cell-Free Binding to Nuclei. If charged rat uterus cytoplasmic receptor in T 7.4 (without EDTA) is transformed by warming at 25° for 30 minutes and is then administered to nuclei at 4° for 1 hour, the receptor recovered from the nuclei sediments solely at 4–5 S in sucrose gradients with 0.4 *M* KCl (Fig. 8) like nuclear receptor found *in vivo.* When the transformation is omitted, however, the recovered receptor sediments solely at 3 S. The 3 S peak found along with the 4 S after 25° incubation is therefore revealed as the result of nuclear binding of receptor not yet transformed.[23] The results are identical when purified nuclei are used.

Final Note

All the receptor forms discussed are artifacts in the sense that they exist under more or less nonphysiological experimental conditions. Almost

[23] G. C. Chamness, A. W. Jennings, and W. L. McGuire, *Nature* (*London*), *New Biol.* **241**, 458 (1973).

any form can be deliberately produced by appropriate manipulation.The interrelationships of these forms and their physiological significance are not yet understood.

Acknowledgments

This work was supported by The National Cancer Institute, The American Cancer Society, and The Robert A. Welch Foundation.

[17] Methods for Extraction and Quantification of Receptors

By WILLIAM T. SCHRADER

I. Introduction

The chick oviduct has been used as a model system for the study of progesterone action in target cells.[1-3] In this tissue, estrogen administration induces growth and differentiation; the estrogen-primed oviduct cells then respond to progesterone by synthesizing the egg protein avidin.[4,5] Studies on the mechanism of this progesterone response have shown that the hormone is first bound to intracellular macromolecules[6,7] and is subsequently transported to the nuclear compartment[8,9] where the hormone–macromolecular complex becomes tightly associated with the

[1] B. W. O'Malley, *Biochemistry* **6,** 2546 (1967).
[2] B. W. O'Malley, W. L. McGuire, P. O. Kohler, and S. G. Korenman, *Recent Progr. Horm. Res.* **25,** 105 (1969).
[3] B. W. O'Malley, W. T. Schrader, and T. C. Spelsberg, *in* "Receptors for Reproductive Hormones" (B. W. O'Malley and A. R. Means, eds.), p. 174. Plenum, New York, 1973.
[4] B. W. O'Malley and P. O. Kohler, *Proc. Nat. Acad. Sci. U.S.* **58,** 2359 (1967).
[5] B. W. O'Malley, G. C. Rosenfeld, J. P. Comstock, and A. R. Means, *Nature (London), New Biology* **240,** 45 (1972).
[6] M. R. Sherman, P. L. Corvol, and B. W. O'Malley, *J. Biol. Chem.* **245,** 6085 (1970).
[7] B. W. O'Malley, M. R. Sherman, D. O. Toft, T. C. Spelsberg, W. T. Schrader, and A. W. Steggles, *Advan. Biosci.* **7,** 213 (1971).
[8] B. W. O'Malley, D. O. Toft, and M. R. Sherman, *J. Biol. Chem.* **246,** 1117 (1971).
[9] B. W. O'Malley, M. R. Sherman, and D. O. Toft, *Proc. Nat. Acad. Sci. U.S.* **67,** 501 (1970).

cell chromatin.[10-13] These studies have been facilitated by use of partially purified progesterone receptors from the oviduct prepared as described below.[14,15]

It is important to keep in mind that there is as yet no true end-point assay for the functional form of a steroid receptor protein in any tissue. At least five parameters have been defined for the progesterone receptor system which may serve as substitutes for such an end-point assay.[16] These parameters should be monitored throughout all manipulations, in order to avoid working with materials which are so altered that they do not represent the starting material. The parameters used in this laboratory are (1) steroid-binding kinetics and specificity,[17] (2) molecular size and charge,[6,18,19] (3) uptake and retention by nuclei,[8,9,20] (4) binding to DNA,[21-23] and (5) binding to chromatin. Whenever a manipulation results in a deviation in one of these parameters, it should serve to prejudice the investigation toward defining the nature of the changes. No step should be adopted into the protocol which results in the isolation of an altered receptor form without a clear realization that the product may be biologically inactive.

II. Tissue Sources

Progesterone receptors can be isolated and identified from chick oviduct of any age or hormonal history. Immature oviducts of average weights of about 10 mg can be pooled for extraction of receptors. It is more convenient, however, to work with immature female chicks injected

[10] T. C. Spelsberg, A. W. Steggles, and B. W. O'Malley, *J. Biol. Chem.* **246,** 4188 (1971).

[11] A. W. Steggles, T. C. Spelsberg, and B. W. O'Malley, *Biochem. Biophys. Res. Commun.* **43,** 20 (1971).

[12] A. W. Steggles, T. C. Spelsberg, S. R. Glasser, and B. W. O'Malley, *Proc. Nat. Acad. Sci. U.S.* **68,** 1479 (1971).

[13] T. C. Spelsberg, A. W. Steggles, F. Chytil, and B. W. O'Malley, *J. Biol. Chem.* **247,** 1368 (1972).

[14] W. T. Schrader and B. W. O'Malley, *J. Biol. Chem.* **247,** 51 (1972).

[15] W. T. Schrader, D. O. Toft, and B. W. O'Malley, *J. Biol. Chem.* **247,** 2401 (1972).

[16] W. T. Schrader, this Vol. [17].

[17] S. G. Korenman, *Endocrinology* **87,** 1119 (1970).

[18] G. A. Puca, E. Nola, V. Sica, and F. Bresciani, *Biochemistry* **10,** 3769 (1971).

[19] W. I. P. Mainwaring and F. R. Mangan, *Advan. Biosci.* **7** (1971).

[20] R. E. Buller and D. O. Toft, *J. Biol. Chem.* (in press).

[21] K. R. Yamamoto and B. M. Alberts, *Proc. Nat. Acad. Sci. U.S.* **69,** 2105 (1972).

[22] D. O. Toft, *in* "Receptors for Reproductive Hormones" (B. W. O'Malley and A. R. Means, eds.), p. 85. Plenum, New York, 1973.

[23] J. D. Baxter, D. G. Rousseau, N. C. Benson, R. L. Garcea, J. Ito, and G. M. Tomkins, *Proc. Nat. Acad. Sci. U.S.* **69,** 1892 (1972).

subcutaneously for 7–21 days with 5 mg per day of diethylstilbestrol in sesame oil. The oviducts grow to 1–2 g each and contain no progesterone. There is no indication that the diethylstilbestrol competes for and masks any progesterone receptor sites. Mature hens can also be used; a typical laying hen oviduct weighing 20–40 g provides a large amount of receptors. However, due to the presence of high titers of endogenous progesterone the concentration of free receptor sites is normally only 10–25% that of the estrogenized immature oviduct. Attempts to free the progesterone sites by treatment of extracts with dextran-charcoal have met with mixed success at best. Hens which have recently stopped laying eggs undergo rapid regression of the oviduct, accompanied by rapid shut-off of progesterone secretion. Thus, recently laying hens having oviducts weighing 5–10 g frequently have high titers of free receptor sites. It must be pointed out, however, that the difficulty of working with hens makes the chick oviduct the tissue of choice.

Receptor preparations are best made from fresh tissue, rather than from stored or frozen sources since the frozen materials yield receptor preparations which tend to be highly aggregated. The progesterone-binding capacity of the tissues is not markedly decreased by freezing, however. No attempt has been made in our laboratory to preserve tissue in liquid nitrogen, nor to extract powdered oviducts.

Progesterone receptors are present in equal titers along the length of the oviduct, including the shell gland. In normal practice the oviduct is cut off anteriorly where it constricts and posteriorly at the broadening, marking the transition to shell gland. It is assumed that the receptors are distributed throughout the oviduct, although to our knowledge no study has been published to measure receptor titers in epithelial vs. stromal or glandular elements.

III. Preparation of Crude Cytoplasmic Receptors

A. Buffers

The standard buffer used is 10 m*M* Tris-HCl, pH 7.4, containing 1 m*M* Na_2EDTA and 12 m*M* 1-thioglycerol (Buffer A). This is supplemented with various amounts of KCl as indicated below. Receptor preparations have also been made using 10 m*M* phosphate buffer with no differences in properties. Thioglycerol is added to stabilize one receptor form obtained at the DEAE-cellulose step[14] (see below). β-Mercaptoethanol can be used equivalently, but glycerol alone (0–30%) is not effective in stabilizing the DEAE eluates.

B. Tissue Removal and Homogenization

Oviducts are cut from chicks freshly killed by cervical dislocation and are obtained by a ventral incision. The process takes less than a minute for each chick. The oviducts should be clipped free of mesentery, blotted, and placed immediately in ice-cold 0.9% NaCl saline solution. The tissue is then blotted, weighed, minced in the cold room, and homogenized using a Polytron (Brinkmann Instruments, PT-10).

It is important to maintain the solution at 0°. This is best done by homogenizing always in ice-water slush rather than crushed ice alone, and by using bursts (5–10 seconds) of the Polytron with cooling between bursts. For preparative studies a few strokes of a large Teflon-glass homogenizer run at low speed in an ice-water bath will improve the yield. The crude homogenate is then centrifuged in a swinging-bucket rotor (Sorvall HB-4) for 10 minutes at 10,000 rpm. The layer of floating lipid is drawn off and the crude low-speed cytoplasmic material decanted. The crude nuclear pellet is stored frozen for preparation of DNA.

C. Preparation of Cytoplasmic Soluble Fraction (Cytosol)

The low-speed supernatant fraction is then centrifuged at 150,000 *g* for 1 hour in a Spinco swinging-bucket rotor (SW-50.1, SW-27, or SW-40 depending on volume) to prepare the cytosol. Another small fat plug is found after this centrifugation and is carefully drawn off in the cold room by aspiration. The cytosol prepared in this manner is largely free of lipid material and gives much cleaner preparations later on. It has been found useful to avoid angle rotors at all times, even for large batches because of the cleanliness of the resultant solutions.

D. Labeling of Cytoplasmic Extracts with Radioactive Steroids

One milliliter of cytosol prepared as described above will contain about 20 mg of protein and will contain about 10 nmoles of progesterone receptor sites. This is about 75 ng of steroid bound at saturation, or 2×10^6 dpm of [^{3}H]progesterone at 50 Ci/mmole.

To saturate these preparations of cytosol with progesterone it is usually convenient to add about a 10-fold excess of [^{3}H]progesterone. This steroid is soluble at about 5 μg/ml and is generally sold as a benzene solution at about 5 μg/ml. To prepare a suitable stock solution, the [^{3}H]progesterone is evaporated under N_2, redissolved in 100% ethanol, diluted 1:10 with Buffer A, and stored frozen. One milliliter of cytosol is then labeled by adding 10 μl of diluted [^{3}H]progesterone. At 0° complete equilibrium requires about 12–18 hours. However, for preparative studies, labeling times of 2–6 hours are usually sufficient.

It is also possible to label the preparations by adding [^{3}H]progesterone to the low-speed supernatant fraction before preparation of the cytosol. However, substantial metabolism of steroids can occur in these perticulate preparations even at 0°. Therefore, labeling of cytosol itself is routinely used. Under these conditions, about 90% of the input steroid is still authentic progesterone after 6 hours. Progesterone complexed to receptor is even more stable; there is essentially no metabolism of the hormone in purified preparations when the receptor-bound material is assayed.

E. Stability of Cytosol Receptors

1. Time

Cytosol receptor preparations generally can be stored at 0° for about 24 hours before the receptor properties dramatically deteriorate. The state of aggregation of the receptor is affected by storage, and sucrose gradient analysis of aged material shows no loss of label from binding sites, but progressively more appearance of receptor at the bottom of the gradient. Gel filtration on Sephadex G-75 also shows no increase in free hormone but a loss of label in the bound fractions, indicative of conversion of the receptor to an altered state which adsorbs to the gel or to the apparatus. Small amounts of detergents (Triton X-100 or deoxycholate, 0.1%) have been tried to prevent this aggregation without success. The aggregation phenomenon is time-dependent. Cytosol has been frozen and thawed repeatedly on the same day with no loss of activity or aggregation, whereas storage in the frozen state caused aggregation within 5 days. The frozen cytosol can be used for steroid-binding competition studies even in the aggregated state. Apparently there is no deterioration of the steroid-binding sites themselves.

2. Temperature

In general, elevated temperature, particularly in the absence of bound steroid, promotes denaturation of progesterone receptors. Bound progesterone labeled to saturation has a half-life as receptor complex of many days at 0°, 6 hours at 23°, and only about 30 minutes at 37°. The process is irreversible at the higher temperatures, as evidenced by the failure of hormone to re-form a complex during subsequent storage at 0°. This denaturation is not merely protease activity, however, since the process has also been observed in highly purified preparations.

A most useful storage method has been simply to lyophilize the labeled receptor preparations. If known volumes are freeze-dried, they can be stored for months and redissolved by adding cold H_2O. Further purification of material stored in this way has shown the receptors to be intact after lyophilization; they have not been tested in the crude state for uptake by nuclei.

3. pH

The progesterone–receptor complexes show a marked pH optimum for stability. They are stable over a broad pH range from pH 7–9, but over half the binding activity is destroyed at pH 6. This is found whether intact progesterone–receptor complexes are lowered at pH 6, or whether nascent receptors are treated at this pH and then assayed for binding activity. The process is also essentially irreversible, since aliquots treated at pH 6 and then incubated at pH 7.4 do not reform the steroid–receptor complex. Finally, if cytosol is brought to pH 4.5, a precipitate forms consisting of ovomucoids. Assay of the precipitate showed no receptor activity in the sedimented material. It should be pointed out that no attempts have been made to raise pH gradually, such as by dialysis which might allow some of the complexes to re-form.

4. Salt and Urea

High concentrations of either KCl or urea cause dissociation of progesterone–receptor complexes. Exposure to 2.0 *M* KCl for 1–2 hours causes a 50% loss of bound progesterone. This is apparently not an irreversible effect since progesterone–receptor complexes precipitated at 50% saturation of ammonium sulfate (2.0 *M*) remain complexed during brief exposure to the salt. Low concentrations of KCl (0.15, 0.3 *M*) are favorable solutes to include in receptor preparations since receptors are maintained in a disaggregated state above about 0.15 *M* KCl.

The relationship between receptor aggregation and the concentration of divalent cations such as Ca^{2+} remains to be established. Stabilization of slowly sedimenting receptor forms by the addition of 4 m*M* $CaCl_2$ has been found to occur with the oviduct progesterone receptor as with other receptors. But these calcium effects have not been correlated with the isolation of the receptor forms reported here.

Urea also dissociates the progesterone–receptor complexes in a reversible process. Exposure of labeled cytosol to 1.0 *M* urea for 6 hours had no effect on the concentration of bound hormone. However, a 50% drop in binding (as determined by chromatography on Sephadex G-75) occurred at 2.0 *M* urea. If the preparation was diluted 10-fold, the [^{3}H]progesterone did not rebind to the receptors. However, if cytosol treated overnight at 3.0 *M* urea was dialyzed to remove the urea, the dialyzed receptor preparation could rebind about 50% of the original concentration of progesterone. Thus the urea denaturant appears to be similar to that observed for other proteins; gradual removal of the agent causes at least some recovery of activity.

F. Characteristics of the Cytosol Receptors

Progesterone receptors prepared in this way sediment on 5–20% sucrose gradients in Buffer A as a 6–8 species showing a broad peak and considerable aggregated material at the tube bottom. If the gradients are run in 0.15 *M* KCl or higher, a sharper peak at about 4 S is seen. This pattern can also be observed if the cytosol is sedimented without added hormone, and each fraction is subsequently assayed for binding sites. Thus the 4 S–8 S relationship is not progesterone-dependent. Furthermore, the pattern can be seen in either unprimed or estrogenized oviduct extracts, and thus is not directly estrogen-dependent either.

The receptors are acidic proteins, having isoelectric points of between pH 4 and pH 4.5. At neutral pH they have an equilibrium constant for progesterone binding of 1×10^{-10} *M* and a half-life of 12 hours under pseudo first-order conditions.[14] The binding constants are summarized in Table I.

IV. Binding-Site Measurements

A. Assays for Bound Hormone

1. *Ammonium Sulfate Precipitation*

If a [^{3}H]progesterone-labeled receptor solution is brought to 50% saturation in ammonium sulfate, the receptor–hormone complexes precipitate and can be collected by centrifugation. This method is convenient because the reagent is soluble rather than particulate (as in the charcoal precipitation method) and because both the pellet (receptor–hormone complex) fraction and supernatant (free hormone) fraction can be counted easily for radioactivity. The method is gentle and can be used with purified receptor preparations. In contrast, the charcoal technique seems to dissociate the complexes or adsorb receptor molecules, especially when used to assay purified preparations.

It is advisable to use polypropylene tubes for steroid receptor assays since the receptors have a tendency to adhere to glass, especially when

TABLE I
APPARENT KINETIC CONSTANTS FOR OVIDUCT RECEPTOR–PROGESTERONE INTERACTION[a]

Receptor fraction	Binding site concentration[b] (nM)	Association rate constant[c] $10^{-5} \times k_a$ (M^{-1} sec^{-1})	Dissociation rate constant[d] $10^5 \times k_d$ (sec^{-1})	Equilibrium constant[e] $10^{11} \times k_d/k_a$ (M)	Equilibrium constant[f] $10^9 \times k_d$ (M)
Crude cytosol	4.6	4.0	1.9	4.8	4.4
Component A	0.7	2.8	1.9	7.0	—
Component B	0.8	6.3	2.4	3.8	—

[a] From Schrader and O'Malley.[14]
[b] Determined from saturation concentration of receptor-bound ^{3}H.
[c] Second-order rate constant obtained from slopes of plots in Fig. 5A.[14]
[d] First-order constant obtained from slopes of plots in Fig. 5B.[14]
[e] Calculated as ratio of rate constants for the dissociation reaction $RH^* \rightarrow R + H^*$.
[f] Calculated from Scatchard plot slope.

purified. This holds true for DNA and chromatin-binding assays as well. Tubes 12 × 75 mm (Falcon Plastics #2053) are suitable.

The protocol used for the ammonium sulfate assay is as follows: 10–100 μl of a receptor preparation is diluted in 0.5 ml with buffer and precipitated by rapid addition of an equal volume of 100% saturated ammonium sulfate in Buffer A. After 30 minutes the tube is centrifuged at 2000 g for 15 minutes. The supernatant fraction is decanted, and an aliquot counted for ^{3}H to determine free steroid in the preparation. The tube is washed with 2 ml of 50% saturated ammonium sulfate in Buffer A, recentrifuged, and the supernatant fraction discarded. The small pellet is dissolved in 700 μl H_2O and counted for ^{3}H to determine the receptor–hormone complex concentration in the preparation.

2. *Charcoal Mixed with Dextran*

Charcoal mixed with dextran has been used extensively for steroid receptor determinations. This method is described in detail elsewhere.[17,24] It is rapid and precise when used to assay crude cytosol receptor levels or when used for steroid competition studies. However, during assays of purified preparations even brief exposure to charcoal causes loss of bound hormone and a concomitant loss of receptor sites in solution. The addition of egg-white protein or bovine serum albumin does not reverse this deleterious effect. The method is thus best used with great caution for assay of purified materials.

[24] S. G. Korenman, this Vol. [4].

3. Sucrose-Gradient Analysis

Sucrose-gradient analysis can be used to determine the binding of hormones to receptors. Due to the time required (9–16 hours) and the fact that it is under nonequilibrium conditions the technique undoubtedly gives estimates of receptor-bound hormone which are too low. Other drawbacks include poor (<50%) recovery of [^{3}H]progesterone when cellulose nitrate centrifuge tubes are used and the small number of samples which can be processed daily. However, since the method offers much more information than simpler methods,[25,26] it is usually included in any receptor study.

4. Gel Filtration

Gel filtration techniques can also be used to assay bound [^{3}H]steroids. Short columns of Sephadex G-75 built in 5-ml disposable syringes are convenient. Two precautions must be mentioned which are described in greater detail elsewhere in this volume.[27] (1) Steroids tend to adsorb to Sephadex and hence radioactivity recovery will not be 100%. This problem is minimized by addition of 0.3 *M* KCl to the eluting buffer. (2) Receptor–hormone complexes adsorb to the gel and also to the filters at the top and bottom of the column. Filters of 200-mesh stainless steel can be used effectively with little loss of receptor during passage through the column. Samples of receptors (about 100 μl) can be resolved into bound and free radioactive fractions by chromatography on 5-cm Sephadex G-75 columns at 0°. About 15 ml of elution buffer is necessary to develop the chromatogram and insure complete elution of all radioactivity. Each fraction is counted for ^{3}H, the elution profile plotted, and the area under the excluded (V_0) and included (V_i) radioactivity peaks determined by cutting out the curve and weighing the peaks on an analytical balance.

B. Scatchard Plot Estimation of K_a and N_{max}

If nascent receptors containing no hormone are to be assayed, it is necessary to construct a Scatchard plot since it is never possible to saturate and detect all the receptor sites in the solution. Saturation is approached asymptotically. Furthermore, in a real system there are competing nonspecific protein–ligand interactions which can bind large quanti-

[25] D. O. Toft and M. R. Sherman, this Vol. [14].
[26] R. G. Martin and B. N. Ames, *J. Biol. Chem.* **236**, 1372 (1961).
[27] M. R. Sherman, this Vol. [18].

ties of the steroid. Thus, it is only possible to determine the saturation value for the number of sites (as is true for V_{max} of an enzyme) by an extrapolation. One method for doing this is the Lineweaver-Burk plot, [bound hormone]$^{-1}$ vs. [free hormone]$^{-1}$. Another method, more useful for protein–steroid interaction, is the Scatchard plot, [bound hormone]/[free hormone] vs. [bound hormone]. A detailed analysis and derivation of this equation is given elsewhere.[28-30]

It is convenient to use 50 μl of cytosol diluted to 500 μl with buffer containing 10^3-10^6 cpm of [^{3}H]progesterone. After overnight incubation, receptors are precipitated with ammonium sulfate as described above. The radioactivity in the pellet and supernatant fractions are converted to molarity and used to construct the Scatchard plot. The plot for a single interacting species of binding activity is a straight line of negative slope with an X-intercept of N_{max}, the molarity of binding in solution, and a slope equal to K_a, the equilibrium constant for the association reaction.

C. Steroid Binding Specificity

This is assayed by a competition method in which [^{3}H]progesterone (10,000 cpm) and varying amounts of unlabeled steroids (0.1 to 100 ng) are added simultaneously to a receptor preparation containing a small number of binding sites. After overnight equilibration, receptor-bound ^{3}H is determined by precipitation with ammonium sulfate as described above. A semilogarithmic plot of percent bound [^{3}H] progesterone vs. the log of the competitor mass added yields a competition curve for each steroid. The amount of steroid necessary to decrease [^{3}H]progesterone binding by 50% is determined from the curves and compared to the amount of authentic unlabeled progesterone required for a similar decrease. The technique of steroid-binding affinity analysis is described in detail elsewhere.[24,31]

D. Rate Constants for Association and Dissociation

For the association reaction of receptor (R) and labeled hormone (H^*) forming complex (RH^*)

$$R + H^* \underset{k_d}{\overset{k_a}{\rightleftharpoons}} RH^* \tag{1}$$

[28] D. Rodbard and Feldman, this Vol. [1].

[29] D. Rodbard, *in* "Receptors for Reproductive Hormones" (B. W. O'Malley and A. R. Means, eds.), p. 289. Plenum, New York, 1973.

[30] G. Scatchard, *Ann. N.Y. Acad. Sci.* **51**, 660 (1949).

[31] D. Rodbard and J. E. Lewald, *Acta Endocrinol: (Copenhagen), Suppl.* **147**, 79 (1970).

At short reaction times, such that dissociation of RH^* (the back reaction) is negligible, the rate equation for this second-order reaction is

$$\frac{dX}{dt} = k_a(T - X)(S - X) \tag{2}$$

where X is the concentration of RH^* complex at time t (the species measured), T is the initial concentration of hormone (total H^* added), S is the initial concentration of receptor-binding sites (concentration of receptor-bound hormone at saturation), and k_a is the association constant in conc^{-1} time^{-1}. In integrated form,

$$\frac{1}{T - S} \ln \frac{T - X}{S - X} = k_a t + \frac{1}{T - S} \ln \frac{T}{S} \text{ for } T > S \tag{3}$$

thus, k_a can be obtained as the slope of a plot of

$$\frac{1}{T - S} \ln \frac{T - X}{S - X} \text{ vs. } t \tag{4}$$

The dissociation of RH^* is studied by preventing the reassociation of receptor and labeled hormone if a dissociation of the complex occurs. This is done by adding a large excess of unlabeled progesterone to the preparation.

Thus the reaction studied is:

$$RH^* + H \underset{k_a}{\overset{k_d}{\rightleftharpoons}} RH + H^* \tag{5}$$

When $H \gg RH^*$, the reaction is pseudo first order and the rate law is

$$\frac{dX}{dt} = k_d(X) \tag{6}$$

In integrated form,

$$\ln X = -k_d t + \ln X_0 \tag{7}$$

where k_d is the dissociation constant in time^{-1} and X_0 is the concentration of RH^* at time = 0. Thus, $-k_d$ is obtained as the slope of a plot of ln X vs. t.

The equilibrium constant for reaction (1) is the ratio of the rate constants:

$$K_d = k_d/k_a \tag{8}$$

It must be stressed that the derived K_d obtained in this way will agree with that obtained by equilibrium methods (i.e., Scatchard plots) only if the reaction proceeds by a pathway adequately defined by the rate equations. This question is treated in greater detail elsewhere.[28,32] More complex behavior will be detected as differences between the two values. This is in fact observed with the case of the progesterone receptor, where the two values of K_d differ by two orders of magnitude, as shown in Table I.

1. Association Rate Constant

To start the binding reaction, a 1.0-ml sample of an unlabeled receptor fraction is mixed with 10 μl of various dilutions of [^{3}H]progesterone in H_2O. The binding reaction is stopped by removing 100-μl aliquots at specific times and adding each to a tube containing 10 μl of unlabeled progesterone (1 μg) in 10% ethanol. The concentrations of bound and free hormone are then determined immediately by the ammonium sulfate assay. The results are plotted as described above, yielding a rate constant from the slope of the line.[33]

2. Dissociation Rate Constant

This is measured in a receptor sample which has been saturated with [^{3}H]progesterone to fill virtually all the receptor sites. To a 1.0-ml sample of receptor, 10 μl of unlabeled progesterone (1 μg) in 10% ethanol is added to start the dissociation experiment. At various times after the addition of unlabeled progesterone, 100-μl aliquots are removed and assayed for bound and free [^{3}H]progesterone by the ammonium sulfate assay. Semilogarithmic plots of bound ^{3}H vs. time give straight lines for the pseudo first-order dissociation reaction. The dissociation rate constant is determined from the slope of the curve.

V. Partial Purification

The initial steps of purification are outlined in Table II. Steps after the DEAE-cellulose procedure are not included, since the protein concentrations were too low to measure. These steps were done using an ammonium sulfate precipitate as starting material.

A. Ammonium Sulfate Precipitation

The crude cytosol receptor preparation can be purified 20–30-fold in about 70–80% yield by precipitation of the receptors at 30% saturation

[32] I. M. Klotz and D. L. Hunston, *Biochemistry* **10,** 3065 (1971).

[33] M. Best-Belpomme, J. Friès, and T. Erdos, *Eur. J. Biochem.* **17,** 425 (1970).

TABLE II
PURIFICATION OF OVIDUCT PROGESTERONE RECEPTOR COMPONENTS[a]

Fraction	Volume (ml)	^3H Bound to receptor[b] (10^{-6} × cpm/ml)	Protein[c] (mg/ml)	Specific activity (10^{-3} × cpm bound/mg protein)	Receptor recovery[d] (%)	Fold purification
Unfractionated total oviduct homogenate	45	0.44	88	5	100	1
Cytosol—105,000 *g* supernatant fraction (S_1)	37	0.39	16.6	23	73	6
30% Sat'd ammonium sulfate precipitate (P_1)	1.5	6.5	10.4	620	49	124
Agarose A-0.5M binding peak, concentrated	25.8	0.15	0.08	1800	19	360
DEAE-cellulose receptor component A	10	0.11	0.03	3700	5	780
DEAE-cellulose receptor component B	10	0.18	0.012	15000	9	3000

[a] From Schrader and O'Malley.[14]
[b] Binding activity determined by Sephadex G-75 or charcoal-dextran assay.
[c] Protein determined by the method of Lowry *et al.* [O. H. Lowry, N. J. Rosebrough, A. L. Farr, and R. J. Randall, *J. Biol. Chem.* **193**, 265 (1951).]
[d] Total bound cpm compared to crude homogenate.

of ammonium sulfate. Saturated ammonium sulfate, neutralized with NH_4OH, is prepared in Buffer A and added dropwise to stirred cytosol in ice until the final solution is 30% saturated in ammonium sulfate. The solution turns cloudy but not opaque, and rigorous removal of lipid during the preparation of the cytosol results in much less visible precipitate at this step. After 30 minutes the white precipitate is collected by centrifugation in a Sorvall HB-4 rotor at 10,000 rpm for 20 minutes. The use of the swinging-bucket rotor enhances recovery and provides a small, compact pellet. The straw-colored supernatant fraction contains any free steroid as well as corticosteroid-binding globulin (CBG) and a few percent of the starting progesterone receptor. Some yield is sacrificed at this step to exclude recovery of CBG in the pellet. This method can be used to precipitate either progesterone–receptor complex or the nascent, uncomplexed receptors. Yields appear to be about the same for both materials.

The tubes containing the pellets are kept rigorously cold, and the inside walls above the pellet rinsed under a stream of cold H_2O, after which the walls are wiped dry. This procedure removes most of the ammonium sulfate and is necessary if the material is to be chromatographed on DEAE-cellulose in good yield. The tubes can be frozen dry at —20° and stored for a period of weeks with no detectable change in hormone-binding activity. This is thus a convenient storage step, as it permits stockpiling of either labeled or nascent receptors.

The receptor pellets are redissolved in Buffer A in ice. After about 30 minutes the denatured material is removed by brief centrifugation. Hormone in the supernatant consists of about 95% receptor complex, with only a small amount of free steroid.

B. Agarose Gel Filtration

Receptors at various stages of purification have been chromatographed on agarose A-0.5M (Bio-Rad Laboratories). The receptors aggregate if run in buffers of low ionic strength and dissociate rapidly if run without agents like 1-thioglycerol. Buffer A containing 0.3 *M* KCl has been found to be effective. Some aggregated material is observed eluting at V_0 even in this buffer. Partially purified preparations show a receptor peak which is broader than expected and the peak sometimes is observed to split, suggesting resolution of multiple forms of the receptor. Gel filtration of receptors is not simple to analyze due to the interaction of receptor with the gel.[27]

Gel filtration of oviduct progesterone receptor has been done in columns 2.6 × 80 cm, as shown in Fig. 1A. Under these conditions about a two- to threefold purification is achieved. The yields of receptor–hormone complex are about 40%, due in part to dissociation of some complexes during the experiment.

The pooled peak fractions of receptor–hormone complex can be dialyzed, diluted 10-fold with Buffer A, or subjected to ultrafiltration to alter salt or volume conditions to suit the next procedure.

C. Dialysis and Ultrafiltration

Dialysis of receptors rapidly removes salts and free hormone, but has little effect on the receptor–hormone complexes themselves. This may be a consequence of progesterone adsorbing to the dialysis bag, thus maintaining an artificially high concentration of free steroid in the bag. Dialysis for a long as 48 hours does not remove bound progesterone if the dialysis is done at 0°.

Ultrafiltration is an effective means of concentrating receptor preparations. In this laboratory several commercial ultrafiltration cells have been used (Amicon) in conjunction with certain membranes (Amicon XM-50). Care must be exercised in the choice of membranes, since the PM and UM series of membranes (including the centrifugation cones) adsorb steroids and result in decreased receptor recovery.

D. DEAE-Cellulose

Receptors are acidic proteins and can be chromatographed on DEAE-cellulose. The resin (Whatman DE-52) is washed and equilibrated in Buffer A by standard procedures. If 10-fold concentrated Buffer A is used for the initial washes, the EDTA concentration should be reduced since it will adsorb to the DEAE and elute during the experiments at 0.15 *M* KCl.

The progesterone receptor preparations can contain salt up to about 0.075 *M* KCl. About 1 ml of resin bed is used for receptors from 500 mg of oviduct tissue. A 10-ml column of DEAE-cellulose is prepared in a 50-ml plastic syringe. The resin bed should be no higher than the diameter of the column. Millipore glass prefilters or stainless-steel screen are used above and below the resin. A receptor solution is applied and washed extensively with Buffer A. If cytosol is chromatographed directly on DEAE, a yellow chromophore band will be observed, and it will slowly wash through the column if enough Buffer A is applied. Elution of the receptors is not begun until this yellow band has been removed.

Free progesterone does not bind to DEAE-cellulose. Therefore the wash is monitored for radioactivity and the elution begins only after the

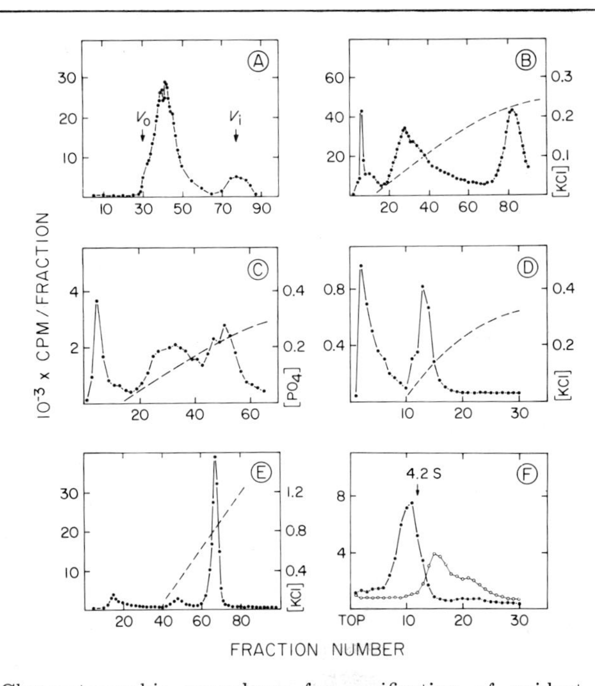

FIG. 1. Chromatographic procedures for purification of oviduct progesterone receptor. [^{3}H]Progesterone was measured in all cases in toluene based scintillation fluid containing toluene–Triton X-100-Spectrafluor (Amersham-Searle) 1000:500:42 (v/v/v) with a counting efficiency of 33%. All procedures were carried out at 0°. Salt concentrations determined using a Radiometer conductivity meter (- - -). (A) Agarose gel filtration using 80 × 2.6-cm column in Buffer A–0.3 *M* KCl. Exclusion limits determined with Blue Dextran 2000 (Pharmacia) and [^{3}H]progesterone. Sample (1.0 ml) chromatographed into 6.0-ml fractions. Fractions 35–45 pooled and concentrated by ultrafiltration over Amicon XM-50 membrane. From Schrader O'Malley.[14] (B) DEAE-cellulose column chromatography of progesterone receptors eluted by KCl gradient elution. Whatman DE-52 exchanger equilibrated in Buffer A was used in a 2.0 × 5.0-cm column. Receptors in Buffer A applied and washed through column before starting elution using 300 ml KCl gradient. Fractions collected were 3.0 ml each. The first peak (Fractions 20–40) was pooled and designated Receptor Component A; the second peak (Fractions 70–90) was pooled and designated Receptor Component B. These could also be collected by stepwise KCl elution at 0.15 *M* and 0.3 *M* KCl, respectively. From Schrader and O'Malley.[14] (C) Hydroxylapatite column chromatography of progesterone receptors. Clarkson Hypatite C was equilibrated in a column (2 × 1.5 cm) with 0.01 *M* Na_xPO_4–0.012 *M* 1-thioglycerol, pH 7.4. Progesterone receptors precipitated with ammonium sulfate and redissolved in the above buffer were applied and washed through the column. Fractions (2.0 ml) were collected. A phosphate gradient (0.01–0.3 *M*) of 100 ml was used to elute the column. Free progesterone eluted in the wash; both peaks consist of receptor–hormone complexes. (D) DNA-cellulose chromatography. Chick oviduct DNA prepared as outlined elsewhere in this volume and coupled to Cellex N-1 (Bio-Rad) by the procedure of Alberts.[34] A 1.0-ml column containing 400 μg

unbound radioactivity falls to a suitable level. When unlabeled receptor preparations are to be chromatographed, a wash of Buffer A five times the bed volume is adequate.

Elution of the receptors is done using Buffer A containing KCl by varying the KCl concentration. Either linear KCl gradient elution or stepwise elution can be used. The total volume of eluate when using gradient elution should be at least 10 times the bed volume of the column. As shown in Fig. 1B, two distinct peaks are eluted. These are identified as receptors A and B, and elute with KCl at 0.12 *M* and 0.22 *M*, respectively. Stepwise elution at 0.15 *M* KCl elutes A, while 0.3 *M* KCl can be used to elute B. The bulk of the protein elutes in a peak with component A.

The two components elute in about equal amounts (assuming equivalent numbers of binding sites on both species). Component A is noticeably less stable than B, and yields of A are frequently poor. The reason for this variability has thus far remained obscure, although it has been observed that component A is absent if thioglycerol is deleted from Buffer A.

The two materials do not interconvert when rechromatographed on DEAE-cellulose after dilution. They are apparently only weakly associ-

---

of adsorbed DNA was equilibrated in Buffer A. Progesterone receptors precipitated with ammonium sulfate and redissolved in 1.0 *M* Buffer A were applied to the column and washed through in 2-ml fractions. A 50-ml KCl gradient was used to elute the bound receptors. The unbound wash fractions contained free progesterone and a receptor component which could be eluted from DEAE-cellulose at 0.22 *M* KCl (Component B). The receptor component bound to DNA and eluting at 0.15 *M* KCl could be eluted from DEAE-cellulose at 0.1 *M* KCl and was thus identified as Component A. (E) Protamine–agarose chromatography. Progesterone receptor was prepared by stepwise elution from DEAE-cellulose at 0.3 *M* KCl (Component B). Salmon sperm protamine (Sigma) was coupled to aminoethyl agarose (Bio-Rad AF101) by the procedure of Cuatrecasas using 1-ethyl-3-(3-dimethylaminopropyl) carbodiimide HCl. The reaction is done at low pH to decrease reactivity of the protamine guanidino groups. The beads are packed in a 10×2.6-cm column and washed exhaustively with Buffer A, 0.1 *M* KCl. The receptor in 0.3 *M* KCl was diluted to 0.1 *M* KCl and applied to the column in a volume of 20 ml and the column was washed with Buffer A, 0.1 *M* KCl, to remove free hormone. Fractions were 2.0 ml each. The column was eluted using a 650-ml KCl gradient in Buffer A from 0.1 to 1.2 *M* KCl. Progesterone receptors eluted at 0.8 *M*. (F) Sucrose-gradient analysis of crude cytoplasmic receptors. Labeled cytosol (200 μl) was layered on 5 to 20% sucrose gradients in Buffer A alone (○—○) or with the addition of 0.3 *M* KCl (●—●). The tubes were centifuged at 45,000 rpm for 16 hours in a Spinco SW 50.1 rotor. Human hemoglobin was used as a 4.2 S marker. The tubes were pierced and 10-drop fractions were collected and counted for ^3H.

ated, since application of cytosol to DEAE-cellulose in Buffer A-0.15 *M* KCl results in elution of component A in the dropthrough fraction and retention of component B on the column.

Stepwise elution with KCl results in preparations of very high activity. Peak fractions may contain upwards of 2 million cpm/ml.

Both components A and B bind progesterone in identical fashion, and thus probably contain the same binding site moiety. They both sediment as 4 S molecules in 5 to 20% sucrose gradients containing greater than 0.15 *M* KCl. In sucrose gradients without KCl, component A is unstable and aggregates whereas component B remains a 4 S species. Finally, both materials cochromatograph on agarose A-0.5M columns with native cytosol receptor.

If the peak radioactive tubes from a KCl gradient elution are pooled, purification of component A is about fourfold and purification of component B is about 20-fold. Yields vary depending on how much of each peak is pooled, but with stepwise elution, yields are typically 70% or more.

E. Hydroxylapatite Chromatography

Progesterone receptors will adsorb readily to hydroxylapatite equilibrated in 0.01 *M* phosphate buffer at pH 7.4. Other salts have little effect on the binding, and the method can thus be readily used to desalt receptor preparations. Partially purified preparations can be eluted stepwise from hydroxylapatite in excellent yields using 0.3 *M* phosphate buffer. A purification of fourfold is achieved by this procedure.

Phosphate gradient elution chromatography of purified progesterone receptor precipitated with ammonium sulfate is shown in Fig. 1C. A 7-ml column was charged with receptor dissolved in Buffer A—0.3 *M* KCl. As in the case of DEAE-cellulose chromatography, two peaks of radioactivity were bound to the column. They eluted at 0.1 *M* and 0.25 *M* phosphate, respectively, again in equal amounts.

F. DNA-Cellulose Chromatography

Cellulose can be used as an inert support for DNA-affinity chromatography. This technique is useful to isolate receptor components which bind to DNA. The DNA is noncovalently coupled to washed, non-ionic cellulose by a drying procedure described by Alberts.[34] Chick DNA has been used, although calf DNA works equivalently. A 5-ml column of

[34] B. Alberts and G. Herrick, Vol. 21 [11].

DNA-cellulose is used to chromatograph receptors from precipitation with ammonium sulfate. The DNA column first is equilibrated in Buffer A, and the receptor preparation applied and washed through the column in the same buffer. About half of the radioactivity fails to bind to the column, although the steroid in this fraction is tightly bound to macromolecules. Subsequent chromatography of this drop-through material on a DEAE-cellulose column showed that this fraction eluted at 0.22 *M* KCl, and thus is identical to receptor component B. This finding is consistent with other studies of DNA binding by progesterone receptors, which demonstrate that component B does not bind to pure DNA.[16]

Elution of the DNA-cellulose column using stepwise KCl elution in Buffer A showed that all the receptor–hormone complex elutes with 0.15 *M* KCl. No receptor-bound radioactivity elutes at 0.5 *M* or 1.0 *M* KCl. Substantial radioactivity elutes if the column is washed finally with 100% ethanol. However, this ethanol peak represents free progesterone bound to the cellulose; this binding occurs with free progesterone alone and does not require DNA on the cellulose.

Gradient elution of a DNA-cellulose column is shown in Fig. 1D. A broad peak is eluted at 0.15 *M* KCl. This broadening may be due to the random nature of DNA attachment to the cellulose producing a multiplicity of conformational states for receptor binding. If material from the peak is chromatographed on DEAE-cellulose, the receptor elutes at 0.1 *M* KCl and thus is component A. This is consistent with other studies showing that component A binds to DNA.

This procedure is useful for purification of receptor component A, as a subsequent step to the DEAE step. Yields of receptor A are high, but the purification has not yet been adequately measured, since only small amounts of progesterone receptors have been chromatographed on DNA-cellulose. The amounts were so small that the radioactivity peaks contained no detectable protein.

G. Protamine-Sepharose Chromatography

The salmon sperm nuclear protein protamine has been used as an affinity chromatography substrate for the isolation of protein kinases.[35] The protamines are coupled through a peptide linkage to an aminoalkyl agarose (Bio-Rad AF 101) according to the method of Cuatrecasas.[36] This protamine-sepharose can be used to chromatograph progesterone receptors by KCl gradient elution. The receptors elute as a minor peak

[35] J. D. Corbin, C. O. Brostrom, C. A. King, and E. G. Krebs, *J. Biol. Chem.* **247,** 7790 (1972).

[36] P. Cuatrecasas, *J. Biol. Chem.* **245,** 3059 (1970).

at about 0.15 *M* KCl and a major peak at about 0.8 *M* KCl. Since the receptors are acidic proteins this process probably represents another type of ion-exchange chromatography involving the strongly basic guanidino groups on the arginine residues. However, it is interesting to note that the progesterone receptor cochromatographs with a protein kinase activity and a cyclic AMP-binding protein.[37]

An elution obtained using this method is shown in Fig. 1E. A 50-ml protamine-agarose column was equilibrated in Buffer A—0.1 *M* KCl. Labeled progesterone receptor in the same buffer was obtained following ammonium sulfate precipitation and DEAE-cellulose chromatography (component B). The receptors were washed through the column in the equilibration buffer, and then the column was eluted using a 600-ml gradient of Buffer A containing KCl from 0.1 *M* to 1.0 *M*. The peak fractions were pooled and assayed for protein. Yield was nearly 90%, and the purification was 66-fold. The receptor peak is eluted at high salt molarity (0.8 *M*) and it is necessary to desalt the material before other assays can be done. This is most easily accomplished by stepwise adsorption and elution from a small hydroxylapatite column, using 0.3 *M* phosphate buffer to elute. This technique also facilitates protein measurements on these highly purified preparations since both Tris and thioglycerol interfere with standard protein assays.

H. Sucrose Gradient Ultracentrifugation

The technique of sucrose-gradient ultracentrifugation was the original definitive procedure used to demonstrate the existence of specific hormone receptor proteins in target tissues.[38] The procedure remains a useful analytical tool and is described in detail elsewhere in this volume.[25] The technique also lends itself to purification steps and is particularly useful in the case of the progesterone receptors. Adenosine-3′,5′-cyclic monophosphate (cyclic AMP) binds strongly to an oviduct acidic protein which copurifies with the progesterone receptor through ammonium sulfate precipitation, gradient elution from DEAE-cellulose, protamine-sepharose, agarose, and hydroxylapatite. However, when a highly purified progesterone receptor preparation is subjected to sucrose-gradient centrifugation in gradients containing 0.3 *M* KCl, the progesterone receptor sediments at about 4 S, whereas the cyclic AMP-binding protein sediments along with protein kinase at about 8 S shown in Fig. 1F.

Most of the proteins of oviduct cytosol sediment in the 4–5 S region of sucrose gradients. Under conditions of low ionic strength (no KCl in

[37] R. K. Keller, W. T. Schrader, and B. W. O'Malley, *J. Biol. Chem.* (in preparation).
[38] D. O. Toft and J. Gorski, *Proc. Nat. Acad. Sci. U.S.* **55**, 1574 (1966).

the gradient) the progesterone receptors aggregate in a complex way and sediment at 6–9 S. Sucrose gradients run under low salt conditions therefore accomplish some degree of purification. This must be balanced against the fact that the method is time consuming (16 hours) and recovery of the materials is tedious (gradient tube fractionation in the cold room).

Finally, yields may be poor. This is especially true for progesterone receptors run in cellulose nitrate centrifuge tubes. Since progesterone strongly adsorbs to this material, pollyallomer centrifuge tubes are used instead.

I. Steroid Affinity Chromatography

A purification step involving the specificity of the progesterone-binding site would be expected to be very efficient at isolating only the receptor molecules. Limited success has been achieved using steroids bound to Sepharose 4B (Pharmacia) according to methods described in detail elsewhere.[36,39,40] The method has been hampered by our inability to achieve reasonable recovery of receptor proteins off of the affinity column.

Deoxycorticosterone-21-hemisuccinate is used as the stationary steroid bound to Sepharose. This steroid binds only half as well as progesterone to the progesterone receptors in steroid competition studies. The steroid-hemisuccinate derivative can be coupled to the Sepharose at a final concentration of a few micromoles per packed milliliter of gel. Nascent progesterone receptors will adsorb to a 1-ml column of this material. The receptors cannot be eluted with KCl. The column is then equilibrated with 10^{-6} *M* [^{3}H]progesterone and allowed to stand overnight at 0°. The first 1 ml of wash is then applied to a 10-cm Sephadex G-75 column, which separates free hormone from that bound to macromolecules. The void volume of this column indeed contains labeled progesterone bound to receptor protein. Yields, however, are much less than 1%, and this system has yet to prove its applicability.

A modification of this technique has been applied successfully to the isolation of calf uterine estrogen receptors.[41] The method has also been used successfully for the oviduct progesterone receptors from chicks, using deoxycorticosterone-21-hemisuccinate coupled to the Sepharose 4B through denatured bovine serum albumin. Such a column binds progesterone

[39] W. Rosner and H. L. Bradlow, *J. Clin. Endocrinol. Metab.* **33,** 193 (1971).
[40] P. Cuatrecasas and C. B. Anfinson, *Ann. Rev. Biochem.* **41,** 259 (1971).
[41] V. Sica *et al., J. Biol. Chem.* **248,** 6543 (1973).

receptors with $K_d = 10^{-10}$ M; a 1.0-ml column will bind receptors from about 10 g of oviduct. Receptors from cytosol are precipitated with ammonium sulfate, redissolved in Buffer A, and incubated overnight at 0° with the affinity resin by shaking in suspension. The slurry is washed with Buffer A containing 0.15 M KCl, and then incubated at room temperature for 30–60 minutes with 10^{-6} M [^{3}H]progesterone. The solubilized receptor-hormone complexes are collected by filtration at 0° and can be isolated from free hormone by any of the column chromatographic steps below. Receptors prepared by this technique are not homogeneous. Analysis of the eluates by SDS-polyacrylamide gel electrophoresis shows several bands of varying molecular weight. Two major bands are seen at 110,000 and 117,000 daltons, which subsequent purification steps show to be the A and B receptor components, respectively.

Details of this method described elsewhere[41] cannot be overstressed. First, the resin must be washed exhaustively with 80% methanol to remove deoxycorticosterone adsorbed noncovalently to the gel. The multipoint attachment of steroid to the backbone afforded by the albumin renders the material much more stable. However, it is advisable to avoid use of cytosol directly. The resin capacity is decreased 70% when crude cytosol is applied, but only a 2% decrease occurs with the ammonium sulfate precipitate. The difference appears to be due to hydrolytic enzymes in cytosol which metabolize the steroid or cause its release from the column. It is not sufficient merely to monitor total deoxycorticosterone bound to the resin as an index of stability, since only about 1% of the bound hormone is accessible to the receptors.

J. Isoelectric Focusing

Partially purified progesterone receptors can be focused on 100-ml ampholyte gradients from pH 3 to pH 10 or from pH 4 to pH 6. However, it is necessary to keep the system rigorously cooled to 0°. It is helpful to prefocus the ampholytes to establish a pH gradient as described elsewhere in this volume by Sherman.[27] The receptors can be separated from the carrier ampholytes by gel filtration, but substantial dissociation of the complexes occurs. The method has been applied to analytical determinations but as yet has not been successfully used to purify the receptors.

K. Polyacrylamide Gel Electrophoresis

Optimal conditions for polyacrylamide gel electrophoresis have not been established to resolve the progesterone receptor proteins. On an analytical scale a modification of the pH 10.2 system of Chrambach[6,42] has been used successfully. Details of gel electrophoresis are discussed elsewhere in this volume.[27] Because the receptors appear to be highly asymmetric it is necessary to use very porous stacking gels with little crosslinking[6] to allow entry to the receptors into the running gel. These gels must be kept rigorously cold; best results require running the gels in a water-cooled gel apparatus in a cold room.

VI. Contaminating Proteins in Progesterone Receptor Preparations

A. Ovalbumin

The cytosol protein fraction from estrogenized oviducts may be about 30–50% ovalbumin.[2] Thus, crude soluble fractions may be treated as "dirty" ovalbumin preparations from which a minor contaminant is to be isolated. Gel electrophoresis patterns show that even after several steps of purification the major visible protein band travels with ovalbumin. Methods should be adopted which optimize the elimination of this protein. The ammonium sulfate step, for example, precipitates very little ovalbumin. This is the reason why this simple method achieves a 30-fold purification. Second, gel filtration steps will separate the lighter ovalbumin from receptors. One method which has not been mentioned previously in this regard is to use an anti-ovalbumin antibody column to remove traces of this protein. Such a material for ovalbumin affinity chromatography has been tried in this laboratory, and the receptors were not retarded by the column. Thus as a polishing step a small anti-ovalbumin column should be very useful for removing traces of this protein.

B. Corticosteroid-Binding Globulin (CBG)

This protein, present in crude oviduct homogenates and cytosol, has an equally high affinity for progesterone as it does for the corticosteroid hydrocortisone and corticosterone.[43]

Since CBG presumably has an entirely different function from that of progesterone receptor, it is necessary to exclude CBG during the purification. Essentially all the CBG arrives in the oviduct by transport through the circulatory system. Thus the first step toward minimizing CBG contamination is to dissect oviducts as free of blood as possible

[42] D. Rodbard and A. Chrambach, *Proc. Nat. Acad. Sci. U.S.* **65**, 970 (1970).
[43] U. Westphal, *Acta Endocrinol. (Copenhagen), Suppl.* **147**, 122 (1970).

and to rinse them in generous amounts of cold 0.9% NaCl. The 30% saturation ammonium sulfate step is very effective at eliminating CBG. Virtually no CBG precipitates at this fractional saturation, and hence the resulting preparation has a 100-fold decreased number of CBG binding sites per unit of protein. The DEAE-cellulose step serves to further eliminate CBG, which washes off the column slowly with Buffer A elution only. Finally, CBG is a smaller protein and chromatographs on agarose gel filtration differently from progesterone receptors. After the ammonium sulfate and DEAE steps, there is no detectable cortisol-binding activity present in the preparations.

C. Other Target-Tissue Steroid Receptor Proteins

Estrogen Receptors

There are specific binding proteins for estradiol in the chick oviduct. This protein has been extremely difficult to study since the estrogen–receptor complexes have short half-lives. Furthermore, oviduct growth is promoted with large doses of estrogen, which will effectively compete for estradiol-binding sites. Nevertheless, all steroid hormone receptors studied to date are remarkably similar and there is no reason to believe the oviduct estradiol receptor is an exception. Thus it is fair to assume that any purification procedure devised for progesterone receptor (with the probable exception of steroid site affinity chromatography) will also copurify the estradiol receptor proteins. This argument could be extended to cover receptors for the glucocorticoids and mineralocorticoids as well.

D. Protein Kinase and Cyclic AMP-Binding Protein

Oviduct cytosol contains several protein kinase activities capable of phosphorylating proteins *in vitro*. At least one of the kinases is inhibited by an acidic protein subunit. Cyclic AMP binds to this subunit, causing its release with concomitant activation of the kinase. The progesterone receptor protein is found associated with the cyclic AMP-binding protein throughout an extensive purification of over 10,000-fold. These steps include separations based on both size and charge. There are roughly three times as many cyclic AMP-binding sites as there are progesterone-binding sites in the cytosol or purified preparation. Thus this material can represent a major contaminant at the end of the procedure. Some manipulations of the system can be done to separate the two binding activities. These involve sucrose gradient ultracentrifugation methods discussed in Section III above.

If a receptor preparation is run on a sucrose gradient in 0.3 *M* KCl, the progesterone receptor is about 4 S while the kinase–cyclic AMP protein complex is about 8 S, and thus can be separated partially. The converse experiment involves adding 10^{-5} *M* cyclic AMP to the preparation and to the gradient as well, but omitting the KCl. Under these conditions, cytosol receptor or the ammonium sulfate material will sediment at about 8 S. The cyclic AMP will cause dissociation of the cyclic AMP-binding protein from the kinase. These two molecules sediment in the dissociated state as about 4 S molecules and thus again can be separated from the progesterone receptor.

[18] Physical–Chemical Analysis of Steroid Hormone Receptors[1]

By MERRY R. SHERMAN

Introduction

Definitive evaluation of the mass, chemical composition and conformation of steroid hormone receptors, and of the changes that may occur on interaction with the steroid, the nucleus, or chromatin components will require milligrams of the purified macromolecules. Our current objective is to estimate the physical parameters from nanogram amounts of impure receptors.

Density gradient centrifugation, one method of analysis, is the subject of several sections of this volume[2-4] and a large proportion of the receptor literature. This chapter deals with three other techniques: gel filtration, gel electrophoresis and isoelectric focusing, and the integration of data from various procedures into a coherent description of the receptor molecule.[5]

[1] This article is dedicated to Professor John T. Edsall, whose classical article, "The Size, Shape and Hydration of Protein Molecules" [*in* "The Proteins" (H. Neurath and K. Bailey, eds.), 1st ed., Vol. 1, Part B, p. 549. Academic Press, New York, 1953], provides the essential background for any investigator of the physical chemistry of steroid receptors.

[2] G. M. Stancel and J. Gorski, this Vol. [15].

[3] W. McGuire, this Vol. [21].

[4] D. O. Toft and M. R. Sherman, this Vol. [14].

[5] Ion-exchange chromatography is used primarily for purification rather than physical characterization of receptors and their subunits, and is discussed elsewhere in this volume (W. Schrader [17]). It should be noted, however, that the effects

The limited availability and unique properties of steroid receptors impose certain restrictions on the application of these techniques. (1) The extremely low concentrations of receptors require highly sensitive methods of detection. Toward this end, specific antibodies to receptors and suitable affinity labels are being developed.[6,7] At present, receptors are detected by noncovalent association with tritiated steroids of high specific activity. The lability of the steroid–protein bonds precludes studies under many conditions of temperature, pH, detergent, and denaturant concentrations. (2) Comparisons with standard proteins are essential in quantitative gel filtration and electrophoresis. The receptors are assumed to be similar in chemical nature and resultant density to these standards. There is no evidence, however, that receptors are not protein conjugates, containing phosphate, carbohydrate, or nucleic acid moieties. (3) The steroid-binding unit may interact with various contaminants in crude preparations, depending on experimental conditions. Different physical methods may therefore reveal different macromolecular complexes, all containing the steroid label. This may prevent a consistent interpretation of data obtained by centrifugation and the techniques described below.

An estimate of a physical–chemical parameter is meaningful only if accompanied by a statement describing the precision of that estimate. The standard deviation of a set of measurements is one method of expressing that precision. Confidence limits provide a way of stating both how close a determination is likely to be to the mean value and the chance of its being that close.[8,9] Analysis of receptors by gel filtration, gel electrophoresis, or any other method necessarily involves a consideration of the statistical factors that determine the uncertainty of the result.

---

of salts on the structure of receptors have profound implications for their behavior on ion-exchange chromatography. A steroid-containing complex may be released from an ion exchanger in a salt gradient either by the disruption of interactions between the receptor and exchanger, or by the salt-induced formation of receptor subunits. The latter process may result in elution of only a portion of the macromolecular complex.

[6] J. Warren, this Vol. [33].

[7] H. Smith, J. R. Neergaard, E. P. Burrows, R. C. Hardison, and R. G. Smith, this Vol. [34].

[8] H. D. Young, "Statistical Treatment of Experimental Data." McGraw-Hill, New York, 1962.

[9] K. A. Brownlee, "Statistical Theory and Methodology in Science and Engineering." Wiley, New York, 1960.

Gel Filtration

Principle. A gel filtration or molecular sieve column is packed with a suspension of porous gel beads or granules. The column contains two solvent phases, one surrounding the beads and one within the gel. Macromolecules passing through the column in a flow of solvent are distributed between the void volume surrounding the granules and the accessible space within the gel interstices. Movement around the granules is less restricted than within the gel. Therefore, molecules that are too large to permeate the granules are eluted most rapidly.[10]

Since the introduction of gel filtration in 1959,[11] there has been widespread agreement that the elution sequence for chemically similar macromolecules is a function of size. There has been no consensus as to which parameter of size is relevant, and numerous formulations of the dependence of elution volume on each parameter have been presented. It is clear that the extent of gel penetration reflects the *geometry* rather than the *mass* (or molecular weight) of a macromolecule. Various approximations to the molecular size and shape have been considered, but the available data have not been sufficiently precise to permit a final choice among them.

Stokes' law[12] defines the frictional resistance (f) to the movement of a sphere of radius R in a medium of viscosity η:

$$f = 6\pi\eta R \tag{1}$$

The Stokes radius (R_S) of a macromolecule is the radius of a sphere that would experience the same frictional resistance as the molecule. The evaluation of this parameter for a receptor, by comparison of its gel filtration pattern with those of well-characterized proteins, thus provides insight into its macromolecular structure. In combination with the sedimentation coefficient and estimates of the partial specific volume and solvation, R_S can be used to estimate the molecular weight and asymmetry.

Gel Selection and Eluent Composition. Clear separation of a steroid–receptor complex from free steroid is achieved by means of a gel of low porosity, which completely excludes the macromolecule. For the structural analysis of a receptor, by contrast, a gel should be selected to *include* its various macromolecular forms. Andrews[13] has outlined the

[10] B. J. Gelotte, *J. Chromatogr.* **3**, 330 (1960).
[11] J. Porath and P. Flodin, *Nature (London)* **183**, 1657 (1959).
[12] G. G. Stokes, *Trans. Cambridge Phil. Soc.* **9**, 8 (1851).
[13] P. Andrews, *Methods Biochem. Anal.* **18**, 1 (1970).

general properties of the three main types of gel filtration media: the cross-linked dextrans ("Sephadex" Pharmacia), cross-linked polyacrylamide beads ("Bio-Gel P" Bio-Rad Laboratories), and agarose ("Bio-Gel A" Bio-Rad; "Sepharose" Pharmacia).[14] The range of molecular size included by a given gel depends on its concentration and degree of cross-linking and is specified by a numerical suffix to the name.

Problems may arise in gel filtration because of the properties of both the steroids and the receptors. All the above gel types adsorb free steroids to varying extents. All apparently contain residual acidic groups which, in media of low ionic strength, may repulse the acidic receptor molecules and cause their exclusion from the gel or premature elution. Charge-free agarose for gel electrophoresis has been prepared by alkaline desulfation and reduction of carboxylic groups.[15] The procedures described are not directly applicable to the beads used for gel filtration, but the study confirms the presence of acidic groups in commercial agarose. Acidic groups in Sephadex may be neutralized by washing the column with 1 *M* aqueous pyridine.[16] The reported presence of zinc in Sephadex[17] may also cause exclusion of receptors, presumably by interaction with sulfhydryl groups and subsequent aggregation.

Steroid-binding studies and ultracentrifugal analyses have demonstrated the dependence of receptor form and stability on ionic strength, and the concentrations of specific divalent cations, chelating agents, polyhydric alcohols, and sulfhydryl-containing reagents. These factors, together with the chemistry of the steroid and of the gel, must therefore be considered in selecting the conditions for gel filtration.

Stabilization of many proteins including steroid receptors by glycerol or other polyhydric alcohols has been discussed elsewhere.[4,18] This stabilization may result from the reduced competition by water molecules for essential hydrogen bonds within the receptors, or from the increased strength of these bonds in environments of low dielectric constant.[19] Regardless of its mechanism, this property recommends the inclusion of glycerol or similar compounds in gel filtration media for receptors and for the standard proteins with which the columns are calibrated. Concentrations of glycerol as high as 30% (w/v) have been used in agarose gel filtration of complexes of [^{3}H]progesterone with chick oviduct recep-

[14] Mention of specific equipment, trade products, or a commercial company does not constitute endorsement over similar products or companies not named.

[15] T. Låås, *J. Chromatogr.* **66**, 347 (1972).

[16] D. Eaker and J. Porath, *Separ. Sci.* **2**, 507 (1967).

[17] R. S. Morgan, N. H. Morgan, and R. A. Guinavan, *Anal. Biochem.* **45**, 668 (1972).

[18] S. L. Bradbury and W. B. Jakoby, *Proc. Nat. Acad. Sci. U.S.* **69**, 2373 (1972).

[19] G. Némethy, I. Z. Steinberg, and H. A. Scheraga, *Biopolymers* **1**, 43 (1963).

tors. The residual amount of bound steroid, assayed by charcoal–dextran treatment of eluted fractions,[20,21] was significantly enhanced by the glycerol.[22] The addition of glycerol, however, has certain unfavorable effects. (1) The increased viscosity of the eluent retards the flow rate through the column. (2) Glycerol,[23] like sucrose,[24,25] interferes in protein induced structural changes in receptors, the effects of the polyhydric determinations by the Lowry and Biuret reactions. (3) In studies of salt-alcohol on the viscosity and dielectric constant, and hence the conductivity of the medium, must be taken into account.[4,26]

The protective effects of sulfhydryl-containing agents on steroid receptors are well known. The inclusion in gel filtration buffers of β-mercaptoethanol, monothioglycerol, or dithiothreitol is therefore recommended. These compounds interfere significantly with the monitoring of eluates for protein by the Lowry method[27,28] or by optical absorption at wavelengths below 230 nm, where sensitivity is highest.[29] Similarly, the addition of EDTA or other chelating agents may enhance receptor stability during filtration, but may interfere in subsequent protein determinations by the Lowry, Biuret,[30,31] or Warburg and Christian methods.[32]

Column Calibration.[33] On the basis of statistical considerations, it is advisable to use about six or seven standards to calibrate the column. The choice of standards for gel filtration is governed by similar factors to those applicable to density gradient centrifugation.[4] It is essential, although difficult, to find standard proteins that are eluted from the

[20] C. A. Nugent and D. B. Mayes, *J. Clin. Endocrinol. Metab.* **26,** 1116 (1966).

[21] S. G. Korenman, this Vol. [4].

[22] M. R. Sherman, S. B. P. Atienza, J. R. Shansky, and L. M. Hoffman, *J. Biol. Chem.* **249,** in press.

[23] M. K. Zishka and J. S. Nishimura, *Anal. Biochem.* **34,** 291 (1970).

[24] B. Gerhardt and H. Beevers, *Anal. Biochem.* **24,** 337 (1968).

[25] D. J. S. Arora, *Can. J. Biochem.* **49,** 139 (1971).

[26] S. Glasstone, "Textbook of Physical Chemistry," 2nd ed., p. 904. Van Nostrand-Reinhold, Princeton, New Jersey, 1946.

[27] C. G. Vallejo and R. Lagunas, *Anal. Biochem.* **36,** 207 (1970).

[28] P. J. Geiger and S. P. Bessman, *Anal. Biochem.* **49,** 467 (1972).

[29] W. J. Waddell, *J. Lab. Clin. Med.* **48,** 311 (1956).

[30] A. R. Neurath, *Experientia* **22,** 290 (1966).

[31] W. W. Ward and R. J. Fastiggi, *Anal. Biochem.* **50,** 154 (1972).

[32] O. Warburg and W. Christian, *Biochem. Z.* **310,** 384 (1942).

[33] Analytical gel filtration in thin layers, as opposed to columns, is a method of potential applicability to steroid receptors [P. Andrews and C. Male, *in* "Chromatographic and Electrophoretic Techniques" (I. Smith, ed.), 3rd ed., Vol. I, p. 823. Heinemann, London, 1969; Z. Wasyl, W. E. Luchter, and W. Bielanski, Jr., *Biochim. Biophys. Acta* **243,** 11 (1971)].

column *before* as well as after the steroid–receptor complex. Only then can the analysis of receptor size be based on interpolation rather than extrapolation of the calibration curve. Many proteins of convenient size are dissociated into subunits by the concentrations of salts or chelating agents in which receptors are studied. This problem may be surmounted by covalently cross-linking the multichain proteins with bifunctional reagents.[34] Other proteins that might serve as markers contain carbohydrate (e.g., γ-globulin) or prosthetic groups that cause deviations from expected chromatographic behavior.[13,35] Proteins with well-established values for the molecular weight (M) are more numerous than those with definitive values for the Stokes radius (R_S). The latter must often be calculated from published values for the diffusion coefficient (D) according to

$$R_S = \frac{kT}{6\pi\eta D} \tag{2}$$

where k is the Boltzmann constant, T is absolute temperature, and η is the solvent viscosity at that temperature. When the value of D in $cm^2\ sec^{-1}$ under standard conditions (20° in water) is used, the equation for the Stokes radius in nanometers reduces to

$$R_S = \frac{2.143 \times 10^{-6}}{D_{20,w}} \tag{3}$$

The parameters of standard proteins used in calibrating the columns are generally obtained under standard conditions, although gel filtration may be conducted in a cold, viscous, high-salt medium. This practice is based on the untested assumption that "environmental" effects on the receptor and standards are analogous.

The parameters of a gel filtration column are the volume of the solvent phase surrounding the gel beads, or void volume (V_0), and the volume of the internal solvent phase (V_i). V_0 is measured by the elution of a solute so large as to be completely excluded from the gel interstices or pores. V_i is determined by subtracting V_0 from the elution volume of a small, uncharged, non-aromatic solute that diffuses freely through the gel.[10] The total volume accessible to most solvent components ($V_0 + V_i$) is less than the total column volume (V_t) by the volumes of the gel matrix and of the water firmly bound to it. The size of a macromolecule determines its distribution between the internal and external solvent phases and hence its elution volume (V_e).

[34] G. E. Davies and G. R. Stark, *Proc. Nat. Acad. Sci. U.S.* **66**, 651 (1970).
[35] D. Schubert, *J. Mol. Biol.* **51**, 287 (1970).

In experiments with [^{3}H]steroid–receptor complexes, the optical density of the eluate may be monitored and recorded continuously before it is collected into fractions for liquid scintillation counting. Available optical equipment includes "Uvicord" absorptiometers (LKB-Produkter AB, Bromma, Sweden; U.S. distributor LKB Instruments, Rockville, Md.), which provide a limited selection of wavelengths, and spectrophotometers that can be operated at 1–4° and are equipped with sample-chamber cooling and flow-through cuvets (e.g., Gilford Model 2400-S; Beckman DB-G with a Hellma Model 178 cell). Colorless proteins can be detected at wavelengths between 236 and 222 nm with sensitivities that are up to eight times higher than at 280 nm, depending on the amino acid composition. Absorbance measurements at the lower wavelengths, however, are more susceptible to errors due to impurities or baseline fluctuations. It is convenient to install an electrical connection or "event marker" from the fraction collector to the recorder, so that data for optical density as well as radioactivity can be expressed as fraction numbers.

When receptors are studied in crude extracts, there is often some radioactivity or a UV absorption peak associated with high molecular weight aggregates or contaminants eluted at V_0. When more purified samples or standard proteins are chromatographed, a solute known to be excluded from the gel, such as Blue Dextran 2000 (Pharmacia) or bacteriophage, must be added as a marker for V_0. A valid indication of $(V_0 + V_i)$ is obtained neither from the elution volume or radioactively labeled water, which has the unique ability to exchange with the gel-bound water, nor by free steroids, which are adsorbed and hence retarded by most gels.[36] This adsorption of steroids may be enhanced by salting out of hydrophobic molecules in eluents of high ionic strength.[16] In experiments with [^{3}H]steroids, the value of $(V_0 + V_i)$ can be measured by adding to the sample a small compound that is neither adsorbed nor repulsed by the gel and is colored or labeled with another isotope, such as [^{14}C]glycerol. $^{45}CaCl_2$ or other charged solutes should be used only with eluates of high ionic strength.[37]

From the column parameters V_0 and V_i, and the elution volume for each standard or steroid-binding component, a distribution coefficient (K_D) may be calculated according to[10]

$$K_D = \frac{V_e - V_0}{V_i} \tag{4}$$

[36] The elution volumes of free [^{3}H]steroids have been incorrectly used for this purpose in our earlier publications and by other investigators.

[37] H.-J. Zeitler and E. Stadler, *J. Chromatogr.* **74**, 59 (1972).

K_D represents the fraction of the *internal solvent volume* accessible to that solute. An alternative expression, K_{av}, denotes the available fraction of the *total internal gel phase*[38]

$$K_{av} = \frac{V_e - V_0}{V_t - V_0} \tag{5}$$

Since V_t is determined from the column dimensions and includes the volumes of the gel and tightly bound water, K_{av} is slightly less than K_D for a given system.

Many of the theoretical and empirical correlations of K_D or K_{av} with the parameters of molecular size have been reviewed by Determann and Michel[39] and by Andrews.[13] Methods utilizing the Stokes radius include:

(1) *Linear correlation of* $K_D^{1/3}$ *with* R_S. This relationship, based on Porath's theoretical calculations of the volume available to spherical molecules in conical pores,[40] has been verified by Siegel and Monty[41] and applied to a testosterone-binding component of human pregnancy serum[42] and to calf uterine estradiol receptors.[43]

(2) *Linear correlation of* $(-\log K_{av})^{1/2}$ *with* R_S. This function is based on Ogston's calculation of the volume available to spherical molecules in a random suspension of rigid rods.[44] A nonlinear form was presented and tested by Laurent and Killander[38] and the linear relationship was confirmed by Siegel and Monty.[41]

(3) *The Renkin equation*[45] *for* K_D *as a function of* (R_S/r). A model for restricted diffusion of the solute within cylindrical channels of radius r in the gel is the basis for this relationship, proposed and tested by Ackers.[46] It has been applied in receptor studies in this laboratory,[47] but is less convenient than the linear correlations.

(4) *Linear correlation of* K_D *with log* R_S. This empirical relationship is probably more widely applicable than the correlation of K_D with log M from which it was derived by Determann and Michel.[39] The range

[38] T. C. Laurent and J. Killander, *J. Chromatogr.* **14**, 317 (1964).
[39] H. Determann and W. Michel, *J. Chromatogr.* **25**, 303 (1966).
[40] J. Porath, *J. Pure Appl. Chem.* **6**, 233 (1936).
[41] L. M. Siegel and K. J. Monty, *Biochim. Biophys. Acta* **112**, 346 (1966).
[42] J. L. Guériguian and W. H. Pearlman, *J. Biol. Chem.* **243**, 5226 (1968).
[43] G. A. Puca, E. Nola, V. Sica, and F. Bresciani, *Biochemistry* **10**, 3769 (1971).
[44] A. G. Ogston, *Trans. Faraday Soc.* **54**, 1754 (1958); A. G. Ogston and C. F. Phelps, *Biochem. J.* **78**, 827 (1961).
[45] E. M. Renkin, *J. Gen. Physiol.* **38**, 225 (1955).
[46] G. K. Ackers, *Biochemistry* **3**, 723 (1964).
[47] M. R. Sherman, P. L. Corvol, and B. W. O'Malley, *J. Biol. Chem.* **245**, 6085 (1970).

of overlap of this function with the theoretical models of Porath[40] and Ackers[46] has been examined by Anderson and Stoddart.[48] Despite the absence of a sound theoretical basis, this relationship has been confirmed by Pagé and Godin[49] and adequately correlates extensive data on standard proteins with which chick oviduct progesterone receptors have been compared.[22]

(5) *Linear dependence of* R_S *on the inverse error function complement* (*erfc*$^{-1}$) *of* K_D. This more recent proposal of Ackers[50] is based on a very general model of the gel microstructure: that the penetrable volumes of any shape are distributed randomly with respect to the sizes of molecules they can accommodate. Use of the resultant correlation requires the availability of tables of the error function complement or its inverse or the corresponding computer programs.[51]

Studies with proteins that are asymmetric or contain non-protein moieties suggest that elution volumes are sensitive to molecular shape and density as well as weight. It is not yet known whether steroid receptors resemble most globular proteins in these properties. In fact, preliminary evidence suggests the contrary. The following methods of column calibration, involving direct correlation of K_D with M, should therefore be applied to receptors with due caution.

(6) *Linear dependence of* $K_D^{1/3}$ *on* $M^{1/2}$. This is based on Porath's[40] model of gel structure [see (1) above] for the case of flexible linear macromolecules, in which the effective radius is proportional to $M^{1/2}$. While native proteins hardly fit this description, this relationship has been confirmed for numerous proteins[52] and has been used to obtain preliminary values of M for progesterone receptors.[47] As pointed out by Anderson and Stoddart,[48] the intercept of this linear plot on the $K_D^{1/3}$ axis is greater than unity.

(7) *Linear correlation of* K_D *with log* M. This relationship is consistent with extensive data on globular proteins.[39] It is related to correlation (4) above, by the additional assumption that the Stokes radii are proportional to a constant fractional power of M for the series of macromolecules under investigation.

Analysis of Results. The Stokes radius is a composite parameter of

[48] D. M. W. Anderson and J. F. Stoddart, *Anal. Chim. Acta* **34,** 401 (1966).

[49] M. Pagé and C. Godin, *Biochim. Biophys. Acta* **194,** 329 (1969).

[50] G. K. Ackers, *J. Biol. Chem.* **242,** 3237 (1967).

[51] A computer subroutine (Fortran) for the error function, from which the error function complement is readily obtained, was developed in connection with previous research [M. M. Rubin (Sherman) and A. Katchalsky, *Biopolymers* **4,** 579 (1966)] and is available on request.

[52] P. Andrews, *Biochem. J.* **91,** 222 (1964).

macromolecular structure that was analyzed by Oncley[53] into the product of (1) the radius ($\bar{R}$) of an anhydrous sphere of the same molecular weight (M) and partial specific volume ($\bar{v}$)

$$\bar{R} = \left(\frac{3M\bar{v}}{4\pi N}\right)^{1/3} \tag{6}$$

where N is Avogadro's number; (2) the radial increment due to a solvation shell

$$\frac{R_{\text{solvated}}}{\bar{R}} = \left(1 + \frac{w}{\bar{v}\rho}\right)^{1/3} \tag{7}$$

where w denotes grams of solvent per gram of protein, and ρ is the solvent density; and (3) the frictional ratio due to asymmetry $(f/f_0)_{\text{shape}}$. The total frictional ratio (f/f_0) is the product of the shape and solvation factors. Thus, the Stokes radius may be expressed as

$$R_S = \left(\frac{3M\bar{v}}{4\pi N}\right)^{1/3} \left(1 + \frac{w}{\bar{v}\rho}\right)^{1/3} \left(\frac{f}{f_0}\right)_{\text{shape}} \tag{8}$$

or

$$R_S = \bar{R}\left(\frac{f}{f_0}\right) \tag{9}$$

It is apparent from Eq. (8) that the value of R_S determined by gel filtration provides only a partial description of the molecule and must be combined with data from other techniques for a complete physical–chemical analysis.

A prerequisite for an integration of data obtained by various methods is the demonstration that, in fact, the same receptor form is being detected in each case. Thus, in order to relate a sedimentation coefficient (s) calculated from density gradient centrifugation to a Stokes radius from gel filtration, the receptor complex isolated from the gradient should be chromatographed with an eluent of the same composition (including the sucrose or glycerol), and the converse experiment performed when possible. Reversible artifacts, such as gel-induced aggregation, may still escape detection by this procedure.

In addition to s and R_S, an independent evaluation of $\bar{v}$ is desirable. Methods of potential applicability include equilibrium centrifugation in gradients of CsCl[54,55] or $NaClO_4$,[56,57] although preferential interactions

[53] J. L. Oncley, *Ann. N.Y. Acad. Sci.* **41,** 121 (1941).

[54] M. Meselson, F. W. Stahl, and J. Vinograd, *Proc. Nat. Acad. Sci. U.S.* **43,** 581 (1957).

[55] J. Vinograd and J. E. Hearst, *Fortschr. Chem. Org. Nautrst.* **20,** 372 (1962).

[56] P. Morin, *Biochimie* **54,** 985 (1972).

[57] J. P. Liautard and K. Köhler, *Biochimie* **54,** 103 (1972).

of the receptor with any constituent of these solutions can alter the value of $\bar{v}$ obtained.[58] Another approach that has been applied to cholinergic receptors[59] involves centrifugation of receptors and standard proteins through sucrose or glycerol gradients made in H_2O and in D_2O or $D_2{}^{18}O$. The value of $\bar{v}$ for the receptor is calculated from the decrease in sedimentation rate due to the heavy water and from the respective densities of the gradient solutions. This method minimizes errors due to preferential interactions with solvent components.[60]

The values for s, R_S, and $\bar{v}$ can then be combined to calculate the molecular weight,

$$M = \frac{6\pi\eta N s R_S}{1 - \bar{v}\rho} \tag{10}$$

The resultant value of M is used in the calculation of $\bar{R}$ by Eq. (6), and the total frictional ratio (f/f_0) by Eq. (9).

The contribution of shape alone may be analyzed by approximating the receptor by an ellipsoid of revolution, with an equatorial semiaxis b and a semiaxis of revolution a. Equations for $(f/f_0)_{\text{shape}}$ for both prolate ($b/a < 1$, cigar shaped) and oblate ($b/a > 1$, flattened globe) ellipsoids were derived by Perrin,[61] and the values are tabulated in books by Svedberg and Pedersen[62] and Schachman.[63] The total frictional ratio, however, cannot be interpreted unequivocally in terms of shape. For each estimate of the solvation, the value of (f/f_0) is compatible with two values of the axial ratio, depending on the type of ellipsoid assumed. This ambiguity was portrayed by Oncley[53] in a contour diagram of (f/f_0) as a function of both the axial ratio and the amount of solvation. The ensemble of data from gel filtration and centrifugation is therefore interpreted with the least bias by presenting a range of axial ratios, for each type of ellipsoid, corresponding to an upper and lower estimate for the solvation.

The diffusion coefficient for the receptor may be calculated from R_S according to Eq. (3). A comparison of the resultant value with published values of $D_{20,w}$ for globular proteins of similar sedimentation coefficients[64]

[58] H. K. Schachman and S. J. Edelstein, *Biochemistry* **5**, 2681 (1966).

[59] J. C. Meunier, R. W. Olson, and J. P. Changeux, *FEBS Lett.* **24**, 63 (1972).

[60] S. J. Edelstein and H. K. Schachman, *J. Biol. Chem.* **242**, 306 (1967).

[61] F. Perrin, *J. Phys. Radium* [7] **7**, 1 (1936).

[62] T. Svedberg and K. O. Pedersen, "The Ultracentrifuge," p. 41. Oxford Univ. Press (Clarendon), London and New York, 1940.

[63] H. K. Schachman, "Ultracentrifugation in Biochemistry," p. 239. Academic Press, New York, 1959.

[64] M. H. Smith, *in* "Handbook of Biochemistry" (H. A. Sober, ed.), Sect. C, p. 10. Chemical Rubber Publ. Co., Cleveland, Ohio, 1968.

provides a rapid assessment of the structural similarity of the receptor to the standard proteins.

Gel Electrophoresis

Principle; Relationship to Gel Filtration. The mobility (m) of a molecule in an electrical field is expressed as the velocity of its motion in cm sec^{-1} divided by the field intensity in V cm^{-1}. In a solution of given ionic strength and dielectric constant, the mobility is proportional to the mean net charge and inversely proportional to a resistance factor which depends on the size and shape of the molecule and viscosity of the medium.[65] Electrophoresis in gels is a tool of high resolving power for the evaluation of macromolecular size as well as charge. Ferguson observed in 1964 that the contributions of size and charge could be distinguished by performing electrophoresis in starch gels of several concentrations under otherwise identical conditions.[66] Chrambach and Rodbard have refined this approach and described in detail its application to the more versatile polyacrylamide gels, in which the acrylamide concentration and degree of cross-linking can be rigorously controlled.[67]

The *microscopic* structure of a gel formed by polyacrylamide, dextran, starch, or agarose may be viewed as a random meshwork of long fibers, if the degree of cross-linking is low, or as an array of beads, when cross-linking is extensive.[44,68] The movement of a macromolecule through such a gel is decreased by its collisions with the fibers or beads. An important distinction exists between the *macroscopic* forms of the gels used for filtration for electrophoresis. In gel filtration, the macromolecules move rapidly around the gel granules or more slowly through the accessible internal volume and are consequently eluted in order of decreasing size. Gel electrophoresis, by contrast, is performed in a solid tube or layer of gel, in which there is no path of migration except through the interstices or pores. Therefore, the largest molecules suffer the most retardation in their electrically driven movement through the gel. Recent investigations of the relationship between molecular retardation in gel electrophoresis and gel filtration[68,69] have led to theoretical and practical advances that should accelerate progress in receptor structure elucidation.

[65] J. T. Edsall and J. Wyman, "Biophysical Chemistry," Vol. 1, p. 506. Academic Press, New York, 1958.

[66] K. A. Ferguson, *Metab., Clin. Exp.* **13**, 985 (1964).

[67] A. Chrambach and D. Rodbard, *Science* **172**, 440 (1971).

[68] D. Rodbard and A. Chrambach, *Proc. Nat. Acad. Sci. U.S.* **65**, 970 (1970).

[69] C. J. O. R. Morris and P. Morris, *Biochem. J.* **124**, 517 (1971).

The molecular sizes of many proteins and their subunits have been determined by electrophoresis in gels of one[70] or several[71,71a] acrylamide concentrations containing sodium dodecyl sulfate. These techniques cannot be applied to a noncovalent steroid–receptor complex, since the steroid would be released from the denatured receptor and would migrate with the ionic detergent.

Estimation of Molecular Radius, Free Solution Mobility, and Valence.[72] Polyacrylamide gels for electrophoresis are characterized by two parameters, T which denotes the total grams of acrylamide monomers plus cross-linking agent/100 ml of solution, and C which represents the amount of cross-linking agent as a percentage (w/w) of the total dry ingredients.[73] With this notation, Ferguson's[66] observation of an exponential decrease of electrophoretic mobility with gel concentration can be expressed

$$\log m = \log m_0 - K_R T \tag{11}$$

where m_0 is the mobility in free solution and K_R is called the retardation coefficient. C must be constant for the series of gels of different T.[74] The mobilities in Eq. (11) are *absolute* mobilities, as defined in the introduction on electrophoresis. Two types of *relative* mobilities are also used. (1) Macromolecular movement through a gel is measured relative to a tracking dye or "front" that is minimally retarded by the gel (m_f).[75] The Ferguson plot for m_f,

$$\log m_f = \log Y_0 - K_R T \tag{12}$$

has the same slope as Eq. (11), while the intercept at $T = 0$ includes the logarithm of the absolute mobility of the dye. The linear dependence of log m_f on total gel concentration has been verified for a wide variety of proteins and experimental conditions[68] and for steroid-binding components of target organs[47] and plasma.[76,76a] (2) Another type of relative mobility compares migration in a gel with that in free solution. Morris[74]

[70] K. Weber and M. Osborn, *J. Biol. Chem.* **244,** 4406 (1969).

[71] G. A. Banker and C. W. Cotman, *J. Biol. Chem.* **247,** 5856 (1972).

[71a] A. R. Ugel, A. Chrambach, and D. Rodbard, *Anal. Biochem.* **43,** 410 (1971).

[72] D. Rodbard and A. Chrambach, *Anal. Biochem.* **40,** 95 (1971).

[73] S. Hjertén, *Arch. Biochem. Biophys., Suppl.* **1,** 147 (1962).

[74] C. J. O. R. Morris, *Protides Biol. Fluids, Proc. Colloq.* **14,** 543 (1967).

[75] The notation m_f for mobility relative to the front has been introduced in place of the widely used symbol R_f to avoid confusion with various definitions of molecular radii.

[76] P. L. Corvol, A. Chrambach, D. Rodbard, and C. W. Bardin, *J. Biol. Chem.* **246,** 3435 (1971).

[76a] J. A. Mahoudeau and P. Corvol, *Endocrinology* **92,** 1113 (1973).

assumed that the ratio (m/m_0) is proportional to the available volume fraction of the gel. Utilizing the notation introduced in the section on gel filtration, this may be expressed

$$m/m_0 = c_1 K_{av} + c_2 \tag{13}$$

where c_1 and c_2 are constants. When c_2 is negligible, Eqs. (11) and (13) may be combined to demonstrate the analogy between the retardation coefficient in gel electrophoresis and $-\log K_{av}$ in gel filtration. Thus, the various methods of correlating the latter function, or other functions of K_D or K_{av}, with molecular dimensions have parallels in the postulated relationships of K_R to the parameters of molecular size.

In early studies of the effect of gel concentration on globular protein mobilities, the retardation coefficient or its equivalent appeared to vary linearly with molecular weight (M).[66,74,77] For molecules with the same partial specific volume $(\bar{v})$, this corresponds approximately to a linear dependence of $(K_R)^{1/3}$ on the geometric mean radius $(\bar{R})$ defined in Eq. (6). In terms of a model for the microscopic structure of the gel, this dependence would be expected for rigid spherical molecules in extensively cross-linked gels (random networks of beads). In more loosely cross-linked gels (random suspensions of long fibers), the volume accessible to rigid spherical molecules would be proportional to the molecular surface area. Thus, $(K_R)^{1/2}$ would be a linear function of $\bar{R}$.

The geometric mean radius is calculated from Eq. (6), which contains the molecular weight and partial specific volume but no terms descriptive of macromolecular conformation or solvation. The use of $\bar{R}$ in electrophoretic analyses of proteins was supported by observations of Ogston and Phelps[44] that partition coefficients for various solutes between buffers and hyaluronic acid networks were insensitive to ionic strength, pH, and the chemical nature of the solute. In extensive electrophoretic analyses of proteins in polyacrylamide gels, however, Rodbard and Chrambach[68,72] as well as Morris[74] have observed variations in the retardation coefficient with pH and ionic milieu. It would therefore seem appropriate to correlate K_R with a more flexible parameter of molecular size than $\bar{R}$. Among the alternatives, neither the Stokes radius nor the radius of a sphere having the same surface area[78] can be excluded on the basis of available data. A further possibility, the radius of gyration of a random coil, would not appear to be applicable to receptors under nondenaturing conditions, in which the steroid label is retained.

Application to Steroid Receptors. In the future, receptors may be de-

[77] J. L. Hedrick and A. J. Smith, *Arch. Biochem. Biophys.* **126**, 155 (1968).

[78] J. C. Giddings, E. Kucera, C. P. Russell, and M. N. Myers, *J. Phys. Chem.* **72**, 4397 (1968).

tectable in gels (1) by autoradiographic[79,80] or semiconductimetric[81] monitoring of [^{14}C]steroid labels of high specific activity, (2) by covalent attachment of fluorescent or radioactive affinity labels,[6] or (3) by immunochemical techniques. Current technology, however, is limited to electrophoresis of the [^{3}H]steroid–receptor complex. The gel is then macerated[82] or sliced[83] and the radioactivity is located by combustion,[84] extraction,[85,86] or dissolution[87] of the gel segments. To demonstrate that the radioactivity coincides with a macromolecular complex, the labeled gel segments may be placed on a fresh gel and again subjected to electrophoresis.

For the analysis of a steroid–receptor complex, the method of gel polymerization and choice of buffer system must be compatible with both the isoelectric point of the macromolecules and the conditions for retention of bound steroid.[88-90] Preparation of a polyacrylamide gel is an oxidative process that is generally catalyzed by persulfate and/or riboflavin under illumination. Polymerization involves the formation of charged and uncharged free radical species, which may interact with receptors and other proteins. Persulfate and other charged polymerization reactants and products may be eliminated by pre-electrophoresis of the

[79] G. Fairbanks, Jr., C. Levinthal, and R. H. Reeder, *Biochem. Biophys. Res. Commun.* **20**, 393 (1965).

[80] P. O. Kohler, W. E. Bridson, and A. Chrambach, *J. Clin. Endocrinol. Metab.* **32**, 70 (1971).

[81] R. Tykva and I. Votruba, *Anal. Biochem.* **50**, 18 (1972).

[82] J. V. Maizel, Jr., *Science* **151**, 988 (1966).

[83] A. Chrambach, *Anal. Biochem.* **15**, 544 (1966).

[84] B. S. McEwen, *Anal. Biochem.* **25**, 172 (1968).

[85] M. Zaitlin and V. Hariharasubramanian, *Anal. Biochem.* **35**, 296 (1970).

[86] Radioactive steroids are eluted from agar gels by xylol/dioxane based scintillation fluid [R. K. Wagner, *Hoppe-Seyler's Z. Physiol. Chem.* **353**, 1235 (1972)].

[87] The procedures for dissolution of acrylamide gels depend on the cross-linking agent. *N,N'*-Methylenebisacrylamide-linked gels are solubilized by heating in "Soluene," a product of Packard Instruments [P. N. Paus, *Anal. Biochem.* **42**, 372 (1971)]; they can be dispersed for counting by maceration in water and addition of "Instagel," also from Packard [S. A. Leon and A. T. Bohrer, *Anal. Biochem.* **42**, 54 (1971)], or dissolved by warming with H_2O_2 [R. W. Young and H. W. Fulhorst, *Anal. Biochem.* **11**, 389 (1965)]. Gels cross-linked with ethylene diacrylate are dissolved by organic or inorganic bases [G. L. Choules and B. H. Zimm, *Anal. Biochem.* **13**, 336 (1965); D. F. Cain and R. E. Pitney, *ibid.* **22**, 11 (1968); P. G. Spear, *ibid.* **26**, 197 (1968)]. Acrylamide gels cross-linked with *N,N'*-diallyl-tartardiamide are dissolved by 2% periodic acid [H. Anker, *FEBS Lett.* **7**, 293 (1970)].

[88] M. D. Orr, R. L. Blakley, and D. Panagou, *Anal. Biochem.* **45**, 68 (1972).

[89] T. M. Jovin, *Biochemistry* **12**, 871, 879, and 890 (1973).

[90] A. Chrambach and D. Rodbard, *Separ. Sci.* **7**, 663 (1972).

gel, before the sample is applied.[91] This process does not remove unreacted acrylamide monomers and other uncharged contaminants that may adversly affect receptors.[92] Regardless of the nature of the polymerization catalyst, or the use of pre-electrophoresis, the sulfhydryl groups of receptors should be protected against oxidation during electrophoresis. In gel filtration, such protection was provided by uncharged sulfhydryl-containing compounds in the buffers, but these reducing agents interfere with gel polymerization. A feasible but inconvenient alternative is to prepare the gel in the absence of sulfhydryl compounds, remove it from the tube or form, permit the reducing agent to enter by diffusion, and replace the gel tightly in its holder. A more practical method utilizes a strong reducing agent, such as thioglycolate, that is negatively charged at moderate pH. Thioglycolic acid is titrated with the common ion of the buffer system to the pH of the upper buffer and layered on the polymerized gel under (or ahead of) the sample. When the current is applied, thioglycolate migrates more rapidly than the macromolecules, eliminating the oxidizing agents in its path.[92]

Adequate temperature control during electrophoresis is essential both for the survival of the steroid–receptor complex and for valid comparison of receptor mobilities with those of standard proteins. Cooling the system will be more difficult if buffers of relatively high ionic strength are used in electrophoresis, as in gel filtration of receptors. Heat dissipation (as well as wall adherence) is facilitated by forming cylindrical gels in thin-walled tubes of Pyrex rather than Plexiglas. The portion of the tubes containing the gel should be completely submerged in the stirred or circulating lower buffer, which is in contact with a jacket of circulating coolant such as aqueous ethylene glycol at 1°. In thin-layer gel electrophoresis, temperature control may be achieved by forming the gel on a glass plate in contact with circulating coolant.

The inclusion of glycerol or unlabeled or tritiated steroid in the gel and electrode buffers may permit the analysis of receptors that have rapid rates of steroid dissociation or are unstable in the absence of bound steroid under the electrophoretic conditions. If labeled steroid is used, as in the density gradients illustrated by Toft and Sherman,[4] the background level must be sufficiently low so that the receptor complex represents a significant radioactive increment. If unlabeled steroid is present during electrophoresis, the optimal conditions for the subsequent exchange of [^{3}H]steroid onto the receptors must be determined. For sufficiently stable receptors, it may be possible to perform electrophoresis

[91] K. Weber and D. J. Kuter, *J. Biol. Chem.* **246**, 4504 (1971).

[92] M. L. Dirksen and A. Chrambach, *Separ. Sci.* **7**, 747 (1972).

on the unliganded form (without bound steroid). The gel may then be exposed to [^{3}H]steroid under conditions favorable to the entry and binding of steroid, but unfavorable to diffusion of the receptor, followed by removal of the free label. This procedure is analogous to standard histochemical techniques. A comparison of data obtained in the presence and absence of steroid may reveal steroid-induced alterations of conformation or effective net charge.

Evaluation of the size and valence of steroid receptors by gel electrophoresis may be facilitated by the following recommendations. Standards should be selected that have well-established values for $\bar{v}$, M, and R_S or the diffusion coefficient [see Eq. (3)]. Data for the free solution mobilities of the standards under various conditions or the titration curves are also desirable. From at least three pairs of values for m_f and T for each standard and binding component, a graph may be constructed according to Eq. (12) from which the values of K_R and Y_0 are determined. In theory, the intercept of this graph at $T = 0$ should depend only on the free solution mobility of a macromolecule. In practice, Y_0 has been shown to vary as a function of the degree of cross-linking of the gels.[93] Calculation of the absolute mobility of the receptor in free solution (m_0) from Y_0 requires knowledge of the absolute mobility of the front. Data for the latter are available for numerous electrophoretic systems[94] and can be calculated for other systems according to the theoretical treatment of Jovin.[59]

Steroid–receptor complexes are more likely to resemble globular proteins than random coils. A dependence of $(K_R)^{1/2}$ rather than $(K_R)^{1/3}$ on molecular radius is therefore expected in gels with long fibers.[68] The choice of correlation functions is then narrowed to a linear dependence of $(K_R)^{1/2}$ on $\bar{R}$, or $(K_R)^{1/2}$ on R_S. Data for the given electrophoretic system and standard proteins may be analyzed in terms of both functions and the deviations from linearity compared. If a dependence of K_R on $\bar{R}$ is assumed, the value of $\bar{R}$ obtained for the receptor can be combined with an estimated or separately determined value for $\bar{v}$ to calculate M by Eq. (6). If K_R is correlated with R_S, the resultant Stokes radius for the receptor may be interpreted in conjunction with sedimentation data as described in the section on gel filtration.

The values of m_0 and molecular radius for the receptor can be combined to estimate the net charge (coulombs/molecule) or valence (number of excess positive or negative charges/molecule in electronic units)

[93] D. Rodbard, C. Levitov, and A. Chrambach, *Separ. Sci.* **7**, 705 (1972).

[94] T. M. Jovin, M. L. Dante, and A. Chrambach, "Multiphasic Buffer Systems Output," P.B. No. 196085 to 196092 and 203016. Nat. Tech. Inform. Serv., Springfield, Virginia, 1970.

in the given ionic conditions.[95] This calculation is simplified by assuming that the valence is less than 10 and that the macromolecule resembles a rigid, nonconducting sphere with uniform surface charge.[72] Electrophoretic data for chick oviduct progesterone receptors[47] and the apparent similarities of many other receptors,[96] however, suggest that these macromolecules may not conform to the simple model described above. In particular, correction terms should most likely be included to account for both molecular asymmetry[95] and a (negative) valence exceeding 10,[97] for experiments at high pH.

In summary, analytical gel electrophoresis demands rigorous technique and statistical evaluation of the results. Chemical interactions of receptors with reactants or products of acrylamide polymerization must be avoided. The sensitivities of the steroid–receptor complex to heat or oxidation of sulfhydryl groups must be respected. It is essential to locate the labeled steroid without loss of the resolution attained by electrophoresis and to verify that the radioactivity coincides with a macromolecular complex. The recommended procedure utilizes gels of several concentrations, in which the mobilities of the complex and of standard proteins are measured relative to a dye. Calculation of the free solution mobility of the complex requires a long extrapolation of the data to zero gel concentration and knowledge of the absolute mobility of the dye in the buffer system used. Receptor dimensions are calculated from the (negative) slope of a graph of the logarithm of mobility vs. gel concentration. This parameter, called the retardation coefficient, varies with the surface area or volume of the macromolecule, depending on whether the gel has long or short fibers. The measured mobilities are proportional to $\exp [-(R + r)^2]$ or $\exp [-(R + r)^3]$, where r is the radius of the fiber and R is the geometric mean or Stokes radius of the macromolecule. For either definition of the radius, the value obtained by gel electrophoresis is based on many assumptions and subject to many limitations. When possible, this value should be corroborated by the application of an independent technique under identical conditions.

Isoelectric Focusing

Principle. Although steroid receptors were referred to above as "acidic macromolecules," like other proteins and ampholytes they contain basic

[95] H. A. Abramson, L. S. Moyer, and M. H. Gorin, "Electrophoresis of Proteins and the Chemistry of Cell Surfaces." Hafner, New York, 1964.

[96] M. R. Sherman, *Advan. Biosci.* **7,** 369 (1971).

[97] M. H. Gorin, *J. Phys. Chem.* **45,** 371 (1941).

as well as acidic groups.[98] The isoelectric point (pI) of an ampholyte is defined as the pH at which it carries no net charge, and therefore does not migrate in an electric field. In isoelectric focusing (IF), also known as electrofocusing, a temporarily linear pH gradient is formed by applying a current to a mixture of low molecular weight synthetic ampholytes, differing in numbers of carboxyl and amino groups, and contained between strongly acidic and basic electrolyte reservoirs. Stabilization of the pH gradient against convection is provided by a solid or granulated gel or by a superimposed density gradient, and by adequate temperature control. Macromolecules may be mixed with the carrier ampholytes or injected into a preformed pH gradient. When the current is applied, they migrate to the position in the gel or column corresponding in pH to their isoelectric points and concentrate into sharp zones. Isoelectric focusing may be performed under conditions in which receptors retain significant amounts of bound [^{3}H]steroid. Subsequent fractionation of the gel or gradient and measurements of the pH and radioactivity in the fractions permit an estimation of the pI of the hormone–receptor complex. The value obtained in this way is an *apparent* pI, since it neglects the presence of residual counterions, carrier ampholytes, or adjacent macromolecules.

Application to Receptors. Excellent recent review articles are available on IF in density gradients,[99-103] in polyacrylamide gels,[99,100,104-107] and in Sephadex.[108] Isoelectric focusing in density gradients is useful as both a preparative and analytical tool, and permits convenient recovery of the fractionated macromolecules. Gel supports have the advantages of preventing disturbance by isoelectric macromolecular precipitates and of facilitating rapid, simultaneous analyses of several samples. The special problems inherent in IF in gels are analogous to those encountered in gel electrophoresis, viz., receptors must be protected from damage by reactants or products of the gelation process. In addition, the rate of migration of a receptor through the gel matrix toward its pI should be optimized by selecting a gel that is mechanically stable but causes mini-

[98] T. Gorell, E. R. DeSombre, and E. V. Jensen, in preparation.
[99] H. Haglund, *Methods Biochem. Anal.* **19,** 1 (1971).
[100] N. Catsimpoolas, *Separ. Sci.* **8,** 71 (1973).
[101] O. Vesterberg, *in* "Methods in Microbiology" (J. R. Norris and D. W. Ribbons, eds.), Vol. 5B, p. 595. Academic Press, New York, 1971.
[102] O. Vesterberg, Vol. 22 [33].
[103] O. Vesterberg, *Protides Biol. Fluids, Proc. Colloq.* **17,** 383 (1970).
[104] C. W. Wrigley, *Protides Biol. Fluids, Proc. Colloq.* **17,** 417 (1970).
[105] C. W. Wrigley, Vol. 22 [38].
[106] N. Catsimpoolas, *Separ. Sci.* **5,** 523 (1970).
[107] G. R. Finlayson and A. Chrambach, *Anal. Biochem.* **40,** 292 (1971).
[108] B. J. Radola, *Biochim. Biophys. Acta* **295,** 412 (1973).

mal molecular sieving.[107,108a] This is done by performing electrophoresis on the receptor in gels of several concentrations (with constant cross-linking), plotting the relative mobilities according to Eq. (12), and extrapolating the results to the gel concentration corresponding to $m_f = 1$.

The design of density gradient IF experiments on receptors involves decisions concerning (1) the analytical or preparative scale, (2) nature of the anticonvectant medium, (3) pI range of the carrier ampholytes, (4) relative position of the anode and cathode, (5) methods of sample preparation and introduction into the column, (6) protection of sulfhydryl groups, (7) duration of focusing, (8) methods of fraction collection and pH determination, (9) detection of bound steroid or steroid-binding activity, and solubilization of isoelectric precipitates of the receptor or hormone–receptor complex. These topics will be briefly discussed.

(1) The first columns that were commercially available for density gradient IF were made in two sizes, 110 and 440 ml, by LKB. These devices require large quantities of expensive carrier ampholytes ("Ampholine," also made by LKB) and must be operated for one to three days to achieve the steady-state. Alternative equipment for density gradient IF permits intermittent inspection of the column contents (ISCO, Lincoln, Neb.) or continuous scanning during focusing.[109] Although receptors labeled with [^{3}H]steroids are detectable only by radioactivity measurements, optical scanning is useful to determine when other proteins in the sample have reached the steady-state. While columns of high resolving power are desirable for final pI determinations, preliminary experiments may be performed on a smaller scale, e.g., 7–14 ml.[110-112] Focusing in flow-through spectrophotometer cuvets[113,114] represents an experimentally more difficult application of IF at the micro-scale.

(2) Column electrofocusing, like density gradient centrifugation, has most often utilized sucrose gradients. Adequate anticonvective support and superior stabilization of receptors or steroid–receptor complexes, however, may be attainable with gradients of up to 75% ethylene glycol[115] or 20 to 70% glycerol.[116]

[108a] A. Chrambach, P. Doerr, G. R. Finlayson, L. E. M. Miles, R. Sherins, and D. Rodbard, *Ann. N.Y. Acad. Sci.* **209,** 44 (1973).

[109] N. Catsimpoolas, *Separ. Sci.* **6,** 435 (1971).

[110] D. L. Weller, A. Heaney, and R. E. Sjogren, *Biochim. Biophys. Acta* **168,** 576 (1968).

[111] L. Osterman, *Sci. Tools* **17,** 31 (1970).

[112] G. N. Godson, *Anal. Biochem.* **35,** 66 (1970).

[113] S. Fredriksson, *Anal. Biochem.* **50,** 575 (1972).

[114] M. Jonsson, S. Pettersson, and H. Rilbe, *Anal. Biochem.* **51,** 557 (1973).

[115] E. Ahlgren, K. E. Eriksson, and O. Vesterberg, *Acta Chem. Scand.* **21,** 937 (1967).

[116] T. Wadström and K. Hisatsune, *Biochem. J.* **120,** 725 (1970).

(3) Ampholytes of a wide pI range (e.g., 3–10) should be used in preliminary experiments, followed by narrower ranges. Although a receptor has an acidic pI, it may be most stable at physiological or higher pH. An analogous situation has hampered IF studies of native corticosteroid-binding globulin.[117,118] When receptors are focused in ampholytes of a narrow pI range (e.g., 4–7), it is advisable to bridge the pH region toward the cathode with carrier ampholytes of intermediate pI (e.g., 7–9), see (5) below.

(4) The steady-state position of a receptor in a pH gradient should not depend on the direction of vertical movement from the point of injection toward its pI. One test of the validity of an IF result is to reverse the positions of the anode and cathode and the respective electrode solutions (acid and base). Under certain conditions, it is more convenient to place the cathode at the top [see (6) below]. The acidic isoelectric points and the enhanced stability of receptors in high concentrations of sucrose or glycerol[4] may also favor this arrangement.

(5) One disadvantage of IF in density gradients, as opposed to gels, is that the density gradient may fail to support macromolecules that are precipitated at their isoelectric points.[100] Since the concentrations of receptors in cellular extracts are extremely low, it is unlikely that copious precipitates would be formed by receptors themselves. On the other hand, receptors or radioactive steroids initially bound to them might become trapped in sedimenting precipitates of contaminating macromolecules. This could result in an incorrect and irreproducible pI for the complex. Receptors should therefore be focused in density gradients only after partial purification, e.g., after centrifugation in density gradients of low ionic strength.[43] Media of high ionic strength are incompatible with IF, since the conductance due to salts would delay the attainment of the steady-state. Gradient centrifugation serves, at the same time, to separate the steroid–receptor complex from unbound steroid. This may also be accomplished by charcoal–dextran treatment[20,21] or by gel filtration.

The method and timing of introducing a sample into the column are factors in the recovery and isolation of the labeled complex. There are several advantages to injecting the sample into the neutral region of a prefocused pH gradient, rather than combining it with the light and/or dense solutions containing the mixed carrier ampholytes before forming the density gradient. The former procedure reduces the time required to focus the receptor, and hence the time available for dissociation of the steroid label. It avoids contact of the receptor with the electrode solutions

[117] U. Westphal, Vol. 15 [41].

[118] H. Van Baelen, M. Beck, and P. De Moor, *J. Biol. Chem.* **247,** 2699 (1972).

or with ampholytes of extremely high or low pI, and precludes the irregular distribution of unbound radioactive steroid in the column that may result from adding the sample to the initial ampholyte solutions. An unbound steroid that does not migrate in the electric field forms a diffuse band around the position of injection (e.g., pH 7) and does not obscure the sharp band of the steroid–receptor complex at lower pH. If the steroid migrates in the field, its position in the ampholyte gradient at the conclusion of focusing should be determined in separate experiments.

To prepare a sample for injection, a calculated amount of the dense reagent (concentrated sucrose solution or glycerol) is added to adjust its density to that of the desired position in the focused ampholyte gradient. With the current turned off, a piece of narrow polyethylene tubing is carefully lowered into the gradient. Two milliliters (of a 110-ml column) are withdrawn into a disposable syringe containing the labeled receptor and are slowly reinjected into the same position in the column. A low molecular weight amphoteric dye may be added to the sample to visualize the initial band position and width before the current is reapplied.[119] After reaching its pI, the dye becomes undetectable as a result of diffusion.

(6) Sulfhydryl groups of receptors may be protected during IF by the presence of dithiothreitol or similar reagents throughout the gradient. Since these compounds become partially dissociated at high pH and migrate toward the anode, additional aliquots may have to be added intermittently at the cathode.[116]

(7) Among the factors that determine the time required for a receptor to reach the steady-state are the applied voltage, the temperature (which should be about 4° but not lower[99]), the steepness of the pH gradient, the distance betwen the position of layering, and the pI, or the dimensions of the column if the sample is uniformly distributed at the start of the experiment. An asymptotic approach to a constant low amperage and the observation of a linear variation in pH over 70 to 80% of the gradient suggest that the carrier ampholytes, but not necessarily the macromolecules, have reached the steady-state. When the receptor has reached its steady-state, neither further focusing nor increased voltage alters the observed pI. It should be noted, however, that receptor bands eluted from the column after different times of focusing may exhibit the same pI, but different elution volumes. This is presumably due to an electroendosmotic effect, by which bands at the steady-state are progressively displaced toward the periphery of the gradient.[109]

(8) Procedures have been investigated for the continuous monitoring

[119] A. Conway-Jacobs and L. M. Lewin, *Anal. Biochem.* 43, 394 (1971).

of pH in a flow-cell during elution of a focused gradient.[120] With available equipment, however, the flow rate must be extremely slow, to avoid pressure on the glass membrane of the electrode and consequent distortion of the pH measurement. The conventional method is to determine the pH on fractions of the gradient, collected from the bottom of the column, or from the top by means of an underlayer of very dense solution. It is preferable to make pH measurements at the temperature of focusing, since temperature corrections calculated for the pH of carrier ampholytes may not be applicable to the receptors. Absorption of CO_2 from the atmosphere may decrease the measured pH of the more basic fractions, but has little effect on the fractions containing receptors. Irreproducible pH measurements on fractions at the bottom of IF density gradients have been observed and were attributed to osmotic effects on the pH electrode.[113]

(9) The potential hazards of simply counting radioactivity in gradient fractions after focusing a labeled steroid-binding component have been analyzed in a careful study of desialylated corticosteroid-binding globulin.[118] First, it should be ascertained whether quenching of radioactivity varies significantly with pH or with the concentration of the dense reagent. Second, the effects of pH (or preferably the isoelectrically fractionated ampholytes) and of the dense compound on the stability of the steroid–receptor complex should be evaluated. It is conceivable, for example, that a complex initially layered at its position of maximal stability (e.g., pH 7 to 8) would become denatured and release most of the bound steroid at around pH 6, although the pI of the macromolecule was 5. This process might be indicated by a diffuse or trailing pattern of radioactivity, rather than a sharp focused zone.

The demonstration that a steroid–receptor complex is intact at its pI may be hindered by the formation of an invisible isoelectric precipitate. In this case, treatment of the labeled gradient fraction with charcoal–dextran and centrifugation might remove the bound as well as the free steroid by trapping the macromolecular precipitate. The soluble form of the complex might be recovered from an isoelectric precipitate by high-speed centrifugation of the gradient fraction (in the absence of charcoal) and resuspension of the pellet in a buffer of higher pH and/or ionic strength.

In conclusion, every attempt should be made to relate the observed isoelectric points to the forms of the receptor characterized by other physical–chemical techniques. Differences between isoelectric points determined in the presence and absence of steroid may reveal effects of the steroid on receptor structure that are not detectable as changes in

[120] M. Jonsson, E. Pettersson, and H. Rilbe, *Acta Chem. Scand.* **23**, 1553 (1969).

parameters of size. Isoelectric focusing, following enzymatic treatment, provides a sensitive tool in the search for non-protein moieties, e.g., phosphate or nucleotides, that may not be involved in steroid binding but may be essential to the function of receptors in steroid-responsive organs.

Acknowledgments

I wish to thank Dr. Andreas Chrambach for valuable discussions and review of the manuscript, Miss Sui Bi Atienza for skillful assistance in experiments with these techniques, and The Paul Garrett Fund, American Cancer Society (Grant PRA-83) and National Institutes of Health (Grant CA-08748) for generous support.

[19] A Filter Technique for Measurement of Steroid-Receptor Binding[1]

By JOHN D. BAXTER,[1a] DANIEL V. SANTI, and GUY G. ROUSSEAU[1b]

Widespread interest in steroid-hormone receptors has stimulated the need for rapid and accurate assay techniques. The method described here depends on the fact that many receptor–steroid complexes bind to diethylaminoethyl (DEAE)-cellulose filters under conditions where free steroids are not retained.[1c] This technique is relatively simple, and several hundred assays may be done daily by one person. It can be performed both with less material and in more dilute solutions of receptor than the charcoal adsorption technique.[2] Nitrocellulose filters, commonly used for studying protein–ligand interactions,[3] cannot be employed because they bind free steroids.[4]

[1] Supported by Grant GM7-17239 from the National Institute of General Medical Sciences and USPHS Grant CA14266 from the National Cancer Institute.

[1a] Recipient of a Career Development Award from the National Institute of Arthritis and Metabolic Diseases, NIH, No. 1-KO4-AM-70528-01.

[1b] Chercheur Qualifié du Fonds National de la Recherche Scientifique (Belgium) and recipient of a Public Health Service International Post-Doctoral Fellowship, No. 1-F05-TW-1725-02.

[1c] D. V. Santi, C. H. Sibley, E. R. Perriard, G. M. Tomkins, and J. D. Baxter, *Biochemistry* **2,** 2412 (1973).

[2] G. G. Rousseau, J. D. Baxter, and G. M. Tomkins, *J. Mol. Biol.* **67,** 99 (1972).

[3] M. Yarus, and P. Berg, *Anal. Biochem.* **35,** 450 (1970).

[4] J. D. Baxter, unpublished observations.

The technique describes dexamethasone binding by specific glucocorticoid receptors in cytosol fractions from cultured hepatoma (HTC) cells.[1,2] The assay has also been used to measure binding of dexamethasone in other glucocorticoid-target tissues and of estrogens and mineralocorticoids by receptors from their respective target tissues.

Materials

[^{3}H]Dexamethasone may be purchased from New England Nuclear Corp., Amersham-Searle, or Schwarz. Nonradioactive dexamethasone may be obtained from Schwarz-Mann. DEAE-cellulose filter paper disks (DE81, 2.4 cm) may be purchased from Whatman Co. Buffer solutions used are: (*A*) 20 m*M* Tricine [*N*-tris-(hydroxymethyl)methyl glycine] 2 m*M* $CaCl_2$, 1 m*M* $MgCl_2$, pH 7.9 at 0°. (*B*) 20 m*M* Tricine, 1.5 m*M* EDTA, pH 7.9 at 0°. A 10-place filter manifold may be obtained from Hoeffer Scientific Instruments. Scintillation fluid contains 0.4% (w/v) Omnifluor (New England Nuclear Corp.), 25% Triton X-100, and 4% H_2O in toluene (CP, Mallinckrodt).

Procedure

The culture and harvest of HTC cells, as well as preparation of cytoplasmic extracts (cytosol) and of steroid solutions, are described elsewhere in this volume.[5] Identical results are obtained when either buffer *A* or buffer *B* is used for preparing cytosol, solutions, and incubations, and for washing filters.

The standard incubation mixture contains, in 0.15 ml of buffers *A* or *B*, 4×10^{-8} *M* [^{3}H]dexamethasone and a limiting amount of receptor sites ($<2 \times 10^{-8}$ *M*; 0.075 ml of cell extract containing about 2 mg protein). A parallel incubation is prepared of identical composition with the addition of 10^{-5} *M* nonradioactive dexamethasone as competitor. The incubations are initiated by adding the receptor and are allowed to proceed for 90 minutes at 0 to 4°.

Filters are soaked in buffer before use, placed on the filter manifold, and freed of excess moisture by applying gentle vacuum. Care should be taken to remove the vacuum before applying the samples and to avoid drying the filter. After incubation, 50-μl portions are uniformly applied to each filter. After at least 1 minute, gentle vacuum is again applied, and the filters are washed 5 times with 1 ml of buffer. Suction is then increased to remove excess moisture, and the damp filters are placed in

[5] J. D. Baxter, S. J. Higgins, and G. G. Rousseau, see this Vol. [20].

5 ml of scintillation fluid. After sufficient time (about 6 hours) for dissolution of steroid, radioactivity is measured.

Analysis of the Data

The amount of steroid specifically bound by glucocorticoid receptor is calculated as follows: radioactivity retained by the filter in the incubation without competing nonradioactive steroid, minus radioactivity retained by the filter in the sample containing the competitor ("background"). The background value corresponds to nonspecific binding of steroid by the filter. Under conditions employed, it is less than 3% of the applied radioactivity. Rationale for this competition technique is described elsewhere in this volume.[5]

Limitations of the Assay

An experiment in which the concentration of dexamethasone was varied with a constant amount of cytosol is shown in Fig. 1. The saturation of specific receptors by dexamethasone at concentrations above 10^{-8} *M* is readily apparent. Nonspecifically bound steroid (also shown) represents a small proportion of total radioactivity retained by the filter (sum

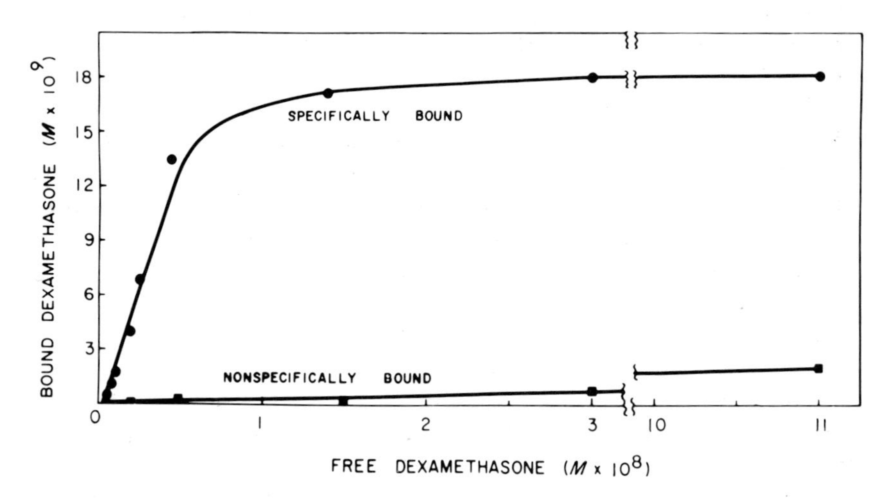

FIG. 1. DEAE-cellulose filter assay of specific binding of [^{3}H]dexamethasone by glucocorticoid receptors in HTC cell cytosol. Nonspecifically bound dexamethasone is also shown. The experiment was performed as described in the text except that the concentration of [^{3}H]dexamethasone was varied. Free dexamethasone concentration is calculated by subtracting the bound from the total steroid concentration in the incubation.

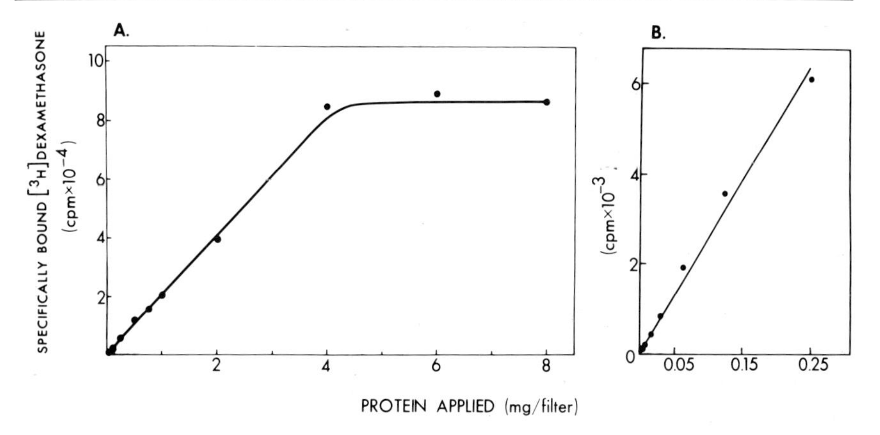

FIG. 2. Linearity of the assay. Specific binding was measured as described in the text except that after binding the amount of reaction mixture applied to the filters was varied (indicated on the abscissa in milligrams of cytosol protein applied).

of both curves). Since this nonspecific binding increases linearly with increasing steroid concentration, specifically bound steroid (which does not increase after the receptors are saturated) becomes a progressively smaller proportion of the total and thus is more difficult to measure accurately. However, this is not a problem at concentrations of [^{3}H]dexamethasone below 10^{-7} *M* (Fig. 1).

The assay is linear below 4 mg of cytosol protein applied to the filter (Fig. 2A); above this concentration, filters are apparently saturated with protein and cannot retain all specifically bound steroid. It should be emphasized that the assay is also linear at low protein concentrations below 0.1 mg (Fig. 2B). At present, the lower limit of linearity appears to be determined only by specific activity of the steroid.

Optimal binding is obtained if the extract is allowed to absorb to the filter for at least 1 minute; longer periods (up to 5 minutes) result in only a slight decrease in specifically bound steroid.

The experimental procedure calls for five 1-ml washes of the filter which is important for reducing background radioactivity (Fig. 3). The background is relatively constant after three 1-ml washes; identical results are obtained with two 5-ml washes. Since dexamethasone–receptor complexes are eluted from DEAE columns by increasing ionic strength,[4] it is not surprising that specifically bound dexamethasone is washed from the filter with buffers of higher ionic strength ($\mu > 0.9$). Retention of specifically bound complex is unaffected by pH of the wash between 7.2 and 8.6; at pH 6.3 and above pH 8.9 the results are erratic. For this

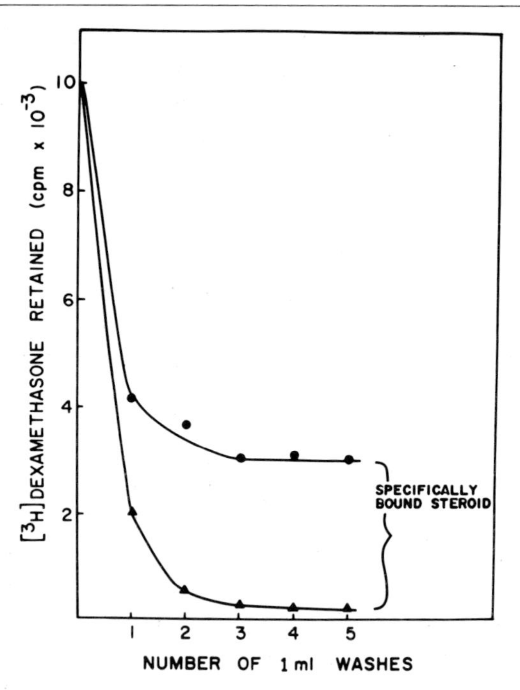

FIG. 3. Influence of washing on [^{3}H]dexamethasone retained by DEAE-cellulose filters. The experiment was performed as described in the text except that the number of washes of the filters was varied. Radioactivity retained from incubations devoid (●) and containing (▲) excess competing nonradioactive dexamethasone is shown. As noted, the difference between these values indicates specific binding.

reason we use buffers at pH 7.2 to 8.3 and favor Tris or Tricine at pH 7.9. The background is slightly higher with phosphate buffer.

The binding measured in cytosol fractions by charcoal adsorption,[2] gel filtration,[2] or the DEAE filter assays[1] is identical. However, the filter assay is not accurate with cytosol preparations in which receptor properties have been altered by heating,[6] suggesting these modified receptor–steroid complexes do not bind to DEAE.

Efficiency of the assay, i.e., probability that the receptor–steroid complex will survive filtration and washing procedures, and thus be measured,[3] is at least 88%. Dexamethasone–receptor binding was measured at low dexamethasone and increasing receptor concentration.[1] Efficiency was then obtained from a double reciprocal plot of the data by determining the percentage of added steroid which is bound at infinite receptor

[6] S. J. Higgins, G. G. Rousseau, J. D. Baxter, and G. M. Tomkins, *J. Biol. Chem.* **248,** 5866 (1973).

concentration. Indication of efficiency can be obtained from the data in Fig. 1, since the initial slope of the curve corresponds to conditions where steroid is limiting. In this case, 86% of total added steroid was specifically retained by the filter, indicating an efficiency similar to that determined above.

Other Applications of the Assay

The filter assay has also been tested in the following systems: dexamethasone binding by glucocorticoid receptors in rat liver and cultured lymphoma cells, estradiol binding by uterine estrogen receptors, and aldosterone binding by mineralocorticoid receptors in rat kidney.

Kidneys and liver are obtained from adrenalectomized rats (250–300 g) maintained on saline solution for at least 3 or 4 days before sacrifice (cervical dislocation). Before removing the organs, animals are perfused with at least 50 ml of ice-cold 0.1 *M* NaCl, 0.025 *M* potassium phosphate, pH 7.6, to chill the tissues and to remove as much blood as possible. The tissues are minced with 1 volume of buffer *A* and homogenized with 8 strokes in a motor-driven (1500 rpm) Teflon-glass homogenizer (Kontes Glass Co). Lymphoma cell cytosol is prepared as described elsewhere.[7] Uteri, obtained from 21-day-old rats, are minced with 1 volume of buffer *B* and homogenized as described above. Homogenates are centrifuged at 100,000 *g* for 2 hours and the supernatant fractions are used for binding studies. Binding is then measured as described for HTC cells and dexamethasone except that buffer *B* is used in experiments with the uterus–estradiol system.

Although not studied as extensively, characteristics of the assay in every case are similar to those of dexamethasone binding by HTC cell receptors. Also, equilibrium constants and concentration of receptor sites are in agreement with those obtained by other methods.[1] Therefore it appears that the DEAE filter assay may be widely adaptable as an accurate and simple assay for measuring steroid–receptor interactions.

[7] W. Rosenau, J. D. Baxter, G. G. Rousseau, and G. M. Tomkins, *Nature (London) New Biol.* **237**, 20 (1972).

[20] Measurement of Specific Binding of a Ligand in Intact Cells: Dexamethasone Binding by Cultured Hepatoma Cells

By JOHN D. BAXTER,[1] STEPHEN J. HIGGINS,[1a] and GUY G. ROUSSEAU[1b]

Hormone and drug receptors have in recent years received considerable attention. It is frequently advantageous to study binding of the effector in intact cells under conditions required for the biological response to occur. Data obtained in this way can often be readily related to the drug or hormone effects. Further, by studying the intact cell, one can avoid problems such as receptor instability, as is observed in cell-free systems, or lack of knowledge of the subcellular site of binding. A suggested approach to this problem is described here using, as an example, binding of dexamethasone by cultured hepatoma (HTC) cells. In these cells, dexamethasone and other glucocorticoid steroids specifically induce the synthesis of tyrosine and aminotransferase and a few other proteins.[1c]

Three procedures are described. (1) Binding of intact cells under conditions in which the biological response occurs.[2] This technique is especially useful when the affinity and association–dissociation rates are unknown or when substantial dissociation of the ligand from its receptors could occur if cells were washed free of the medium in which they are incubated. (2) Binding under the same conditions followed by chilling the cells to slow the rate of receptor–ligand complex dissociation and washing them to reduce the amount of nonspecifically bound ligand. This technique is useful either when dissociation of specifically bound material is slow or when the cells can be treated (as by chilling) to decrease the dissociation. (3) Subcellular localization of the ligand specifically bound in the intact cell, again taking advantage of a slow dissociation at 0°.

For all these procedures, the competition principle is employed in which incubations of cells with [^{3}H]ligand are performed in parallel to

[1] Recipient of a Research Career Development Award from the National Institute of Arthritis and Metabolic Diseases, NIH, No. 1-KO4-AM-70528-01.

[1a] Recipient of a Damon Runyon Memorial Fund Fellowship, No. DRF-630.

[1b] Chercheur Qualifié du Fonds National de la Recherche Scientifique (Belgium) and recipient of a Public Health Service International Postdoctoral Fellowship, No. 1-FO5-TW-1725-02.

[1c] G. M. Tomkins, T. D. Gelehrter, D. Granner, D. Martin, Jr., H. H. Samuels, and E. B. Thompson, *Science* **166**, 1474 (1969).

[2] J. D. Baxter and G. M. Tomkins, *Proc. Nat. Acad. Sci. U.S.* **65**, 709 (1970).

incubations with [^{3}H]ligand plus excess nonradioactive ligand. The latter inhibits specific, but not nonspecific, binding of the [^{3}H]ligand and thus allows determination of specific binding.

Materials

Cultured hepatoma cells are grown as described elsewhere in loosely capped flasks in spinner culture at 37°, to a density of 2–10 $\times$ 10^5 cells/ml in Swimm's 77 medium (Grand Island Biological Co., New York), buffered with 0.05 *M* Tricine, pH 7.6 to 7.8, and supplemented by 0.002 *M* glutamine, 0.5 g/liter $NaHCO_3$, and 10% of either calf serum or a mixture (1:1) of fetal calf and calf serum.[3]

Dexamethasone (obtained from Schwarz-Mann) is prepared as a 0.005 *M* solution in ethanol. Dexamethasone ([1,2,4-^{3}H] or [1,2-^{3}H]) may be purchased from Amersham-Searle, Schwarz Bioresearch, or New England Nuclear Corp. and its purity is verified by thin-layer chromatography as described elsewhere.[2] Induction medium is the same as that in which the cells are grown except that it lacks serum. Homogenization buffer is 0.02 *M* *N*-Tris-(hydroxymethyl)methyl glycine (Tricine), 0.002 *M* $CaCl_2$, 0.001 *M* $MgCl_2$, pH 7.8 at 0°. Phosphate-buffered saline is 0.1 *M* NaCl, 0.025 *M* potassium phosphate, pH 7.6 at 0°. [^{14}C]Inulin may be obtained as sterile, pyrogen-free solution (50 μCi/10 ml, 1–3 mCi/g) from New England Nuclear Corp. Scintillation fluid is toluene (Mallinckrodt, CP) with 8 g/liter Butyl-PBD (Ciba) or 4 g/liter Omnifluor (New England Nuclear Corp), and 10% Biosolv (Formula BBS-3, Beckman Instrument Co.), or 250 ml/liter Triton X-100, 710 ml/liter toluene, 40 ml/liter water, and 4 g/liter of Omnifluor.

Binding by Intact Cells

Principle. Cells are incubated with [^{3}H]ligand. At the end of the incubation [^{14}C]inulin is added, cells are centrifuged, and the supernatant medium is removed. The cells are resuspended and assayed for radioactivity and protein. ^{3}H Radioactivity reflects ligand associated with cells plus that in incubation medium trapped in the pellet. Since inulin is excluded from the cells, ^{14}C radioactivity reflects the amount of extracellular fluid in the cell pellet. ^{3}H Radioactivity corresponding to that amount of fluid is then subtracted from total ^{3}H radioactivity in the cell pellet, yielding the amount of cell-associated [^{3}H]ligand.

Procedure. Sterility must be maintained unless the experiment is of

[3] A. Hershko and G. M. Tomkins, *J. Biol. Chem.* **246,** 710 (1971).

short duration (e.g., less than 2 hours). A solution of [^{3}H]dexamethasone in induction medium is made and the benzene is removed by evaporation in a stream of air. One has to take into account the solubility of the steroid. There is no problem at a concentration of dexamethasone below 5×10^{-5} *M*.

Cultured hepatoma cells are harvested by centrifugation at room temperature (600 *g*, 5 minutes). The pellet is immediately resuspended in induction medium. The reaction is started by adding the [^{3}H]dexamethasone-containing medium to the cell suspension. To have enough radioactivity in the cell pellet, the final incubation mixture should contain at least 30 ml of cells at 10^{6}/ml. This cell density should not be exceeded in experiments lasting longer than 45 minutes. Higher densities may be used in shorter experiments. The cells are kept in suspension in loosely capped bottles at 37° during the incubation by shaking or stirring (100–125 rpm). After 30 minutes (by which time binding is maximal), [^{14}C]inulin is added to the reaction mixture. Ordinarily, 0.075 ml/30 ml incubation will suffice. The cells are then collected by centrifugation at room temperature at 800 *g* for 3 minutes. The supernatant medium is immediately removed and assayed for ^{3}H and ^{14}C. The pellet is resuspended by agitation after addition of 0.5 to 1 ml (per 3×10^{7} cells) of phosphate-buffered saline, and portions (ordinarily 0.2 ml) are assayed for cell density (hemacytometer), radioactivity, and protein.[4]

Several points deserve comment. First, time of incubation required for reaching equilibrium should be determined. As a rough indication of minimal time which may be required, one can refer to the time course of the biological response. For instance, induction of tyrosine aminotransferase by dexamethasone in HTC cells can be detected within 30 minutes and is maximal by 16 hours. In fact, even at low concentrations, dexamethasone binding by HTC cells is maximal by 30 minutes. Second, one has to make sure inulin is actually excluded from the cells under the experimental conditions. In the case of HTC cells, this was determined by comparing the volume of distribution of inulin with that of trypan blue which is excluded from the cells as verified by light microscopy. Probably any compound which is generally excluded from cells such as sucrose, albumin, or dextran would suffice. Third, with HTC cells and dexamethasone, the amount of cell-associated ^{3}H reflects cell-associated dexamethasone, since there is little or no metabolic conversion of this steroid by these cells. In other cases, it should be determined whether the radioactive material in the pellet and the incubation medium are the

[4] O. H. Lowry, N. J. Rosebrough, A. L. Farr, and R. J. Randall, *J. Biol. Chem.* **193**, 265 (1951).

original materials or metabolites. This illustrates an advantage of examining binding in intact cells: If specific binding is found and is related to the biological response, chemical nature of the specifically bound material can indicate the nature of the authentic biological effector.

Analysis of the Data. Data may be analyzed as follows: (1) Concentration of free steroid in the incubation medium is determined from ^{3}H cpm in the supernatant medium. (2) The cell-associated steroid is obtained from the difference between ^{3}H cpm in the pellet and that due to extracellular fluid trapped in the pellet. The latter equals the ^{14}C cpm in the pellet times the ratio of ^{3}H to ^{14}C in the supernatant medium.

Figure 1 shows results of such an experiment. The logarithmic scale is used to show the binding obtained over a wide range of steroid concentrations. This binding is linearly related to the free steroid at higher concentrations and this line is extrapolated at the lower concentrations (dotted line) where there is a superimposed hyperbolic curve which merges into the linear one between 10^{-8} and 10^{-7} *M* free steroid. These data suggest that the steroid associates with at least two distinct classes

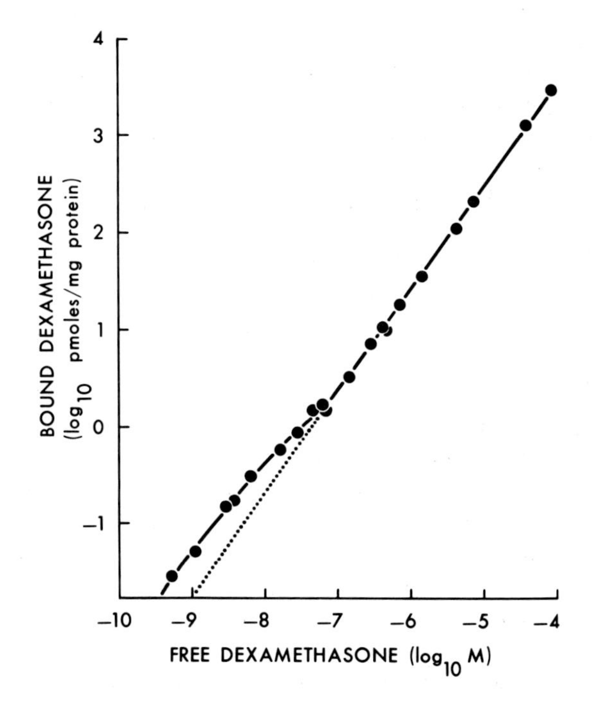

FIG. 1. Association of dexamethasone with intact HTC cells determined as described in the text. Data reprinted (with permission from the publisher) from J. D. Baxter and G. M. Tomkins.[2]

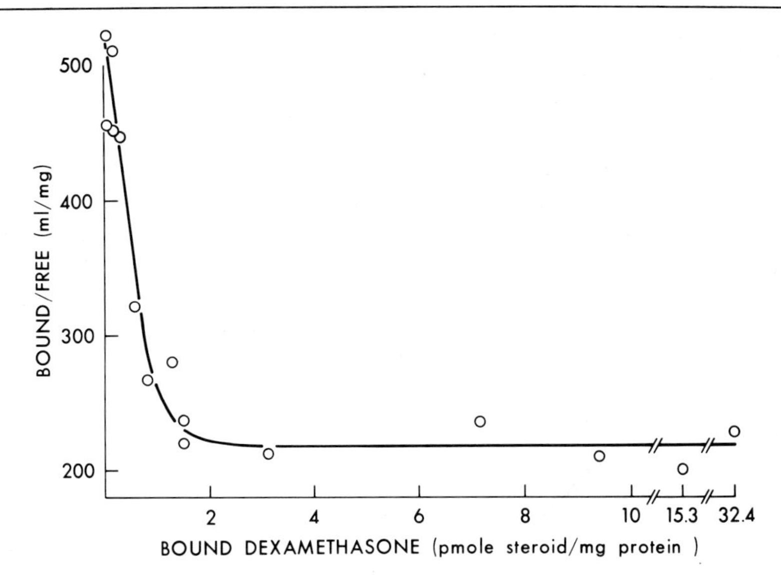

FIG. 2. Scatchard analysis[5] of the data shown in Fig. 1.

of sites. This is supported by a Scatchard[5] analysis (Fig. 2) which suggests the presence of a component of high affinity and low capacity ("specific" component, corresponding to the steep part of the curve), and of a "nonspecific" component of much higher capacity and much lower affinity, corresponding to the part of the curve which is nearly parallel to the abscissa. So far, the terms "binding" and cell association have been used synonymously for convenience. Other experiments do indicate that the specific component is due to the glucocorticoid receptor proteins. However, the foregoing data do not prove steroid binding since cell association could, for example, result from active transport or from different steroid solubility. Affinity and number of sites of the high-affinity component may be computed from the slope and intercept on the abscissa of the steep portion of the curve, providing one corrects for contribution of the second compartment. (Otherwise the affinity would be underestimated and the number of sites would be overestimated.)

Measurement of Specific Binding by Competition Analysis. From the data discussed above, the specific binding component becomes saturated at low concentrations of free steroid (below 10^{-7} M), whereas only a small proportion of the low-affinity sites are saturated, even at 10^{-5} M dexamethasone. Therefore, specific binding can be estimated by measuring,

[5] G. Scatchard, *Ann. N.Y. Acad. Sci.* **51**, 660 (1949).

at low concentrations of [^{3}H]dexamethasone (e.g., 10^{-8} *M*), the difference in bound steroid between cells incubated in the absence ("total" bound steroid) and in the presence ("background binding") of an excess (10^{-6} to 10^{-5} *M*) of nonradioactive dexamethasone. The latter effectively competes with [^{3}H]dexamethasone for binding to specific sites of high affinity and limited capacity, but does not inhibit binding of [^{3}H]dexamethasone to the low-affinity sites. The validity of this technique can be assessed experimentally (Fig. 3). Cultured hepatoma cells were incubated with 3×10^{-9} *M* [^{3}H]dexamethasone and increasing amounts of nonradioactive dexamethasone. It can be seen that binding of [^{3}H]dexamethasone is progressively inhibited at increasing concentrations of competitor until a plateau ("background") is reached. Although there is increasing nonspecific binding of the steroid between 10^{-6} and 10^{-5} *M*, no significant further competition occurs because the nonspecific binding capacity (Fig. 1) is large and relatively nonsaturated.

Thus, for determining specific binding by competition, two cell incubations are performed for each [^{3}H]steroid concentration, one without and one with a large excess of nonradioactive steroid as a competitor. The amount of specific binding then equals the difference in cell-associated radioactivity in each pair of incubations ("total" minus "background").

Techniques described above apply without making any assumptions of the rate of dissociation of the receptor–steroid complex. It is therefore useful even if the rate of dissociation is rapid. Its disadvantages lie in

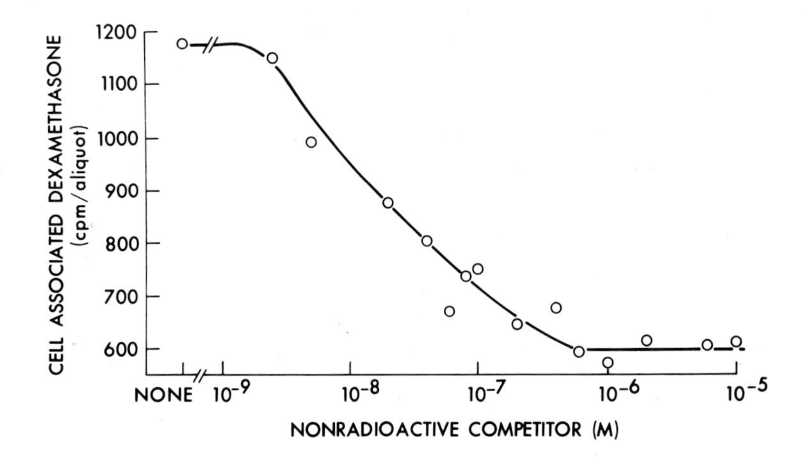

FIG. 3. Competition by nonradioactive dexamethasone for the binding of 3×10^{-9} *M* [^{3}H]dexamethasone (4 Ci/mmole) by intact HTC cells. Details are given in the text.

the facts that the cell pellet is contaminated by extracellular radioactivity and the "nonspecific" ("background") binding is high. In fact, radioactivity from these sources increases linearly with increasing concentrations of [^{3}H]steroid (shown for the "nonspecific" binding in Fig. 1), so that at higher incubation concentrations (above 5×10^{-8} *M*), the specifically bound ^{3}H represents a trivial proportion of the total and thus cannot be accurately determined by the competition technique.

Binding by Intact HTC Cells Followed by Chilling and Washing

Since nonspecifically bound dexamethasone dissociates from HTC cells much more rapidly than specifically bound steroid, the latter can be estimated in a simpler way, without resorting to a double isotope assay. This involves thorough washing of the cells after incubation with steroid. This removes most of the steroid nonspecifically bound or trapped in the cell pellet without substantially affecting the specifically bound steroid.

Procedure. Cultured hepatoma cells are incubated with a low concentration of [^{3}H]dexamethasone without or with nonradioactive competitor as described above. After incubation, the cells are centrifuged at 800 *g* for 3 to 5 minutes and the supernatant medium is removed. The pellets are immediately resuspended, using a pipet or a Vortex mixer, in at least 3 ml of phosphate-buffered saline per 30 ml of cells at 10^6 cells/ml. The cells are pelleted and washed again. The final cell pellet is resuspended and assayed, and specifically bound steroid is determined by difference as described above.

For quantitative experiments in which it is desirable to retain all the specifically bound steroid, the cells should be washed at 0 to 4°. At room temperature, however, only a small amount of specifically bound steroid is lost, whereas the background has become small enough (Table I and Fig. 4) to allow a clearer demonstration of specific binding.

Subcellular Localization of Specifically Bound Dexamethasone Using Competition Analysis

The competition technique may also be used to determine the subcellular localization of specifically bound steroid. After incubation with [^{3}H]dexamethasone and washing with cold phosphate-buffered saline as described above, the cells are disrupted and fractionated (for example by differential centrifugation). Radioactivity in the various cell fractions is then measured. An example of such an experiment is shown in Table

TABLE I
SPECIFIC BINDING OF DEXAMETHASONE IN SUBCELLULAR FRACTIONS

Fraction	Dexamethasone binding (cpm)			Specific (% of total)
	A	B	Specific (A − B)	
Crude nuclei (800 *g* pellet; after washes)	141,980	4,270	137,710	51.0
Purified nuclei	122,150	830	121,320	45.0
Large granules (10,000 *g* pellet)	8,817	2,281	6,536	2.5
Microsomes (105,000 *g* pellet)	2,110	1,871	239	0.1
Cytosol (105,000 *g* supernatant)	201,300	78,250	123,050	46.0

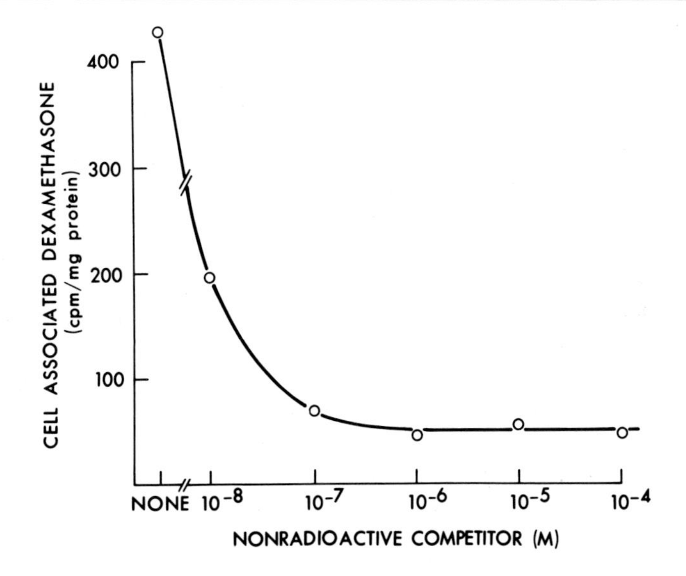

FIG. 4. Competition by various concentrations of nonradioactive dexamethasone for the binding of 5×10^{-9} *M* [^{3}H]dexamethasone (4 Ci/mmole) by intact HTC cells. Binding was measured after washing the cells as described in Table I.

I. It can be seen that most of the steroid specifically bound in intact HTC cells at 37° was localized in the nuclear and cytosol fractions.

Cultured hepatoma cells (40 ml at 8.2×10^6 cells/ml) were incubated in growth medium at 37° for 45 minutes with 2.5×10^{-8} *M* [^{3}H]dexamethasone (22 Ci/mmole) (cells A) or with 2.5×10^{-8} *M* [^{3}H]dexamethasone plus 2.5×10^{-5} M nonradioactive dexamethasone (cells B). The cells were harvested by centrifugation, washed twice with cold phosphate-buffered saline, and were then disrupted at 0° in homogenization buffer (10 ml) using a Teflon-glass tissue grinder (Kontes Glass Co.). The homogenate was centrifuged at 800 g for 10 minutes. The pellet was washed twice with homogenization buffer (10 ml), resuspended in 4.5 ml homogenization buffer containing 0.25 M sucrose, and portions assayed for radioactivity ("crude nuclei"). The remainder was centrifuged (60,000 g for 60 minutes) through 1.7 M sucrose to yield "purified nuclei." The supernatant medium from the initial centrifugation was centrifuged at 10,000 g for 30 minutes to yield a pellet ("large granules") and a supernatant medium which was further centrifuged at 100,000 g for 60 minutes to produce a pellet ("microsomes") and cytosol. Portions of the cytosol and the pellets (resuspended in water) were assayed for radioactivity, and the dexamethasone bound specifically in each fraction was determined by taking the difference between the radioactivities of fractions from cells A and cells B.

[21] Quantitation of Estrogen Receptor in Mammary Carcinoma

By William L. McGuire

Experiments in this laboratory with rat mammary tumors[1-3] and preliminary data in human patients[4] indicate that tumors which contain appreciable cytoplasmic estrogen receptor (R) regress after endocrine ablation therapy whereas tumors without R fail to respond. The cytoplasmic R in human mammary tumors has properties similar to that from rat uterus,[5] rat pituitary tumor,[6] or rat mammary tumor.[7] It has been suggested that assays of human tumors for R might lead to valid prognos-

[1] W. L. McGuire and J. A. Julian, *Cancer Res.* **31,** 1440 (1971).
[2] W. L. McGuire, J. A. Julian, and G. C. Chamness, *Endocrinology* **89,** 969 (1971).
[3] W. L. McGuire, K. Huff, A. Jennings, and G. C. Chamness, *Science* **175,** 335 (1972).
[4] E. V. Jensen, G. E. Block, S. Smith, K. Kyser, and E. R. DeSombre, *Nat. Cancer Inst., Monogr.* **34,** 55 (1971).
[5] G. C. Chamness and W. L. McGuire, *Biochemistry* **11,** 2466 (1972).
[6] W. L. McGuire, M. DeLaGarza, and G. C. Chamness, *Endocrinology* **93,** 810 (1973).
[7] W. L. McGuire and M. DeLaGarza, *J. Clin. Endocrinol. Metab.* **36,** 548 (1973).

tication regarding endocrine therapy.[8] I have assayed 200 human tumors for R and would like to emphasize the following points. (1) The concentration of R in both primary and metastatic human mammary carcinoma varies over a wide range. Thus, the assay must be quantitative. (2) Nonspecific binding of radioactive estradiol ([^{3}H]E_2) may be considerable so it is essential to have independent checks and controls to insure specificity of the observed binding.

Sample Preparation

Tissues are removed, trimmed of fat and normal tissue, and dropped into liquid nitrogen. If the specimen needs to be shipped long distance to the assay laboratory, it is packed in dry ice, shipped, and stored in a Revco ultra-low freezer at —75° until assay. Tissues are thawed at 4° and homogenized in a motor-driven Duall glass homogenizer at low speed in TED buffer [0.01 *M* Tris-HCl, 0.0015 *M* EDTA, 0.5 m*M* dithiothreitol (DTT) pH 7.4], 1.2 ml per gram minced tissue. The homogenate is centrifuged at 40,000 rpm (102,000 *g*) for 50 minutes to obtain the supernatant cytosol fraction. Protein is quantitated by the method of Lowry.[9]

Sucrose Gradients

All sucrose gradients are prepared in TED buffer as described in an accompanying paper in this volume.[10] Samples for gradient analysis are prepared by incubating 250 μl of cytosol with 1.0 pmoles [^{3}H]estradiol-17β 96 Ci/mmole for 60 minutes at 4°. Control cytosols are preincubated with 100 pmoles nonradioactive estradiol-17β prior to adding the labeled estradiol. After the incubation the nonbound estradiol is removed by treatment with dextran-coated charcoal (DCC) and the treated cytosol is applied to the sucrose gradient.

Saturation Analysis

Earlier data were obtained by a minor modification[11] of Korenman's method.[12] We have since made extensive comparisons of a variety of

[8] E. V. Jensen, E. R. DeSombre, and P. W. Jungblut, *in* "Endogenous Factors Influencing Host Tumor Balance" (R. W. Wissler, T. L. Dao, and S. Wood, Jr., eds.), pp. 15 and 68. Chicago Univ. Press, Chicago, Illinois, 1967.

[9] O. H. Lowry, N. J. Rosebrough, A. L. Farr, and R. J. Randall, *J. Biol. Chem.* **193**, 265 (1951).

[10] G. C. Chamness and W. L. McGuire, this vol. [16].

[11] W. L. McGuire, *J. Clin. Invest.* **52**, 73 (1973).

[12] S. G. Korenman and B. A. Dukes, *J. Clin. Endocrinol. Metab.* **30**, 639 (1970).

charcoal assays using a wide range of incubation times and temperatures, with and without dithiothreitol, and find the following procedure far superior to those currently reported in the literature (detailed comparisons to be reported elsewhere). Cytosol is prepared in TED buffer as described above and 200 μl ($\sim$2.0 mg protein/ml) is incubated with increasing quantities of [^{3}H]E_2 (0.015–0.2 pmoles) in duplicate for 18 hours at 4°. One-half milliliter of DCC suspension (0.25 g% Norit A, 0.0025 g% Dextran Grade C, Mann Research Labs, in 0.01 *M* Tris-HCl, pH 8.0) is then added for an additional 30 minutes at 4° with vigorous shaking. The mixture is then centrifuged for 10 minutes at 2000 *g* and the supernatant radioactivity is quantitated in a modified Bray's scintillation cocktail.

Illustrative Examples

Of all the available methods for demonstrating R, sucrose gradient centrifugation is the most specific. The presence of an 8–10 S peak of estradiol radioactivity unequivocally indicates the presence of R in a tumor cytosol. In Fig. 1 a representative human breast cancer cytosol-[^{3}H]E_2 sucrose centrifugation is displayed. The specific 8–10 S peak can be seen near the bottom of the gradient and the 4–5 S peak near the top. Whereas a 8–10 S peak always represents specific R, a 4–5 S peak could represent nonspecific [^{3}H]E_2 binding, so a preincubation of an identical cytosol with a 100-fold excess of unlabeled estradiol is always included. Figure 1 shows that the unlabeled estradiol has predictably eliminated all the 8–10 S peak of [^{3}H]E_2, but in addition has eliminated the majority of the 4–5 S peak. Therefore, in this particular cytosol, the majority of the 4–5 S peak represents specific [^{3}H]E_2 binding. Figure 2 reveals a tumor cytosol which lacks an 8–10 S peak but has an appreciable 4–5 S peak. In this instance the 4–5 S peak is unaffected by nonradioactive estradiol preincubation. Hence, all the estradiol binding observed in this tumor cytosol is nonspecific and emphasizes the need for proof of specificity in measuring R.

The addition of DCC to a preincubated mixture of cytosol and [^{3}H]E_2 quantitatively removes the nonbound [^{3}H]E_2 and leaves the bound [^{3}H]E_2 in the supernatant for measurement. Since the [^{3}H]E_2 left in the supernatant may be bound to R or to nonspecific proteins, it is important to incubate cytosols with very low quantities of [^{3}H]E_2 ($<1 \times 10^{-9}$ *M*) to minimize nonspecific binding. Figure 3 shows a representative DCC binding curve of human mammary carcinoma cytosol. The bound [^{3}H]E_2 is plotted as a function of the total [^{3}H]E_2 input. Usually, this method of data presentation is not adequate to accurately determine the

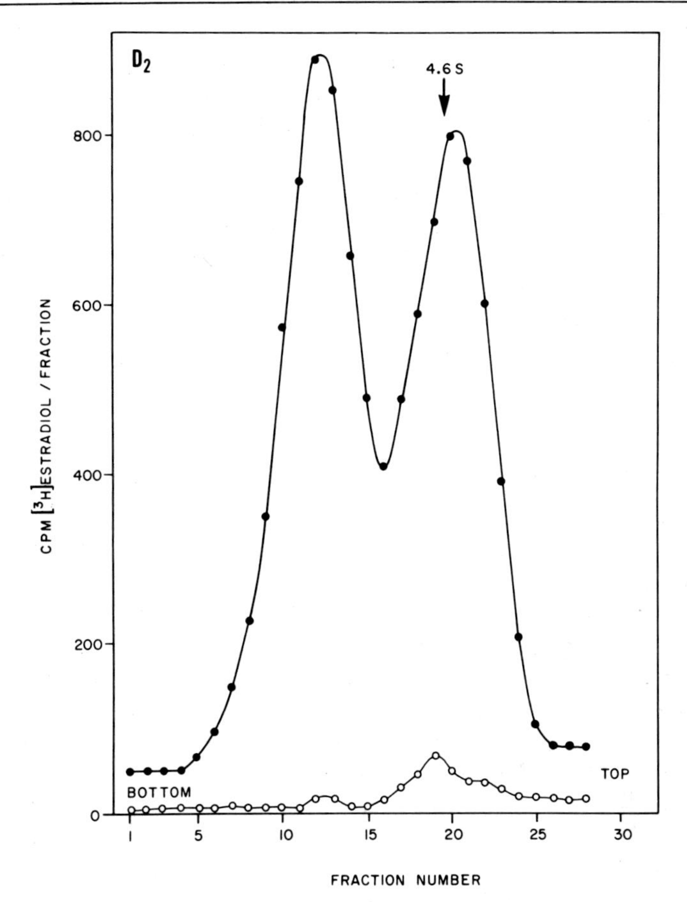

FIG. 1. Sucrose gradient centrifugation of human mammary carcinoma cytosol with (○-○-○) and without (●-●-●) nonradioactive estradiol preincubation as described in the text.

saturation value, thus a Scatchard analysis[13] is always performed as in Fig. 4. The data in Fig. 3 have been replotted, and the linear relationship that results indicates one class of binding sites. It is important to emphasize that at higher dose levels of [^{3}H]E_2, nonspecific binding becomes appreciable and the Scatchard plot is no longer linear. In this cytosol there are 53 fmoles of estradiol binding sites/mg cytosol protein. Another advantage of the Scatchard plot is that an estimate of the dissociation constant (K_d) is readily obtained. This, in turn, is a reflection of the specificity of the cytosol–[^{3}H]E_2 interaction since specific R–[^{3}H]E_2 interactions yield K_d's much lower than 1×10^{-9} M whereas nonspecific

[13] G. Scatchard *Ann. N.Y. Acad. Sci.* **51, 660 (1949).**

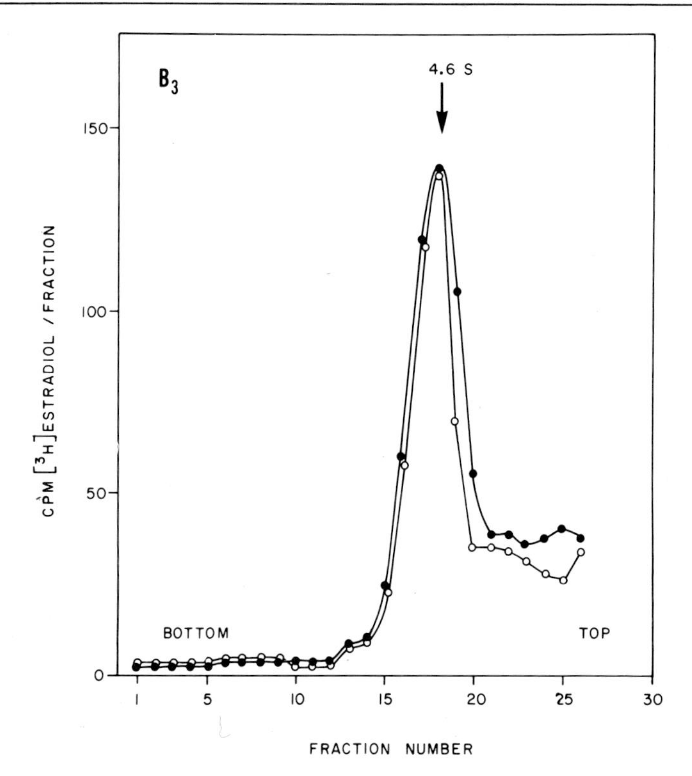

FIG. 2. Sucrose gradient centrifugation of human mammary carcinoma cytosol with (○-○-○) and without (●-●-●) nonradioactive estradiol as described in the text.

$[^3H]E_2$ interactions yield K_d's several orders of magnitude higher (low affinity binding). Figure 4 reveals the K_d of this particular cytosol to be 2.6×10^{-10} *M*. Using both the DCC assay and sucrose gradient centrifugation, we have evaluated 200 human breast cancers and find a range of values from 0 to 1100 fmoles R/mg cytosol protein. The number of sites fall in a rather continuous spectrum, and the 8–10 S peak is present in cytosols containing a concentration of R greater than 9 fmoles/mg of cytosol protein.[11]

Comment

I have tried other methods of measuring human tumor R such as gel filtration and protamine precipitation[7] but find them less satisfactory be-

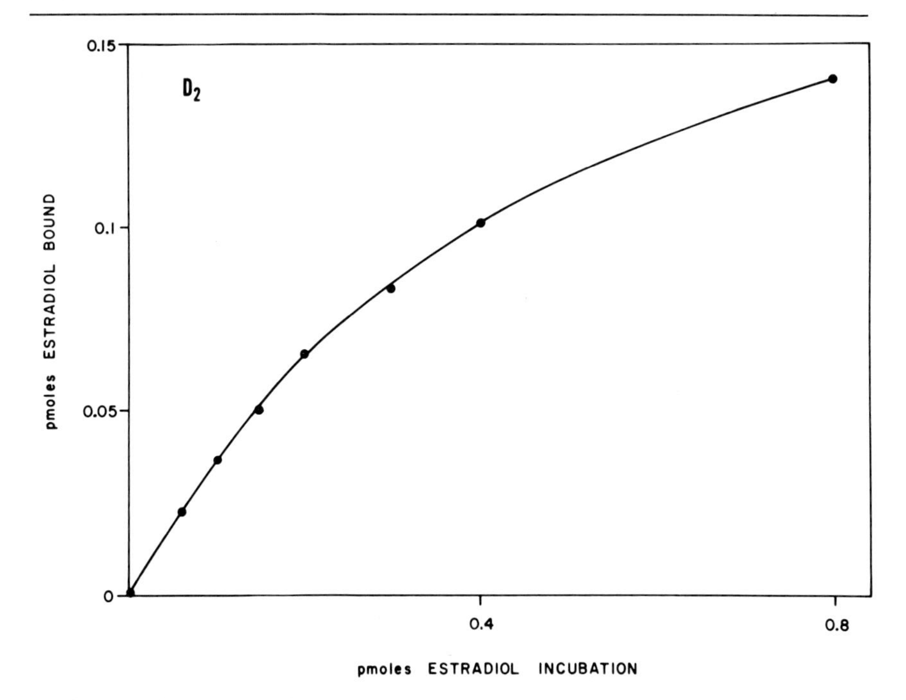

FIG. 3. Binding of radioactive estradiol to human mammary carcinoma cytosol as determined by the dextran-coated charcoal technique described in the text.

cause of either the cumbersome requirement of running the many columns needed for a quantitative assay or the high nonspecific background in the protamine assay. The popular *in vitro* tissue slice uptake method is not quantitative and requires additional controls to insure specificity.[4,8]

The DCC method gives accurate quantitative data, and sucrose gradient centrifugation with and without nonradioactive estradiol preincubation gives ideal specificity proof. It could be argued that the sucrose gradient method is expensive and unnecessary since the K_d derived from the DCC assay is sufficient proof of the specificity of the interaction. This may well turn out to be true, but at present the gradient centrifugation is the only method available to evaluate 4–5 S binding. Steggles and King reported that a 4–5 S estradiol-binding protein distinct from the usual 8–10 S or salt-derived 4–5 S form was present in the uteri of mature rats.[14] Furthermore, this 4–5 S form disappeared after ovariectomy or hypophysectomy and thus could be very important in breast tumor response to endocrine ablation. Thus until improved methods such as radio-

[14] A. W. Steggles and R. J. B. King, *Biochem. J.* **118**, 695 (1970).

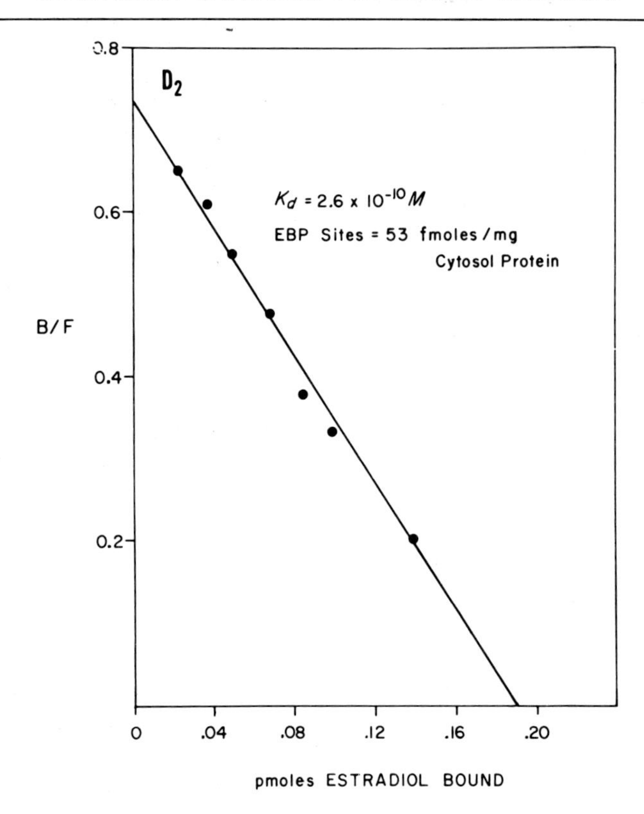

FIG. 4. Scatchard analysis of the data in Fig. 3.

immunoassay become available, I recommend both the DCC assay and sucrose centrifugation on each sample.

Acknowledgments

This work was supported by The National Cancer Institute, The American Cancer Society, and The Robert A. Welch Foundation.

[22] Methods for Assessing Hormone-Receptor Kinetics with Cells in Suspension: Receptor-Bound and Nonspecifically Bound Hormone; Cytoplasmic-Nuclear Translocation

By ALLAN MUNCK[1] and CHARLES WIRA

General

The methods described in this chapter have been developed specifically for studying kinetic aspects of glucocorticoid–receptor interactions in intact lymphocytes. We have applied them mainly to suspensions of rat thymus cells,[1a–4] and have also found them applicable to rat lymphocytes from lymph nodes and spleen and to human circulating lymphocytes from patients with chronic lymphocytic leukemia. Most of these methods can probably be adapted for studying binding of other hormones and related substances to other types of cells in suspension, though optimum conditions must be determined for each case. As described separately in this volume, modifications of these methods can also be used to study kinetics of sugar uptake by cells in suspension.

For concreteness, we assume throughout that the hormones dealt with are labeled with ^{3}H and are assayed by scintillation counting in an appropriate fluid such as Bray's solution,[5] quenching being compensated for where necessary. Furthermore, we assume that the hormones either are not significantly metabolized by the cells or can be separated from their metabolites prior to assay.

Cell Concentrations. The concentration of cells in a suspension is conveniently expressed by the cytocrit V (milliliters of packed cells per milliliters of cell suspension[6]), which can be determined by standard microhematocrit procedures. For the types of experiments described here we have generally used initial values of V in the range of 0.2 to 0.4.

Incubation Conditions. Lymphocytes have been studied as suspensions in Krebs-Ringer bicarbonate buffer. Other biological media can be employed equally well; but if they contain proteins or other macromolecules,

[1] Supported in part by Research-Grant AM-03535 and Research Career Development Award AM-K3-16.

[1a] A. Munck and T. Brinck-Johnsen, *Proc. Int. Congr. Horm. Steroids, 2nd, 1966* Int. Congr. Ser. No. 132, p. 472 (1967).

[2] A. Munck and T. Brinck-Johnsen, *J. Biol. Chem.* **243,** 5556 (1968).

[3] C. Wira and A. Munck, *J. Biol. Chem.* **245,** 3436 (1970).

[4] A. Munck and C. Wira, *Advan. Biosci.* **7,** 301 (1971).

[5] G. A. Bray, *Anal. Biochem.* **1,** 279 (1960).

[6] A. Munck, *J. Biol. Chem.* **243,** 1039 (1968).

it may be necessary to determine what fraction of the hormone in the medium is free in solution and what fraction is bound to the macromolecules. At least with steroid hormones, it is likely that only the free hormone molecules interact the cells.

Cell-Bound Hormone. This expression will be used rather loosely to refer to the hormone molecules that sediment with the cells under conditions to be described. Some of these molecules are not "bound" in the strict sense of forming a complex with a cell structure or macromolecule, but are free in solution in the cell water. In fact, where a hormone can rapidly pass through the cell membrane, as is probably the case with cortisol in lymphocytes at 37°, the concentrations of free hormone inside and outside the cell will be very similar. Such free intracellular hormone will form part of the so-called "nonspecifically bound" hormone referred to below.

Kinetics of Hormone Dissociation and Association[2,4]

Rates of dissociation are determined by first bringing a cell suspension to an equilibrium (or a steady state; the distinction is not easy to make experimentally) with a given concentration of free hormone in the medium, then diluting the suspension rapidly and measuring at intervals the cell-bound hormone.

The concentration of free hormone in the medium, in moles per liter, is symbolized by (S). Cell-bound hormone, in moles per liter packed cell volume, is symbolized by (Sc). Equilibrium values are denoted by a subscript e, as in $(\mathrm{S})_e$ and $(\mathrm{Sc})_e$. For convenience, binding data is normalized by dividing by $(\mathrm{S})_e$.

Equilibrium or Steady-State Binding. A cell suspension of concentration V(0.2–0.4) is incubated with radioactive hormone for sufficient time (10 minutes at 37°; 1–2 hours at 3°, with cortisol and lymphocytes) to reach a steady value of (Sc). A 50-μl aliquot of the suspension is counted to give the value T (counts per minute). The suspension is centrifuged for about 2 minutes in a Beckman Microfuge, and a 50-μl aliquot of the supernatant is counted to give S (counts per minute). If L is the cell-bound hormone in 50 μl packed cells (in counts per minute), then the equation $\mathrm{T} = \mathrm{S}(1\text{-}\mathrm{V}) + \mathrm{LV}$ accounts for the total hormone in 50 μl of cell suspension, S(1-V) being that which is in the medium, and LV being that which is cell-bound. Since $(\mathrm{Sc})_e/(\mathrm{S})_e = \mathrm{L}/\mathrm{S}$, it follows that

$$\frac{(\mathrm{Sc})_e}{(\mathrm{S})_e} = \frac{\mathrm{T} - \mathrm{S}(1\text{-}\mathrm{V})}{\mathrm{VS}} \tag{1}$$

This quantity, which in effect is a distribution coefficient for distribution of hormone between cells and medium, gives a measure of total cell-bound hormone at equilibrium with a given hormone concentration $(S)_e$ and serves as an initial or zero time-point for dissociation.

Dissociation. A cell suspension equilibrated as above, from which values of V, T, and S have been obtained, is rapidly diluted 50-fold into buffer at whatever temperature dissociation is to be measured. Beginning immediately after dilution samples are removed and centrifuged to obtain cell pellets for assay. The nominal time for each sample is the time centrifugation is started.

A practical, rapid procedure that allows sampling every 2 minutes, with the first sample 1 minute after dilution, is the following. Starting about 40 seconds before the nominal sampling time, 0.80 ml of the diluted suspension is drawn into a 1-ml long-tip pipet calibrated to the tip. A Propipette or similar pipetting device is almost essential. Duplicate 0.40-ml samples are delivered into polyethylene Beckman Microfuge tubes, the tip of the pipet being drawn out of the tube to avoid overflow as the volume is released. The tubes are then centrifuged for 105 seconds in a Beckman Microfuge. Supernatants are immediately removed by suction, leaving the cell pellets in the tubes. The tubes should subsequently be kept horizontal or inverted to prevent any remaining supernatant from draining to the pellets. The next step should be carried out within an hour or so, before the pellets have time to dry out and become stuck to the polyethylene. Each tube is cut off 1 or 2 mm above the pellet. The resulting tip is inverted over the end of a bent 1.5-in. No. 22 syringe needle that is attached to a 5-ml syringe partly filled with counting fluid. While held on the needle inside a counting vial against the shoulder of the vial, the tip is flushed to wash out the cells, and is then left with the cells in the vial. The vial is counted to give C, the counts per minute of cell-bound hormone in 0.4 ml of dilute suspension at time t after dilution. Duplicate values of C should rarely differ by more than 5%. A 100-μl sample of the dilute suspension is counted to give D (counts per minute), from which the dilution factor is calculated at 2T/D.

Division of the original cell concentration, V, by this dilution factor gives the cell concentration in the dilute suspension, so 0.4VD/2T is the milliliters of packed cells (2–3 μl under the present conditions) in 0.4 ml of dilute suspension. The counts of cell-bound hormone per milliliter of packed cells is therefore C divided by the above expression, or 2CT/0.4VD. This quantity, when divided by 1000S/50 (the counts of hormone per milliliter of medium before dilution) gives

$$(Sc)/(S)_e = CT/4VDS \tag{2}$$

Since $(S)_e$ is simply a constant normalizing factor, $(Sc)/(S)_e$ is a measure of the amount of cell-bound hormone at time t after dilution. As mentioned, the value of $(Sc)/(S)_e$ for $t = 0$ is $(Sc)_e/(S)_e$, given by Eq. (1).

The various steps in this method have been studied[2] with the cortisol–lymphocyte system to ascertain that: (a) exchange of hormone between cells and medium stops within 15 seconds of the beginning of centrifugation; (b) there is no significant loss of hormone by adsorption to glassware or centrifuge tubes; (c) contamination of the cell pellet by supernatant or by hormone adsorbed to the tube is negligible; and (d) for dilution at 37°, cooling of the dilute suspension caused by use of pipets that are at room temperature produces negligible error. Similar controls should be made with each new system. A lower limit to the size of the cell pellet, hence an upper limit to the dilution factor, is imposed by the requirement that centrifugation should bring exchange between cells and medium to a halt. For example, with a 500-fold dilution dissociation appears to continue after centrifugation, and erratic results are obtained. Cooling prior to or after centrifugation may extend these limits. Alternatively, where dissociation is intrinsically slow, the same size of pellet can be obtained from a more dilute solution by centrifuging a larger volume (see below).

Example: Quantitation of Receptor-Bound and Nonspecifically Bound Hormone; Cell Permeability.[2-4] Results obtained by these procedures are given somewhat schematically in Fig. 1, which shows the time course of equilibration and dissociation from thymus cells of cortisol, a glucocorticoid that acts directly on thymus cells to produce metabolic effects, and cortisone, a closely related inactive steroid. At the equilibrium concentrations used, $(S)_e = 6$ nM, cortisol gives an equilibrium value, $(Sc)_e/(S)_e$, of about 3, and cortisone of about 2. Less polar steroids such as progesterone may give values as high as 50.

Considering only the dissociation curves obtained by dilution into buffer (KRB) at 37°, it can be seen that cortisone dissociates rapidly, almost reaching a new equilibrium value by 1 minute. Most of the bound cortisol also dissociates rapidly. But cortisol in addition exhibits a fraction that dissociates relatively slowly, with a velocity constant for dissociation that is shown by an analysis of the dissociation curve into exponential components to be about 0.3 min^{-1}. A variety of criteria[2-4] (specificity for glucocorticoids, dissociation constant in the physiological concentration range, etc.) indicate that this slowly dissociating "specific" fraction consists of receptor-bound cortisol.

The cortisol–receptor complex can be extracted,[3,4] and in isolation it has a velocity constant for dissociation at 37° very close to that found with intact cells. Since in intact cells the complex formed at 37° (as in

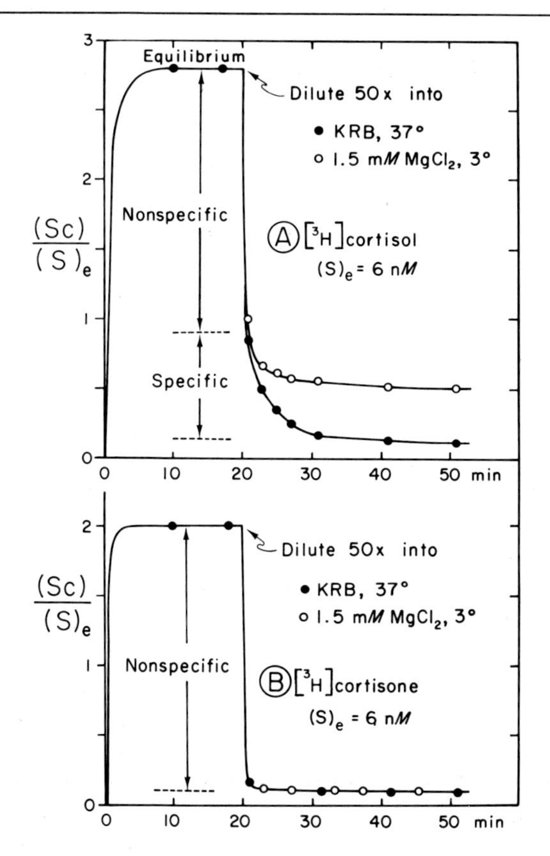

FIG. 1. Typical experiments illustrating equilibration of [^{3}H]cortisol and [^{3}H]cortisone with rat thymus cells in suspension, and dissociation following dilution. (A) [^{3}H]Cortisol at a concentration $(S)_e = 5$ nM is equilibrated for 20 minutes at 37° with cells suspended in Krebs-Ringer bicarbonate buffer (KRB). At 20 minutes the suspension is diluted 50-fold into either KRB at 37° or into 1.5 mM $MgCl_2$ at 3°, and the amount of cell-bound [^{3}H]cortisol measured at intervals (1, 3, 5, 7, etc., minutes) after dilution as described in the text. (B) Similar experiments with [^{3}H]cortisone at a similar initial concentration.

Fig. 1A) turns out to be located largely in the nucleus, the coincidence of dissociation rates leads to the conclusion that in the intact cell there is no significant permeability barrier to cortisol between the receptor in the nucleus and the outside of the cell.

If this conclusion is valid it follows that (a) the concentration of free hormone inside lymphoid cells is the same as that outside, so there is no need to postulate the existence of a system for transporting glucocorticoids into cells, a role occasionally ascribed to receptors; (b) kinetic measurements of specific hormone interaction with intact cells reflect directly the properties of the hormone–receptor complex *in situ.*

The amount of receptor-bound hormone present prior to dilution can be determined by analyzing the dissociation curve into exponential components and extrapolating back to 0 minute the specific or receptor-bound component.[2] With the aid of a computer, with a least-squares criterion for best fit, this procedure is objective, convenient, and reproducible. It requires, however, a full dissociation curve, carried over 40 minutes or so, and a minimum initial volume of cell suspension of about 0.25 ml. A more rapid procedure, requiring fewer cells, is described in the next section. Alternatively, a reasonable quantitative estimate of the amount of receptor-bound hormone is given by the 1-minute point on the 37° dissociation curve.

The rapidly dissociating fraction of bound cortisol (or cortisone) is referred to as "nonspecific" because it is found with all steroids tested, regardless of hormonal activity. It accounts almost entirely for the pronounced equilibrium binding of nonpolar steroids and undoubtedly consists partly of molecules that are weakly bound to various cell structures and partly of molecules that are free in the cell water.

Kinetic Separation of Nonspecific from Receptor-Bound Hormone.[2,4] If dissociation of cortisol is carried out by 50-fold dilution into buffer at 3°, the nonspecifically bound fraction dissociates almost completely within about 40 minutes. Since the receptor-bound fraction takes many hours to dissociate at 3°, such a preliminary 40-minute dissociation at 3° yields cells containing almost exclusively receptor-bound hormone.

A dissociation that is initiated by 50-fold dilution into buffer at 37°, but that is stopped by cooling the dilute suspension to 3° after 30 seconds, will similarly result in almost complete dissociation of nonspecifically bound steroid and a suspension of cells retaining only receptor-bound hormone. Cooling of 10 ml of dilute suspension from 37° to 3° can be achieved within about 45 seconds by pouring the dilute suspension into a 500-ml Erlenmeyer flask precooled in ice water and swirling; similarly, 2 ml of dilute suspension can be rapidly cooled by pouring it into a precooled 50-ml Erlenmeyer. When this procedure is followed by a measurement of cell-bound hormone it becomes a rapid, convenient method for measuring receptor-bound hormone, giving results in good agreement with those obtained by exponential analysis of the 37° dissociation curve.[7] For this method 40 μl of incubated cell suspension are diluted into 2 ml buffer at 37°. After 30 seconds the dilute suspension is cooled to 3° as described above. Between 10 and 20 minutes later (the exact time is not

[7] C. Wira and A. Munck, *J. Biol. Chem.* in press.

important, since receptor-bound hormone dissociates slowly at 3°) duplicate 0.4-ml aliquots, kept at 3°, are centrifuged in a Beckman Microfuge and the resulting pellets counted as described under Dissociation. These values represent receptor-bound hormone.

Association.[2] The rate of association of hormone to the receptors in intact cells is measured by adding hormone to a cell suspension at whatever temperature is to be studied, and at various intervals thereafter diluting aliquots of the suspension 50-fold into buffer at 37°. The amount of receptor-bound hormone is then determined by one of the methods described in the preceding two sections. A value of S is obtained at each dilution time. A single value of T can be determined after equilibrium is attained. From these values an association curve can be constructed.

The velocity constant for association of receptor-bound cortisol in thymus cells determined this way is in reasonable agreement with the value calculated from the dissociation constant and the velocity constant for dissociation at 37°.

Competition.[2] Cells are equilibrated simultaneously with labeled hormone (at a concentration well below the dissociation constant of the hormone–receptor complex) and with a known concentration of the unlabeled competing substance to be studied. Receptor-bound hormone is then determined by one of the procedures described above. Comparison of this magnitude with that obtained with various concentrations of unlabeled hormone in the absence of competing substance gives a measure of the receptor-binding affinity of the competing substance relative to the hormone.

Kinetics of "Cytoplasmic"-Nuclear Translocation[3,4,7]

Rapid Isolation of Lymphocyte Nuclei.[3] Dilution of an aliquot of a thymus cell suspension (V = 0.2–0.4) into 50 times its volume of 1.5 m*M* $MgCl_2$ leads within less than 5 minutes to disruption of all the cells, with release of cytoplasmic contents but reasonable preservation of nuclear structure. The suspension of nuclei (from which if necessary membrane fragments may be removed) can be treated in the same way as described above with cell suspensions. In particular, the procedure for obtaining a dissociation curve with intact cells can be applied to the suspension of nuclei, the values of C and (Sc) in Eq. (2) referring then to nuclear-bound rather than cell-bound hormone. Centrifugation of the dilute suspension for 105 seconds in the Beckman Microfuge leads to formation of a nuclear pellet containing 70–80% of the DNA in the intact cells.

Cell disruption occurs rapidly whether the $MgCl_2$ solution is at 37°

or at 3°. A clear indication of disruption at 3° is that, following equilibration with cortisol or cortisone, already 1 minute after dilution the pellet obtained contains almost no nonspecifically bound steroid (Fig. 1). By contrast, when dilution is into buffer at 3°, then as mentioned above, the nonspecifically bound cortisol takes about 40 minutes to dissociate.

Nuclear and "Cytoplasmic" Receptor-Bound Hormones.[4,7] As can be seen in Fig. 1A, after cells that have been incubated at 37° with cortisol are disrupted in $MgCl_2$ at 3°, the nuclei retain a fraction of cell-bound cortisol. This fraction can be shown to correspond to the receptor-bound cortisol seen in the intact cells and is therefore ascribed to a nuclear cortisol–receptor complex.[3] It generally amounts to 50–100% of the receptor-bound fraction measured with intact cells.

If the initial incubation is carried out at 3° (for about 90 minutes) rather than at 37°, no such nuclear receptor-bound fraction is found; the intact cells, however, still exhibit a receptor-bound fraction, measurable by the methods described earlier. The receptor-bound cortisol present under these conditions is referred to for convenience as "cytoplasmic," although its location is not firmly established.

A rapid method for measuring nuclear receptor-bound hormone is to dilute 20 μl of the cell suspension into 1 ml of 1.5 mM $MgCl_2$ at 3°, 10–20 minutes later to centrifuge duplicate 0.4-ml aliquots of dilute suspension for 105 seconds at 3° in the Beckman Microfuge, and then to measure the hormone in the nuclear pellets as described earlier. This procedure, which is best carried out in a cold room, can easily be combined with simultaneous measurements of total receptor-bound hormone in the cell by the method described above, employing dilution for 30 seconds into buffer at 37° and rapid cooling. The difference between nuclear and total receptor-bound hormone can be taken to represent cytoplasmic receptor-bound hormone.

Rapid Measurement of "Cytoplasmic" Receptor-Bound Hormone. If a cell suspension that has been preincubated with labeled hormone is diluted into 5 times its volume of 1.5 mM $MgCl_2$ at 0–3° and centrifuged, the lysate in the resulting supernatant is found to contain most of the cytoplasmic hormone-receptor complex.[4,8] Nuclei and nuclear-bound complexes sediment in a gellike mass. Bound hormone in the supernatant can then be assayed by a conventional charcoal adsorption method.[8] In the following procedure considerable simplification is achieved by combining the two steps, cell lysis and charcoal adsorption, into a single step.

Dextran-coated charcoal in suspension is prepared by stirring 0.1 g Norit-A charcoal (Fisher) and 0.01 g Dextran type 60C (Sigma) in 10 ml

[8] P. A. Bell and A. Munck, *Biochem. J.* **136,** 97 (1973).

of 1.5 mM $MgCl_2$ for one hour at room temperature. A 20-μl aliquot of the cell suspension to be assayed is added to 100 μl of the dextran-charcoal suspension which has been precooled to 0–3° in a plastic centrifuge tube of the type used in the Eppendorf Microcentrifuge 3200. After vigorous mixing for about 5 seconds the tube is left at 0–3° for about 15 minutes. During this period the cell membranes break to release cytoplasmic hormone-receptor complex, and free hormone is adsorbed by the charcoal. The tube is then centrifuged for 2 minutes in the Eppendorf Microcentrifuge 3200 to sediment simultaneously the charcoal, the cell nuclei, and membrane fragments. Presence of cell fragments if anything improves removal of charcoal, and contamination of the supernatant by charcoal specks is rare. An aliquot of the supernatant is counted to determine the amount of bound hormone.

Rate of Translocation of Receptor-Bound Hormone from Cytoplasmic to Nuclear Form.[7] When a cell suspension that has been incubated at 3° for 90 minutes with cortisol is warmed to 37° over a period of 45 seconds, it is found that by 1 minute there is almost complete transformation of receptor-bound cortisol from cytoplasmic to nuclear form.[4] The kinetics of this rapid process can be determined accurately by the following simple and general method, which permits warming of cells for time intervals as short as 1 second.

A suspension of lymphocytes (V = 0.3–0.35) is incubated with hormone for about 90 minutes at 3°. A 20-μl aliquot of the suspension is placed in the bottom of a cold 15-ml graduated conical polypropylene centrifuge tube (Nalgene). At time = 0 second, 0.20 ml of buffer at 41° is rapidly added on top of the 20-μl aliquot, and the tip of the centrifuge tube is immersed in a bath at 41°. If the warming period exceeds 30 seconds, then at 30 seconds the tube is transferred to a 37° bath. With this schedule of warming the temperature to which the cells are exposed (as recorded with a thermistor with a time constant of a fraction of a second) rises instantaneously to about 37°, drops in 10 seconds to 35° or so, and by 30 seconds has returned to 37°. After the required interval of warming the suspension is rapidly cooled by adding to it 10 ml of 1.5 mM $MgCl_2$ at 3° and mixing briefly by inverting the tube. For a zero time-point, 0.2 ml of buffer at 3° instead of at 41° is added. The time required for the interior of a free-floating cell to come to temperature equilibrium with the medium can be calculated to be of the order of milliseconds, so it can be assumed that in this procedure the temperatures of the medium and the cells at any instant are identical.

Addition of the $MgCl_2$ solution simultaneously cools, dilutes, and disrupts the cells with release of nonspecifically bound hormone. The total dilution of the original 20 μl is 500-fold. After 10–20 minutes the resulting nuclei are sedimented into pellets by centrifuging the tubes at 1800 *g* in a swinging bucket centrifuge for 7 minutes at 3° (an angle-head is unsatisfactory as the nuclei tend to stick to the wall of the tube). The supernatant is decanted, and the tubes are left upside down for a few minutes, after which they can be transferred to normal room temperature. The tips of the tubes are cut off above the nuclear pellets (at about the 0.1 ml mark); for this purpose a sharp carving knife is useful. Tips and nuclei are then dropped into scintillation vials with fluid and counted. Brief, vigorous shaking of the vials is sufficient to disperse the nuclear pellets.

As measured by this method, the half-time for translocation at 37° of receptor-bound cortisol in thymus cells from cytoplasmic to nuclear form is about 30 seconds.

Section IV

Nuclear Receptors for Steroid Hormones

[23] Estrogen Interaction with Target Tissues; Two-Step Transfer of Receptor to the Nucleus

By E. V. JENSEN, P. I. BRECHER, M. NUMATA, S. SMITH, and E. R. DESOMBRE

Introduction

Studies from many laboratories, summarized in detail elsewhere,[1-3] have established the pattern by which estrogenic hormones interact with receptor substances in estrogen-responsive or "target" tissues. On entering the uterine cell, estradiol associates with the binding unit of an extranuclear receptor protein, activating it to undergo temperature-dependent conversion to a new form that can be distinguished from the native receptor by its sedimentation properties in salt-containing sucrose gradients. This transformation of the receptor protein is accompanied by translocation of the estrogen–receptor complex to the nucleus, where it associates with chromatin and in some way initiates or accelerates nuclear reactions, especially RNA biosynthesis.

This chapter describes experimental conditions for studying the interaction of tritiated estrogens with uterine horns and other excised tissues *in vitro* and for demonstrating the estradiol-induced transformation of the receptor in uterine cytosol and the binding of the transformed complex to isolated nuclei.

Estrogen Uptake by Tissues *in Vitro*

Experimental Systems. Various *in vitro* systems have been employed by different investigators to study the interaction of radioactive estrogens with excised tissues. The main features of some of these are listed in Table I. Although most procedures employ a small concentration of glucose in the incubation medium and in some cases oxygen is supplied, experience, in our laboratory and elsewhere,[4] has shown that the presence of glucose has little effect on the incorporation of estradiol by uterine tissue and essentially the same uptake is observed when exposure is carried out in an atmosphere of nitrogen as in air or oxygen.

[1] H. G. Williams-Ashman and A. H. Reddi, *Annu. Rev. Physiol.* **33**, 31 (1971).
[2] E. V. Jensen and E. R. DeSombre, *Annu. Rev. Biochem.* **41**, 203 (1972).
[3] E. V. Jensen and E. R. DeSombre, *Science* **182**, 126 (1973).
[4] G. M. Stone and B. Baggett, *Steroids* **5**, 809 (1965).

TABLE I
SYSTEMS FOR ESTROGEN UPTAKE *in Vitro*

Estrogen (n*M*)	Medium					Reference
	ml	Buffer[a]	pH	Additions[b]		
0.5	Flow	KR-phosphate	7.3	G 0.1%	O_2	c
0.1	5	KR-phosphate	7.4	G 0.1%	Air	d
0.1	300	KR-phosphate	7.3	G 0.1%	Air	e
0.14	4	KR-phosphate	7.4	G 0.1%	Air	f
0.14	4	KR-phosphate	7.2	G 0.1%	Air	g
0.5	25	KR-phosphate		G 0.1%	Air	h
2.0	4	KR-phosphate	7.4	G 0.25% + BPA 2%	Air	i
2.5	3	KR-phosphate	7.4	BPA 2%	Air	j
0.03–0.13	10	KR-bicarbonate	7.3	G 0.2%	Air	k
0.1–1.5	50	Tris-EDTA	7.4		Air	l
2–20	2.5	Eagle-HeLa culture medium			95% O_2 – 5% CO_2	m
0.4–18,000	Flow	KR-bicarbonate	7.4	G 0.1%	95% O_2 – 5% CO_2	n

[a] KR = Krebs-Ringer.
[b] G, glucose; BPA, bovine plasma albumin.
[c] P. W. Jungblut, E. R. DeSombre, and E. V. Jensen, *in* "Hormone in Genese und Therapie des Mammacarcinoms" (H. Gummel, H. Kraatz, and G. Bacigalupo, eds.), p. 109. Akademie-Verlag, Berlin, 1967.
[d] G. M. Stone and B. Baggett, *Steroids*, **5**, 809 (1965).
[e] E. V. Jensen, E. R. DeSombre, D. J. Hurst, T. Kawashima, and P. W. Jungblut, *Arch. Anat. Microsc. Morphol. Exp.* **56**, Suppl. 3–4, 547 (1967).
[f] S. Sander, *Acta Pathol. Microbiol. Scand.* **75**, 520 (1969).
[g] F. James, V. H. T. James, A. E. Carter, and W. I. Irvine, *Cancer Res.* **31**, 1268 (1971).
[h] R. H. Wyss, R. Karsznia, W. L. Heinrichs, and W. L. Herrmann, *J. Clin. Endocrinol. Metab.* **28**, 1824 (1968).
[i] H. Johansson, L. Terenius, and L. Thorén, *Cancer Res.* **30**, 692 (1970).
[j] L. Terenius, *Acta Endocrinol. (Copenhagen)* **57**, 669 (1968).
[k] L. H. Evans and R. Hähnel, *J. Endocrinol.* **50**, 209 (1971).
[l] T. A. Musliner, G. J. Chader, and C. A. Villee, *Biochemistry* **9**, 4448 (1970).
[m] G. Giannopoulos and J. Gorski, *J. Biol. Chem.* **246**, 2524 (1971).
[n] L. Tseng, A. Stolee, and E. Gurpide, *Endocrinology* **90**, 390 (1972).

For studying estradiol–receptor interaction *in vitro*, we have favored exposing excised tissue to estradiol concentrations (0.1 n*M*) which approximate the blood levels resulting from a physiological dose of hormone in the immature rat,[5] using a sufficient volume of solution (250–400 ml)

[5] E. V. Jensen, E. R. DeSombre, D. J. Hurst, T. Kawashima, and P. W. Jungblut, *Arch. Anat. Microsc. Morphol. Exp.* **56**, Suppl. 3–4, 547 (1967).

that the hormone concentration is not depleted significantly by tissue uptake. In this concentration range, nonspecific binding, shown by all tissues investigated, is less pronounced than at higher estradiol concentrations, and estradiol incorporation into uterine tissue is of the same order of magnitude as that observed *in vivo*.[6] A simple system for this purpose is described.

Procedure for Tissue Uptake. Exposure of excised tissues to tritiated estradiol is carried out in pH 7.3 Krebs-Ringer-Henseleit glucose buffer (KRH) containing the following reagents for each liter: $Na_2HPO_4 \cdot 7H_2O$, 1.73 g; KH_2PO_4, 0.2 g; NaCl, 8 g; KCl, 0.2 g; glucose, 1.0 g; $CaCl_2$, 0.05 g; and $MgCl_2$, 0.5 ml of 1 *M* solution. All components except $CaCl_2$ and $MgCl_2$ are dissolved in somewhat less than the final volume of distilled water, previously deionized by passage through mixed bed ion-exchange resin. The $CaCl_2$ is then added; when this has dissolved, the $MgCl_2$ solution is added and the mixture diluted to mark. Unused buffer is stored at 2°; if it is to be kept for periods longer than 2 weeks, penicillin (10^6 units/liter) is also added.

[6,7-^{3}H]Estradiol of high specific activity is stored in methanol at —20° at a concentration less than 2 μg/ml to minimize radiochemically induced deterioration. An aliquot portion of this solution is evaporated at room temperature in a stream of nitrogen, and the residue taken up in sufficient deionized water to give a 1 μM solution. A trace of ethanol may be used to dissolve the film of estradiol before the water is added, but this is not necessary. For each 100 ml of KRH buffer, in a beaker of appropriate size, a 10-μl portion of the estradiol solution is added; the beaker is covered with a watch glass, placed in a water bath at the desired temperature (usually 37°), and the contents allowed to come to temperature equilibrium while stirring with a Teflon-coated magnetic bar. Duplicate 200-μl aliquots of the resulting 0.1 n*M* estradiol solution are taken before and after the tissue incubation, and the initial and final estradiol concentrations determined by liquid scintillation counting.

Tissues to be investigated are rapidly trimmed of fat or other extraneous material. Rat uterine horns are slit lengthwise to expose both surfaces to hormone solution; mammary tumors and larger organs, such as liver and kidney, are manually cut into sections 0.5 mm thick, using a Stadie-Riggs tissue slicer. Specimens are kept in cold saline or KRH buffer until all are ready.

Tissue segments or slices are added to the hormone solution at zero time, using at least 10 ml of 0.1 n*M* [^{3}H]estradiol in KRH for each 10 mg of tissue. At appropriate intervals, specimens are removed with a for-

[6] E. V. Jensen and H. I. Jacobson, *Recent Progr. Horm. Res.* **18**, 387 (1962).

ceps or a glass rod, rinsed with a little cold KRH buffer, blotted gently with filter paper, and either pooled for homogenization, dissolved in a tissue solubilizer, or lyophilized and the dry residue weighed and combusted in a Packard Model 300 tissue oxidizer to furnish tritiated water for scintillation counting. Five to six specimens usually are taken at each time point; this is especially important with tumor slices so that variations due to nonhomogeneity of the tissue can be minimized.

Specific vs. Nonspecific Binding. When a target tissue, such as rat uterus, is exposed to tritiated estradiol in the *in vitro* system described above, the uptake observed is a combination of specific binding to the receptor system plus the nonspecific binding that estradiol shows with target and nontarget tissues alike. The two types of binding are readily differentiated by comparing the uptake in the presence and absence of an inhibitor of specific binding, such as Parke-Davis CI-628 (1-[2-(*p*-[α-(*p*-methoxyphenyl)-β-nitrostyryl]phenoxy)ethyl]pyrrolidine monocitrate)[7] or nafoxidine (U-11,100A = 1-[2-(*p*-[3,4-dihydro-6-methoxy-2-phenyl-1-naphthyl]phenoxy)ethyl]pyrrolidine hydrochloride).[8] For such studies, a 10 mM solution of the antagonist is prepared in 95% ethanol (6.37 mg/ml for CI-628; 4.62 mg/ml for nafoxidine), and 10 μl of this solution added for each 100 ml of KRH–estradiol solution before the tissue specimens are introduced. As shown in Fig. 1, the presence of 10 μM PD CI-628 reduces the binding by rat uteri in 0.1 nM estradiol to the level observed with diaphragm but has no effect on the nonspecific binding seen with either uterus or diaphragm.[9]

The amount of nonspecific binding shown by whole tissues in the *in vitro* system depends both on the hormone concentration and on the structure of the steroid being studied. Certain derivatives of estradiol, such as estradiol 3-methyl ether or 17α-ethynylestradiol 3-methyl ether (mestranol), substances which do not bind to the receptor protein of uterine cytosol and which are not incorporated as such by rat target tissues *in vivo*,[10] show marked *in vitro* incorporation by uterine tissue comparable to that of estradiol (Fig. 2). That this binding is of a nonspecific type is indicated by the similar large uptake of these ethers by diaphragm

[7] M. R. Callentine, R. R. Humphrey, S. L. Lee, B. L. Windsor, N. H. Schottin, and O. P. O'Brien, *Endocrinology* **79**, 153 (1966).

[8] G. W. Duncan, S. C. Lyster, J. J. Clark, and D. Lednicer, *Proc. Soc. Exp. Biol. Med.* **112**, 439 (1963).

[9] E. V. Jensen, H. I. Jacobson, S. Smith, P. W. Jungblut, and E. R. DeSombre, *Gynecol. Invest* **3**, 108 (1972).

[10] E. V. Jensen, H. I. Jacobson, J. W. Flesher, N. N. Saha, G. N. Gupta, S. Smith, V. Colucci, D. Shiplacoff, H. G. Neumann, E. R. DeSombre, and P. W. Jungblut, *in* "Steroid Dynamics" (G. Pincus, T. Nakao, and J. F. Tait, eds.), p. 133. Academic Press, New York, 1966.

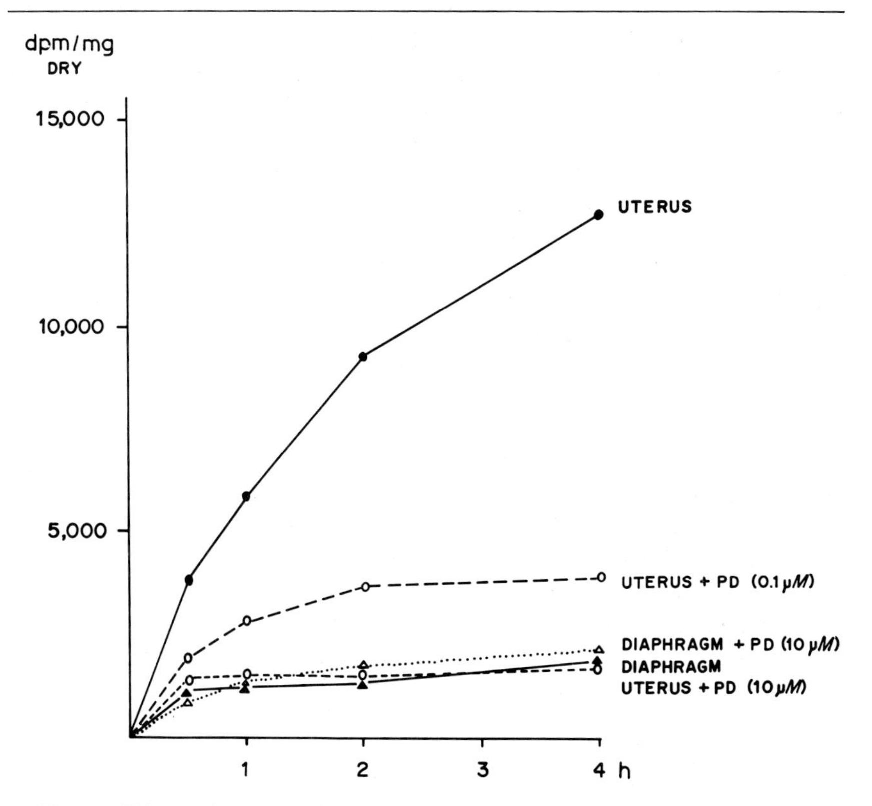

FIG. 1. Tritium levels in slit uterine horns and hemidiaphragms of immature rats after stirring in 0.12 nM [6,7-^{3}H]estradiol (57 Ci/mmole) at 37° in Krebs-Ringer-Henseleit glucose buffer, pH 7.3, in the presence and absence of an estrogen antagonist, Parke-Davis CI-628. Each point is the median value of five specimens. From E. V. Jensen, H. I. Jacobson, S. Smith, P. W. Jungblut, and E. R. DeSombre, *Gynecol. Invest.* 3, 108 (1972).

tissue and the fact that the uterine uptake is not reduced significantly by the presence of CI-628 or nafoxidine. Sulfhydryl blocking reagents, such as organic mercurials which also prevent specific binding of estradiol by uterine tissue,[5] do not abolish the uptake of the 3-methyl ethers (Fig. 2).

Estrogen-Induced Receptor Transformation

The estrogen-dependent conversion of the estrogen-binding unit of the receptor protein from its native to an activated form can be demonstrated by warming a mixture of uterine cytosol and tritiated estradiol to temperatures between 20 and 37° and comparing the sedimentation of the estra-

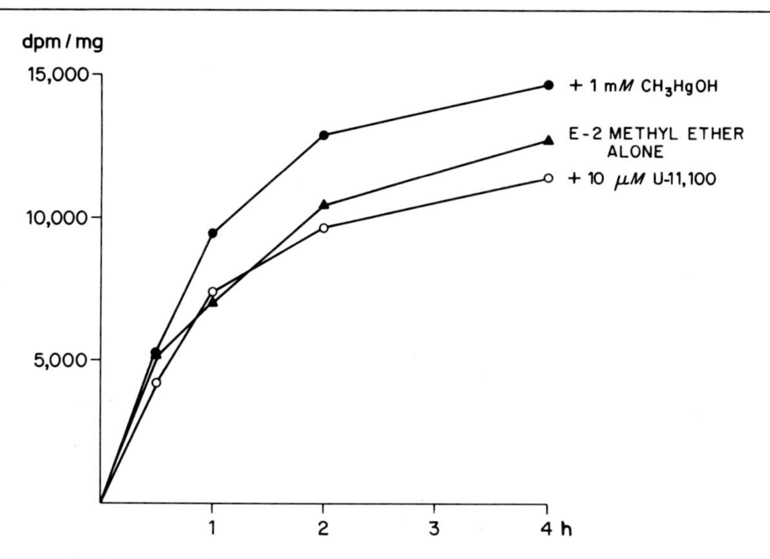

FIG. 2. Uptake of radioactivity by immature rat uteri from 0.08 n*M* [6,7-^{3}H]estradiol 3-methyl ether (48 Ci/mmole) in presence and absence of 10 μ*M* nafoxidine (U-11,100) or 3 m*M* methylmercurihydroxide. From E. V. Jensen, H. I. Jacobson, S. Smith, P. W. Jungblut, and E. R. DeSombre, *Gynecol. Invest.* **3**, 108 (1972).

diol–receptor complex with that of the complex in unheated cytosol, using salt-containing sucrose gradients.[11,12] The conversion takes place readily at pH conditions between pH 7.5 and 8.5 and is faster at the higher pH. The presence of EDTA in the medium retards but does not prevent transformation, and 150 m*M* KCl exerts a significant catalytic effect. The reaction is strongly temperature-dependent, but because decomposi-
receptor transformation.
tion of the receptor protein also takes place as the temperature is raised, we have found that 25° is a favorable temperature at which to carry out

In a typical experiment, illustrated by Fig. 3, minced calf endometrium (5 g) is homogenized in four volumes of cold 0.32 *M* sucrose in 10m*M* Tris, pH 7.4, using a Polytron PT-20 with each 10-second homogenization period followed by a 50-second cooling interval to keep the temperature below 2°. The homogenate is centrifuged at 2° for 30 minutes at 246,000 g. A 5-ml portion of the cytosol fraction is made 5 n*M* in [6,7-^{3}H]estradiol by addition of 25 μl of a 1 μ*M* aqueous solution the hormone and then incubated for 2 hours at either 25 or 0° or for

[11] E. V. Jensen, M. Numata, P. I. Brecher, and E. R. DeSombre, *in* "The Biochemistry of Steroid Hormone Action" (R. M. S. Smellie, ed.), p. 133. Academic Press, New York, 1971.

[12] M. Geschwendt and T. H. Hamilton, *Biochem. J.* **128**, 611 (1972).

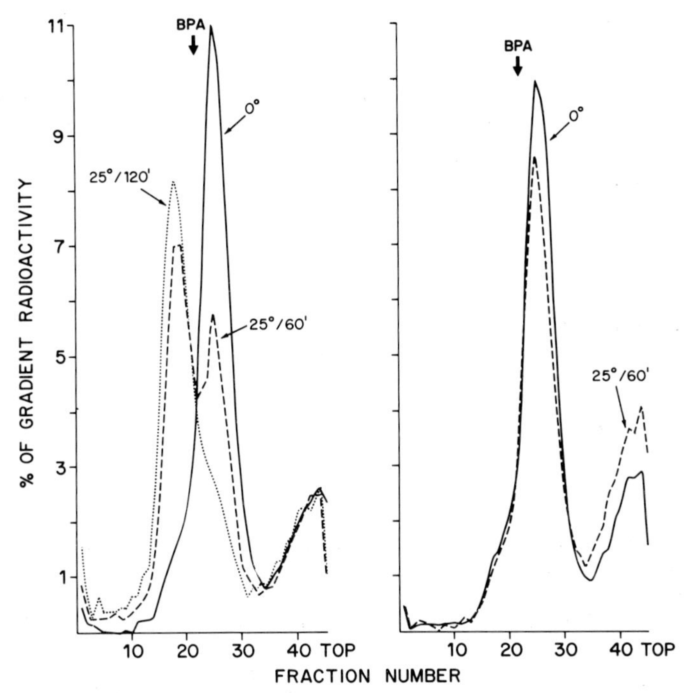

FIG. 3. Sedimentation patterns in salt-containing sucrose gradients of mixtures of tritiated estradiol and calf endometrium cytosol incubated as described in the text. Left panel: [^{3}H]estradiol present during incubation; right panel: [^{3}H]estradiol added after incubation.

1 hour at 25° followed by 1 hour at 0°. Portions of the cytosol without estradiol are similarly incubated, after which they are made 5 n*M* with tritiated estradiol. After dilution with an equal volume of 10 m*M* Tris buffer, pH 7.4, a 200-μl aliquot portion of each preparation is layered on a 4.4-ml preformed 5 to 20% sucrose gradient, prepared in 10 m*M* Tris buffer, pH 7.4, containing 400 m*M* KCl and 1 m*M* EDTA, and centrifuged at 2° for 16 hours at 308,000 *g* using cellulose nitrate tubes in a Spinco SW-56 rotor. After centrifugation the tubes are punctured at the bottom and successive 100-μl fractions collected for scintillation counting. A 0.5% solution of bovine plasma albumin (BPA) containing 6 m*M* tritiated estradiol is used as a marker for the 4.6 S position.

As illustrated in Fig. 3, the conversion of the estradiol receptor complex from the native to the transformed form does not take place appreciably in the cold under conditions of low ionic strength. It proceeds readily at 25, reaching completion in between 1 to 2 hours with calf uterus and somewhat more rapidly with rat uterus. In the absence of the estrogenic hormone, warming the cytosol does not cause transformation of the receptor protein.

Binding of Transformed Complex to Nuclei

The formation in uterine nuclei of a 5.2 S estradiol–receptor complex, extractable with 0.3 *M* or 0.4 *M* potassium chloride and indistinguishable from that obtained after estradiol administration *in vivo,* takes place on exposure of excised uterine tissue to estradiol in the *in vitro* system described above or on incubating uterine nuclei with estradiol in the presence but not the absence of the cytosol receptor protein. Whole uterine homogenate may be warmed with tritiated estradiol at 25 to 37°, or purified uterine nuclei may be recombined with the high-speed supernatant fraction for incubation with the hormone.

In a typical experiment, illustrated in Fig. 4, calf endometrium cytosol, prepared in 0.32 *M* sucrose as described in the previous section, is made 5.6 n*M* in [6,7-^{3}H]estradiol and incubated for 45 minutes either at 0° (A and B) to maintain the receptor in its native form or at 25°

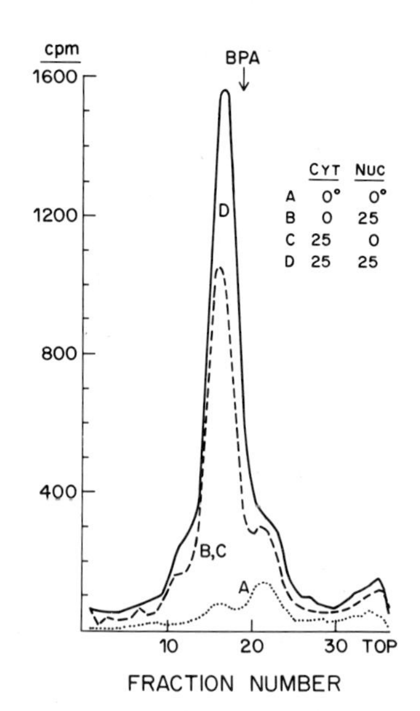

FIG. 4. Sedimentation patterns on salt-containing sucrose gradients of KCl extracts of purified calf endometrium nuclei incubated with heated and unheated estradiol–cytosol mixtures as described in the text. From E. V. Jensen, S. Mohla, T. Gorell, S. Tanaka, and E. R. DeSombre, *J. Steroid Biochem.* 3, 445 (1972).

(C and D) to effect its transformation. Equal portions of the nuclear fraction from the same homogenate, after purification by centrifugation in 2.2 *M* sucrose,[13] are suspended in either the heated or the unheated estradiol–cytosol mixture, using a ratio of cytosol to nuclei comparable to that of the original homogenate (the recovery of purified nuclei from the 2.2 *M* sucrose is usually about 30%). The nuclear suspensions are incubated for 60 minutes at either 0° (A and C) or 25° (B and D), after which the nuclei are separated by centrifugation at 2° for 10 minutes at 1,000 *g*, washed once by centrifugation in 0.32 *M* sucrose, and resuspended in 1 ml of 400 m*M* KCl in 10 m*M* Tris, pH 8.5. After standing for 30 minutes at 2° with occasional agitation (Vortex mixer), the nuclei are removed by centrifugation for 20 minutes at 10,000 *g*, and a 200-μl portion of each KCl extract (containing in CPM: A, 1,520; B, 7,430; C, 7,450; D, 11,200) is layered on a 3.6-ml salt-containing 5 to 20% sucrose gradient for sedimentation analysis, as described for Fig. 3 (except for the use of 3.8 ml polyallomer rather than cellulose nitrate centrifuge tubes).

Figure 4 indicates that the transformed and native forms of the estradiol–receptor complex differ markedly in their affinities for purified uterine nuclei. To generate an extractable 5.2 S complex in the nucleus, the native complex, which shows little binding in the cold (A), requires incubation at temperatures where transformation can accompany interaction with the nucleus (B). In contrast, the pretransformed complex shows substantial uptake both in the cold (C) or at 25° (D). Thus, receptor transformation appears to be a requisite for nuclear binding of the estradiol–receptor complex and probably represents the temperature-dependent process associated with nuclear incorporation in whole uterine tissue.

[13] J. Chauveau, Y. Moule, and C. Rouiller, *Exp. Cell. Res.* **11**, 317 (1956).

[24] Techniques for Monitoring the Distribution of the Estradiol-Binding Protein Complex between Cytoplasm and Nucleus of Intact Cells[1]

By DAVID WILLIAMS and JACK GORSKI

A number of observations have led to an hypothesis in which the initial step in the stimulation of uterine growth by estradiol (E) is the

[1] Supported by grants HD 4828 and HD 00181 from the National Institutes of Health and Ford Foundation Training Grant 700-0333.

interaction of the hormone with specific binding proteins or receptors (B) in the cytoplasm of uterine cells.[1a] The estradiol-binding protein complex (EB) thus formed is subsequently transferred to the cell nucleus. The methods to be described here have been employed to examine the formation of EB and its distribution between cytoplasmic and nuclear fractions of immature rat uteri.

Consider as an example the situation at the conclusion of an *in vitro* incubation of the uterus with [^{3}H]E or an *in vivo* treatment with the hormone. The labeled hormone is present in several states within the tissue: (1) as EB complexes in cytoplasmic and nuclear fractions (specifically bound E); (2) as complexes with proteins other than the specific estradiol-binding proteins (nonspecifically bound E or EP); and (3) as E free within the cells, in extracellular spaces, and partitioned into lipid components of the tissue (trapped E). Similarly, when a particular amount of EB is present, a defined amount of unfilled specific binding sites will also be present within the uterine cells. The distribution of E among these components will be a function of E concentration, incubation time, temperature, the quantity of each component present, and the relative affinity each component exhibits for E. For an examination of EB as a function of time or its distribution between cytoplasmic and nuclear fractions, several requirements must be met. (1) It must be possible to stop the formation of EB and preserve the EB formed to that point. (2) Methods must be available to separate cytoplasmic and nuclear fractions. (3) Methods must be available to assay bound E in these fractions. (4) It must be possible to distinguish EB from EP. These points will be considered in turn.

Stopping the Reaction

Principle

The rate of estrogen uptake by the intact uterine cell at 0° is approximately 1% that observed at 37°.[2] It is possible, therefore, to stop the formation of EB within the cell by rapidly lowering the temperature and preventing estrogen from associating with B. The dissociation of E from the EB complex proceeds with a $t_{1/2}$ of 20–30 hours at 0°.[3] Consequently, little of the EB complex present within the tissue will be lost through dissociation during the preparation of EB for subsequent assay.

The slow dissociation of E from EB at 0° also permits one to eliminate

[1a] J. Gorski, D. Toft, G. Shyamala, and A. Notides, *Recent Progr. Horm. Res.* **24,** 45 (1968).

[2] D. Williams and J. Gorski, *Biochemistry* **12,** 297 (1973).

[3] B. M. Sanborn, B. R. Rao, and S. G. Korenman, *Biochemistry* **10,** 4955 (1971).

the situation in which excess [^{3}H]EB complexes are formed during homogenization of the tissue. This problem appears to arise because [^{3}H]E which is present within the tissue as trapped hormone at 37° is brought into contact with unfilled specific binding sites in the cytoplasmic fraction during homogenization at 0°. Since EB is readily formed in cell-free extracts at 0°, the formation of [^{3}H]EB during homogenization can lead to an overestimate of the quantity of cytoplasmic EB that was present at the conclusion of the *in vitro* incubation or *in vivo* hormone treatment. This additional [^{3}H]EB may cause an overestimation of actual cytoplasmic [^{3}H]EB by as much as 10-fold. The extent of [^{3}H]EB formation during homogenization is a reflection of the quantity of estradiol present as trapped hormone and the quantity of unfilled specific binding sites remaining at the conclusion of the experimental treatment. It has not been possible to eliminate this problem by altering the *in vitro* incubation conditions or washing the tissue extensively before homogenization. On the other hand, the long $t_{1/2}$ for dissociation of [^{3}H]EB permits homogenizations to be carried out in the presence of excess unlabeled estradiol. The unlabeled hormone eliminates the formation of additional [^{3}H]EB during homogenization but has little effect on the [^{3}H]EB formed prior to homogenization.[4]

Procedures

As uteri are removed from the hormone-treated animal or the *in vitro* incubation flask, they are placed in incubation medium or a balanced salt solution at 0°. The distribution of EB between cytoplasmic and nuclear fractions is stable at 0° for at least 2 hours. The uteri are washed several times in iced homogenization buffer and homogenized in buffer containing excess unlabeled estradiol. For *in vitro* incubations at [^{3}H]E concentrations up to 2×10^{-8} M, unlabeled estradiol at 5×10^{-7} to 1×10^{-6} M is sufficient to block formation of [^{3}H]EB during homogenization.

Comments

The temperature sensitivity of the binding process in the intact cell is probably due to the inability of E to enter uterine cells at 0°, although other explanations have not been ruled out.[2,4] It is not known whether the uptake of other estrogens is similarly sensitive to temperature.

The rate constant for the dissociation of E from the EB complex appears to be even more temperature sensitive than the rate constant for the association reaction.[3] It is imperative, therefore, to carry out all operations at 0° after the addition of excess unlabeled E. An increase

[4] D. Williams and J. Gorski, *Biochem. Biophys. Res. Commun.* **45**, 258 (1971).

in the temperature to even 15° during subsequent centrifugation procedures will cause extensive exchange to occur.

The use of excess unlabeled E during homogenization is contingent on the $t_{1/2}$ for dissociation being long relative to the time required to fractionate the tissue and assay [^{3}H]EB. Whether this procedure can be used with other estrogens must be determined on the basis of their dissociation rates from their respective hormone-binding protein complexes.

Preparation of Cytoplasmic and Nuclear Fractions

Principle

For an analysis of EB distribution within the uterine cell, two criteria are desirable: (1) complete cell breakage and separation of cytoplasmic and nuclear fractions, and (2) a nuclear fraction consisting of purified nuclei in quantitative yield. The muscular nature of the uterus makes the preparation of purified nuclei in high yields very difficult, however, such that it is usually necessary to choose a procedure which either satisfies the first criterion above or provides purified nuclei in low to moderate yields. The nature of the question under investigation will dictate the method to be used.

Method 1: Preparation of Purified Uterine Nuclei[5,6]

All operations are carried out at 0°. Uteri are homogenized in 0.25 *M* sucrose–0.01 *M* Tris-HCl, pH 7.4 (25°), with 1–5 uteri/ml buffer. Homogenization is with a motor-driven all-glass tissue grinder at low speed for one to two 15-second grinding periods. Centrifugation of the homogenate at 600 *g* for 10 minutes yields a supernatant and a nuclear sediment. The supernatant can be treated as in Method 2 to yield the cytosol fraction. The nuclear sediment is suspended in 5.0 ml of 2.2 *M* sucrose, 0.003 *M* $CaCl_2$, 0.01 *M* Tris-HCl, pH 7.4 (25°), and centrifuged at 47,000 *g* (average) for 30 minutes in a swinging bucket rotor (Beckman SW-39) to yield the purified nuclear fraction.

Advantages. This nuclear preparation consists of intact clean nuclei as judged by phase microscopy. Electron microscopy of these nuclei shows a slight contamination with collagen fibers, but no other cytoplasmic contaminants are evident.

Disadvantages. With the gentle homogenization required to maintain the integrity of the nuclei, cell breakage and, hence, the recovery of specific binding sites is not complete. Recoveries of DNA in the purified

[5] G. Shyamala Harris, *Nature (London), New Biol.* **231**, 246 (1971).

[6] G. Shyamala Harris, personal communication.

nuclear preparations range from 30 to 45%. It is not known whether this procedure selects for nuclei from particular uterine cell types or gives a uniform sampling from all cell types.

This procedure avoids the use of detergents to remove the outer nuclear membrane.[7] The purified nuclei can be washed with detergent to accomplish this, but it should be emphasized that the effects of various detergents on nuclear EB are not known. We have determined that exposure to Tween 80 for 1–3 hours at concentrations of 0.2–0.5% has no effect on cytoplasmic EB as judged by gel filtration on Sephadex G-25.[8] On the other hand, Triton X-100 and deoxycholate at these concentrations reduce cytoplasmic EB by 60–80%. The effects of these detergents on nuclear EB have not been studied.

This procedure does not employ polyamines in the buffers to reduce the loss of RNA polymerase during the nuclear preparation.[9] Spermidine at concentrations up to 0.005 *M* does not solubilize nuclear bound EB or affect its subsequent extraction with 0.4 *M* KCl (see section Assay of Bound Hormones below).[8]

An alternative procedure for preparing uterine nuclei employs homogenization with a Virtis homogenizer and does not employ the 2.2 *M* sucrose centrifugation step.[10]

Method 2: Preparation of a Crude Nuclear Fraction[3]

Care is taken to carry out all operations on ice or at 0°. Immature rat uteri are homogenized in 0.01 *M* Tris-HCl, 0.0015 *M* EDTA, pH 7.4 (25°), containing 1×10^{-6} *M* E with 1–5 uteri/ml buffer. Homogenization is with a motor-driven all-glass tissue grinder (Kontes) for three to four 15-second grinding periods with 30-second cooling periods between. To minimize heating, the homogenizer is submerged in an ice-water or salt-ice-water bath. Centrifugation of the homogenate at 800 *g* for 20 minutes yields the low speed supernatant (LSS) which can be centrifuged at 226,000 *g* (average) for 45 minutes to yield the high speed supernatant (HSS). The 800 *g* pellet is washed vigorously (with a vortex mixer) four times in homogenization buffer (5 ml), each wash followed by centrifugation at 800 *g* for 10 minutes to yield the washed nuclear fraction.

Advantages. This method provides complete cell breakage and complete recovery of tissue DNA in the washed nuclear fraction. If uteri are incubated at various E concentrations, the sum of the specific binding

[7] S. Penman, I. Smith, and E. Holtzman, *Science* **154**, 786 (1966).

[8] D. Williams and J. Gorski, unpublished observations.

[9] S. P. Blatti, C. J. Ingles, T. J. Lindell, P. W. Morris, R. F. Weaver, F. Weinberg, and W. J. Rutter, *Cold Spring Harbor Symp. Quant. Biol.* **35**, 649 (1970).

[10] J. Barry and J. Gorski, *Biochemistry* **10**, 2384 (1971).

sites found in the HSS and the washed nuclear fraction are constant and equal to the total specific binding capacity extractable from the tissue[11]; this indicates the quantitative recovery of estradiol-binding proteins in these two fractions. Similarly, the equilibrium distribution of EB between the HSS (or LSS) and the washed nuclear fraction[12] is indistinguishable from the autoradiographic distribution of [^{3}H]E between cytoplasm and nuclei after the *in vivo* administration of the hormone.[13] This result gives assurance that this method provides an accurate reflection of the filled binding site distribution in the uterine cell.

Disadvantages. The washed nuclear pellet consists of intact nuclei, broken nuclei, nuclear debris, and other cellular components which sediment at 800 *g*. This fraction is obviously not a suitable nuclear preparation with which to ask about the subnuclear distribution of EB.

Assay of Bound Hormone

Principle

The measurement of EB in the HSS (or LSS) and washed nuclear fraction requires a means of distinguishing bound from unbound hormone. The slow dissociation of E from EB permits the use of nonequilibrium methods for the physical separation of E from EB. Among those methods which have been employed are gel filtration,[14] sucrose gradient centrifugation,[15] charcoal adsorption of unbound E,[4] and hydroxylapatite (HAP) adsorption of the EB complex.[16] The HAP method described here provides a rapid, reliable, and sensitive assay for the routine analysis of many samples.[17]

Assay of Cytoplasmic EB

All operations are carried out on ice or at 0°. BioGel HT hydroxylapatite (BioRad Industries) is washed with 0.05 *M* Tris-HCl, 0.001 *M* KH_2PO_4 (pH 7.2, 0°) until the pH of the wash is 7.2 at 0°. The volume of the slurry is adjusted such that 1.0 ml slurry contains 0.6–0.7 ml packed HAP. High speed supernatant (HSS) ($>$100,000 *g* supernatant fraction) or LSS (800 *g* supernatant fraction (0.05–2.0 ml) is added to the HAP slurry (0.50 ml) and allowed to stand on ice for 10–20 minutes with several mixings. Washing buffer (0.05 *M* Tris-HCl, pH 7.3, 25°)

[11] G. Giannopoulos and J. Gorski, *J. Biol. Chem.* **246**, 2524 (1971).
[12] D. Williams and J. Gorski, *Acta Endocrinol.* (*Copenhagen*), *Suppl.* **168**, 420 (1972).
[13] W. E. Stumpf, *Endocrinology* **83**, 777 (1968).
[14] G. A. Puca and F. Bresciani, *Nature* (*London*) **218**, 967 (1968).
[15] D. Toft and J. Gorski, *Proc. Nat. Acad. Sci. U.S.* **55**, 1574 (1966).
[16] T. Erdos, M. Best-Belpomme, and R. Bessada, *Anal. Biochem.* **37**, 244 (1970).
[17] D. Williams, Ph.D. Thesis, University of Illinois, Urbana (1972).

(4 ml) is added, the sample mixed, and centrifuged at 800 g for 2 minutes. The wash is discarded, the HAP pellet resuspended in washing buffer (5 ml), mixed vigorously with a vortex mixer, and centrifuged as above. This washing procedure is repeated three more times. The washed HAP pellet is extracted with 100% ethanol (4 ml) for 15 minutes at room temperature, the sample centrifuged, and the ethanol extract assayed for radioactivity after addition to a toluene-based scintillation fluid. This extraction procedure transfers greater than 97% of the radioactivity to the scintillation vial.

Comments. Adsorption of EB to HAP is maximal by 5 minutes and stable for at least 90 minutes. Adsorption of EB to HAP is linear with increasing quantities of HSS or LSS up to at least 1.2 uterine equivalents of EB per 0.3 ml packed HAP. Identical results are obtained with HSS and LSS. Assay of EB in HSS by HAP or gel filtration on Sephadex G-25 gives identical results.

With 0.3 ml packed HAP the adsorption of one uterine equivalent of HSS or nuclear extract (see below) is independent of volume between total assay volumes of 0.55–2.5 ml, independent of Tris concentrations of 0.01–0.05 M, and independent of KCl concentrations up to and including 0.4 M. Unbound E does not significantly adsorb to HAP under the conditions of the assay. Estradiol bound to nonspecific sites also adsorbs to HAP, necessitating the use of competitive methods to correct for contributions from this component (see section Distinguishing EB from EP).

Assay of Nuclear EB

The preparation of the washed nuclear fraction separates unbound E from nuclear bound hormone present as EB and EP. Consequently, nuclear bound E can be determined with an ethanol extraction of the washed nuclear pellet as described above for the HAP pellet. EB can then be distinguished from EP as described below. Alternatively, EB can be extracted from the washed pellet or a purified nuclear preparation with 0.01 M Tris-HCl, 0.001 M EDTA, 0.4 M KCl, pH 8.5, for 1 hour at 0°.[14] Subsequent centrifugation at 226,000 g for 45 minutes yields the high speed nuclear extract which is suitable for further examination of EB. EB in this extract can be assayed with HAP as described above without altering the KCl concentration.

Distinguishing EB from EP

Principles

With the possible exception of sucrose gradient centrifugation under conditions where the 8 S binding protein is separated from the bulk of uterine proteins,[15] all the methods noted above measure bound E irrespective of whether it is EB or EP. Although the use of nonequilibrium assay methods minimizes the contribution of EP to the total bound E, EP is never zero and it may be as large as EB at high E concentrations. The quantity of E present as EB can be distinguished from that present as EP by taking advantage of the fact that the nonspecific P binding sites are a low affinity binding system present in very large numbers while the specific B binding sites are a high affinity binding system present in very limited numbers.[18] Hence, the specific B binding system saturates at E concentrations that are still in the linear range of the nonspecific P binding system. The quantity of [^{3}H]EP present at any [^{3}H]E concentration can be determined with a parallel sample which contains a 100-fold excess of unlabeled E in addition to the [^{3}H]E. In this parallel sample the limited number of specific B sites contain E and [^{3}H]E in a 100:1 ratio, such that EB complexes contribute only a few percent of the radioactivity they have in the sample containing only [^{3}H]E. On the other hand, the unlimited number of P sites contain 100 times as much E at a specific radioactivity equal to 1% of that in the sample containing only [^{3}H]E; hence the EP complexes contain the same quantity of radioactivity in both samples. The difference between the samples gives the quantity of [^{3}H]EB present.

Procedures

For any sample at a particular [^{3}H]E concentration, a parallel sample containing [^{3}H]E and a 100-fold excess of unlabeled E is also incubated. The samples are then treated by the methods described above. The values of EP determined for cytoplasmic and nuclear fractions are subtracted from the total bound E in each fraction to yield the quantities of EB present.

Comments

For this procedure to be valid, EP must behave as a linear function of the E concentration within the employed concentration range. When working with immature rat uteri *in vitro*, EP present in cytoplasmic and washed nuclear fractions behaves linearly up to E concentrations of at least 2×10^{-6} *M*.[17]

It should be noted that assayable EP is an operational quantity which may differ in magnitude depending on the method used for assay. In con-

[18] D. J. Ellis and H. J. Ringold, *in* "The Sex Steroids" (K. W. McKerns, ed.), p. 73. Appleton, New York, 1971.

trast to the situation with EB, the quantity of EP remaining after a nonequilibrium assay may be only a small fraction of that present during the tissue incubation.

Since the nonspecific P sites are present in great excess over the specific B sites, the equilibrium quantity of EP will be attained much faster than the equilibrium quantity of EB. During brief incubations, therefore, EP will constitute a greater part of the total bound hormone than it will at equilibrium.

The use of E in a 100-fold excess is chosen simply for flexibility over a large part of the saturation range. When working in the lower 10% of the saturation curve, a higher ratio of E to [^{3}H]E can be employed to ensure that EB complexes contribute little to the value obtained for EP. Similarly, when working at [^{3}H]E concentrations appreciably greater than 10^{-8} *M*, a lower ratio of E to [^{3}H]E can be employed to ensure that the hormone stays in solution.

[25] [^{3}H]Estradiol Exchange Assay for the Determination of Nuclear Receptor–Estrogen Complex[1]

By J. H. CLARK, J. N. ANDERSON, and E. J. PECK, JR.

Introduction

The association of the receptor–estrogen complex (RE) with nuclear binding sites is considered to be of fundamental importance in the mechanism of action of estrogen. We have developed a method for the determination of the concentration of RE in the nuclear fraction of estrogen target tissues which relies on the exchange of [^{3}H]estradiol with estradiol which is bound to the nuclear fraction.[1a] This method permits an evaluation of the concentration of nuclear RE under *in vivo* conditions and as a function of nonlabeled estrogenic compounds.

Assay Method

The [^{3}H]estradiol exchange assay can be used for the measurement of nuclear RE in the uterus, pituitary, and hypothalamus. The method as used for the analysis of uterine tissue will be described initially and

[1] This work was supported by grants from The National Institutes of Health (HD 04985) and The Research Corporation, Atlanta, Ga., USA.

[1a] J. N. Anderson, J. H. Clark, and E. J. Peck, Jr., *Biochem. J.* **126**, 561 (1972).

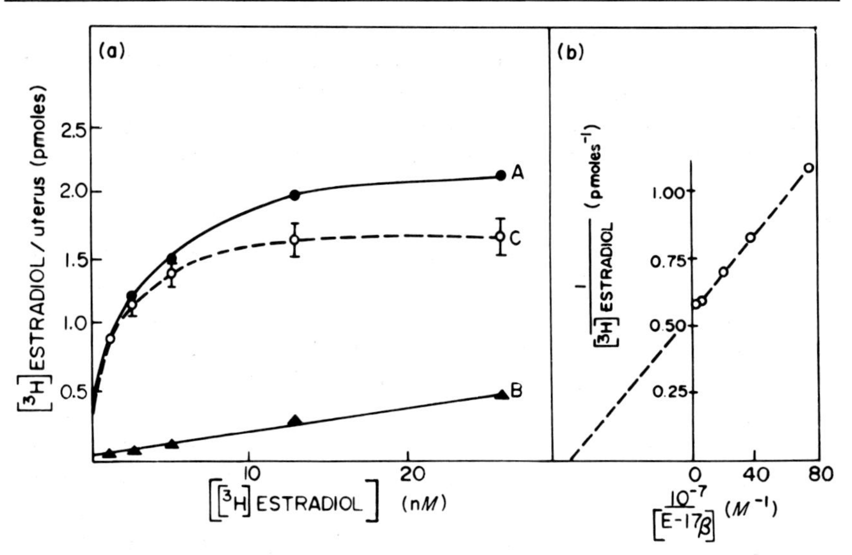

FIG. 1. The determination of the quantity of nuclear RE by the [^{3}H]estradiol exchange assay. (a) This experiment is presented as an example of the use of the [^{3}H]estradiol exchange assay for the determination of the quantity of nuclear RE following an injection of estradiol. Immature rats, age 21–23 days, were injected subcutaneously with 2.5 μg of estradiol. One hour after the administration of hormone the rats were sacrificed and the uterine nuclear fraction was prepared as described in Methods. This fraction was incubated for 1 hour at 37° with either [^{3}H]estradiol alone (A) or [^{3}H]estradiol plus DES at 100 times the concentration of [^{3}H]estradiol (B). The dashed line (C) was obtained by subtracting B from A. Each point represents the mean of 6 determinations with 3 rats per determination. (b) A double-reciprocal plot of the data represented by (C) of Part 1a. The apparent K_d^R is 1.3×10^{-9} M and the number of binding sites per uterus is 1.8 pmoles.

the necessary modifications for the analysis of the hypothalamus or pituitary will be noted.

The uterus is stripped of adhering fat and mesentery and placed in cold 0.9% NaCl. All subsequent procedures should be performed at 4° unless otherwise stated. The tissue is washed in 10 mM Tris buffer, pH 7.4, that contains 1.5 mM EDTA (TE buffer) and subsequently homogenized in 1.5 ml of TE (wet weight/volume, 20 mg/ml) in an all glass Kontes homogenizer using a motor-driven pestle. The nuclear fraction is obtained by centrifugation of the homogenate at 800 g for 10 minutes. This fraction is washed three times with 3 ml of TE buffer and the nuclear pellet is mixed thoroughly each time with a vortex mixer. Each wash is followed by centrifugation at 800 g for 10 minutes. The washed nuclear pellet is rehomogenized and suspended in TE buffer to a concentration of 0.5 uterus/ml (immature rat uterus). Portions (0.5 ml) of this suspension are dispensed into two series of tubes, A and B, containing 0.2 ml of buffer. Series A should contain various concentrations of [^{3}H]estradiol (1–25 nM) and is used to determine the total amount of [^{3}H]estradiol exchange. Series B should contain the same concentration of [^{3}H]estradiol as in A plus a 100-fold excess of diethylstilbestrol (DES) and is used to determine the nonspecific [^{3}H]estradiol exchange. The nuclear fractions are incubated with shaking at 37° for 1 hour. Following incubation, the nuclear fraction is washed three times with 3 ml of TE buffer as described above. The nuclear fraction is extracted with 3 ml of 100% ethanol and the extract is added to 10 ml of scintillation fluid [95.5% toluene, 0.45% 2-5-diphenyloxazole, 0.05% 1,4-bis-(5-phenyloxazol-2-yl) benzene] for the determination of radioactivity. The amount of [^{3}H]estradiol that is bound in the nuclear fraction in the absence (series A) and presence (series B) of DES represents total and nonspecific binding, respectively (Fig. 1). The [^{3}H]estradiol that is bound in a specific, DES-competable manner (Fig. 1a, C) may be determined by subtracting the nonspecific binding (Fig. 1a, B) from the total binding (Fig. 1a, A). The maximum quantity of nuclear RE and the dissociation constant for RE, K_d^R, are obtained from a double reciprocal plot of the data for the specifically bound [^{3}H]estradiol (Fig. 1a, C). A representative plot is shown in Fig. 1b.

The procedure for the measurement of nuclear RE in the hypothalamus or pituitary is essentially the same as that used for the analysis of uterine tissue with these modifications: The tissue is homogenized in a glass homogenizer, with a Teflon pestle (0.11–0.15 mm clearance). The nuclear fraction is obtained and washed as above. The washed nuclear pellets are resuspended in TE buffer to a concentration of 3 hypothalami/ml (wet weight, 26 mg/immature hypothalamus) or 3 pituitaries/ml (wet weight, 2.2 mg/immature pituitary). One-milliliter portions of this suspension are dispensed into two sets of tubes, series A and B as above. The concentrations of [^{3}H]estradiol should range from 0.46–3.66 nM. The remainder of the procedure is identical to that used for the uterus.

To determine the number of picomoles of receptor–estrogen complex present in the nuclear fraction under any experimental situation, either a saturation curve such as Fig. 1 or a concentration of estradiol equivalent to $K_d^R \times 10^3$ must be employed. Since either experiment is expensive in experimental animals and/or labeled estradiol, the assay may be performed at a concentration of 13 nM [^{3}H]estradiol for uterine tissue and 3.6 nM [^{3}H]estradiol for hypothalamic and pituitary tissue. These concentrations are 10 fold greater than the apparent K_d^R for the receptor of these tissues (uterus, 1.3 ± 0.16 nM; hypothalamus, 0.39 ± 0.19 nM;

pituitary, 0.35 ± 0.06 nM) and should be sufficient to saturate approximately 91% of the specific binding sites. Under our conditions, approximately 95% of the specific binding sites are measured at these concentrations. The difference between observed and predicted degrees of saturation are reflected in the means obtained for maximal binding sites ± the standard error of the mean for each tissue (uterus, 60.0 ± 5.9; pituitary, 14.3 ± 0.23; hypothalamus, 0.47 ± 0.06 fmole/mg wet weight).

Conclusion

The [^{3}H]estradiol exchange assay can be used to determine the number of specific nuclear binding sites in estrogen-sensitive tissues and has permitted us to examine the influence of endogenous and exogenous estrogen on the concentration of nuclear RE and the concentration-time-response parameters of estrogen action.[2,3] This assay is of special significance in view of the recent demonstration that approximately 90–95% of the estrogen which is specifically bound in the uterus is that which is present in the nuclear fraction.[4]

[2] J. H. Clark, J. N. Anderson, and E. J. Peck, Jr., *Science* **176**, 528 (1972).

[3] J. N. Anderson, J. H. Clark, and E. J. Peck, Jr.,, *Biochem. Biophys. Res. Commun.* **48**, 1460 (1972).

[4] D. Williams and J. Gorski, *Biochem. Biophys. Res. Commun.* **45**, 258 (1971).

[26] A Technique for Differential Extraction of Nuclear Receptors[1]

By D. Marver and I. S. Edelman

Introduction

Mineralocorticoid-specific binding proteins have been isolated and characterized in target tissues of adrenalectomized rats injected with [^{3}H]aldosterone.[2-7] *In vivo,* the binding of aldosterone to these receptors[8]

[1] Supported by USPHS, National Heart and Lung Institute, Grant Nos. HL-06285 and HL-05725.

[2] D. Marver, D. Goodman, and I. S. Edelman, *Kidney Int.* **1**, 210 (1972).

[3] J. W. Funder, D. Feldman, and I. S. Edelman, *J. Steroid Biochem.* **3**, 209 (1972).

[4] T. S. Herman, G. M. Fimognari, and I. S. Edelman, *J. Biol. Chem.* **243**, 3849 (1968).

[5] G. E. Swaneck, L. L. H. Chu, and I. S. Edelman, *J. Biol. Chem.* **245**, 5382 (1970).

[6] D. D. Fanestil and I. S. Edelman, *Proc. Nat. Acad. Sci. U.S.* **56**, 872 (1966).

[7] G. E. Swaneck, E. Highland, and I. S. Edelman, *Nephron* **6**, 297 (1969).

[8] The term "receptor" is used only to imply high-affinity, specific binding sites.

is an early event, occurring well before the physiological response. Maximum nuclear labeling of [^{3}H]aldosterone in the rat kidney occurs 10 minutes after intravenous injection of steroid, although the physiological change in urinary Na^+/K^+ levels is not demonstrable for about an hour and reaches a peak response in 3–4 hours.[9] In this chapter, we describe the methods for labeling of the aldosterone receptors and differential extraction of aldosterone–receptor complexes from the nucleus.

Preparation of Labeled Complexes

In Vivo. Aldosterone-binding proteins can be labeled *in vivo* by injecting adrenalectomized rats intravenously or subcutaneously with physiological doses of [^{3}H]aldosterone (0.1–0.5 μg/100 g body wt., 50 Ci/mM). To correct for nonspecific binding, control animals receive the same dose of [^{3}H]aldosterone plus a 100-fold excess of unlabeled aldosterone. With intravenous administration, animals are sacrificed in 2 to 10 minutes, depending on whether maximal cytosol labeling (2 minutes) or maximal nuclear labeling (10 minutes) is desired. Subcutaneous injections result in a somewhat lengthened time course, with maximal cytoplasmic and nuclear labeling occurring at about 15 to 30 minutes, respectively.[6]

In Vitro: Slices. Kidney slices from adrenalectomized rats are labeled with [^{3}H]aldosterone by incubation in a complete salt media at 25° with 10^{-10} to 10^{-8} M [^{3}H]aldosterone. The kidneys are removed, decapsulated, and transferred immediately to a solution composed of: NaCl = 135, KH_2PO_4 = 5, Tris-HCl = 5, $MgCl_2$, = 0.5, $CaCl_2$ = 1, glucose = 5 (all in mM), pH 7.4, at 0°. The kidneys are sliced (275 μm) in a McIlwain tissue slicer,[10] collected on Nytex,[11] and briefly rinsed with cold media. Erlenmeyer flasks containing 20 ml incubating media and 10^{-10} to 10^{-8} M [^{3}H]aldosterone $\pm$ a 100-fold excess of unlabeled steroid are placed in a rotary shaking water bath. At zero time, slices from one to two kidneys are transferred to each flask and incubated for 5 to 60 minutes. Within this time period, cytoplasmic labeling reaches a plateau in 15 minutes, while nuclear labeling rises steadily for an hour or more. At the end of the incubation period, the slices are recollected on Nytex, rinsed with ice-cold buffer, and immediately transferred to homogenizing media (0.25 M sucrose, 3 mM $CaCl_2$) at 0°.

In Vitro: Cell-Free Recombination Studies. Cell-free mixtures of nuclei and cytosol labeled with [^{3}H]aldosterone can be used to generate

[9] G. M. Fimognari, D. D. Fanestil, and I. S. Edelman, *Amer. J. Physiol.* **213**, 954 (1967).

[10] Brinkmann Instrument Co., Burlingame, Calif.

[11] Nylon cloth, 132 mesh, John Stanier & Co., Manchester, England.

both low salt- and high salt-extractable nuclear complexes. For cell-free incubations, cytosol obtained from the 19,000 *g* isolation step or the 105,000 *g* isolation step (details in next section) is diluted with spectroquality glycerol to a final concentration of 25% (v/v). [^{3}H]Aldosterone (10^{-10} to 10^{-8} *M*) ± a 100-fold excess of unlabeled aldosterone is added and allowed to stand at 0° for 30 minutes. The crude nuclear 600 *g* pellet (details in next section) is washed twice with 0.25 *M* sucrose, 3 m*M* $CaCl_2$ using volumes at least equivalent to the decanted crude cytosol volume, by resuspending and recentrifuging at 600 *g* for 10 minutes. After washing, the pellet is suspended in a small amount of 0.25 *M* sucrose, 3 m*M* $CaCl_2$ and divided between a number of 25-ml Erlenmeyer flasks so that each flask contains the nuclei from 1–2 kidneys. Labeled cytosol is added to the nuclear suspension at 0°. For optimal binding, the cytosol added per flask should be equivalent to the amount obtained from 1–2 kidneys and the proportion of cytosol volume to nuclear suspension volume in the incubating flask should be $\geqq 4:1$. At zero time, the mixtures are placed in a 25° shaking water bath, incubated 15 to 30 minutes, and then poured into iced centrifuge tubes and spun at 600 *g* for 10 minutes (0°) to separate nuclei and cytosol.

Preparation of Purified Cytosol and Nuclear Fractions

All procedures are at 0–4°. Kidneys or kidney slices are homogenized in 0.25 *M* sucrose, 3 m*M* $CaCl_2$ (~1 g tissue/4 ml sucrose) using a Teflon-glass homogenizer. A crude nuclear and cytoplasmic fraction is obtained by centrifugation of the homogenate at 600 *g* for 10 minutes. The supernatant is decanted and recentrifuged at 19,000 *g* for 15 minutes and then 105,000 *g* for 1 hour. A separation of bound and free aldosterone in the final 105,000 *g* supernatant is made by filtering 1-ml aliquots over 0.28 × 15 cm G-50 fine Sephadex[12] columns equilibrated with 0.1 *M* Tris-HCl, 3 m*M* $CaCl_2$, pH 7.2 (the pH of this Tris buffer can be 7.2–8.0 for both cytosol and nuclear complexes). Following filtration, 0.5–1.0 ml fractions of the void volume of each G-50 column is added to 15 ml of dioxane counting solution[13] and assayed for radioactivity in a liquid scintillation counter. The protein content of the void volume is estimated from the ratio of the O.D. at 260 and 280 nm[14] and the final results expressed as moles of aldosterone bound per milligram protein. "Specific"

[12] Pharmacia, Piscataway, New Jersey.

[13] 10 g 2,5-diphenyloxazole (PPO), 0.5 g 1,4-bis-2-(4-methyl-5-phenyloxazolyl) benzene (dimethyl-POPOP), and 80 g napthalene in 428 ml ethylene glycol monomethyl ether, 428 ml *p*-dioxane, and 143 ml xylene.

[14] O. Warburg and W. Christian, *Biochem. Z.* **310,** 384 (1942).

binding is then calculated as the difference between total binding and binding in the presence of a 100-fold excess of unlabeled aldosterone.

To purify nuclei obtained from *in vivo*, slice, or recombination studies, the 600 *g* pellet is resuspended twice in 0.25 *M* sucrose, 3 m*M* $CaCl_2$ and recentrifuged each time at 600 g for 10 minutes. The pellet from 2–4 kidneys is then mixed with 20 ml of 2.2 *M* sucrose, 3 m*M* $CaCl_2$ and layered over 10 ml of this same solution. This mixture is centrifuged at 56,000 *g* in a Spinco 30 rotor[15] for 1 hour, using polycarbonate 30-ml tubes. The 2.2 *M* sucrose and cellular debris are decanted and excess sucrose removed from the walls of the centrifuge tube with cotton-tipped applicator sticks. The sedimented nuclei are gently resuspended in 1.0 ml of 0.1 *M* Tris-HCl 3 m*M* $CaCl_2$, pH 7.2, and transferred by Pasteur pipet to 3.0 ml of the same Tris buffer. The suspension is allowed to stand 10 minutes at 0° and then spun at 10,000 *g* for 10 minutes. The aldosterone-bound protein released into the supernatant of this fraction is referred to as the "nuclear Tris-soluble complex." The supernatant Tris extract is decanted and the pellet resuspended in 2 ml of 0.1 *M* Tris-HCl buffer, recentrifuged at 10,000 *g* for 10 minutes, and the supernatant discarded. The pellet is then extracted with 1–2 ml of 0.4 *M* KCl, 0.01 *M* Tris, pH 7.2. This mixture is incubated at 0° for 10 minutes and then centrifuged at 10,000 *g* for 10 minutes. The resultant supernatant contains the majority of the residual [^{3}H]aldosterone nuclear counts, and the protein:steroid conjugate obtained in this fraction is referred to as the "chromatin-bound aldosterone complex."[5] For concentration of both fractions, an equal volume of saturated ammonium sulfate is added to the fraction at 0°. After 30 minutes the mixture is spun at 19,000 *g* for 10 minutes and the supernatant discarded. The precipitate is resuspended in 0.5/1.5 ml of either 0.1 *M* Tris-HCl or 0.4 *M* KCl, 0.01 *M* Tris-HCl, pH 7.2, and any insoluble material removed by centrifugation at 19,000 *g* for 10 minutes.

To assay binding, aliquots of the Tris-soluble and chromatin-bound nuclear fractions are added to dioxane counting solution and protein estimated spectrophotometrically by the O.D. at 260 and 280 nm. Results are expressed as moles of aldosterone bound per milligram protein minus nonspecific binding determined in the presence of a 100-fold excess of unlabeled aldosterone.

Characterization of Nuclear Aldosterone Complexes on Glycerol Density Gradients

The nuclear binding proteins can be further characterized by their sedimentation properties on glycerol density gradients. Aliquots of

[15] Spinco Corp., Palo Alto, Calif.

nuclear low and high salt extracts can be layered directly on gradients or concentrated first with ammonium sulfate as described above. Linear gradients made from starting concentrations of 10 and 34% spectroquality glycerol (v/v) which contain in addition, either 0.1 *M* Tris, pH 7.2, or 0.4 *M* KCl + 0.01 *M* Tris, pH 7.2. [^{3}H]Aldosterone, at a final concentration of 10^{-9} *M*, may also be added as a constant background to stabilize bound complexes during centrifugation. Either 12.5 or 4 ml gradients can be used, the larger gradient spinning at 283,000 *g* for 40 hours (rotor SB 283) and the smaller at 405,000 *g* for 16 hours (rotor

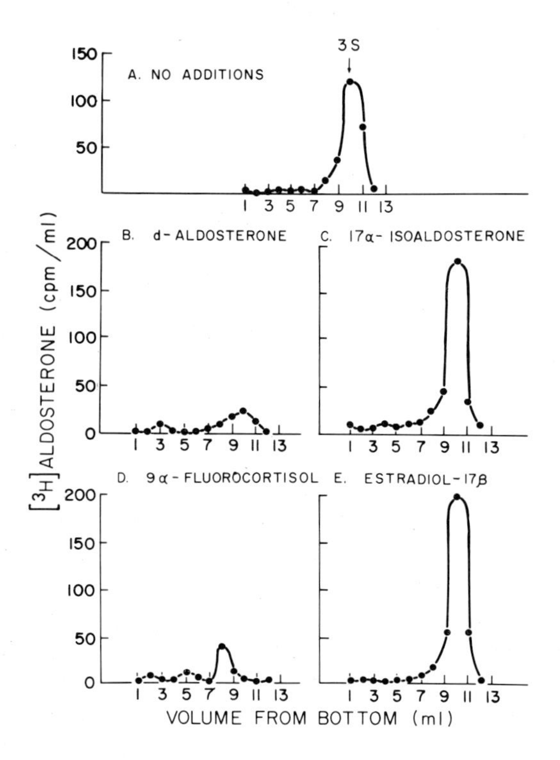

FIG. 1. Glycerol density gradients of Tris-soluble nuclear extracts labeled *in vivo*. Adrenalectomized 150 g rats were injected with 2.6×10^{-10} moles [^{3}H]aldosterone and either diluent (Panel A), or 2.6×10^{-9} moles *d*-aldosterone (Panel B), or 2.6×10^{-9} moles 17α-isoaldosterone (Panel C), or 2×10^{-8} moles 9α-fluorocortisol (Panel D), or 2×10^{-8} moles estradiol-17β (Panel E). The 10–34% glycerol density gradients also contained 0.1 *M* Tris-HCl, 3 m*M* $CaCl_2$, pH 8.0, and 1.6×10^{-9} *M* [^{3}H]aldosterone with or without a 10- or 78-fold excess of the appropriate competitor. The 12.5-ml gradients were centrifuged 283,000 *g* for 40 hours. Following centrifugation, 1-ml fractions were collected and filtered through Sephadex G-50 columns to remove free [^{3}H]aldosterone. (From Marver *et al.*[2])

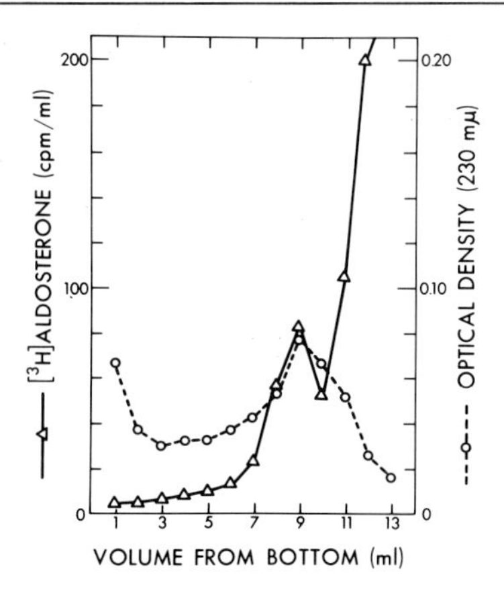

Fig. 2. Glycerol density gradient of the chromatin-bound [^{3}H]aldosterone complex labeled *in vivo*. Adrenalectomized 150–200 g rats were injected with 2.6×10^{-10} moles [^{3}H]aldosterone. Chromatin was isolated from the nuclear pellet following extraction with 0.1 *M* Tris-HCl, 3 m*M* $CaCl_2$, pH 8.0, by the method of K. Marushige and J. Bonner [*J. Mol. Biol.* **15**, 160 (1966)]. A 0.3 *M* KCl extract of chromatin was layered on a 10–35% 12.5 ml, glycerol density gradient containing 0.3 *M* KCl–0.01 *M* Tris-HCl, pH 8.0, and centrifuged 283,000 *g* for 40 hours. Following centrifugation, 1-ml aliquots were collected and assayed directly for radioactivity and O.D. at 230 mm. [From G. E. Swaneck, L. L. H. Chu, and I. S. Edelman, *J. Biol. Chem.* **245**, 5382 (1970).]

SB 405).[16] Following centrifugation, the base of the tube is punctured with a #26 needle and aliquots collected. The fractions may be counted directly and assayed for 260:280 protein, or if [^{3}H]aldosterone was used to stabilize binding, the fractions are filtered through G-50 columns to removed unbound steroid.[2]

Gradients of Tris soluble fractions yield a specific 3 S peak as shown in Fig. 1 (whether the gradients contain low or high salt). The gradient pattern of the chromatin-bound fraction isolated after *in vivo* labeling and centrifuged in the presence of 0.4 *M* KCl is shown in Fig. 2. In high salt (0.4 *M* KCl) gradients, the chromatin-bound species migrates as a 4 S complex.[5] In low salt gradients (0.1 *M* Tris or less) however, this complex aggregates and pellets to the bottom of the centrifuge tube.

Properties of Nuclear Complexes

Isolated nuclear and cytoplasmic aldosterone-specific complexes are susceptible to degradation by sulfhydryl reagents and proteases such as pronase, trypsin, and chymotrypsin, but not by nucleases. In the absence

[16] International Centrifuge Corp., Needham Heights, Mass.

TABLE I
TEMPERATURE DEPENDENCE OF [^{3}H]ALDOSTERONE BINDING IN RAT RENAL CYTOSOL AND NUCLEAR FRACTIONS[a]

Temperature (°C)	$\times 10^{-15}$ moles/mg protein		
	Cytosol	Tris-soluble	Chromatin-bound
0	22.5	19.0	2.1
25	31.4	28.4	12.9
0/25	71.7%	66.9%	16.3%

[a] Adrenalectomized rat kidney slices were incubated at either 25° or 0° for 1 hour in 5.2×10^{-9} *M* [^{3}H]aldosterone with or without 5.2×10^{-6} *M* *d*-aldosterone. The quantity of [^{3}H]aldosterone bound has been corrected for nonspecific labeling.

of glycerol, at least 70% of the bound aldosterone is dissociated from the complexes in 24 hours at 0°. Glycerol at a 20–25% final concentration is able to preserve a minimum of 75% of this activity over the same time period at 0°. The steroid-labeled complexes are also heat labile and may be irreversibly dissociated by incubating isolated fractions at 37° for 30 minutes. *In vitro* incubations have provided more information regarding the properties of the nuclear complexes. The transfer of labeled complexes from cytosol to nuclei is temperature-dependent. Table I shows the effect of incubating rat kidney slices at 0° rather than 25°. While the amount of labeled cytosol and Tris-soluble complex isolated from the 0° incubation was 72% and 67% that of the 25° incubation, the chromatin-bound fraction had only 16% of the labeling seen at 25°.

[27] Steroid Hormone–Receptor Interactions with Nuclear Constituents

By WILLIAM T. SCHRADER, SUSAN H. SOCHER, and RICHARD E. BULLER

I. Introduction

Over the past decade considerable evidence has accumulated to support the concept that steroid hormones act to alter the expression of genes for specific proteins in target cells. This regulation appears to occur in the nucleus, where radioactive steroids administered *in vivo* rapidly accumulate tightly associated with the chromatin. The steroids are found first bound to cytoplasmic receptor proteins and subsequently these complexes appear in the nuclei. In order to study these nuclear processes more fully, it has been necessary to develop or adopt procedures for observing nuclear uptake and steroid–receptor binding *in vitro*.

The chick oviduct has been studied extensively as a model system for steroid hormone action.[1-3] In this tissue, estrogens induce growth and differentiation of the immature oviduct and induce the synthesis *de novo* of the egg white proteins ovalbumin and lysozyme. Progesterone administration to the oviduct results in the induction of a new protein, avidin. This progesterone response proceeds through the early events outlined above. Since stable receptor proteins can be isolated for this steroid,[4-6] the intranuclear accumulation mechanism for progesterone could be readily studied in this system.

A. Preparation of Progesterone Receptors

Detailed methods for isolation and partial purification of chick oviduct progesterone receptors appear elsewhere in this volume.[7] For routine assays of chromatin acceptor sites and nuclear binding, crude cytoplasmic extracts are prepared in 0.01 *M* Tris-HCl, 0.025 *M* KCl, pH 7.4 (Buffer A). Buffers containing EDTA can be used for chromatin binding studies, which show no demonstrable cofactor requirement, but EDTA should be avoided in nuclear studies since it will chelate some of the Mg^{2+} in the nuclear preparation.

High-speed cytoplasmic extracts (cytosol) from oviduct or nontarget tissues will contain about 20 mg protein/ml if homogenization is done in 4 ml buffer/g tissue. The cytosols should be diluted to about 10 mg protein/ml for the incubations. Tritiated hormone ([1,2-^{3}H]progesterone, 40–50 Ci/mmole) should be added 1 to 2 hours before the incubations begin in order to have freshly labeled preparations. It is important to add subsaturating doses of progesterone (10^{-8} *M*) so that only small

[1] B. W. O'Malley, *Biochemistry* **6**, 2546 (1967).

[2] B. W. O'Malley, W. L. McGuire, P. O. Kohler, and S. G. Korenman, *Recent Progr. Horm. Res.* **25**, 105 (1969).

[3] B. W. O'Malley, T. C. Spelsberg, W. T. Schrader, F. Chytil, and A. W. Steggles, *Nature (London)* **235**, 141 (1972).

[4] M. R. Sherman, P. L. Corvol, and B. W. O'Malley, *J. Biol. Chem.* **245**, 6085 (1970).

[5] B. W. O'Malley, M. R. Sherman, D. O. Toft, T. C. Spelsberg, W. T. Schrader, and A. W. Steggles, *Advan. Biosci.* **7**, 213 (1970).

[6] W. T. Schrader and B. W. O'Malley, *J. Biol. Chem.* **247**, 51 (1972).

[7] W. T. Schrader, this Vol. [17].

amounts of free hormone remain in solution. This helps to lower the background in the chromatin and nuclear-binding studies. Alternatively, the preparation can be labeled for 6 to 12 hours with saturating amounts of hormone (10^{-7} *M*) and then chromatographed on a short Sephadex G-25 column (10 × 2.6 cm). The void volume fractions will contain labeled receptors essentially free of uncomplexed steroid. Any of the chromatographic steps used to purify the progesterone receptors also separate free hormones from receptor complexes. Thus, partially purified preparations can be used immediately after they are made.

Since several of the receptor purification steps involve use of KCl elution, it is frequently necessary to dilute the salt concentration to suit the required assay conditions. Nuclear uptake studies can be performed over the widest range of KCl concentration: The assay described here uses 0.025 *M* KCl during the incubation and wash steps before extraction at 0.4 *M* KCl. Thus, receptor preparations should be diluted to 0.025 to 0.15 *M* KCl before incubation. Chromatin binding is optimal at 0.15 *M* KCl.[8] Since variations in KCl above or below this value can cause net changes in the total mass of protein bound to chromatin, receptors should not be added to chromatin until the KCl concentration in the receptor preparation has been lowered to less than 0.15 *M* KCl. DNA–receptor interactions dissociate at about 0.15 *M* KCl.[7] Therefore, to maximize receptor binding but minimize effects of other proteins the receptor–DNA studies are done at 0.1 *M* KCl. Receptor separations should be adjusted to a suitably low KCl concentration before combining with the DNA sample.

One of the most illuminating results of progesterone receptor purification has been the resolution of the receptors into two classes by DEAE-cellulose chromatography.[5,9,10] Both classes bind progesterone but only one binds to DNA.[7] The other receptor class binds only to chromatin. These receptor preparations were assayed for binding to DNA and chromatin as discussed in the following sections. Immediately after elution of the labeled complexes from the ion-exchange column, the receptors were assayed for binding activity. Receptor component A was eluted stepwise with 0.15 *M* KCl and was diluted to 0.1 *M* KCl for DNA-binding studies. Receptor B, which eluted stepwise at 0.3 *M* KCl, was diluted to 0.15 *M* KCl for chromatin-binding analysis.

[8] T. C. Spelsberg, A. W. Steggles, and B. W. O'Malley, *J. Biol. Chem.* **246**, 4188 (1971).

[9] W. T. Schrader, D. O. Toft, and B. W. O'Malley, *J. Biol. Chem.* **247**, 2401 (1972).

[10] B. W. O'Malley, W. T. Schrader, and T. C. Spelsberg, *in* "Receptors for Reproductive Hormones" (B. W. O'Malley and A. R. Means, eds.), p. 174. Plenum, New York, 1973.

The two different receptor components can be used to analyze DNA and chromatin binding independently. Since the binding activities are mutually exclusive, it is possible to study chromatin binding separately from DNA interactions by use of component B. By choosing component A, the DNA binding can be studied independently as well.

B. Isolation of Nuclei

Similar methods of isolation are used for the preparation of nuclei for both the nuclear- and chromatin-binding assays.[8] For the nuclear-binding experiments, fresh tissue is routinely used to prepare nuclei. However, chromatin can be prepared from either fresh or frozen tissue. The tissues to be assayed are removed from the chicks and rinsed in cold 0.9% NaCl. They can be stored frozen at —20°.

All steps for the isolation of nuclei are performed at 4°. The desired tissue is weighed, minced, and placed in 5 to 10 volumes (w/v) of Buffer B (0.025 *M* KCl, 0.002 *M* $MgCl_2$, 0.05 *M* Tris-HCl, pH 7.5) containing 0.5 *M* sucrose. The tissue is initially homogenized either with a Pt-10 Polytron (Brinkman) using four 5-second bursts at 45 V or with 10 strokes of a loose-fitting Teflon-glass tissue homogenizer (Glenco Instruments). These homogenizers are selected so that the Teflon pestle will slowly drop down the glass homogenizer under gravity when dry. Further homogenization is accomplished by 5 strokes in a Teflon-glass homogenizer turned by an electric drill at 500 rpm. A Teflon-glass homogenizer is used in all subsequent steps for resuspension. The initial homogenate is centrifuged at 6000 *g* for 10 minutes. The crude nuclear pellets are resuspended in 4 to 10 volumes of Buffer B containing 0.5 *M* sucrose and passed through 4 layers of cheesecloth to remove tissue fragments. This homogenate is centrifuged at 6000 *g* for 10 minutes. The nuclear pellets are then resuspended in 10 volumes of Buffer B containing 2.0 *M* sucrose, and following homogenization the sucrose concentration is adjusted to 1.75 to 1.8 *M* sucrose. (During homogenization, care must be taken not to create a vacuum below the Teflon pestle; this results in broken nuclei and low yields.) The resuspended nuclei are centrifuged at 25,000 *g* for 20 minutes. At this step in the isolation, the nuclei are sufficiently purified for the nuclear-binding assay. They are free of cytoplasmic contaminants but have intact nuclear membranes.

For the preparation of nuclei for chromatin studies, the nuclear pellets collected through heavy sucrose are gently homogenized in 5 volumes of Buffer B containing 0.5 *M* sucrose and 0.2% Triton X-100. Triton is used to obtain purified nuclei without ultracentrifugation[11] and to reduce pro-

[11] S. Panyim and R. Chalkley, *Biochemistry* **8**, 3972 (1969).

tease activity in stored chromatin.[12] The homogenate is then passed through organza cloth (100 mesh) and centrifuged at 6000 *g* for 10 minutes to pellet the purified nuclei. It is advisable to check the nuclei microscopically throughout the isolation process using aceto-orcein stain or phase-contrast microscopy.

C. Isolation of Chromatin

Purified nuclear pellets are resuspended by hand homogenization in 10 to 20 volumes of 0.08 *M* NaCl + 0.02 *M* EDTA, pH 6.3. The homogenate is centrifuged at 12,000 *g* for 5 minutes, and the chromatin pellets are rehomogenized twice more in the same solution. At this stage, the preparation should be checked microscopically for complete nuclear breakage. Additional washes with 0.08 *M* NaCl + 0.02 *M* EDTA, pH 6.3, may be required to break all nuclei. When nuclear lysis is complete, the chromatin is washed with 10 volumes of 0.35 *M* NaCl and centrifuged at 12,000 *g* for 5 minutes. The chromatin pellets are resuspended in 10 volumes of 0.002 *M* Tris-HCl, 0.0001 *M* Na_2EDTA, pH 7.5, and centrifuged at 20,000 *g* for 10 minutes. The chromatin is then allowed to swell in 20 volumes of the same buffer. The chromatin is passed through organza cloth and centrifuged at 20,000 *g* for 20 minutes. The chromatin pellets are resuspended in a small volume of the above buffer to give a final DNA concentration of 0.5 to 1 mg/ml. The chromatin suspension is analyzed for histone, nonhistone protein, DNA, and RNA concentration.[13,14] Chromatin can be stored at −20° for short periods of time.

In experiments requiring sheared chromatin, the chromatin is sheared in a French Pressure Cell at 12,000 psi. After shearing, the chromatin solution is centrifuged at 3,000 *g* for 10 minutes and the pellet discarded. This procedure solubilizes 94 to 98% of the DNA.

D. Preparation of Animal Cell DNA

The procedure is a modification of the method of Marmur[15] with the addition of a polysaccharide hydrolysis step using α-amylase and the omission of the isopropanol precipitation procedure.

Reagents

Solution A: 0.25 *M* Sucrose, 0.001 *M* $CaCl_2$, 0.05 *M* Tris-HCl, pH 8.0

Solution B: 0.15 *M* NaCl, 0.1 *M* Na_2EDTA, pH 8.0

[12] J. R. Tata, M. J. Hamilton, and R. D. Cole, *J. Mol. Biol.* **67**, 231 (1972).

[13] K. Marushige and J. Bonner, *J. Mol. Biol.* **15**, 160 (1966).

[14] T. C. Spelsberg and L. S. Hnilica, *Biochim. Biophys. Acta* **228**, 202 (1971).

[15] J. Marmur, *J. Mol. Biol.* **3**, 208 (1961).

Solution C: 10% Sodium dodecyl sulfate (SDS)
Solution D: 5.0 *M* $NaClO_4$
Solution E: Chloroform–isoamyl alcohol, 24:1 (v/v)
Solution F: 95% Ethanol (cold)
Solution G: Pancreatic ribonuclease A (Worthington Biochemicals code RAF) 2 mg/ml in 0.15 *M* NaCl, pH 5.0
Solution H: α-Amylase (Worthington Biochemicals code AA) 20 mg/ml in 0.1 × SSC
Solution I: Pronase (Calbiochem) 5.0 mg/ml self-digested for 2 hours at 37° in 0.1 × SSC
10 × SSC: 1.5 *M* NaCl, 0.15 *M* Na citrate, pH not adjusted
0.1 × SSC: 0.015 *M* NaCl, 0.0015 *M* Na citrate, pH 7.0

All steps should be done in ice or a cold room at 4° whenever possible. This is essential at the early steps, until the first chloroform–isoamyl alcohol extraction. The procedure can be stopped overnight at any of these chloroform additions. At the completion of this procedure the DNA should meet or exceed the following criteria of purity: (a) spectrophotometric absorbance ratios, $A_{260}/A_{280} > 1.9$, $A_{260}/A_{230} > 2.0$; (b) protein concentration[16] should be less than 3% of the DNA concentration[17]; (c) DNA content determined by the diphenylamine assay should be greater than 93% of the DNA content determined by UV absorbance (1.0 μg DNA/ml has an $A_{260} = 0.022$); (d) melting temperature T_m should be within 2° of a purified salmon sperm DNA standard (87°); and (e) hyperchromicity should be not less than 93% of that found for purified salmon sperm DNA standard (36%).

Procedure. Fresh or frozen tissues have been processed. Mince the tissue with scissors in the cold and then add two volumes (2 ml/g tissue) of solution A and homogenize with several strokes of a Teflon-glass homogenizer turned at low speed by an electric drill. Tissues which are difficult to disrupt by this process can be broken up by several 10-second bursts of a Brinkman Polytron set at 45 V. Cool in ice between bursts. The crude homogenate is then centrifuged at 1500 *g* for 10 minutes in a Sorvall HB-4 rotor to pellet nuclei. Discard the supernatant fraction. Resuspend the pellet in 5 to 10 volumes of solution B and then add 1 volume of solution C per 9 volumes of suspension to bring the solution to 1% in SDS. Shake briefly. The solution will become viscous. One volume of solution D is added per 4 volumes of suspension to bring the solution to 1.0 *M* in sodium perchlorate. Shake vigorously for 60 to 90

[16] O. H. Lowry, N. J. Rosebrough, A. L. Farr, and R. J. Randall, *J. Biol. Chem.* **193,** 265 (1951).
[17] K. Burton, *Biochem. J.* **62**, 315 (1956).

seconds. This step concludes the disruption of nuclei and the exposure of chromatin material. Subsequent steps are used to strip proteins from the DNA.

Chloroform–isoamyl alcohol extractions are used to remove most of the protein. Place the suspension obtained above in a flask with a ground-glass stopper. Add an equal volume of solution E and shake well for 30 minutes on a wrist-action shaker. Separate the resulting emulsion by centrifuging in the Sorvall HB-4 rotor at 16,000 g for 10 minutes. Draw off the upper (aqueous) phase containing the DNA and place it in a clean glass-stoppered flask. A Pasteur pipet, cut off at the constriction, is convenient for this purpose. Avoid removing any of the interface material. Repeat this extraction twice, saving only the upper phase each time. The interface should become smaller each time. If substantial interface material collects after the third extraction, repeat an additional time.

Pour the aqueous DNA-containing phase into a beaker and add two volumes of solution F to precipitate the DNA. Spool out the DNA using several sealed Pasteur pipets. Press out as much liquid as possible; then transfer the pipet to a closed flask containing 0.1 $\times$ SSC. Choose a volume that will result in a DNA concentration of about 1 mg/ml. Redissolve the DNA by gently oscillating the flask for 24 to 72 hours. The time required will vary with the tissue and source of the DNA.

Add 1 volume of 10 $\times$ SSC per 10 volumes of DNA solution to bring the solution to 1 $\times$ SSC. Next, add 24 μl of solution G and 2.5 μl of solution H per 1.0 ml of DNA solution to bring the solution to 50 μg/ml in pancreatic RNase A and α-amylase. Incubate at 37° for 30 minutes. Next, add 10 μl of solution I per 1.0 ml of DNA solution to bring the solution to 50 μg/ml in pronase. Incubate at 37° for 60 minutes. These enzymatic steps digest most of the tenaciously associated contaminants in the DNA. Subsequent steps are required to separate the DNA from the enzymes themselves.

Perform the chloroform–isoamyl alcohol extraction procedure two more times as outlined above. After the second extraction, precipitate the DNA with 95% ethanol, spool it out, and redissolve as above in 0.1 $\times$ SSC. Readjust to 1 $\times$ SSC by adding 10 $\times$ SSC as described above.

Perform a third chloroform–isoamyl alcohol extraction and again precipitate the DNA, spool it out, and redissolve in 0.1 $\times$ SSC at about 1 mg/ml.

Store the purified DNA in this buffer frozen at −20° or lower. The DNA is stable for months under these conditions. The yield of DNA prepared by this procedure varies with the type of tissue and the organism of origin. For chick oviduct or chick liver, the yield is about 200 μg of DNA/g of tissue.

II. Receptor Binding to Purified Nuclei

A. Conditions

1. Buffers

Purified nuclear pellets prepared as described above are resuspended in 8 to 12 ml of receptor preparation using a loose fitting Teflon-glass tissue homogenizer. The final buffer and salt concentrations should be approximately the same as Buffer A. Other ions or minor variations from this formula have not been found to interfere with the uptake reaction.[18]

2. Time and Temperature

In contrast to studies using crude nuclear preparations,[19,20] the optimal conditions for use of purified nuclei appear to be incubation of nuclei and receptors at 25° for 120 minutes. The rate of uptake is decreased below this temperature and is nearly zero at 0°. At incubation temperatures of 37°, the initial uptake rate is very high but falls after 10 to 15 minutes to low values. This failure of uptake parallels a heat-dependent destruction of the cytosol receptors themselves, and thus is attributed in part to a falling concentration of receptor complexes.

The uptake reaction continues at 25° for about 120 minutes and then plateaus. Over this time interval there is a progressive loss of cytosol-binding sites concomitant with their transfer to the nuclear fraction, but only about a 10% decay of total receptor sites in the assay. Pretreatment of either cytosol or nuclei by warming to 25° does not capacitate the uptake reaction to occur more rapidly at 0°.

3. Requirements for Steroids

Purified nuclei from chicks not exposed to progesterone should not contain receptor proteins. If nuclei from immature estrogen-primed chicks are isolated as described above and treated with 0.4 *M* KCl to extract receptors, the extract will contain a small number of receptor sites. These are presumably cytoplasmic contaminants and can be detected by the addition of [^{3}H]progesterone to label the sites followed by separation of receptor complexes from free hormone as described elsewhere in this

[18] R. E. Buller, unpublished.

[19] B. W. O'Malley, M. R. Sherman, and D. O. Toft, *Proc. Nat. Acad. Sci. U.S.* **67,** 501 (1970).

[20] D. O. Toft and M. R. Sherman, this Vol. [14].

volume.[7] From a nuclear pellet containing 1 mg DNA, about 4000 cpm of bound [^{3}H]progesterone can be extracted. If progesterone-labeled oviduct cytosol is used, about 60,000 cpm/mg DNA are extracted after incubation under the same conditions. The reaction requires intact progesterone–receptor complexes, since incubation of nuclei with labeled liver cytosol or labeled Tris buffer (both lacking receptors) results in extracts containing less than 4000 cpm/mg DNA. The liver cytosol generally results in even lower activity than the buffer control, an effect which can be mimicked by adding 10 mg/ml bovine serum albumin to the buffer. The liver cytosol proteins presumably lower the buffer blank value by adsorbing the free [^{3}H]progesterone and preventing the steroid from binding to perinuclear receptors.

B. Protocol for Nuclear Binding Experiments

1. Incubation

Purified nuclei are prepared as pellets in centrifuge tubes, with each tube containing nuclei from about 1 g of oviduct. To these pellets in ice, add about 10 ml of receptor preparation with the receptor sites labeled with [^{3}H]progesterone. Resuspend the pellets using a rubber policeman or Teflon pestle from a homogenizer swirled in the tube. The tubes are then set out at room temperature for the desired time of 0 to 2 hours. An unincubated control kept at 0° should be included. The tubes should be agitated periodically during the incubation.

2. Extraction

At the end of the incubation, the tubes are centrifuged at 20,000 *g* at 0° in a Sorvall SS-34 rotor and the supernatant fraction decanted. An aliquot of this can be assayed for bound progesterone as an index of receptor stability during the experiment. The nuclear pellet is resuspended in 20–30 ml Buffer B containing 1.5 *M* sucrose and centrifuged at 31,000 *g* for 20 minutes in the SS-34 rotor. This step removes material loosely associated with the nuclear membranes. Examination of nuclei at this point in the experiments by light and electron microscopy showed them to be intact with the nuclear membranes slightly damaged and with little contamination. A second wash of the nuclear pellets is done with 20 to 30 ml of Buffer B to remove excess sucrose so that nuclear extracts can be layered on 5 to 20% sucrose gradients. The supernatant fraction from this wash is discarded.

The nuclear pellets are then treated at 0° for 60 minutes with 0.4 *M* KCl, 0.0015 *M* Na_2EDTA, 0.01 *M* Tris-HCl, pH 7.5, to extract the intranuclear receptors.[19,21] The nuclei are resuspended in 2 to 3 ml of this buffer and allowed to stand in ice for 1 hour. The nuclei are pelleted by centrifugation at 31,000 *g* for 10 minutes in a Sorvall SS-34 rotor. After extraction with KCl, the nuclei are significantly damaged and clumped. They are then washed in 10 ml of ice-cold 0.3 *N* $HClO_4$.

3. *Assays for Bound Receptors and DNA*

The nuclear pellets are resuspended in 1.0 ml of 0.3 *N* $HClO_4$, heated at 90° for 30 minutes, cooled in ice, and centrifuged at 31,000 *g* to pellet the nuclear debris.[8] The supernatant fraction containing the DNA is assayed for DNA by the diphenylamine reaction.[17]

The supernatant fraction containing the receptors is kept at 0° and assayed for progesterone receptor activity by sucrose-gradient ultracentrifugation. This method is described in detail elsewhere in this volume.[20] An aliquot (200 μl) of nuclear extract is layered on a 5 to 20% sucrose gradient containing 0.01 *M* Tris-HCl, pH 7.4, 0.001 *M* Na_2EDTA, 0.012 *M* 1-thioglycerol, 0.3 *M* KCl. The gradients are centrifuged at 48,000 rpm for 16 hours in a Spinco SW-50.1 rotor. The tube is pierced, and fractions are collected from the bottom and counted for radioactivity. The size of the 4 S receptor peak of radioactivity is an index of the extent of nuclear binding of progesterone–receptor complex.

Results of a sample experiment are shown in Fig. 1A. Nuclei from estrogen-primed chick oviducts were prepared as outlined above and incubated with progesterone-labeled oviduct cytosol, or with unlabeled preparations of oviduct cytosol, liver cytosol, or buffer alone. After 2 hours, 0.4 *M* KCl extracts were prepared from each sample. The extracts from nuclei incubated without progesterone were labeled at this step with radioactive progesterone equivalent to that present in the labeled extract. After an additional 60-minute incubation at 0°, aliquots (200 μl) of each 1.0 ml extract were layered on sucrose gradients for analysis. The figure demonstrates that a significant 4 S receptor peak appeared only in the sample derived from nuclei incubated with labeled oviduct cytosol. The nuclear extract from the incubations of oviduct cytosol without progesterone showed no intranuclear receptors, since progesterone is required for nuclear retention of the complexes. Control incubations of nuclei with buffer alone or with unlabeled liver cytosol showed that intranuclear receptors were present in very low concentrations.

[21] B. W. O'Malley, D. O. Toft, and M. R. Sherman, *J. Biol. Chem.* **246**, 1117 (1971).

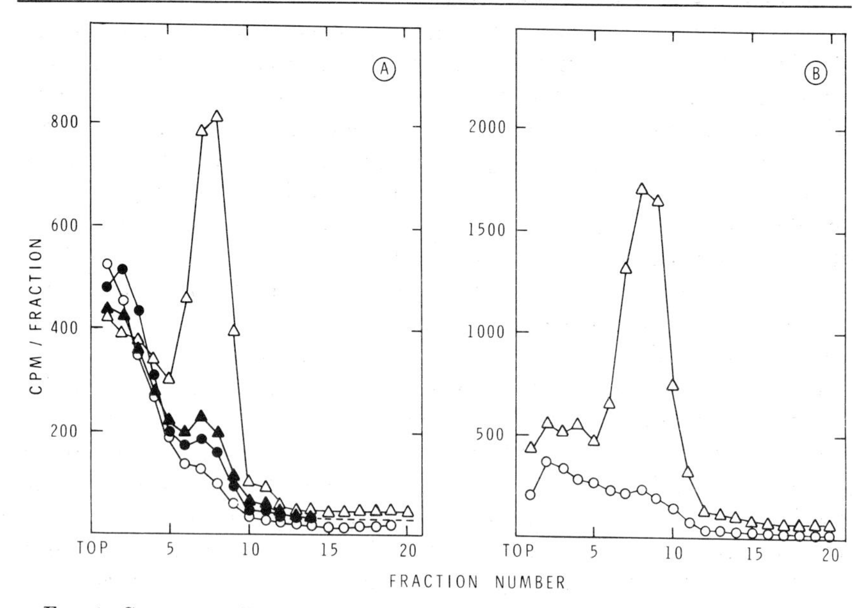

FIG. 1. Sucrose-gradient ultracentrifugation analysis of chick nuclear extracts. Nuclei were purified from oviduct or spleen. Nuclei from about 1 g tissue each were incubated with appropriate receptor preparations as described in the text at 25° for 60 minutes. Nuclei were then washed twice and resuspended in 2.0 ml of 0.01 *M* Tris-HCl, 0.0015 *M* Na_2EDTA, 0.4 *M* KCl, pH 7.5, for 60 minutes at 0° to extract intranuclear receptors. Identical 200-μl aliquots of each extract were centrifuged as described in the text on 5 to 20% sucrose gradients in Buffer D containing 0.3 *M* KCl. Gradients were run in a Spinco SW 50.1 rotor at 48,000 rpm for 16 hours. (A) Requirement for oviduct progesterone–receptor complex. Oviduct nuclei were combined with: oviduct cytosol labeled with 10^{-8} *M* [^{3}H]progesterone (△——△) or with unlabeled oviduct cytosol (▲——▲), unlabeled liver cytosol (●——●), or unlabeled Buffer A (○——○). Extracts from the unlabeled tubes were labeled with [^{3}H]progesterone so that all extracts contained equivalent radioactivity. After an additional 60 minutes at 0° for progesterone to bind to any free receptors, the extracts were layered on gradients as described above. (B) Requirement for oviduct nuclei. Oviduct cytosol, labeled with 10^{-8} *M* [^{3}H]progesterone, was combined with equivalent amounts (as nuclear DNA) of purified nuclei from oviduct (△——△) or spleen (○——○).

Similar experiments have also been performed using heterologous cytosol, plasma, or buffer to which 10^{-7} *M* [^{3}H]progesterone had been added. In such experiments, not included in the figure, extracts again contained no receptors. Thus the nuclear-binding system described here shows an absolute requirement for progesterone receptor complex.

The nuclear extracts can also be assayed for bound progesterone by more rapid procedures. Any of the methods used to assay for progesterone receptors[7] can be used. Of particular interest is the precipitation of free steroid bound to dextran-coated charcoal.[22] In this assay, radioactivity remaining in the supernatant fraction is receptor-bound steroid. Contact with the charcoal should be kept brief, particularly when the method is used to assay nuclear binding of purified receptors. Steroid-competition studies done using this method showed the nuclear 4 S receptors to be specific progesterone-binding proteins indistinguishable from the cytosol receptors, and presumably identical to them.

If nuclear-binding studies are carried out using nuclei from several different tissues of the chick, target-tissue specificity can be observed. Nuclei from oviduct cells bind two- to tenfold more receptors than do nuclei from other tissues, as shown in Fig. 1B. If equivalent amounts of oviduct and spleen nuclei are incubated separately with labeled oviduct cytosol, KCl extracts assayed on sucrose gradients show a 4 S receptor peak about 1600 cpm in amplitude for the oviduct nuclei but only a 200 cpm shoulder is seen in the spleen experiment. These studies correlate well with the results of chromatin-binding experiments described in Section III, showing that target-tissue nuclear specificity resides in the chromatin.

4. *Adaptation to Small Samples*

The methods outlined above demonstrate receptor-dependent hormone binding in oviduct nuclei. Receptor-hormone complexes can be extracted from the nuclei in sufficient quantities to permit their further characterization by various chromatographic methods. However, for direct studies of the kinetics of receptor-nuclear interaction, this method requires an unnecessarily large nuclear pellet for each assay point. In order to process large numbers of tubes, it is desirable to scale down the assay to use smaller nuclear pellets and less receptor. For this purpose nuclei suspended in Buffer B containing 0.5 *M* sucrose can be dispensed (50–100 μg nuclear DNA per sample) into 12 × 75 mm tubes. The nuclei are collected as pellets by a brief centrifugation at 1500 *g* at 0°. They are then resuspended for the assay in 0.5 ml of receptor preparations containing the desired concentrations of receptors or salts. Incubations are carried out as described above, followed by wash steps using 2 ml of wash buffer (Buffer B in 0.5 *M* sucrose) and centrifugation at 1500 *g*. After the wash, nuclear-bound receptors are assayed by extracting the pellets with 1.0 ml of ethanol at room temperature for 15 minutes. Control experiments have shown that values for nuclear-bound receptor ob-

[22] S. G. Korenman, *J. Clin. Endocr. Metab.* **28**, 127 (1968).

tained by salt extraction correlate nicely with those values using ethanol. This procedure allows the processing of several hundred samples per assay.

III. Receptor Binding to Chromatin

For the chromatin-binding assay, a constant amount of chromatin (40–50 μg DNA) from target and nontarget tissues is incubated with increasing amounts of receptor.[8] The receptor sources which have been used in these studies are freshly prepared cytosol, ammonium sulfate precipitated receptor, and the binding components A and B prepared by chromatography on DEAE-cellulose.[7] The chromatin is incubated with the receptor for 1 to 2 hours at 4° in a final volume of 0.5 ml containing 0.15 *M* NaCl, 0.01 *M* Tris-HCl, 0.001 *M* Na_2EDTA, pH 7.4. These conditions appear to be optimal for the binding of receptor to chromatin. At an incubation temperature of 4°C, binding is complete in 60 minutes and the receptor–chromatin complex is stable for several hours. At higher incubation temperatures the initial rate of binding is high but falls to low levels in a short period of time. A salt concentration of 0.15 *M* NaCl was chosen since this results in high levels of receptor binding to chromatin. The use of lower salt concentrations results in nonspecific binding of cytosol proteins to chromatin. Incubation at higher ionic strengths results in no or low levels of receptor binding to chromatin. During the incubation the tubes are gently shaken every 10 minutes since in the presence of 0.15 *M* NaCl the chromatin precipitates. The incubation is performed in either disposable glass tubes (10 × 75 mm) or polypropylene tubes (Falcon #2053). The use of polypropylene tubes reduces the problem of chromatin sticking to glass. At the end of the incubation the tubes are centrifuged at 1200 *g* for 10 minutes. The sedimented chromatin is then washed with 2 ml of 0.15 *M* NaCl, 0.01 *M* $MgCl_2$, 0.01 *M* Tris-HCl, pH 7 (Buffer C), and recentrifuged. The pellets are then resuspended in 2 ml of Buffer C and collected under vacuum on Millipore filters (0.45 μm pore size). The incubation tube is washed twice with 2 ml of Buffer C and each filter is washed with 15 ml of Buffer C. The filters are dried and counted in 5 ml of toluene-based POP-POPOP scintillation solution with a counting efficiency of 45%. Alternatively, if polypropylene tubes are used, the bottom of the tube containing the chromatin pellet is cut off, placed in a scintillation vial, and dried in an oven at 90° for 2 hours. The vials are then shaken vigorously and counted in 5 ml of scintillation fluid.

In each experiment, solutions with the radioactively labeled receptor but no chromatin are treated similarly to monitor background adsorption of receptor. An additional control for nonspecific adsorption and trapping is to add receptor to chromatin and centrifuge immediately with no incubation time. This control contains all the elements of the binding reaction and is preferable to the one in which chromatin is omitted.

When the chromatin is collected on filters, the filters are removed from the vials and air-dried after counting. Each filter is cut up and the DNA is hydrolyzed by heating in 0.5 ml of 0.3 *N* $HClO_4$ for 30 minutes at 90°. The filter hydrolyzates are analyzed for DNA by the diphenylamine reaction. Purified calf thymus DNA is spot-dried on filters to serve as a standard. The amount of radioactivity per filter is corrected for the amount of DNA per filter. The results are expressed as the amount of labeled steroid bound per milligram of chromatin DNA.

Dehistonized chromatin (acidic protein–DNA complexes) or pure DNA have also been used in hormone-binding studies.[8,23] A modification of the above procedure for chromatin is used since dehistonized chromatin and pure DNA do not sediment at low centrifugal forces. In this method the low-speed centrifugations and washes with Buffer C are eliminated. These steps are replaced by a 24-hour centrifugation at 120,000 *g* at 0–4°. After centrifugation, the supernatant fraction is discarded and the nucleic acid pellet is resuspended in 100 μl of Buffer C and spot-dried on Millipore filters. The filters are counted and analyzed as described above.

Using these methods, steroid hormone receptors are found to bind preferentially to target-tissue chromatin. This has been demonstrated for the progesterone receptor in the chick oviduct,[24] the estradiol receptor in the rat uterus,[25] and the dihydrotestosterone receptor in the rat prostate.[25,26] The binding of the hormone receptor appears to be tissue specific; the highest levels of binding are observed when target tissue cytosol is incubated with target-tissue chromatin, as shown in Fig. 2. Lower levels of binding are observed when target-tissue chromatin is incubated with nontarget cytosol or free hormone; the incubation of nontarget tissue chromatin with target-tissue cytosol also results in low levels of hormone binding.

The binding of steroid hormone receptors to chromatin has been extended to determine which component or components of chromatin are

[23] T. C. Spelsberg, A. W. Steggles, F. Chytil, and B. W. O'Malley, *J. Biol. Chem.* **247,** 1368 (1972).

[24] A. W. Steggles, T. C. Spelsberg, and B. W. O'Malley, *Biochem. Biophys. Res. Commun.* **43,** 20 (1971).

[25] A. W. Steggles, T. C. Spelsberg, S. R. Glasser, and B. W. O'Malley, *Proc. Nat. Acad Sci. U.S.* **68,** 1479 (1971).

[26] W. I. P. Mainwaring and B. M. Peterken, *Biochem. J.* **125,** 285 (1971).

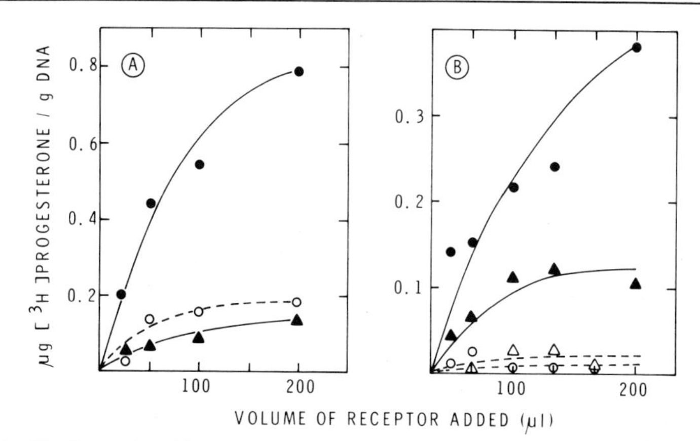

FIG. 2. Binding of oviduct progesterone receptor complexes to chick chromatin. Chromatins were prepared from purified nuclei as described in the text. Chromatin (40–50 μg DNA) was suspended in the indicated receptor preparations in a final volume of 0.5 ml containing 0.01 *M* Tris-HCl, 0.001 *M* Na_2EDTA, 0.15 *M* NaCl, pH 7.4, for 2 hours at 4°. Chromatin was pelleted and washed by centrifugation at 1000 *g*, and then resuspended in Buffer C and collected on Millipore filters. (A) Oviduct chromatin was combined with indicated amounts of labeled oviduct cytosol (●——●) or with diluted chick serum corticosteroid-binding globulin preparation labeled with 10^{-8} *M* [^{3}H]progesterone (○– –○). Labeled oviduct cytosol was also combined with chick heart chromatin (▲——▲). (From O'Malley *et al.*[10]) (B) Partially purified oviduct progesterone receptor complexes were resolved into two fractions by stepwise KCl elution from DEAE-cellulose and were assayed for chromatin binding activity. Both receptor fractions were adjusted to contain 50,000 cpm/ml of progesterone–receptor complex. The receptor component eluting at 0.15 *M* KCl (component A) was combined with oviduct (○– –○) or spleen (△– –△) chromatin. The receptor component eluting at 0.3 *M* KCl (component B) was also combined with oviduct (●——●) or spleen (▲——▲) chromatin. (From Schrader *et al.*[9])

responsible for the tissue specificity observed. These studies utilized the technique of dissociation and reconstitution of chromatin. The methods used have been described[8,23,27,28] elsewhere. Essentially the technique involves the selective dissociation of protein fractions from chromatin by the use of 2 *M* NaCl and 5 *M* urea and the reconstitution of the dissociated proteins by reducing the ionic strength through gradient or stepwise dialysis. In the chick oviduct system, such studies have shown that the acidic proteins, not the histones, play an important role in the specific binding of the progesterone–receptor complex to oviduct chromatin.[3,8,23]

[27] I. Bekhor, G. M. Kung, and J. Bonner, *J. Mol. Biol.* **39**, 351 (1969).
[28] R. S. Gilmour and J. Paul, *J. Mol. Biol.* **40**, 137 (1969).

IV. Receptor Binding to Purified DNA

A. Conditions

1. Ionic Strength

Progesterone receptors will bind to purified DNA in solutions of low ionic strength. The binding is nearly zero in 0.2 *M* KCl and reaches a peak at 0.1 *M* KCl. At molarities less than this, the receptors tend to aggregate with themselves. Since the methods used to detect DNA binding involve separation of DNA complexes from free receptors on the basis of size, such receptor aggregations must be avoided. Thus, all DNA-binding experiments have been performed in Buffer D (0.01 *M* Tris-HCl, 0.001 *M* Na_2EDTA, 0.012 *M* 1-thioglycerol, pH 7.4) containing 0.1 *M* KCl.

2. Time and Temperature

All binding experiments have been done at 0°. Since the purified receptor components used in the assays are very labile to heat, no study has been done at elevated temperatures.

Even in dilute solutions, receptor–DNA binding is very rapid, reaching equilibrium in about 10 minutes. This rapid association rate is much faster than that seen for receptor binding to chromatin, which requires about 2 hours at 0°. Receptor–DNA complexes have a half-life of several hours at 0°, and thus can be chromatographed by gel filtration or sucrose-gradient ultracentrifugation.

These characteristics are those of a strong interaction between the receptor and the DNA. The measured equilibrium constant is 3×10^{-10} *M*, measured by the Scatchard plot.[10,29] The gel filtration method outlined below was used. The rapidity of the DNA-binding process also suggests that the mechanism does not select specific, unique gene sequences in the DNA but rather involves binding to DNA regions which occur frequently in the structure, such as regions of similar base composition.

The binding reaction is saturable, with about one receptor site per 10^6 nucleotide pairs. This amount is much less than the amount of binding to DNA exhibited by organic dyes or metal ions and therefore presumably involves some degree of nucleotide specificity. If one assumes that an oviduct nucleus contains 3×10^{-12} g DNA, then there are about 2500 binding sites for receptor component A in the DNA of an oviduct cell.

[29] W. T. Schrader and B. W. O'Malley, unpublished.

3. Incubations

From a practical standpoint, it is necessary to use 50–300 μg of DNA in order to recover DNA–receptor complexes labeled with a few thousand cpm of [^{3}H]progesterone. Purified chick oviduct DNA, prepared as described above, is precipitated in 95% ethanol, spooled out of solution, and redissolved in Buffer D containing 0.1 *M* KCl. The DNA solution should be made at about 1 mg DNA/ml. Receptor preparations at various stages of purification are used in Buffer D. The receptor solutions are adjusted to 0.1 *M* KCl before addition to the DNA.

Polypropylene tubes are used for the incubations to diminish adsorption and subsequent loss of receptors on glass tubes. Three-hundred micrograms of DNA (300 μl of stock solution) are added to receptor solutions containing varying amounts of labeled progesterone receptor. The receptor solution volume is brought to 700 μl with Buffer D, 0.1 *M* KCl.

The solutions are mixed and allowed to stand in ice for up to 2 hours. Since the binding reaction appears to be complete in much less time than this, the assay is generally started after 1 hour.

4. Binding-Site Requirements

It has been of interest to examine whether receptor–DNA interactions occur with both native and denatured DNA and whether the receptors require an occupied progesterone-binding site for DNA binding to occur. These studies have been done by varying the additions to the binding reaction tubes.

In the case of DNA, both assays require the use of a very high molecular weight DNA and the separation of receptor–DNA complexes on the basis of their size difference from uncomplexed receptors. Therefore, it has not been possible to use sheared DNA in these studies. Native and denatured DNA's both bind progesterone receptors extensively but their relative affinities have not been determined in this system.

Receptor requirements for binding can be examined either directly by adding a particular preparation to the DNA or indirectly by competition studies in which an excess of receptors must be present to compete for available DNA sites. For these studies 100 μg of DNA is used, to which 100–300 μl of labeled receptor preparation is added containing about 100,000 cpm of bound [^{3}H]progesterone. It is important to prepare fresh receptors containing only small amounts of free steroid so that the free hormone will not occupy sites on competing receptors. If a purified or lyophilized receptor preparation is used, free hormone can be eliminated by chromatographing the receptors on a short Sephadex G-25 (coarse) column and collecting the void volume fractions. Competing receptors have been either nascent receptor preparations (no progesterone present on the steroid-binding sites) or receptors charged with unlabeled progesterone. These are added to the labeled preparation and immediately incubated with the DNA. Results in this laboratory have shown that nascent receptors do not compete equivalently for DNA sites. Under these conditions receptors saturated with unlabeled progesterone do compete as expected.

A second way to study receptor-binding requirements is simply to incubate nascent receptors with DNA, chromatograph, and collect the DNA-containing fractions, and then add [^{3}H]progesterone and assay after additional incubation in the cold for labeled progesterone bound to receptors by any of the assays described elsewhere in this volume.[7] Under these conditions no receptors are found bound to the DNA, confirming the requirement for progesterone to permit receptor–DNA interaction.

B. Binding Assay Using Sucrose–Gradient Ultracentrifugation

DNA prepared as described above has a very high average molecular weight, and has a sedimentation coefficient of over 14 S. The receptor proteins, on the other hand, sediment at about 4 S. Thus, by laying a DNA–receptor mixture on a 5 to 20% sucrose gradient containing the appropriate salt concentration, it is possible to separate free receptor (4 S) fractions from the more rapidly sedimenting DNA–receptor complex fraction (> 14 S) which quickly reaches the bottom of the centrifuge tube.

Sucrose gradients are prepared and handled as described elsewhere in this volume.[20] The gradients are linear 5 to 20% sucrose in Buffer D, 0.1 *M* KCl. Samples to be analyzed are prepared as described in the previous section and layered (200 μl) on the gradients which are centrifuged in a Spinco SW 50.1 rotor at 45,000 rpm for 16 hours. The tubes are pierced and fractions allowed to drip out of the tube bottom. The radioactive receptor peak at about 4 S appears about one-third of the distance down the gradient, while DNA–receptor complexes are detected by cutting off the centrifuge tube bottom and counting it for labeled receptor.

The DNA effect is twofold. First, the normal 4 S receptor peak is diminished in size due to binding of labeled receptors by the DNA. Second, these labeled receptors appear at the tube bottom bound to DNA. By comparison of the radioactivity in the 4 S region to that at the tube bottom, an analysis is possible of the receptor–DNA binding. If increasing amounts of DNA are incubated with a constant receptor concentration, the 4 S peak height will progressively decrease in size. This can be done using either chick DNA's or DNA's from other vertebrate species. Under these conditions, chick liver DNA and calf thymus DNA give binding curves identical to that obtained using chick oviduct.[10,29]

The method is semiquantitative, due to several sources of loss in the experiment. First, the ultracentrifugation requires 16 hours to run, which is over one half-life for the receptor–hormone complex itself. Thus, substantial amounts of breakdown of the receptor–steroid complexes occur, reducing the effective concentration of this species for the DNA interaction. Second, recovery of receptors from sucrose gradients is significantly less than 100%, due to binding to the tube walls and the piercing apparatus. Finally, it has not been possible to recover receptors from the tube bottom in a quantitative manner. Any excess fluid adherent to the tube bottom will be counted as DNA–receptor complex although the fluid may have fallen to the bottom after fractionation was completed. Since receptors aggregate, especially in the impure state, there is always some heavy material at the tube bottom resulting from this aggregation, not from DNA interaction.

This technique is nevertheless quite advantageous, however, because the sucrose-gradient technique permits analysis of the hydrodynamic integrity of the receptor fraction being tested. A typical experiment is shown in Fig. 3. Two labeled receptor fractions prepared by DEAE-cellulose chromatography were analyzed for their ability to bind to DNA. The left panel shows that component A, the less acidic receptor, could bind to DNA and be displaced from its 4 S sedimentation position. On the other hand, component B, which is the more acidic receptor component, was not affected by the presence of the DNA. Thus there are two receptor forms in the chick oviduct, one binding to DNA (component A) and the other which does not bind to DNA (component B). Note that the sucrose-gradient technique demonstrates that both receptors are otherwise normal 4 S forms and the profiles do not indicate any other sort of DNA or buffer effects on the preparations.

There is an additional advantage of this technique, namely, that it will show DNA binding to occur even with crude cytoplasmic preparations. The column chromatographic method outlined below is more readily quantitative, but crude cytoplasmic extracts aggregate so badly that they do not chromatograph cleanly on agarose columns without KCl concentrations of 0.3 *M* or higher. Therefore, the sucrose-gradient method is certainly the first one to employ when investigating a crude receptor preparation for evidence of DNA binding.

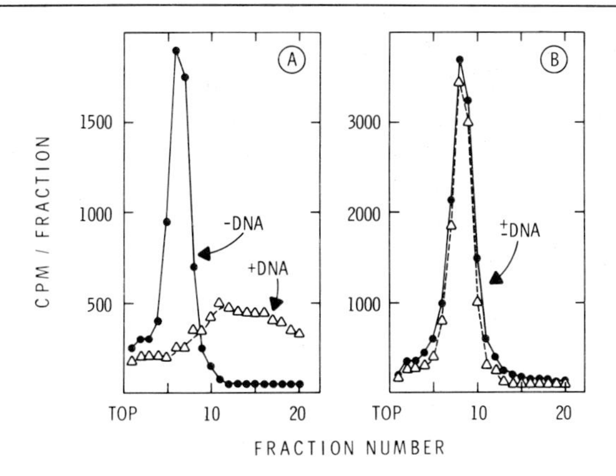

FIG. 3. Sucrose-gradient ultracentrifugation analysis of progesterone receptor-DNA binding. Labeled progesterone receptor complexes were prepared by stepwise DEAE-cellulose elution into two fractions. These were incubated with chick DNA as described in the text. About 50,000 cpm of either receptor fraction was combined with 100 μg DNA in 0.1 *M* KCl. Aliquots of the incubations (200 μl) were centrifuged in 5 to 20% sucrose gradients containing 0.1 *M* KCl in Buffer D for 16 hours at 45,000 rpm. (A) Receptor component A, eluted at 0.15 *M* KCl from DEAE-cellulose, was incubated alone (●——●) or in the presence of chick DNA (△--△). (B) Receptor component B, eluted at 0.3 *M* KCl from DEAE-cellulose, was incubated alone (●——●) or in the presence of chick DNA (△--△). (From Schrader *et al.*[9])

C. DNA Binding Assay Using Agarose Gel Filtration Chromatography

A second method of separating receptors from DNA–receptor complexes on the basis of their size differences is to chromatograph the mixture on a gel filtration column.[10,30] Porosity of the beads can be chosen so that DNA–receptor complexes are eluted in the void volume of the column and the receptors alone elute farther back toward the exclusion volume. Agarose beads made by Bio-Rad or Pharmacia are to be preferred over Sephadex dextran beads, since there is less interaction between receptors and the agarose, thus improving the recovery and resolution of the column.

An agarose A-15 M column 40 × 1.5 cm has been used to separate DNA complexes from receptors. Coarse bead size (50–100 mesh) is preferable, since flow rates are correspondingly higher. The column is equilibrated in Buffer D, 0.1 M KCl and a 1.0 ml receptor–DNA sample applied and eluted in the same buffer. About 50 fractions (2 ml) should be collected and assayed for [^{3}H]progesterone. The radioactivity eluting

[30] J. D. Baxter, G. G. Rousseau, N. C. Benson, R. L. Garcea, J. Ito, and G. M. Tomkins, *Proc. Nat. Acad. Sci. U.S.* **69**, 1892 (1972).

with the void volume fractions (V_0) represents DNA–receptor complexes; free receptors elute next, followed finally by free uncomplexed steroid, as shown in Fig. 4. The areas under the peaks can be cut out of graph paper and weighed to measure total binding in the three regions.

Agarose A-15 M is preferable to more porous beads such as A-50 M, since the A-15 will resolve receptor–steroid complexes from free steroid alone.[10] If a preparation is initially free of uncomplexed steroid, presence of a peak of free hormone in the elution profile provides an index of receptor stability during the experiment. Since the columns require about 3 hours to complete, there is always some dissociation of receptor–hormone complexes, as well as dissociation of the complexes with DNA. However, by varying the rate of elution of the columns for identical receptor–DNA complexes, we have found that these complexes have half-lives much longer than the time required for chromatography. Thus, although the method is not rapid enough for rate kinetics measurements it can be used for equilibrium measurements of the sort detailed in Section IV, A.

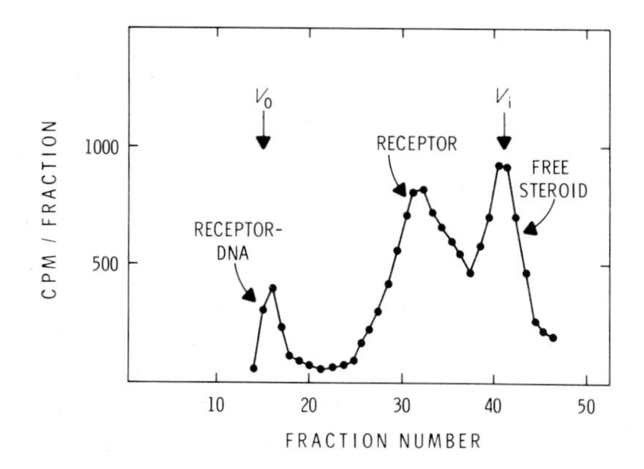

FIG. 4. Agarose A-15 M column chromatography of progesterone receptor–DNA complexes. A 1.5 × 40-cm column was prepared and equilibrated in Buffer D–0.1 *M* KCl. Calibration standards for the column were as follows: High molecular weight chick DNA eluted at the void volume (V_0) as determined in control chromatograms by DNA absorbance at 260 nm. Labeled progesterone receptors chromatographed as a peak in fraction 31, while free [^{3}H]progesterone was eluted in the included volume (V_i). When chick DNA was combined with a labeled progesterone receptor preparation as described in the text, a 1.0-ml sample eluted as shown. Fractions (2.0 ml) were collected and assayed for [^{3}H]progesterone (●——●). Radioactivity in the DNA region (V_0) was totaled and corrected to determine molarity of steroid–receptor complexes bound to DNA in the original sample. Similarly, radioactivity eluted in the receptor peak region was totaled to determine the molarity of progesterone–receptor complex in the sample.

The two methods differ in one important detail. In the sucrose-gradient method the measured quantity is the receptor itself, whereas the column method allows more accurate and sensitive detection of DNA–receptor complexes. Thus, the sucrose method lends itself to studies in which a small amount of receptor is combined with varying (or large) amounts of DNA, as in DNA-competition studies. The column method, on the other hand, lends itself to studies in which a constant amount of DNA is combined with varying amounts of receptors. When used together the two techniques permit analysis of receptor–DNA interactions under conditions in which the receptor:DNA ratio varies from very low values (sucrose-gradient method) to very high values (column method).

[28] Evaluation of Androgenic Compounds by Receptor Binding and Nuclear Retention

By SHUTSUNG LIAO and TEHMING LIANG

In certain androgen-sensitive tissues, 17β-hydroxy-5α-androstan-3-one [5α-dihydrotestosterone (DHT)] is believed to be an active form of androgen.[1,2] In rat ventral prostate, DHT is bound to a cytoplasmic protein, which has been named β-protein,[3] in a highly specific manner.[3-7] This β-protein appears to be an androgen receptor that may play a central role in the androgen action.[1-10] The receptor protein can be retained tightly by the prostate cell nuclei only in a DHT-bound form. The androgen-dependent translocation of the specific cytoplasmic protein to a specific nuclear "acceptor" site is believed to be an essential step in the regulation of certain nuclear functions in androgen-sensitive cells. On this basis, one can use β-protein binding and nuclear retention to assay and characterize various potentially active androgens and anti-

[1] S. Liao and S. Fang, *Vitam. Horm. (New York)* **27,** 17 (1969).
[2] J. D. Wilson and R. E. Gloyna, *Recent Progr. Horm. Res.* **26,** 309 (1970).
[3] S. Fang and S. Liao, *J. Biol. Chem.* **246,** 16 (1971).
[4] S. Fang, K. M. Anderson, and S. Liao, *J. Biol. Chem.* **244,** 6584 (1969).
[5] W. I. P. Mainwaring and B. M. Peterken, *Biochem. J.* **125,** 285 (1971).
[6] S. Liao, T. Liang, and J. L. Tymoczko, *J. Steroid Biochem.* **3,** 401 (1972).
[7] S. Liao, *in* "Biochemistry of Hormones" (H. Kornberg, D. C. Phillips, and H. V. Rickenberg, eds.). MTP Int. Rev. Sci., Oxford (in press).
[8] J. L. Tymoczko and S. Liao, *Biochim. Biophys. Acta* **252,** 607 (1971).
[9] S. Liao, T. Liang, and J. L. Tymoczko, *Nature (London)* **241,** 211 (1973).
[10] S. Liao, J. L. Tymoczko, T. Liang, K. M. Anderson, and S. Fang, *Advan. Biosc.* **7,** 155 (1971).

androgens.[6,10] The procedure described below is based primarily on the ability of various test compounds to compete with [^{3}H]DHT for binding to β-protein or for retention by prostate cell nuclei. The method is especially useful if these compounds are available only in minute quantities (< 1 μg).

Receptor Binding Assay

Rats (adult males weighing 300 to 400 g) are castrated 18 to 20 hours before they are killed, so that levels of endogenous androgens are minimized. Castration is via the scrotal route, with the use of ether anesthesia. Ten animals are killed by cervical dislocation, the ventral prostates (5 to 8 g) are dissected free of their capsules, and their combined volume is measured by displacement in several milliliters of ice-cold Medium A (0.32 *M* sucrose containing 3 m*M* $MgCl_2$ and 20 m*M* Tris-HCl buffer, pH 7.0). Th prostate tissue is then minced and homogenized in 3 volumes of Medium A in an all glass Potter-Elvehjem homogenizer at 0–2°. The homogenate is centrifuged at 600 *g* for 10 minutes. The supernatant is separated from the "nuclear pellet" and centrifuged again at 100,000 *g* for 60 minutes. Granular $(NH_4)_2SO_4$ (enzyme grade, 243 mg/ml cytosol) is added to the cytosol slowly over 10 minutes. The mixture is allowed to stand at 0° for 20 minutes and is then centrifuged at 10,000 *g* for 10 minutes. The protein precipitate is dissolved in 0.5 ml of Medium B (1.5 m*M* EDTA and 20 m*M* Tris-HCl buffer, pH 7.5) containing 0.4 *M* KCl. The resulting solution is passed through a column (1×30 cm) of Sephadex G-25 equilibrated with the same buffer. The fractions containing the protein peak at the position of the excluded volume are pooled (β-protein fraction[3]; 3 to 5 mg protein/ml).

A test compound (at concentrations desired, usually 0.03 to 3 pmole) and [1,2-^{3}H]DHT (0.015 μCi/0.3 pmole) are first dissolved in a small volume (< 5 μl) of ethanol and then added to a small glass tube which contains 0.1 ml of Medium B containing 0.4 *M* KCl. The freshly prepared β-protein preparation (0.2 ml) is added to the contents of the tube. After mixing, the tube is allowed to stand at 0° for 20 minutes and then kept frozen at −20°. The frozen mixture can be stored for at least 2 weeks. Samples can be taken out, thawed, and analyzed by gradient centrifugation as ultracentrifuges become available. Over 60 samples can be prepared in one day. With two ultracentrifuges for gradient centrifugation, more than 10 compounds, each at several different concentrations, can be handled in a week.

Gradient centrifugation is performed by layering 0.2 ml of the sample on the top of a sucrose or glycerol gradient (5 to 20% linear in Spinco SW-56 rotor tube) containing 0.4 *M* KCl, 1.5 m*M* EDTA, 20 m*M*

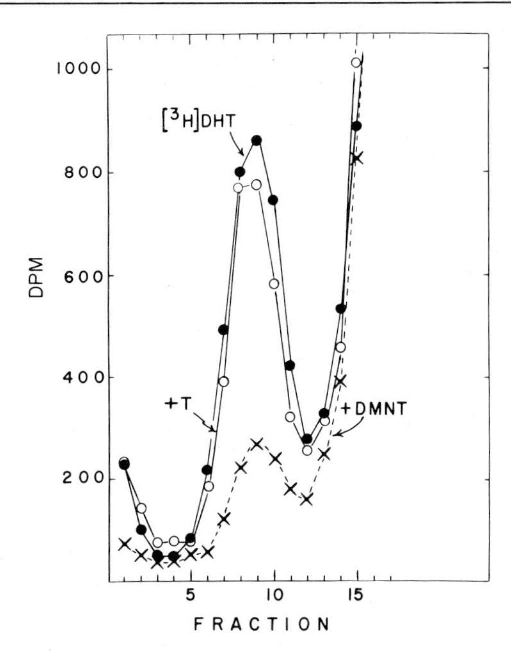

FIG. 1. Effect of nonradioactive steroids on formation of [^{3}H]DHT–β-protein complex. The β-protein fraction (1.5 mg of protein/0.5 ml of medium B) was mixed with 0.012 μCi of [^{3}H]DHT (final concentration 0.5 nM) and 1.5 nM of testosterone (T) or 7α,17α-dimethyl-19-nortestosterone (DMNT). The mixture was kept at $-20°$ for 20 hours. After thawing, 0.2 ml of the sample was analyzed by gradient centrifugation as described in the text.

Tris-HCl buffer, pH 7.5. The tube is centrifuged at 56,000 rpm for 18 hours at 0 to 2°. Fractions (0.2 ml each) are collected and numbered starting from the bottom of the tube, and the radioactivity is measured in a scintillation counter. The [^{3}H]DHT-receptor complex sediments with a sedimentation constant of about 3 to 3.5 S, which is somewhat slower than that of bovine albumin ($s_{20,w}$ = 4.6 S).[11] The radioactivity associated with the protein peak (Fig. 1) is computed as *3 S-protein-bound DHT* (*in cpm*), and used for comparison.

Nuclear Retention Assay (with Tissue Incubation System)

Ventral prostates from rats castrated 3 days earlier are minced into pieces of about 1 mm^3. These are suspended in Medium A, so that 3 ml of the medium contains about 1 g of the minced tissues. A test compound (0.01 to 2.0 nmole) and [1,2-^{3}H]DHT (5 μCi/0.1 nmole) are dissolved in 2 ml of Medium A and placed in a glass bottle (diameter 5 cm, height

[11] R. G. Martin and B. N. Ames, *J. Biol. Chem.* **236**, 1372 (1961).

5 cm). The minced tissues (3 ml each) are then allocated by a wide mouth pipet. The bottle is flushed with a gas mixture containing 95% O_2 and 5% CO_2 and is tightly covered. The bottle is then incubated at 37° in a water-bath shaker. The maximum retention of radioactive steroid by cell nuclei can be achieved within 20 to 40 minutes. After incubation, tissues are separated from the medium by a brief centrifugation at 1,000 *g* for 3 minutes, washed with 8 ml of medium A three times, and homogenized in 3 volumes of medium A. Then they are mixed with 34 ml of Medium C (2.2 *M* sucrose containing 1 m*M* $MgCl_2$, 20 m*M* Tris-HCl buffer, pH 7.0). The mixture is layered on the top of 5 ml of Medium C, which is placed at the bottom of a centrifuge tube beforehand. The tube is centrifuged in a SW-27 rotor at 22,000 rpm for 60 minutes in a L2-65B Spinco centrifuge at 0–2°. The nuclear sediments are suspended in 3 ml of Medium A containing 0.25% of Triton X-100. Cell nuclei are recollected by centrifugation at 2,000 *g* for 10 minutes and washed once with 5 ml of Medium A and once with 5 ml of Medium B.

One milliliter of ice-cold 10% trichloroacetic acid is added to release radioactive steroids from nuclei. The mixture is centrifuged at 10,000 *g* for 10 minutes. The radioactivity in the supernatant and DNA content in the chromatin aggregates are measured. The nuclear-bound radioactivity is expressed in terms of cpm/mg DNA and used for comparison. In some experiments, to ascertain that the nuclear steroids are indeed in a protein-bound form, 0.4 ml of Medium B containing 0.6 *M* KCl is added to the nuclei washed with Medium A. They are mixed gently and allowed to stand at 0° for 40 minutes with occasional stirring. The mixture is then centrifuged at 14,000 *g* for 20 minutes. The clear nuclear extract is then removed from the tube and analyzed by sucrose gradient centrifugation.

For DNA measurement, the nuclear sediment should be washed once with 5 ml of 10% trichloroacetic acid so that residual sucrose is removed. Three milliliters of 10% trichloroacetic acid are added, and the tube is heated for 15 minutes at 90°. After cooling on ice and centrifugation, the hydrolyzed DNA in the supernatant is measured by the diphenylamine test.[12]

Data Analysis

For the purpose of comparison, we derived the "relative competition index" in the following manner. If

$$Y = \frac{\text{3 S-protein-bound cpm in the presence of a competitor}}{\text{3 S-protein-bound cpm in the absence of a competitor}},$$

[12] K. Burton, *Biochem. J.* **62**, 315 (1956).

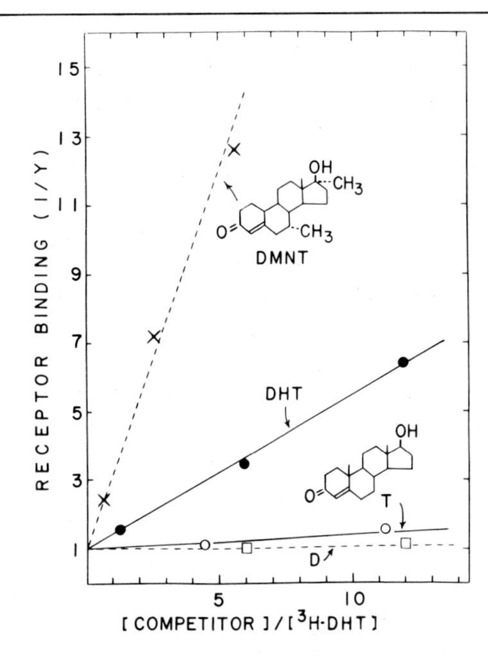

FIG. 2. RC-plot for DHT, testosterone (T), androstenedione (D), and 7α,17α-dimethyl-19-nortestosterone (DMNT) in the receptor-binding assay.

or

$$Y = \frac{\text{nuclear-bound cpm in the presence of a competitor}}{\text{nuclear-bound cpm in the absence of a competitor}},$$

then

$$Y = \frac{[^3\text{H-DHT}]}{[^3\text{H-DHT}] + a\,[\text{competitor}]},$$

where a is a factor (competition index, CI) characteristic of a competing steroid. This factor can be used to compare the relative ability of various steroids to compete with DHT for β-protein binding. By plotting (RC-plot) $1/Y$ as a function of [competitor]/[^{3}H-DHT], one can obtain a from the slope. Theoretically, a is 1 for nonradioactive DHT. However, due to the presence of undefined amounts of endogenous DHT in the β-protein fractions, it is necessary to determine a for DHT (a_{DHT}) experimentally, while a for a competitor (a_{comp}) is measured at the same time (Figs. 2 and 3). The relative competition index (RCI) for a competitor is defined as

$$\text{RCI}_{\text{comp}} = \frac{a_{\text{comp}}}{a_{\text{DHT}}}$$

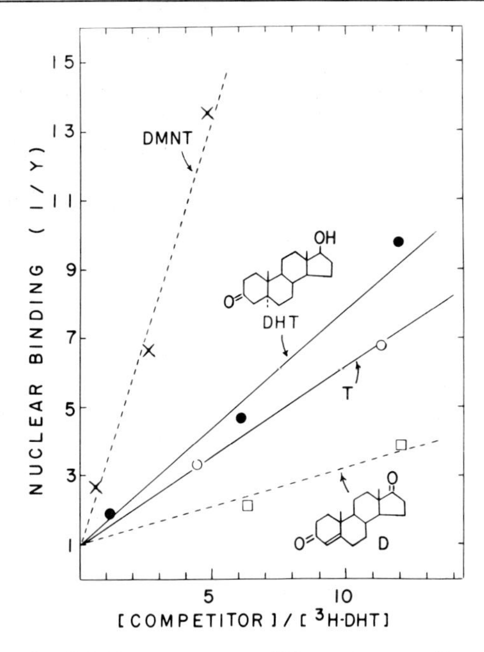

FIG. 3. RC-plot for DHT, testosterone (T), androstenedione (D), and 7α,17α-dimethyl-19-nortestosterone (DMNT) in the nuclear retention assay, using the tissue-incubation system.

Relative competition index values for some steroids we have estimated are shown in Table I. The deviation of the RCI values in different experiments is generally about 20%. A compound with an RCI of 1 is expected to bind to β-protein receptor with an affinity equivalent to that of DHT. A compound with an RCI higher than 1 is expected to be biologically more active than DHT. A low RCI in the β-protein-binding assay for a potent androgen (like testosterone) that also shows a high RCI in the nuclear retention assay with the tissue-incubation system suggests that the biological activity of the compound is dependent on its transformation to an active form, or on a process not dependent on β-protein binding. These assay methods[13] may be useful in the studies of the interaction of a receptor protein and an androgen and in the search for potent androgens as well as anti-androgens.[14]

[13] S. Liao, T. Liang, S. Fang, E. Castañeda, and T. C. Shao, *J. Biol. Chem.* **248**, 6154 (1973).

[14] The work presented in this article was supported by grants AM-09461 and HD-0711 from the United States National Institutes of Health.

TABLE I
RELATIVE ANDROGENICITIES AND RELATIVE COMPETITION INDEXES OF SOME REPRESENTATIVE STEROIDS[a,b]

Steroids	Androgenic activity on rat prostate	RCI nuclear binding[c]	Receptor binding[d]
5α-Dihydrotestosterone (DHT)	1.0	1.0	1.0
5β-Dihydrotestosterone	0.1	0.0	0.0
Testosterone	0.4	0.7	0.1
Androstenedione	0.1	0.2	0.0
7α,17α-Dimethyl-19-nortestosterone	5.7	3.6	3.5
7α,17α-Dimethyl-19-nor-5α-DHT	0.3	0.4	1.0
Estradiol-17β	0.0	0.1	0.0
Cortisol	0.0	0.0	0.0

[a] DHT taken as 1.0.
[b] See Liao and Fang.[13]
[c] In tissue incubation system.
[d] With β-protein fraction.

[29] Methods for Assessing Steroid-Receptor Effect on RNA Synthesis in Isolated Nuclei

By C. RAYNAUD-JAMMET, M. M. BOUTON, M. G. CATELLI, and E. E. BAULIEU

The injection of 5 μg of estradiol to the immature or castrated rat results, 2 to 3 hours later, in a marked increase in RNA biosynthesis,[1] as shown by the increased incorporation of radioactive precursors into uterine RNA. This accrued synthesis may be explained by the increase in the activity of the so-called nucleolar polymerase A or I, which catalyses the synthesis of ribosomal type RNA. This enzyme can be assayed in a low salt medium[2,3] and is insensitive to α-amanitin.[4]

This relatively late increase in polymerase activity is dependent on

[1] G. C. Mueller, A. Herranen, and K. Jervell, *Recent Progr. Horm. Res.* **8**, 95 (1958).
[2] T. H. Hamilton, C. C. Widnell, and J. R. Tata, *J. Biol. Chem.* **243**, 408 (1968).
[3] J. Gorski, *J. Biol. Chem.* **239**, 889 (1964).
[4] C. Raynaud-Jammet, F. Biéri, and E. E. Baulieu, *Biochim. Biphys. Acta* **247**, 355 (1971).

the prior synthesis of a protein,[5] "Key Intermediary Protein" (KIP),[6] itself deriving from a short-lived RNA synthesized by an α-amanitin sensitive extranucleolar polymerase (B or II), in all likelihood mRNA.[7] This may be substantiated by the appearance after 15 minutes of one or two distinct types of RNA with a molecular weight compatible with a messenger nature.[8]

On the other hand, 30 to 60 minutes after exposure of an atrophic uterus to estradiol either *in vivo* or *in vitro*, an induced protein peak (IP) appears, which may be accounted for by a single or very limited number of proteins.[9] The formation of this protein peak depends on an actinomycin D, α-amanitin sensitive, process, very likely due to the early synthesis of its mRNA.[6,10-12] Whether or not it may be KIP is beyond the scope of this chapter.[6,12]

The synthesis of RNA thus appears to be an early response of the target tissue to estradiol. In fact, an increase in nuclear RNA has been described at 2 minutes.[13]

To elucidate the mechanism of early accrued RNA synthesis and of increased RNA polymerase activity, receptor–nucleus interactions were studied in order to show their effect on RNA biosynthesis in the absence or presence of estradiol. It is well known that the cytosol, presumably cytoplasmic, receptor acquires under the action of estradiol a particular affinity for the nucleus to which it adheres or into which it penetrates.

For such a study it is necessary to have at one's disposal a cell-free system which responds to estradiol and to which the hormonal receptor may be added at will. The RNA polymerase activity of isolated nuclei from calf and rat uterus was measured. The experimental conditions were chosen in such a way as to obtain an increase in this activity *in vitro* under the influence of estradiol, in other words, conditions which alter neither the receptor nor the interactions between the estradiol receptor complex and nuclei. This was obtained by eliminating tissue incubation and by introducing estradiol as soon as possible, during the homogeniza-

[5] J. A. Nicolette and G. C. Mueller, *Biochem. Biophys. Res. Commun.* **24**, 251 (1966).

[6] E. E. Baulieu, A. Alberga, C. Raynaud-Jammet, and C. R. Wira, *Nature (London). New Biol.* **236**, 236 (1972).

[7] C. Raynaud-Jammet, M. G. Catelli, and E. E. Baulieu, *FEBS Lett.* **22**, 93 (1972).

[8] C. R. Wira and E. E. Baulieu, *C.R. Acad. Sci.* **274**, 73 (1972).

[9] A. Notides and J. Gorski, *Proc. Nat. Acad. Sci. U.S.* **56**, 230 (1966).

[10] A. B. De Angelo and J. Gorski, *Proc. Nat. Acad. Sci. U.S.* **66**, 693 (1970).

[11] R. F. Mayol and S. A. Thayer, *Biochemistry* **9**, 2484 (1970).

[12] E. E. Baulieu, C. R. Wira, E. Milgrom, and C. Raynaud-Jammet, *Acta Endocrinol. (Copenhagen), Suppl.* **168**, 396 (1972).

[13] A. R. Means and T. H. Hamilton, *Proc. Nat. Acad. Sci. U.S.* **56**, 1594 (1966).

tion step (experiment type I). It was thus possible to confirm the fundamental role of the receptor in a cell-free reconstituted system (experiment type II).[14]

Methods

Preparation of Nuclei

From Calf Endometrium.[14] Calf uteri are excised at the slaughterhouse and transported to the laboratory in crushed ice. The endometrium is separated from the myometrium by scraping with a scalpel, weighed, suspended in 3 volumes of a medium consisting of 0.32 M sucrose, 3 mM $MgCl_2$, shaken, and centrifuged at 700 g for 10 minutes at 0°. The residue obtained is weighed and homogenized in 3 volumes of homogenization medium = x ml in a Teflon-glass Potter. The homogenate is filtered through two layers of gauze, diluted 1.6-fold with the homogenization medium, further diluted with distilled water to a final concentration of 0.25 M, and then carefully poured on top of a layer of homogenization medium (which volume is equal to the volume of the homogenate diluted with the homogenization medium). The residue obtained on centrifuging at 700 g for 10 minutes is suspended in x ml of 2.4 M sucrose, 1 mM $MgCl_2$, transferred into a Spinco tube, and recentrifuged at 50,000 g for 1 hour at 0°. The pellet is resuspended in 0.25 M sucrose, 1 mM $MgCl_2$ in a Teflon-glass Potter to give a suspension containing approximately 100 μg of DNA/0.1 ml.

The preparation of the nuclei lasts nearly 3 hours. Their purity is verified by phase-contrast microscopy. The yield, as estimated on the basis of the DNA recovered, is approximately 50%. One-tenth of a milliliter of the final suspension contains about 100 μg of DNA and 50 μg of protein.

From Rat Uterus. Twenty-two-day-old immature rats or 150–200 g adult rats castrated 15 days previously are killed by decapitation. The uteri are immediately excised, weighed, finely cut up with scissors, and then crushed in a glass-glass Potter in 6 volumes of 2.2 M sucrose, 1 mM $MgCl_2$. The homogenate is diluted twice with this medium, placed in an ultracentrifuge tube to which is added a further quantity of medium not exceeding two-thirds of the diluted homogenate volume (the homogenate moves up to the surface), and centrifuged at 50,000 g for 60 minutes at 0°. This method of ultracentrifuging over a layer of 2.2 M sucrose has the disadvantage of giving rather low yields of nuclei (40% recovery of DNA). According to latest experiments, these yields may be increased

[14] C. Raynaud-Jammet and E. E. Baulieu, *C.R. Acad. Sci.* **268**, 3211 (1969).

by omitting the addition of sucrose prior to centrifuging. The floating cell debris and supernatant are discarded, whereas the nuclear pellet is suspended in a Teflon-glass Potter in 0.32 M sucrose, 3 mM $MgCl_2$ to give a suspension containing approximately 100 μg of DNA/0.1 ml.

Assay of RNA Polymerase Activity in the Isolated Nuclei

Calf Endometrium. Nuclear RNA synthesis is measured on a 0.5-ml volume containing: 50 μmoles Tris-HCl buffer (pH 8), 2.5 μmoles $MgCl_2$, 10 μmoles cysteine, 3 μmoles NaF, 35 μmoles KCl, 0.3 μmoles GTP and UTP, 0.5 μmoles ATP, 1 μCi [^{3}H]CTP (Schwarz: a 20 Ci/mmole preparation diluted with cold CTP to a final radioactivity concentration of 1 μCi/10 mμmoles), and variable amounts of nuclear suspension containing 50 to 150 μg of DNA.

The above are incubated for 15 minutes at 37°. Since under these "low salt" conditions the incorporation of [^{3}H]CTP reaches a plateau after 10 to 15 minutes (Fig. 1), 15 minutes were chosen for single experimental points under the various conditions.

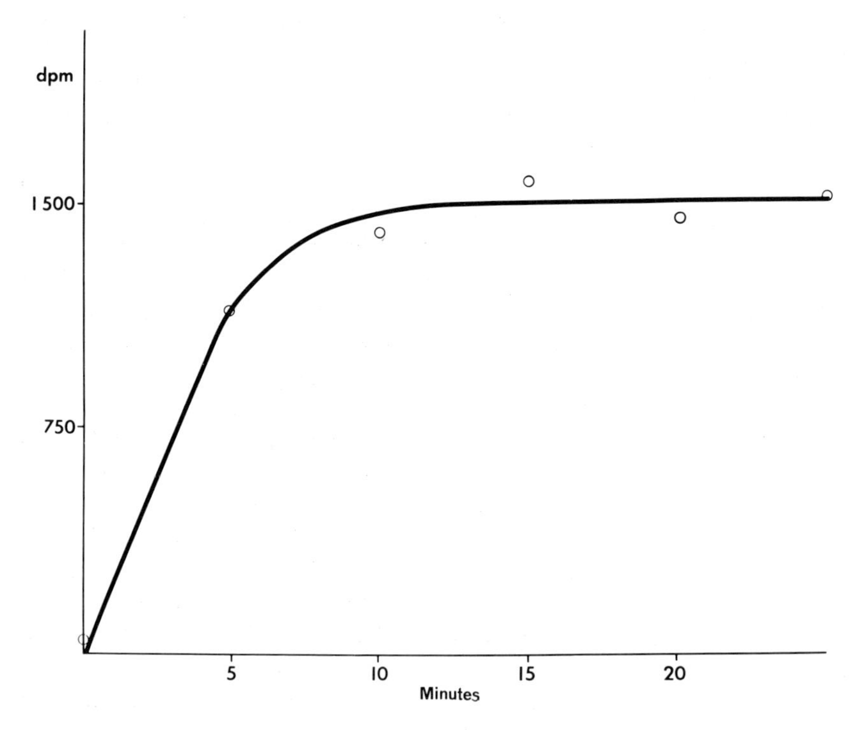

FIG. 1. Incorporation of [^{3}H]CTP into RNA in isolated nuclei from calf endometrium.

Rat Uterus. Ribonucleic acid polymerase activity has been assayed in low and high salt containing media.

Low salt: The incubation medium is identical to that used for calf endometrium, except for a final 25% v/v concentration of glycerol which is added to stabilize polymerase activity.

High salt: $(NH_4)_2SO_4$ is added to the above medium. (Final concentration = 0.25 *M*.)

The incubation procedure is carried out as for calf endometrium.

Determination of the Radioactivity Incorporated into RNA

Calf Endometrium. Following addition of 2 ml of an aqueous solution containing 400 μg of yeast RNA activator (Sigma type XI) at 0°, the reaction is stopped with 2.5 ml of chilled 1 *N* $HClO_4$. The insoluble precipitate formed is recovered by filtration on Whatmann GF/B glass-paper under vacuum, washed twice with 1 ml of 0.5 *N* chilled $HClO_4$, twice with chilled distilled water, and twice with 1 ml absolute ethanol, then placed in Packard vials and dried under vacuum. Hyamine hydroxide (1 ml) is added to each filter and left to react for 12 hours at room temperature. Each filter is then dispersed by hand-stirring in 15 ml of 0.5% PPO, 0.03% POPOP in toluene to measure RNA radioactivity. Control measurements are carried out in each case on reaction mixture containing no nuclei (20% of the values obtained for nuclei containing 100 μg of DNA) and systematically subtracted from the experimental values. A linear relationship is obtained and may be extrapolated to zero (Fig. 2a). The amount of nuclei is expressed in terms of DNA assayed by the method of Burton for a nuclear suspension volume equal to that used for the determination of RNA polymerase activity.

Rat Uterus. Following incubation, 0.5 ml of 1 *N* $HClO_4$ are added at 0°. The precipitate formed is washed successively with 1 ml of 0.5 *N* $HClO_4$ (three times), of ethanol RP (once), of ethanol: chloroform (1:2, v/v) (once), of ethyl ether (once), and centrifuged between each washing at 700 *g* for 10 minutes at 0°. Following the last washing, the precipitate is dried in a dessicator *in vacuo,* hydrolyzed with 1.5 ml of 0.3 *N* KOH for 1 hour at 37°, and then acidified with $HClO_4$ to a final concentration of 0.5 *N*. Following centrifugation at 700 *g* for 10 minutes at 0°, a 1.5-ml sample of the supernatant is dissolved in 15 ml of Bray's scintillation fluid (this contains per liter: 60 g naphthalene, 4 g PPO, 200 mg POPOP, 20 ml ethylene glycol, 100 ml methanol RP, dioxane) and counted.

Ribonucleic acid is assayed on the remaining supernatant by spectrometry at 260 nm. The results are expressed in dpm/mg RNA. Control measurements are carried out in each case on reaction mixture containing no nuclei. Blank measurements have been made on incubation medium containing 0.12 *M* EDTA, but the values obtained were so high that they were subsequently ignored. Blanks obtained by eliminating a nucleotide in the incubation medium gave values virtually equal to those for blanks without nuclei.

Results

Experiment Type I. Necessity of an Extranuclear Factor

The batch of calf endometrium tissue is divided into two halves: One half is treated as described above, but in the presence of 1 n*M* estradiol. This concentration corresponds to a physiological estradiol concentration in the cellular medium and is added to all the media used throughout the experiment, that is from homogenization of the endometrial tissue to the incubation of nuclei in the presence of [^{3}H]CTP. The other half is treated in the absence of estradiol. The results indicate that estradiol activates RNA synthesis whatever method is used: simple precipitation or phenol extraction (Fig. 2). If, however, SDS is not used during extraction or if heating to 60° is omitted, no increase in specific activity of RNA is recorded, thus implying the insertion of the neoformed RNA into a structure which requires denaturation or cleavage. Addition of 20 μg/ml actinomycin D to the incubation medium completely inhibits the effect of estradiol on RNA synthesis.

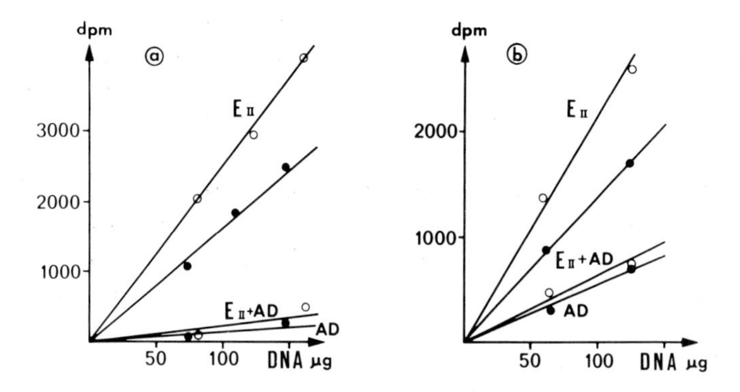

FIG. 2. Influence of the addition of estradiol during the isolation of nuclei from calf endometrium. (a) Direct precipitation of RNA; (b) phenol extraction.

This is the first time that a response of the uterine tissue to estradiol has been detected *in vitro*. Failure by other investigators to detect such a response may be explained by experimental conditions which disrupt the interaction between estradiol and a functional extranuclear factor which could be the cytoplasmic receptor. Since this receptor is extremely fragile, sucrose gradient ultracentrifugation was used to check whether it remained intact in the endometrial preparations. When the receptor is damaged, the nuclei do not respond to the hormone.

Experiment Type II. The Effect of a Cytosol Component, Presumably Receptor

Calf Endometrium. A batch of endometria, divided into two halves, is used to prepare, on the one hand, nuclei to which estradiol is not added and, on the other hand, a soluble extract obtained by homogenizing in 0.01 *M* Tris-HCl, 0.001 *M* EDTA buffer (pH 7.4) and centrifuging at 105,000 *g* for 60 minutes at 0° in a Spinco fitted with an SW 65 rotor. The presence of the undamaged cytosol receptor is checked systematically by sucrose gradient ultracentrifugation. The nuclei are incubated for 1 hour at 0° with estradiol 1 n*M* soluble extract prepared from the same weight of endometrium and then centrifuged at 700 *g* for 10 minutes at 2°. After resuspension of these nuclei in a Teflon-glass Potter in 0.25 *M* sucrose, 1 m*M* $MgCl_2$, the biosynthesis of RNA is measured (Fig. 1). Control measurements are carried out on nuclei in Tris EDTA (pH 7.4) (N), either in the presence of estradiol ($N + E_{II}$ 1 n*M*), of cytosol (N + C), of cytosol heated to 60° for 60 minutes ($N + C_h$) or of heated cytosol and estradiol ($N + C_h + E_{II}$ 1 n*M*). The need for the simultaneous presence of estradiol and intact soluble extract is obvious (Fig. 3).

These results establish first that it is possible to induce a biosynthetic response to a hormone added *in vitro* to an acellular system and second that a relationship exists between one (or more) element(s) of the cytoplasm and nuclear activation of the cells specifically responsive to this hormone.

Whether such experiments can be regularly reproduced is difficult to establish. Nuclei incubated with cytosol in the absence of estradiol often give rise to increased incorporation of [^{3}H]CTP into RNA as compared to nuclei incubated with Tris EDTA. This could mask a response to estradiol.

Rat Uterus. A homogeneous batch of rats is divided into two groups. One group is used to prepare nuclei, the other to prepare a soluble extract obtained by homogenizing in a glass-glass Potter in 0.01 *M* Tris-HCl, 0.0015 *M* EDTA, 0.15 *M* KCl, pH 7.4, and centrifuging at 105,000 *g*

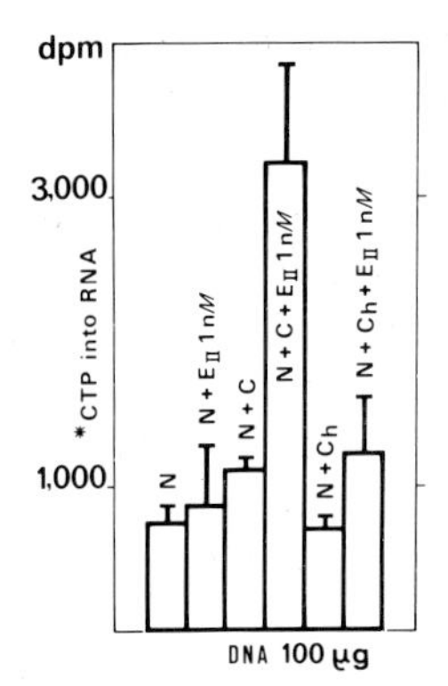

FIG. 3. Action of estradiol on the "reconstituted" cell-free system. The limits represent the standard error of the mean as determined in 4 replicate measurements.

for 90 minutes at 0°. The supernatant is incubated with 5 n*M* estradiol either for 30 minutes at 25° or overnight at 4° and stirred slowly. Controls are carried out by incubating Tris EDTA with or without estradiol and supernatant without estradiol. Following incubation, a volume of nuclear suspension prepared from the same weight of uterus is added to the corresponding supernatant volume. This is reincubated for 30 minutes at 25°, then centrifuged at 700 *g* for 10 minutes at 0°. The nuclei are resuspended in a Teflon-glass Potter in a 0.32 *M* sucrose, 3 m*M* $MgCl_2$ volume such that each 0.1 ml of suspension contains approximately 100 μg of DNA. Ribonucleic polymerase activity is assayed.

ACTIVITY AS MEASURED IN A LOW SALT MEDIUM. The RNA polymerase activity of nuclei preincubated with supernatant in the absence of estradiol is higher than that of control nuclei in nearly all experiments. The addition of estradiol further increases RNA polymerase activity, the response being identical to that obtained for calf endometrium. The reproducibility of the experiments is yet again difficult to control.

ACTIVITY AS MEASURED IN A HIGH SALT MEDIUM. Ribonucleic acid polymerase activity in control nuclei is two to three times higher than in a low salt medium.

The increases in activity recorded for the supernatant and for supernatant plus estradiol are similar to those obtained in a low salt medium, although the reproducibility of the results cannot be definitely asserted. The increase in radioactivity recorded for the supernatant is effectively due to an increase in RNA synthesis since the addition of 20 μg/ml actinomycin D brings down the uptake level to that of control nuclei measured in the presence of actinomycin D.

The original experiments[14] have been recently reproduced and developed[15-18]; in particular the organ specificity of both the cytosol and the nuclei has been demonstrated, and also evidence has been presented that a specific conformational change of the cytosol receptor by estradiol may be a prerequisite for an increased RNA polymerase activity.[16] It seems pertinent to discuss the rapid and large increase of the low salt polymerase activity, as well as the difficulty in obtaining reproducible results. Such a rapid increase, easily reaching 200% in a few minutes, does not correspond to a similar augmentation of the polymerase A activity *in vivo*. Indeed the very early increase of RNA which has been described[13] may correspond to an activation of polymerase B activity which could have been overlooked *in vitro*[19,20]. Finally, it is not impossible that the *in vitro* reconstituted system of nuclei and hormone cytosol complex provides an interesting tool for studying some specific interaction of various pieces of the cellular machinery, especially the receptor, even if these pieces are not necessarily organized as *in vivo*. In other words, even if some molecular characteristics could be worked out with such a system, it might not reproduce the overall psychological mechanism.

[15] M. Arnaud, Y. Beziat, J. C. Guilleux, A. Hough, D. Hough, and M. Mousseron-Canet, *Biochim. Biophys. Acta* **232**, 117 (1971).

[16] S. Mohla, E. R. DeSombre, and E. V. Jensen, *Biochem. Biophys. Res. Commun.* **46**, 661 (1972).

[17] G. C. Mueller, B. Vonderhaar, U. H. Kim, and M. Le Mahieu, *Recent Progr. Horm. Res.* **28**, 1 (1972).

[18] M. M. Bouton, unpublished results.

[19] S. R. Glasser, F. Chytil, and T. C. Spelsberg, *Biochem J.* **130**, 947 (1972).

[20] M. M. Bouton, unpublished results.

Section V

Purification of Receptor for Steroid Hormones

[30] Purification of Estrogen Receptors. I

By GIOVANNI ALFREDO PUCA, ERNESTO NOLA, VINCENZO SICA, and FRANCESCO BRESCIANI

I. Definition of Estrogen Receptors

The word *receptor* was originally used by Ehrlich[1] to indicate a cell component able to recognize and interact with a specific molecule of the cellular environment and included the concept that the informational content of the incoming molecule was expressed through the interaction with its cellular receptor. This full and original meaning of the word still holds among pharmacologists.[2] Therefore, when speaking of *estrogen receptors* one must preliminarily state that this term refers to cellular components of which, at present, we know are proteins able to recognize and specifically bind estrogenic molecules with high affinity (in excess of 10^9 liters/mole at $+4°$). We have no definite information yet as to whether these proteins also fulfill the second part of Ehrlich's proposition, i.e., whether the estrogenic ligand acts by inducing some modification of these proteins which, in turn, enables them to activate the cell machinery so as to produce the final estrogen effects. *Estrogen receptors* is thus used here not in the strictest sense of the word but as synonymous with *high affinity, cellular estrogen-binding proteins* (*EB-proteins*).

II. Different Forms of Estrogen Receptors

In 1971, at the Schering Workshop on Steroid Hormone Receptors in Berlin,[3] 13 different EB-protein forms were claimed to exist. Since then, however, it has become increasingly clear that this high multiplicity of forms is in part apparent—the result of precarious identification based on sucrose-gradient analysis of crude cellular extracts under variable conditions. For calf uterus, when characterization is carried out after partial purification, we find[4-7] (1) a native EB-protein which at low ionic strength exists as a heavier soluble state as well as insoluble aggregates

[1] P. Ehrlich, *Festschr. von Leyden* p. 76 (1902).
[2] R. J. Wurtman, *Science* **159**, 1261 (1968).
[3] *Advan. Biosci.* **7**, (1971).
[4] G. A. Puca, E. Nola, and F. Bresciani, *Res. Steroids* **4**, 319 (1970).
[5] G. A. Puca, E. Nola, V. Sica, and F. Bresciani, *Advan. Biosci.* **7**, 97 (1971).
[6] G. A. Puca, E. Nola, V. Sica, and F. Bresciani, *Biochemistry* **10**, 3769 (1971).
[7] G. A. Puca, E. Nola, V. Sica, and F. Bresciani, *Biochemistry* **11**, 4157 (1972).

TABLE I
PHYSICAL CHARACTERISTICS OF PARTIALLY PURIFIED ESTROGEN-BINDING PROTEINS FROM CALF UTERUS[a]

	Native		
	Low salt ($<0.1\ M$)[b]	High salt ($>0.2\ M$)[b]	Derived
1. $s_{20,w}$	8.6 ± 0.05	5.3 ± 0.1	4.5 ± 0.01
2. Stokes radius (Å)	67	54	33
3. Molecular weight	238,000	118,000	61,000
4. f/f_0	1.65	1.67	1.25
5. I.P.	6.2	—	6.6, 6.8, (7.0)

[a] Data from Puca *et al.*[4-7] Sedimentation coefficients are average ± S.E. of a number of independent measures by sucrose gradient centrifugation. Stokes radii were obtained by reverse gel filtration on calibrated columns. Molecular weight and frictional ratio were calculated assuming a partial specific volume $\bar{v} = 0.725$ cm^3/g. Isoelectric points were measured by electrofocusing. I.P. of native EB-protein in high-salt could not be assessed due to incompatibility of high ionic strength with electrofocusing.

[b] Between 0.1 and 0.2 M salt (KCl, NaCl) both the 8.6 S and 5.3 S states are present. The ionic strengths refer to 2–4° temperatures; at physiological temperature the ionic strength required for the 5.3 S state to exist is expectedly lower.

and (2) an EB-protein which derives from the former by the action of an enzymatic factor of cytoplasm and is stable also at low ionic strength.

The physical properties of these proteins purified from calf uterus are shown in Table I.[4-7] At about physiological, or higher, ionic strength (>0.1–$0.2\ M$ salt) the native EB-protein sediments at 5.3 S on a sucrose gradient and shows a M.W. of 118,000, as derived from sedimentation coefficient and Stokes radius measured by chromatography on a calibrated column of Sephadex G-200; at lower ionic strength (<0.2–$0.1\ M$ salt) it sediments instead at 8.6 S[8] and weighs 240,000 as measured by the same methods. Both states are rather asymmetric (f/f_0, 1.65–1.67) and we know that at least the 8.6 S form is acidic also, at low ionic strength, and especially when the concentration of —SH groups is low, substantial EB-protein aggregation occurs. While the 8.6 S reverses into 5.3 S when the concentration of salt is increased (with both forms present between 0.1 and 0.2 M salt), the aggregates do not; in fact, to date they have resisted dissociation by any means. The M.W. ratio of about two for the 8.6 S and 5.3 S states (240,000:118,000) as well as the tendency

[8] This 8.6 S form corresponds to the "8 S" of other authors.

toward aggregation of the 5.3 S form strongly suggest that the 8.6 S state is the still soluble complex of two 5.3 S proteins. Because salt concentration in uterus is at least 0.15 *M*,[9] it may be that the 8.6 S state, as well as larger aggregates, are artifacts resulting from homogenization of the tissue in low-salt solutions (0.01 *M* salt), a customary procedure used from the earliest experiments toward identification of estrogen-binding cell components.[10,11]

By the action of an enzyme, the Receptor transforming factor (RT-factor),[5,7] a distinct EB-protein derives from the native one. The RT-factor is present in cytoplasm and is Ca^{2+}-dependent. Ionic strength in the physiological range or higher is also required for production of this distinct EB-protein, possibly because the RT-factor can use only the 5.3 S form as substrate. The RT-factor, a protein of at least 100,000 M.W., is probably a protease, but certainly not trypsin or an enzyme of the trypsin group, and formation of the derived EB-protein appears to be a case of limited hydrolysis. The derived EB-protein sediments at 4.5 S[12] on sucrose gradients, independent of ionic strength, weighs just about one-half (61,000) as much as the parent 5.3 S form and is definitely less asymmetrical (f/f_0, 1.25 against 1.67 of parent 5.3 S). Formation of the 4.5 S EB-protein is thus consistent with splitting of 5.3 S into two fragments of about equal length. Whether or not both the fragments resulting from the splitting of the 5.3 S bind estrogen (with the derived EB-protein then actually being a mixture of the two fragments) cannot be answered on the basis of present information. It is true that the 4.5 S EB-protein is not homogeneous at electrofocusing (double isoelectric point, see Table I) but other explanations of this lack of homogeneity also exist.[6] It is an important feature of the 4.5 S that it has lost the tendency to aggregate and is stable also in low-salt media (0.01 *M*).

The above-described EB-proteins are found in cytoplasm, more specifically in cytosol (cytoplasmic soluble fraction), and formation of a complex with estrogen occurs spontaneously on addition of estrogen to cytosol. On the other hand, incubation of subcellular uterine nuclear fraction with estrogen does not give rise to any extractable EB-protein. Only nuclei prepared from uterine tissue preexposed to estrogen, either as an intact tissue or an homogenate, will yield an EB-protein–estradiol complex. When the crude nuclear extract is analyzed as such on a sucrose

[9] D. F. Cole, *J. Endocrinol.* **7**, 12 (1950).

[10] G. P. Talwar, S. J. Segal, A. Evans, and O. W. Davidson, *Proc. Nat. Acad. Sci. U.S.* **52**, 1059 (1964).

[11] D. Toft and J. Gorski, *Proc. Nat. Acad. Sci. U.S.* **55**, 1574 (1966).

[12] This 4.5 S EB-protein corresponds to the "4 S" of other authors.

gradient, the complex sediments at 5–6 S[13,14] This 5–6 S form[15] is, however, labile, and after partial purification invariably ends as a form sedimenting at 4.5 S and indistinguishable from the 4.5 S EB-protein from cytoplasm not only in sedimentation rate but also in all other molecular parameters and properties which were investigated by methods suited to partially purified preparations.[4,5] This finding suggests that the RT-factor produced 4.5 S EB-protein is the only cytoplasmic form able to move to the nucleus, and only as a complex with the hormone.[7]

III. Basic Guidelines for Purification

From the aforementioned properties of cellular EB-proteins and of RT-factor, the basic principles for preparation of either one of the different forms of this type of protein may be derived. When preparation of éither state (5.3 S or 8.6 S) of native EB-protein is required then one should not go through procedures which result in activation of the RT-factor, with resultant loss of the desired protein. This may be easily accomplished by avoiding the use of buffers with Ca^{2+} and complexing endogenous Ca^{2+} by EDTA. When, instead, the 4.5 S EB-protein by cytoplasm must be prepared, pretreatment of cytosol must aim at the opposite result and therefore both sufficient Ca^{2+} (4 m*M*) and KCl or NaCl (final molarity 0.2 or higher) must be added to cytosol to obtain efficient formation of 4.5 S.

When EB-proteins are prepared from cytosol, preliminary exposure of tissue to estrogen either *in vivo* or *in vitro* is not required; in fact, EB-proteins may be prepared estrogen-free from prepubertal or ovariectomized animals.[6] On the contrary, receptor protein from nuclei is obtained only when the tissue *in vivo* or the tissue (or tissue homogenate) *in vitro* was preexposed to estrogen and is obtained only as a protein–estrogen complex. Preparation of 4.5 S EB-protein from nuclei does not require that the nuclear fraction or nuclear extract be treated under conditions which activate the RT-factor.

After these preliminary operations, standard methods for purification of proteins may be applied. However, one must take into account the many factors detrimental to estrogen receptors (Table II). These factors either destroy binding activity for the specific ligand or because they

[13] E. V. Jensen, D. J. Hurst, E. R. DeSombre, and P. W. Jungblut, *Science* **158,** 385 (1967).

[14] G. A. Puca and F. Bresciani, *Nature* (*London*) **218,** 967 (1968).

[15] Usually, this complex is referred to as nuclear "5 S", but actually it sediments closer to 6 S than to 5 S in most runs.

TABLE II
SOME FACTORS DETRIMENTAL TO ESTROGEN RECEPTORS[a]

1. Binding activity is irreversibly destroyed by:
   a. Proteases
   b. +65° for 5 minutes[b]
   c. Acidic pH
   d. Iodination
   e. Cold ethanol or Ether
   f. 1.5–5 *M* Urea
   g. 1.5–2.5 *M* Guanidine HCl

2. Loss of native EB-protein (8.6 S or 5.3 S) by irreversible aggregation occurs easily and is strongly favored by low concentration of SH groups, freezing, precipitation by $(NH_4)_2SO_4$, storage. The derived 4.5 S EB-protein is stable and does not aggregate even in hypoionic media.

3. Some protection from inactivation by heat, acidic pH, freezing, and total protection from inactivation by I_2 is afforded by excess estradiol-17β.

[a] Data from Puca *et al.*[6,7]

[b] On the contrary, aspecific binding of 17β-estradiol (for instance by albumin) is unaffected or increased after heating at +65° and this furnishes a quick method of discrimination.

produce irreversible aggregation of the proteins. Loss by aggregation is insignificant in the case of the 4.5 S EB-protein.

Assay of estrogen-binding activity at various stages of purification may be carried out either by the charcoal method[16] or by the gel exclusion method described later in this chapter. It must be stressed, however, that due to the tendency of estradiol-17β to bind aspecifically to virtually every protein in solution, measurements of binding activity of a preparation must always be accompanied by an estimate of the binding constant in order to be acceptable. The association constant can be easily estimated by using either the charcoal or the exclusion chromatography method. A rapid preliminary check of the specific estrogen-binding activity of a given preparation can be carried out by taking advantage of the finding[6] that heating at 65° for 5 minutes destroys all specific activity, while aspecific binding is unaffected or increased.

IV. Sources of Estrogen Receptors

In principle, any estrogen target tissue can be used. Preferably the tissue should be obtained from prepuberal or ovariectomized animals, in

[16] S. G. Korenman, this Vol. [4].

order to avoid high concentrations of endogenous estrogen which cause a low content of estrogen-free receptors. Indeed, once formed, estrogen–receptor complexes dissociate very slowly under conditions which are harmless for receptor-binding ability, and labeling with added tritiated estradiol-17β is weak.

The most convenient source of estrogen receptors for purification purposes is calf uterus. Fresh calf uterus can be obtained with relative ease from the local slaughter house and combines the advantages of being an organ richest in estrogen-free EB-proteins with those of being of considerable size and rather readily homogenized. Cow uterus has a lower concentration of estrogen-free EB-proteins (due to endogenous hormones) and, in addition, is very tough because of a substantial component of fibrous connective tissue. Calf pituitary, human uterus and breast cancer tissue, and rat and mouse uterus have also been used as sources of estrogen receptors.

In general, fresh tissue is recommended. This is a prerequisite when the 4.5 S EB-protein must be prepared because freezing inactivates the RT-factor. Frozen tissue can be used for preparation of native forms 5.3 S and 8.6 S, but uteri should be cleaned of parametrial tissue before freezing. Freezing is done best by dipping the tissue in liquid nitrogen and storing at —80°, a procedure which results in no loss of binding activity for months. Storage at —20° will produce slight but significant loss of activity (10–15% per month).

V. Collection and Preparation of Tissue

In general, the tissue is excised rapidly and immediately cooled to 0°. Calf uteri are collected at the local slaughter house as soon as the animal is killed and kept in separate plastic bags buried in crushed ice. Once in the laboratory cold room (4°) they are carefully cleaned of parametrial connective tissue before any further processing. In order to obtain the highest concentration of estrogen-free EB-proteins, calf uterine horns weighing more than 20 g (after being cleared of parametrium) are discarded. Human tissue is collected immediately after surgical excision and further treated as described for calf uterus.

VI. Purification of Estrogen Receptors from Cytosol

The procedure will be described as applied to calf uterus. It has also been applied with satisfactory results to calf pituitary, human uterus and breast cancer tissue, and rat and mouse uterus. Molecular characteristics of EB-proteins from human and rodent tissues may be different from those of calf.

A. Preparation of Cytosol[5–7]

All operations are carried out at 4°, either in a cold room or in refrigerated centrifuges. Uterine horns are minced twice in a meat grinder and weighed in a plastic container. Four milliliters of ice-cold TEN buffer, pH 7.5,[17] to which 1 m*M* dithioerythritol was added immediately before use, are added to each gram of ground uterus and the tissue is homogenized by means of an Ultraturrax homogenizer (Janke and Kunkel, Model TP 18/2) in six runs of 15 seconds each, with 60-second intervals between runs. Temperature is continuously monitored and kept below 4°. The homogenate is centrifuged at 105,000 g for 1 hour. Dithioerythritol is very effective in reducing irreversible aggregation of EB-proteins.[7] Even at 20 m*M* concentration, this thiol compound does not interfere with binding of estradiol-17β to receptor proteins, nor does it dissociate the complex once formed, or modify the molecular characteristics of EB-proteins.[7] Cytosol thus prepared usually contains 5–7 mg of protein/ml. At this stage, the cytosol may either be labeled by addition of 6,7-*tert*-estradiol-17β (spec. act. 40 Ci/m*M*), generally to a radioactive hormone concentration of 10^{-8} *M*, or further purification carried out without preliminary labeling if estrogen-free receptors are desired. Centrifugation of the above cytosol preparation on a sucrose gradient in low-salt buffers (0.1 *M* or lower KCl or NaCl) shows bound radioactivity sedimenting at about 8.6 S; if high-salt buffers (0.2 *M* or higher KCl or NaCl) are used, bound radioactivity sediments instead at about 5.3 S. RT-factor is present and stable in this cytosol, but inactive because of EDTA complexing endogenous Ca^{2+}.

B. Purification of Native Receptor (8.6 S or 5.3 S, Depending on Ionic Strength)[5,7]

Principles. RT-factor is present but inactive because of addition of EDTA to cytosol. After the $(NH_4)_2SO_4$ purification step (20% saturation) the RT-factor remains in the supernatant while the EB-protein precipitates, and thus the possibility of loss of native EB-protein due to RT-factor activity is definitely eliminated.

Procedure. The procedure is basically the same for both the 5.3 S and the 8.6 S forms, except that high-salt buffers (KCl or NaCl molarity higher than 0.2, usually 0.4) are used for preparation of 5.3 S, while low-salt buffers (KCl or NaCl molarity less than 0.1, usually 0.01) are used for the 8.6 S. In general, purification of form 8.6 S is much less efficient because at low ionic strength a substantial amount of EB-protein is lost

[17] TEN buffer, pH 7.5: Tris-HCl, 10 m*M*; EDTA, 1.5 m*M*; NaN_3, 1 m*M*.

due to formation of aggregates. One can easily obtain 8.6 S by lowering the ionic strength of purified solutions of 5.3 S.

The purification procedure to be described consists of only two steps: (1) $(NH_4)_2SO_4$ precipitation (20% saturation) of protein from cytosol, and (2) exclusion chromatography on Sephadex G-200 columns. The double electrofocusing step which is successful when applied to the 4.5 S EB-protein after the exclusion chromatography stage (see below) does not improve purification in the case of the 8.6 S form and cannot be applied at the ionic strength required for the 5.3 S to exist. The same is true of other separation methods which have been tried (see Comments for more details). All operations are carried out at 0–4°, either in a cold room or in refrigerated centrifuges.

STEP 1. One-quarter of a milliliter of a saturated and neutralized solution of $(NH_4)_2SO_4$ for each milliliter of cytosol is added at once. Ten minutes later the mixture is centrifuged at 19,000 *g* for 10 minutes. After discarding the supernatant, the sediment is redissolved by stirring or vibrating for 30 minutes in the cold in a volume of high-salt TENK buffer,[18] pH 7.5, containing dithioerythritol (5 m*M*) added just before use, equivalent to $\frac{1}{10}$ of the original cytosol volume. In case of large preparations, it is efficient to apply the Ultraturrax at 50–70 V for a few seconds to the sediment in TENK buffer. Insoluble material is sedimented by either (1) centrifugation at 105,000 *g* for 90 minutes or at 310,000 for 30 minutes; and (2) centrifugation at 105,000 for 4 hours on a cushion of 3 *M* sucrose or at 310,000 for 90 minutes. Protein concentration of the supernatant after either treatment (1) is about 2.4–4.5 mg/ml, with about 70% recovery of specific binding activity in cytosol. After either treatment (2) protein concentration of the supernatant is only about 1.5–2 mg/ml, with about 50% recovery of binding activity of cytosol. Either one of the treatments in (2) is preferred when the next purification step is carried out in a low-salt buffer (in order to obtain 8.6 S) for reasons which will become apparent later. Sucrose-gradient centrifugation of the redissolved precipitate shows bound hormone sedimenting either at 8.6 S or 5.3 S, depending on ionic strength of the medium. Despite the presence of dithioerythritol, the rate of EB-protein aggregation in the redissolved $(NH_4)_2SO_4$ precipitate is rapid and therefore it is important to pass as soon as possible to the next purification step.

STEP 2. Exclusion chromatography of the redissolved $(NH_4)_2SO_4$ precipitate is carried out on a Sephadex G-200 column of appropriate size. Sample volume should not exceed 3% of the total volume of the column.

[18] TENK buffer, pH 7.5: Tris-HCl, 0.1 *M*; EDTA, 1.5 m*M*; NaN_3, 1 m*M*; and KCl, either 0.4 *M* (high-salt buffer) or 0.05 *M* (low-salt buffer).

The column is fitted with an upward-flow adaptor. The upward flow is generated by a peristaltic pump which is set to give the preliminarily measured flow rate produced by 15 cm hydrostatic pressure in the same column. However, while applying the sample and until the effluxed volume is equivalent to about $\frac{1}{10}$ the total volume of the column, we keep the flow going 2–3 times faster than the above value. This method gives a better separation of the excluded material from the included EB-protein, possibly because of a rapid initial dissociation of the two fractions. Chromatography is carried out in either low-salt or high-salt TENK buffer. Typical results are shown in Fig. 1. In both cases, in addi-

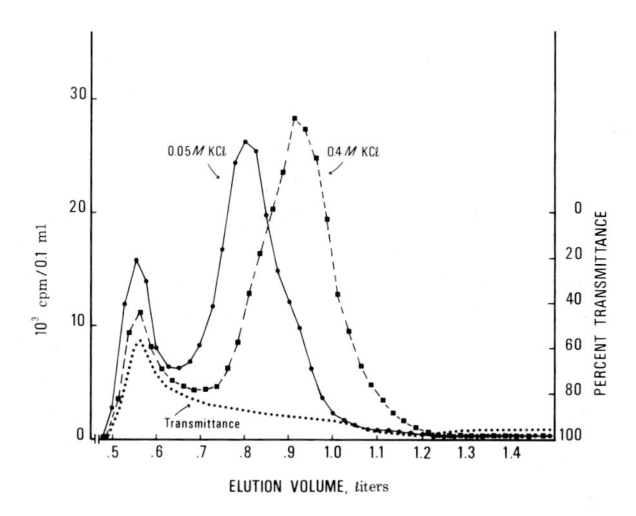

FIG. 1. Preparative exclusion chromatography of native estrogen receptor in either low-salt (0.05 *M* KCl) or high-salt (0.4 *M* KCl) buffer. Continuous line: $(NH_4)_2SO_4$ precipitate from 6,7-*tert*-estradiol-17β labeled cytosol (158 g of fresh calf uteri) was prepared and redissolved in 60 ml of buffer as described in text. After sedimentation of aggregates (310,000 *g* for 90 minutes), the 60 ml sample was applied to a column of Sephadex G-200 (5×97 cm; V_t, 1900 ml) in low-salt TENK-buffer, pH 7.5. Inverted flow rate = 22 ml/hour. The smaller, earlier peak is tritiated estradiol-17β bound to excluded material; the larger, later emerging peak is radioactive estradiol bound to protein sedimenting at 8.6 S on a sucrose density gradient in the same buffer. Broken line: Same preparation as above except that starting material was 140 g of fresh calf uteri and chromatography was carried out in high-salt TENK buffer, pH 7.5. The initial peak is radioactive hormone bound to aggregates, while the larger second peak is hormone bound to protein sedimenting at 5.3 S in the same high-salt buffer. Dotted line: Transmittance at 280 nm. Association constant of 8.6 S and 5.3 S hormone–protein complexes is in excess of 10^9 liters/mole. Electrofocusing pattern of the protein-tritiated hormone complex is shown in Fig. 2.

tion to a small excluded peak there is a major peak of bound tritiated estradiol-17β. When the low-salt buffer (0.05 M KCl) is used, the major peak emerges at about 0.42 the total volume of the column. In high salt buffer (0.4 M KCl) the major peak emerges later, at about 0.48 the total volume of the column. Of course these elution values may vary in absolute terms, from column to column, depending on packing, flow rate, gel particle size, volume of the sample to volume of the particular column, and so on. However, the relative difference between high-salt and low-salt is always found. Sucrose-gradient analysis in the same buffer as used for chromatography shows that the low salt major peak is 8.6 S and that the high-salt major peak is 5.3 S. In both cases, the material in the excluded peak rapidly sediments to the bottom of the gradient.

It is important to warn that the excluded peak (aggregates) may present a problem; especially if cytosol preparation is old and/or clarification of resuspended $(NH_4)_2SO_4$ precipitate is insufficient and/or dithioerythritol buffers are old and/or chromatography is carried out in low-salt buffer. Under these conditions the peak of aggregates may be very large and overlap substantially with the included EB-protein peak. Therefore, when using low-salt chromatography, previous clarification of the redissolved $(NH_4)_2SO_4$ precipitate must be carried out by either one of treatments (2) (see above) in order to reduce aggregates to a minimum. The specific activity of EB-protein preparations after step 2 is between 2 and 5×10^6 dpm of 6,7-*tert*-estradiol-17β (40 Ci/mmoles) per milligram of protein, with recovery varying from about 25% (low-salt chromatography) to about 40% (high-salt chromatography) of the original binding activity of cytosol. These values of specific activity and recovery refer to the pooled central fractions of the "included peak." Electrofocusing of the above preparations, necessarily in low-salt buffer, always shows a single sharp peak of EB-protein at pH 6.2 (Fig. 2).

STORAGE. The pooled chromatographic fractions corresponding to the peak of interest are best stored without attempts toward increasing protein concentration. Such attempts, either by $(NH_4)_2SO_4$ reprecipitation or dialysis under reduced pressure or filtration through Amicon membranes, always result in substantial loss of EB-protein, due especially to aggregation. Native EB-protein stored in high-salt TENK buffer, pH 7.5, made 5 mM in dithioerythritol and to which estradiol-17β is added in excess, shows no significant loss of activity after two months at 0–4°. Higher storage temperatures result in appreciable loss of activity, and freezing in substantial loss of protein due to aggregation. Storage of low-salt preparation as such is not recommended because, even with the addition of dithioerythritol and excess estradiol-17β, aggregation takes place at a

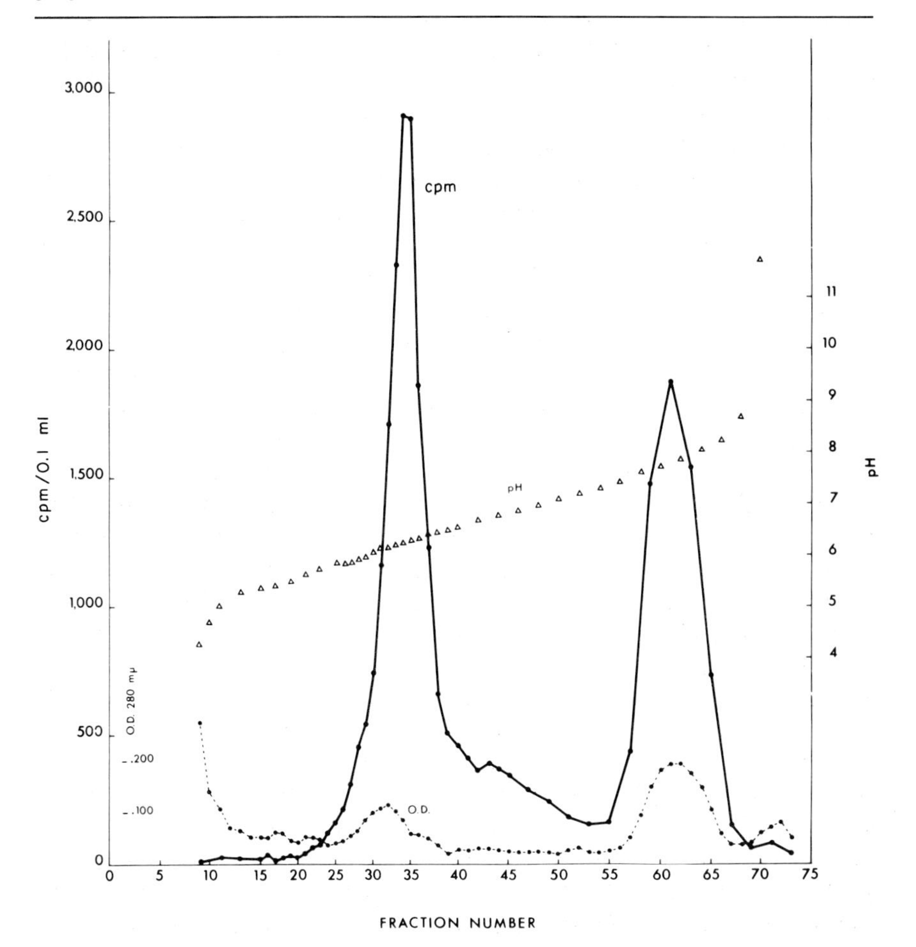

FIG. 2. Electrofocusing pattern of native receptor after the exclusion chromatography purification step. Electrofocusing is carried out in the LKB 110-ml electrofocusing column. Positioning of the sample is accomplished as described in text. The sharp central radioactive peak (pH 6.2) is 6,7-*tert*-estradiol-17β bound to EB-protein. The radioactive peak on the right is free hormone. The O.D. peak on the right is dithioerythritol. From Puca *et al.*[6]

rapid rate. Addition of salt, at least 0.2 M KCl or NaCl, is required for hindering aggregation.

COMMENTS. The use of higher $(NH_4)_2SO_4$ concentration (30% saturation) to precipitate EB-protein (step 1) results in about 10–15% higher yield of specific binding activity but at the cost of more than halving the specific activity. The use of 20% saturation also ensures that the redissolved precipitate (step 1) is completely free of RT-factor, which is still soluble at 20% saturation of $(NH_4)_2SO_4$; indeed, no formation of 4.5 S EB-protein occurs in the redissolved precipitate after addition of Ca^{2+} (4 mM) and increase of ionic strength (0.4 M KCl).[7]

Attempts at purification of the EB-protein preparation further than step 2 have proved essentially unsuccessful. At electrofocusing, most proteins in the preparation focus at about pH 6.0, almost coincident with the peak of native EB-protein (Fig. 2), and preparative electrofocusing produces no appreciable increase of specific activity while resulting in significantly lower recovery due to aggregation.[19] DEAE-cellulose chromatography has also been applied[19] but no improved purification was achieved: The EB-proteins are eluted between 0.15 and 0.30 M KCl together with most of the other proteins in the preparation; furthermore, recovery from the column is poor. CM-cellulose at neutral pH does not retain the EB-proteins, and at acidic pH the EB-protein is quite unstable and binding activity is lost.[19] Phosphocellulose does not retain EB-proteins, while they cannot be eluted once bound to hydroxylapatite.[19]

C. Purification of Derived Receptor (4.5 S, Independent of Ionic Strength)[5–7]

Principles. Ionic strength, Ca^{2+} concentration, and pH of cytosol are set so as to favor optimal activity of RT-factor, with resulting formation of the 4.5 S EB-protein from the parent 5.3 S EB-protein. The 4.5 S is stable at low ionic strength and has lost all tendency to aggregate, which makes it much easier to deal with. The purification procedure consists of 4 steps: (1) pretreatment of low-salt cytosol; (2) $(NH_4)_2SO_4$ precipitation; (3) exclusion chromatography, and (4) preparative electrofocusing. Specific activity of the final purified preparation is up to two orders of magnitude higher than that achieved for native estrogen receptor.

Procedure. Except when specified, all operations are carried out at 0–4°, either in a cold room or in refrigerated centrifuges.

STEP 1. KCl as salt is added to either labeled or unlabeled cytosol up to 0.5 M. After 15 minutes, $CaCl_2$ as a 0.2 M solution is further added dropwise up to 4 mM final concentration. The pH of the cytosol, usually 7.0–7.1, is slowly adjusted to pH 8.0 by careful addition of 0.1 N NaOH. These conditions are optimal for RT-factor activity.[7] After 60 minutes of incubation at 4°, sucrose gradient analysis of the cytosol on either

[19] G. A. Puca, E. Nola, V. Sica, and F. Bresciani, unpublished observations.

low or high-salt gradient shows bound 6,7-*tert*-extradiol-17β to sediment at about 4.5 S.

STEP 2. Finely powdered $(NH_4)_2SO_4$, slowly and with stirring, is then added in the amount of 0.113 g/ml of cytosol. After standing for an additional 60 minutes, the mixture is centrifuged at 19,000 g for 10 minutes. Redissolving of the precipitate in TENK buffer, pH 7.5, and elimination of insoluble material is carried out as described for native receptor.

STEP 3. This step is carried out as already described for native receptor. The pattern of elution of the 4.5 S EB-protein is the same in either low-salt or high-salt TENK pH 7.5 buffer. A typical chromatographic pattern in low-salt (Fig. 3) consists of an excluded, small peak corresponding to aggregates, a just included peak corresponding to nontransformed 8.6 S, and a last, large peak emerging at 0.68–0.70 the total volume of the column, this last peak is the 4.5 S EB-protein, as can be shown by sucrose-gradient analysis on either low-salt or high-salt gra-

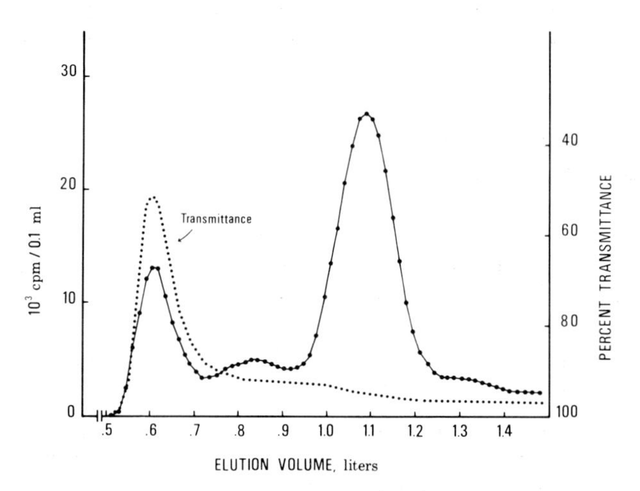

FIG. 3. Preparative exclusion chromatography of derived estrogen receptor (4.5 S EB-protein). Continuous line: $(NH_4)_2SO_4$ precipitate from 6,7-*tert*-estradiol-17β labeled cytosol (172 g of fresh calf uteri), in which RT-factor had been activated (1 hour) by Ca^{2+} (4 mM) at 0.4 M KCl concentration, was redissolved in 65 ml buffer as described in text. After sedimentation of aggregates (310,000 g for 30 minutes) the 65-ml sample was applied to a column of Sephadex G-200 (5×91.5 cm; V_t, 1780 ml) in low-salt TENK-buffer, pH 7.5. Inverted flow: 22 ml/hour. The smaller, earlier emerging peak is tritiated estradiol-17β bound to "excluded" material; the larger, later emerging peak is radioactive estradiol bound to protein sedimenting at 4.5 S on either low-salt or high-salt sucrose gradients. Dotted line: Transmittance at 280 nm. Association constant of the protein–hormone complex is in excess of 10^9 liters/mole. Electrofocusing pattern of the protein–tritiated hormone complex is shown in Fig. 4.

dients. When only fractions in the central part of the large peak are pooled, typically the specific binding activity ranges from 5 to 12 $\times 10^6$ dpm of 6,7-*tert*-estradiol-17β (40 Ci/mM) per milligram of protein. Recovery is of the order of 30% of the activity in cytosol. The 4.5 S EB-protein has no tendency to aggregate and at this stage of purification may be stored at 4° for months without loss of activity, best in the presence of estradiol-17β and dithioerythritol (1–5 mM).

STEP 4. The 4.5 S EB-protein preparation from step 3 may be further purified by DEAE-cellulose chromatography followed by electrofocusing, but recovery is very poor. Better results are obtained when, instead, the material is immediately double electrofocused (Fig. 4). Contrary to native receptor, the 4.5 S EB-protein focuses as a two-peak band at a definitely higher pH (6.6–6.8) than most other proteins in the preparation (5.9–6.0). On the other hand, free radioactivity possibly released during the experiment focuses at about pH 8.0, together with O.D. due to dithioerythritol. These facts can be exploited to eliminate the bulk of contaminating protein and free hormone by selecting and pooling the central fractions of the 4.5 S peak in the electrofocusing. These pooled fractions are then concentrated under vacuum, dialysed against buffer solution to eliminate most of the Ampholine and sucrose, and refocused. This second time there is no detectable O.D. under the band of bound 6,7-*tert*-estradiol-17β. Satisfactory results, however, are obtained only by using selective positioning of sample in the electrofocusing columns, as described in Comments. In a purification experiment in which we started from 840 g of fresh calf uterus, the pooled fractions of the 4.5 S peak in the second electrofocusing contained 1.2 mg of protein with a specific activity of 160 $\times 10^6$ dpm of bound 6,7-*tert*-estradiol-17β (40 Ci/mM) per milligram. The recovery was 11% of binding activity in original cytosol. Assuming a ligand: 4.5 S EB-protein molar ratio of 1:1, this final preparation is at least 30% pure. On gel electrophoresis this preparation gives four bands of which only one shows estrogen-binding activity.[19]

STORAGE. In contrast to native EB-protein, the 4.5 S receptor does not aggregate, even at low ionic strength, and can be concentrated and stored at 2–4° in low-salt solutions, pH 7.5, containing 1 mM dithioerythritol, without appreciable loss of activity over several months.

Comments. ELECTROFOCUSING WITH SELECTIVE POSITIONING OF SAMPLE. Due to the high sensitivity of receptor-binding activity at pH below 7.0 or above 9.0, the following method of positioning the sample in the electrofocusing column is recommended to avoid losses.[6] Either 110 ml or 440 ml standard LKB electrofocusing columns are used, according to the amount of material to be electrofocused. A 5–50% sucrose gradient with 1% ampholine, pH 5–8, is prepared by the LKB gradient former

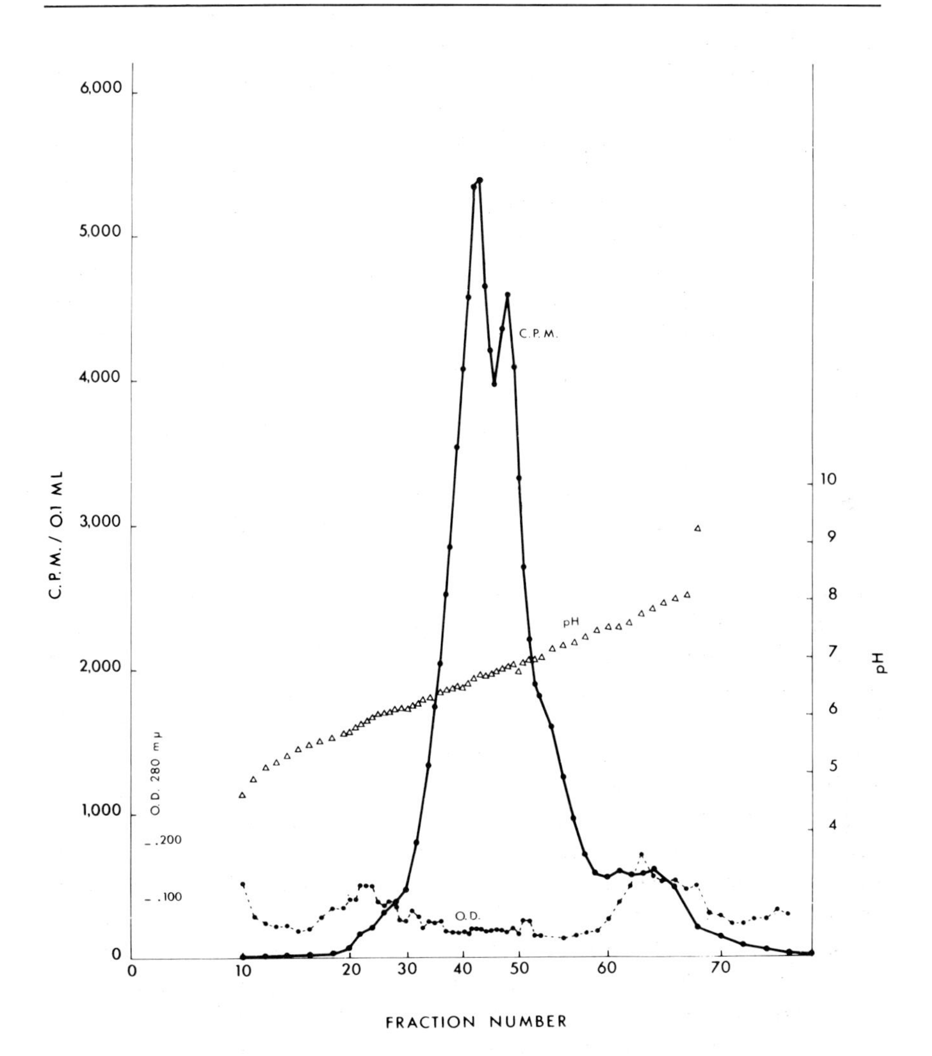

FIG. 4. Electrofocusing pattern of derived estrogen receptor (4.5 S) after the exclusion chromatography purification step. Electrofocusing is carried out in the 110-ml LKB column. Position of sample is accomplished as described in text. The protein–tritiated hormone complex focuses as a double-spiked band (pH 6.6 and 6.8) with a shoulder (pH 7.0) on the right; this pattern is typical and reproducible. Note that the left bump in the O.D. profile, due to contaminating protein, is definitely shifted with respect to the EB-protein band. Also, free hormone (bump on the right in the radioactive profile) and dithioerythritol (bump on the right in the O.D. profile) are off the EB-protein peak. These are advantageous features which can be exploited for purification by double preparative electrofocusing, as described in text. From Puca *et al.*[6]

device. Temperature of the water running through the double-jacketed column is kept at 2–3°, throughout the run. The pH gradient is preformed by applying voltage and keeping it until the milliamperage falls to a constant value (0.8 and 2.4 mA at 800 V for the 110- and 440-ml columns, respectively). Power is turned off and a capillary tubing is introduced slowly from the top into the gradient, down to the upper third of the column, in the pH 7–8 range of the gradient. Ten milliliters 110-ml column) or 40 ml (440-ml column) of the gradient are slowly sucked off by means of a peristaltic pump. The fraction of gradient sucked off is mixed with the sample and pushed back into place in the column through the same tubing by the same pump. Samples of up to 3 ml for the 110-ml column and of up to 14-ml for the 440 ml column can be applied by the above procedure with good results. Power is turned on again and the increasing amperage is kept within 3.5 mA (110-ml column) or 10.5 mA (440-column) by decreasing voltage. Generally after 6–7 hours (110-ml column) or 10–11 hours (440-ml column) the initial amperage at 800 V is reached again. The gradient is now collected in the cold room in appropriate fractions, with O.D. of effluent being monitored by an UV spectrophotometer connected with a recorder. The pH of the fractions is measured in the cold room and radioactivity measured on small aliquot parts of the fractions by liquid scintillation. The fractions containing the bound radioactive hormone are pooled, dialysed against buffer pH 7.5 to remove Ampholine and sucrose, and concentrated by dialysis under vacuum.

The above electrofocusing procedure avoids destruction of receptor protein and gives reproducible patterns when two conditions are fulfilled: (1) Applied preparations of receptor must be purified at least through the gel exclusion step; and (2) efficient cooling of the column (2–3°) is maintained throughout the run.

D. Purification of Estrogen Receptors from Nuclear Fraction[4,14,20,21]

As stated in the Introduction, the estrogen–protein complex from the nuclear fraction sediments at about 5–6 S ("5 S") in the stage of crude nuclear extract. The 5–6 S complex, however, is labile and even after mild purification procedures becomes a 4.5 S sedimenting EB-protein, which is indistinguishable from the cytoplasmic 4.5 S also in Stokes radius, molecular weight, isoelectric pattern (electrofocusing), and all other properties which were investigated by methods suited to partially

[20] G. A. Puca and F. Bresciani, *Nature* (*London*) **223**, 745 (1969).

[21] G. A. Puca, E. Nola, and F. Bresciani, *Atti Accad. Naz. Lincei, Cl. Sci. Fis. Mat. Natur., Rend.* **46**, 72 (1969).

purified preparations.[4,5] Whether the labile, faster sedimenting state in crude nuclear extract has a physiological meaning or it is simply the consequence of aspecific interaction with other molecules in the nuclear extract is still an open question. It is also to be remembered that virtually no receptor protein can be extracted from nuclei of tissue not previously exposed to estrogen, and this explains the requirement of preliminary incubation with estrogen (see later) when the tissue used as source of receptor derives from prepubertal or ovariectomized animals.

Principles. The tissue used as source is exposed to radioactive estradiol-17β. The nuclear fraction is prepared and extracted with high salt at alkaline pH. The extract contains the specific estrogen–receptor complex, which is then further purified by $(NH_4)_2SO_4$ precipitation, exclusion chromatography, and electrofocusing.

Procedure. Except when specified, all operations are carried out in the cold room or in refrigerated centrifuges at +4°. The procedure is described as applied to calf uterus. Fresh uterine horns are cut into strips with scissors. The strips of tissue are washed in saline and then incubated at 37° in Krebs-Ringer phosphate buffer, pH 7.4, containing 10^{-9} *M* 6,7-*tert*-17β (40 Ci/mmole) and kept in suspension by gentle magnetic stirring. The ratio of tissue to incubation medium (w/v) is 0.02. After 60 minutes the uterine strips are rapidly washed with cold saline, blotted dry, and homogenized in 3 times their volume of ice-cold TEN buffer, pH 7.5, containing 1 m*M* dithioerythritol, as described for preparation of cytosol. The homogenate is centrifuged at 3000 *g* for 10 minutes to sediment nuclei; the sediment is resuspended by the Ultraturrax homogenizer (3 runs of 10 seconds each at 70–90 V with 30-second intervals between runs) in about 5 times its volume of fresh homogenization medium and centrifuged again. This operation is repeated 3 times and the final sediment (nuclei heavily contaminated by myofibrils) typically contains no less than 70% of the radioactivity taken up by the uterine strips. Cell autoradiography has given uncontroversial evidence that the radioactive estradiol-17β is concentrated in nuclei not in myofibrils.[22] The final nuclear sediment is suspended in 1.5 times its volume of 0.4 *M* KCl in 10 m*M* Tris-HCl, pH 8.5, containing 1 m*M* dithioerythritol by means of the Ultraturrax homogenizer (6 runs of 15 seconds each at 115 V, with 30-second intervals between runs) and magnetically stirred for 60 minutes before centrifugation at 105,000 *g* for 60 minutes. The decanted supernatant, or nuclear extract, typically contains 50 to 70% of the radioactivity originally present in the initial nuclear fraction. The nuclear extract is processed through $(NH_4)_2SO_4$ precipitation, exclusion chroma-

[22] W. E. Stumpf, *Endocrinology* **85**, 31 (1969).

tography, and electrofocusing as described for the 4.5 S EB-protein from cytosol. However, due to the higher pH of the nuclear extract (8.5) a slightly higher concentration of $(NH_4)_2SO_4$ (0.16 g/ml) is recommended to obtain satisfactory recovery of receptor–estrogen complex.

Storage. The purified nuclear receptor does not aggregate and can be stored at 2–4° in TEN buffer, pH 7.5, containing 1 m*M* dithioerythritol without significant loss of activity over several months.

Comments. The use of a higher than 0.4 *M* of KCl as well as a second extraction of the high speed nuclear pellet decreases the specific activity of the protein extract and increases the amount of aggregated receptor. A pH lower than 8.5 during extraction results in significantly lower solubilization of EB-protein from nuclei.

E. Assay of Estrogen-Binding Activity by Exclusion Chromatography[6]

This assay is based on exclusion chromatography experiments described elsewhere,[14,20] and measures *specific* estrogen-binding activity (i.e., K_{ass} of binding $>10^9$ liters/mole at $+4°$) in preparations spanning from crude cytosol to preparations of the highest achieved purity. It is carried out as follows. The sample (0.1–0.3 ml) to be tested, containing 0.01–0.1 mg of protein, is brought to 1 ml with low-salt TENK buffer, pH 7.5, which contains 3×10^5 dpm of 6,7-*tert*-estradiol-17β (40 Ci/mmole) per milliliter and is colored with Blue Dextran. The mixture is incubated at 4° for 5–6 hours, a sufficient time to reach equilibrium. After incubation, separation of free from protein-bound hormone is accomplished by chromatography on Sephadex G-25 at 4°, under standard conditions of column size (ϕ = 1.2 cm; total volume 20 ml) and flow rate (40 ml/hour). The *bound* hormone emerges associated with the first macromolecular peak, which is excluded by the gel (void volume) while *free* hormone is eluted as a peak emerging after the internal volume of the column. The excluded peak is easily identified by the color (Blue Dextran); there is no adsorption of estradiol to the dye. If the amount of bound hormone is in excess of $\frac{1}{10}$ of that free, the test is repeated after dilution of the sample in order to ensure saturation of binding sites. Under the above described conditions of chromatography, aspecific hormone binding by macromolecules other than receptors is negligible, including binding by albumin (K_{ass} ~10^5 liters/mole). The method furnishes a linear response for binding activity up to 3×10^4 dpm of 6,7-*tert*-estradiol-17β bound. Association constants between estrogen receptors and estradiol-17β can be estimated with the above method, using a range of estradiol-17β concentrations. Affinity for other cold ligands can also be measured indirectly from the study of competitive inhibition of hot estradiol-17β binding by the cold ligand under consideration.

A quick test of the quality of binding, preliminary to measurement of K_{ass}, consists in heating the preparation under study at +65° for 5 minutes. The above heat treatment completely destroys binding activity due to estrogen receptor while, on the contrary, aspecific binding is unaffected or increased.[6]

[31] Purification of Estrogen Receptors. II

By EUGENE R. DESOMBRE and THOMAS A. GORELL

Introduction

A number of unique and separate forms of the estrogen receptor have been identified in uterine tissue (Table I). These can basically be separated into 3 types by their sedimentation behavior: (1) the native cytosol receptor consisting of a core binding unit sedimenting near 4 S in high salt but associating either with other 4 S core units or with nonbinding moeities to sediment in the 8 S region when the salt is removed. (2) The stable 4 S form of the receptor, found along with the 8 S cytosol receptor in adult uterus[1] and obtained from immature uterine cytosol (prepared with EDTA) on treatment with salt and calcium,[2,3] sediments as a 4.5 S complex with estradiol in both high- and low-salt media. On the basis of recent experiments,[4] its formation appears to derive from the action of a calcium-dependent macromolecular factor present in uterine cytosol (prepared in EDTA) on other forms of the receptor. (3) 5 S Forms of the receptor complex obtained on extraction of uterine nuclei from animals treated with the hormone *in vivo*, from uteri or uterine homogenates treated with estradiol at 25° or 37° *in vitro*,[5] by warming uterine cytosol with estradiol,[5] or salt precipitation of uterine cytosol prepared in the

[1] R. J. B. King, J. Gordon, J. Marx, and A. W. Steggles, *in* "Basic Actions of Sex Steroids" (P. O. Hubinont, F. Leroy, and P. Galand, eds.), p. 21. Karger, Basel, 1971.

[2] E. R. DeSombre, G. A. Puca, and E. V. Jensen, *Proc. Nat. Acad. Sci. U.S.* **64**, 148 (1969).

[3] E. R. DeSombre, J. P. Chabaud, G. A. Puca, and E. V. Jensen, *J. Steroid Biochem.* **2**, 95 (1971).

[4] G. A. Puca, E. Nola, V. Sica, and F. Bresciani, *Biochemistry* **11**, 4157 (1972).

[5] E. V. Jensen, P. I. Brecher, M. Numata, S. Smith, and E. R. DeSombre, this Vol. [23].

TABLE I
FORMS OF ESTROGEN RECEPTORS IN IMMATURE UTERUS

Source	Treatment	Sedimentation behavior[a]		Name
		Low salt	High salt	
1. Cytosol (Tris/EDTA)	None	8 S	3.8 S	Native
	+Ca(4 m*M*), KCl (400 m*M*)	4.5 S	4.5 S	Stable 4 S
2. Cytosol (Tris)	None	8 S	3.8 S	Native
	+Estradiol, warmed 25°	8 S + aggregates	5.3 S	Cytosol 5 S
	Ammonium sulfate precipitation, 0° (± estradiol)	8 S + aggregates	5.5 S	Low temp. 5 S
3. Nuclear[b] extract	None	8 S + aggregates	5.2 S	Nuclear 5 S

[a] Analysis on sucrose density gradients in: Low salt = 10 m*M* Tris, 10 m*M* KCl, 1 m*M* EDTA, pH 7.4; high salt = 10 m*M* Tris, 400 m*M* KCl, 1 m*M* EDTA, pH 7.4. Sedimentation coefficients based on the sedimentation of bovine albumin, 4.6 S.

[b] Extracts of uterine or endometrial nuclei with 10 m*M* Tris, 400 m*M* KCl, 1 m*M* EDTA, pH 7.4 buffer, after estrogen administration *in vivo*, to whole tissue *in vitro*, or to homogenates appear indistinguishable.

absence of EDTA.[6] Like the native cytosol complex from which it is derived, the 5 S complexes all give rise to an 8 S complex (as well as some aggregates) on sedimentation analysis in low ionic strength.

Although the most frequently cited differences among these receptor complexes depend on their sedimentation behavior, the various preparations of 5 S receptor complex can be differentiated from the native cytosol complex both in their interaction with nuclei and effects on nuclear RNA polymerase activity.[5-7] Insofar as they have been studied, all these forms of the estrogen receptor appear to possess about the same affinity for the estrogenic hormones.[8]

We will describe in detail the purification procedures for the stable 4 S receptor complex of the cytosol and the nuclear 5 S receptor complex

[6] E. R. DeSombre, S. Mohla, and E. V. Jensen, *Biochem. Biophys. Res. Commun.* **48**, 1601 (1972).

[7] E. V. Jensen, P. I. Brecher, M. Numata, S. Mohla, and E. R. DeSombre, *in* "Advances in Enzyme Regulation" (G. Weber, ed.), Vol. 11, p. 1. Pergamon Press, New York, 1973.

[8] E. V. Jensen and E. R. DeSombre, *Annu. Rev. Biochem.* **41**, 203 (1972).

and finally describe a method to obtain a 5 S complex from the cytosol, which appears to be similar to the nuclear 5 S complex. For the preparation of both types of receptor from cytosol, the procedures can be carried out either in the presence or absence of the hormone to obtain the receptor complexes or the naked receptor proteins, respectively. In the procedures described, tritiated estradiol is employed both for the convenience of following the hormone binding during the purifications and because the receptor complex is less labile than the naked receptor. If the procedure is carried out in the absence of the hormone, one of the several available methods for quantitating the amount of receptor capacity in the fractions[8] is employed. The preparation of the nuclear 5 S receptor complex requires addition of estrogen.

Methods designed to purify significant amounts of the estrogen-binding proteins (estrogen receptors) of estrogen target tissues must cope with the very small amounts present (several milligrams of receptor per kilogram of uterus). Initial attempts to purify an estrogen receptor involved affinity chromatography[9] using a column containing estradiol linked to cellulose by an azobenzyl grouping. As was subsequently observed by others employing estrogens linked to other adsorbents, these columns resulted in an apparent adsorption of the receptor from target tissue extracts but no receptor could be eluted in a form which was still able to bind the hormone.[10]

Because of the minute amounts of estrogen receptors in target tissue extracts, it is convenient to employ a concentration step early in the purification procedure. Unfortunately, many of the precipitation methods attempted with cytosol of target tissues have given rise to significant amounts of nonspecific aggregation (often irreversible) of the receptor with other proteins, hampering further purification procedures. Purification of the cytosol receptor of uterus was therefore facilitated by the observation[2] that treatment of uterine cytosol (prepared from an homogenate using a Tris-EDTA buffer) with calcium ions in the presence of salt yields a "stabilized" 4 S binding unit which does not aggregate or even revert to the native low-salt form (8 S) when salt is removed.

[9] P. W. Jungblut, R. I. Morrow, G. L. Reeder, and E. V. Jensen, *Abstr. 47th Meet. Endocrine Soc.* p. 56 (1965).

[10] However, antibodies to material eluted from the column in an inactive form (i.e., no longer able to bind the hormone) were able to precipitate receptor complexed with estradiol. P. W. Jungblut, I. Hatzel, E. R. DeSombre, and E. V. Jensen, *Colloq. Ges. Physiol. Chem.* **18,** 58 (1967). A recent paper reports the successful application of affinity chromatography to purification of the cytosol estrogen receptor [V. Sica, I. Parikh, E. Nola, G. A. Puca, and P. Cuatrecasas, *J. Biol. Chem.* **248,** 6543 (1973)].

Buffers and Reagents

TE, 10 m*M* Tris, 1.5 m*M* EDTA, pH 7.4
TKC, 100 m*M* Tris, 400 m*M* KCl, 1 m*M* $CaCl_2$, pH 7.4
$TK_{10}E$, 10 m*M* Tris, 10 m*M* KCl, 1 m*M* EDTA, pH 7.4
$TK_{30}E$, 10 m*M* Tris, 30 m*M* KCl, 1 m*M* EDTA, pH 7.4
$TK_{400}E$, 10 m*M* Tris, 400 m*M* KCl, 1 m*M* EDTA, pH 7.4 or 8.5
TK_{400}, 10 m*M* Tris, 400 m*M* KCl
$T_{100}K_{400}$, 100 m*M* Tris, 400 m*M* KCl
E*2, [6,7-^{3}H]Estradiol-17β, 57 or 5.7 Ci/mmole

Saturated ammonium sulfate, enzyme grade ammonium sulfate saturated at 2° and titrated to pH 7.2

$CaCl_2$, 200 m*M* $CaCl_2$, pH 7.4
3.5 *M* KCl, pH 7.4

Sucrose gradient solutions, high salt are 5–20% sucrose in $TK_{400}E$, pH 7.4 buffer, low salt are 10–30% sucrose in $TK_{10}E$, pH 7.4 buffer, each made either with automatic gradient former (Beckman Instruments) or manually layered in 5 steps of equal volumes and allowed to equilibrate in the cold

Disc Gel Electrophoresis

Acrylamide and *N,N'*-methylene bisacrylamide (Canalco, Inc.)
Tris-glycine, 3.0 g Tris, 14.4 g glycine, distilled H_2O to 1 liter, pH 9
Tracking dye, 0.005% bromphenol blue, aqueous
Amido Black, 0.5% in 7% acetic acid
Destaining solution, 7% acetic acid
Coomassie brilliant blue, 0.2%, in methanol : acetic acid : distilled H_2O (5:1:5)
Destaining solution, 7.5% acetic acid, 5% methanol

Stable 4 S Receptor of Cytosol

All procedures are carried out at 0–4°. The procedure described is conducted on a scale intended to yield purified receptor complex that can be accumulated for structure determination and antibody preparation. Since our usual experience indicates that calf uteri contain from 1 to 4 mg of receptor per kilogram of fresh uterus (based on a molecular weight of about 75,000 for the stable 4 S), it is necessary to process very large amounts of tissue for these purposes. The procedures can be scaled down with no significant changes evident,[2,3] with the advantage that in general a somewhat better purification is achieved with particle free cytosol[3] (high speed) more readily obtained on a small scale. The scale-limiting feature of the procedure described is the gel filtration step; using the sectional columns (available from Pharmacia Fine Chemicals, Inc.), this limitation can readily be alleviated by adding more sections (16 liters each) to increase the scale. Adequate separations on Sephadex G-200 have been obtained using sample (prepared as described) corresponding to 2 to 4% of the G-200 bed volume. The purification data (Table II) are based on the amount and specific activity of the original cytosol receptor.

Step 1. Cytosol Preparations

Fresh immature calf uteri (less than 20 g/uterus), stored in ice no more than 3 days, are trimmed free of connective tissue and minced through a meat grinder. The mince (2.5 kg) is divided into 125-g portions in 1-liter beakers. To each beaker, 500 ml TE buffer, pH 7.4, is added and the tissue homogenized (Polytron PT35 at setting 4 to 5) using five 10-second periods of homogenization, each followed by cooling periods in the ice bath. The cytosol is obtained after low speed centrifugation of the homogenate (30 minutes at 9500 g, GC-3 rotor, Sorvall).

Step 2. Ammonium Sulfate Precipitation

The cytosol (9620 ml) is divided into 4 portions of about 2500 ml each in 6-liter Ehrlenmeyer flasks and 10^{-5} M tritiated estradiol (E*2) added (2 to 3 ml/liter) to give a final concentration of 2 to 3×10^{-8} M. Aliquot samples of the cytosol containing E*2 are saved for sedimentation analysis of 8 S receptor capacity on low-salt sucrose density gradients and protein analysis.[11] After 15 minutes of incubation, 3.5 M KCl (400 ml/liter) and 200 mM $CaCl_2$ (20.4 ml/liter) are added during 15 minutes to give final concentrations of 1 M KCl and 4 mM $CaCl_2$; stirring is continued for an additional 45 minutes to insure maximal binding of estradiol to the receptor and maximal conversion of the receptor to the stable 4 S form. Saturated ammonium sulfate (250 ml/liter of cytosol mixture) is added, the first half fairly rapidly during 10 minutes, the last half more slowly during 20 minutes, bringing the suspension to 20% of saturation in ammonium sulfate. After stirring for an additional 30 minutes, the suspension is centrifuged 20 minutes at 9500 g and the supernatant carefully decanted and drained. Subsequent portions of the ammonium sulfate precipitates are collected by centrifugation in the same

[11] O. H. Lowry, N. J. Rosebrough, A. L. Farr, and R. J. Randall, *J. Biol. Chem.* **193**, 265 (1951).

TABLE II

PURIFICATION OF CYTOSOL STABLE 4 S RECEPTOR COMPLEX

	ml[a]	Total DPM[b]	Total protein (mg)	Specific activity[b] (DPM/mg protein)	Recovery % ^3H	Purification factor
Cytosol	9620	164.9×10^8 [c]	88,023	187,000[c]	100.0	—
Ammonium sulfate precipitation	960	107.5×10^8	4,032	2,670,000	65.2	14.3
G-200 Pool	9559	31.0×10^8	459	6,760,000	18.8	36.2
DEAE Pool	82.2	7.92×10^8	14.7	53,960,000	4.8	288
Disc electrophoresis	—	—	—	>80,000,000[d]	2.4[d]	>425

[a] Corrected for material used for assay or for other reasons not carried on to subsequent steps.

[b] Specific activity based on tritiated estradiol, sp. act. 57 Ci/mmole. On large scale preparation the specific activity is usually diluted 10-fold to reduce radiochemical decomposition in the highly purified product.

[c] Based on the total cytosol receptor found by sedimentation analysis. The total DPM added to the cytosol is generally 2 to 3 times the receptor capacity. The receptor capacity varies considerably from one preparation to another. This figure would correspond to a high capacity, equivalent to about 10 mg of receptor for the 2.5 kg of uterus. Most capacities correspond to 1–3 mg receptor/kg of uterus.

[d] At least two-thirds of the staining protein is removed by this step with a ^3H recovery of about 50%.

bottles to reduce the number of precipitates to be resuspended. After the last centrifugation, the final portions of supernatant fluid are removed by pipet. Using a total volume of TKC buffer equal to 10% of the original cytosol volume, these precipitates are resuspended, initially by Pasteur pipet, finally by gentle rehomogenization, and the suspension magnetically stirred for 60 minutes. The resuspended 0–20% ammonium sulfate precipitate is then clarified by centrifugation for 60 minutes at 78,000–90,000 *g*.

Step 3. Gel Filtration on Sephadex G-200

The clarified, redissolved ammonium sulfate precipitate is applied by upward flow to a sectional column (three 16-liter sections, Pharmacia Fine Chemicals, Inc.) containing Sephadex G-200 in $TK_{30}E$ buffer and elution conducted by pumped upward flow with $TK_{30}E$ buffer at a flow rate of about 4 liters/hour. The fractions, about 900 ml each, are collected and assayed for tritium and absorbance at 280 nm. Almost all the 280 nm absorbing material, excluded from the gel, elutes at the void volume of the column, tailing somewhat, but also containing some of the tritiated estradiol (as irreversibly aggregated receptor).[12] The stable 4 S receptor complex is included in the gel and elutes as a symmetrical peak of tritium with no associated peak of protein. Although the K_{av} ($K_{av} = V_E - V_0 / V_T - V_0$, where V_T is the total volume of the column, V_0 is the void volume, and V_E is the elution volume of the receptor complex) of smaller G-200 columns, was reproducibly 0.5,[2,3] with the sectional columns and their more rapid elution, the K_{av} is reduced somewhat, but the stable 4 S complex is still adequately separated from the peak of proteins in the void volume. The fractions comprising the stable 4 S receptor complex (usually about 10) are combined to give the G-200 pool (9 to 10 liters volume).

Step 4. DEAE-Cellulose Chromatography

A column of DEAE-cellulose (Whatman DE 52, 200 ml bed volume) is prepared in $TK_{30}E$ buffer. With those batches of DEAE-cellulose which give rise to 260 nm absorbing impurity on elution with 0.5 *M* KCl, the

[12] It is not unequivocally established as to how stable the macromolecular factor required for stable 4 S formation is on storage of whole uteri at 0° for the several days needed to accumulate 2 to 3 kg. There is suggestive evidence that the stability of this factor under these storage conditions may be less than the stability of the receptor itself. This would be consistent with larger amounts of V_0 tritium on large scale preparations using 0° stored uteri.

column is washed with 1 *M* KCl until no more 260 nm impurity is seen, then washed with large quantities of $TK_{30}E$ buffer. The pool from the G-200 column is then loaded onto the DEAE-cellulose column at a flow rate of 150–200 ml/hour by gravity flow. During the loading, some 20–25% of the tritium is lost in the effluent. However, our experience with large scale membrane concentration of G-200 pools prior to loading on DEAE-cellulose shows even greater losses on concentration (due mainly to membrane build-up of protein and tritium) than are found on direct loading of the G-200 pool. After loading, the DEAE-cellulose column is washed with an additional 2 column volumes of $TK_{30}E$ buffer and then the elution is effected with a 600-ml linear gradient of KCl, 30 to 400 m*M*, in 10 m*M* Tris, 1 m*M* EDTA, pH 7.4 buffer. Depending on the amount of protein applied and the size of the column, the stable 4 S receptor complex elutes with 0.1 to 0.15 *M* KCl, along with the smaller of the 2 peaks of 280 nm absorbance, tailing somewhat, but clearly separated from the larger peak of 280 nm absorbance which elutes with 0.2 to 0.3 *M* KCl.[3]

Step 5. Acrylamide Disc Gel Electrophoresis[13]

Samples of DEAE-cellulose purified receptor complex containing 100–200 μg of protein are made to about 7% sucrose (w/v) and layered on 12 identical gel columns containing 1.0 ml separating gel (7% acrylamide) and 0.3 ml stacking gel.[14,15] In most cases, 1.0 ml of receptor solution is used in 6 in. gel tubes available for the analytical disc electrophoresis apparatus (Canalco, Inc.). The buffer (Tris-glycine, pH 9) is carefully layered on top of the sample in sucrose and the tubes placed in the electrophoresis apparatus with ice water cooling. The electrophoresis is conducted in the cold room at 2.5 mA/gel until the tracking dye has reached the bottom of each gel. When the electrophoresis is completed for a gel, the tube is removed from the apparatus, the gel is removed from the gel tube, and the separating gel is sliced into 32–33 slices (about 1.6 mm each, Canalco slicer) except for one gel which is stained with amido black for 1 hour and destained electrophoretically to indicate the

[13] Despite the good recoveries (70–85% of the total tritium, most as receptor complex) found on extraction of purified receptor complex from slices of analytical acrylamide gels, we have been uniformly unsuccessful in the use of preparative acrylamide disc gel electrophoresis apparatus, used either by elution or by extracting sliced preparative gels. Therefore, the procedure used, as outlined here, employs large numbers of analytical gel columns, used preparatively.

[14] L. Ornstein, *Ann. N.Y. Acad. Sci.* **121,** 321 (1964).

[15] B. J. Davis, *Ann. N.Y. Acad. Sci.* **121,** 404 (1964).

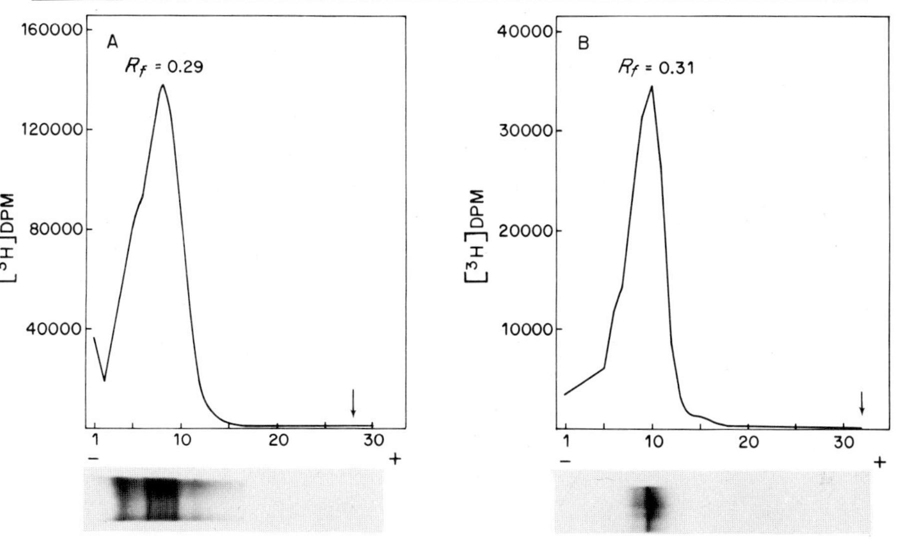

FIG. 1. Disc electrophoretic purification of stable 4 S receptor complex of cytosol (A) Elution of like slices (about 1.6 mm) of ten separating gels with 1.0 ml $TK_{30}E$ buffer for 72 hours after electrophoresis of DEAE-cellulose purified receptor complex 148 μg protein/gel) at pH 8.8, 2.5 mA/gel in the cold. Tritium elution pattern corresponds to the DPM for one gel. The R_f is calculated from the migration of the tritium peak relative to the tracking dye (arrow). Migration is from left to right. Stained gel at the bottom is from an additional gel run at the same time, stained with amido black and destained electrophoretically. (B) Rerun of a portion of the pooled tritium peak fractions of A under the same conditions. DPM determined after combustion of the individual slices of one gel. Gel stained with amido black.

protein bands. Like slices from the 11 sliced gels are combined and eluted for 72 hours in the cold with $TK_{30}E$ buffer (0.1 ml buffer/slice). After 72 hours of elution, the buffer is removed from the slices and assayed for tritium (Fig. 1A). The 3 fractions corresponding to the peak of tritium are combined (generally consisting of 40–60% of the total tritium applied to the gels) and analyzed by sedimentation analysis and analytical disc gel electrophoresis to establish the integrity of the purified stable 4 S receptor complex (Fig. 1B).

Nuclear 5 S Receptor

Introduction

As detailed elsewhere[5] one can readily convert the native cytosol–receptor complex (3.8 S in high salt) into the 5 S complex by simply warming the cytosol–estradiol mixture. In contrast to the native cytosol 4 S complex, this 5 S complex is readily taken up by purified nuclei as well as by lysed nuclei, and after extraction from nuclei it appears identical to the nuclear receptor obtained on administration of the hormone *in vivo.* In the procedure described, nuclear sediment, preextracted with 0.4 *M* KCl buffer, is used as a specific adsorbent for the cytosol 5 S receptor from which it is then solubilized with smaller amounts of total protein by extraction with 0.4 *M* KCl buffer. The purification data presented are based on the amount and specific activity of the original cytosol receptor (Table III).

Procedure. All procedures are carried out at 0–4° unless otherwise mentioned.

Step 1. Preparation of Nuclear Extract

Fresh immature calf uteri (1.6 kg) are finely minced through a meat grinder. After addition of 6400 ml of 10 m*M* Tris buffer, pH 7.4, and homogenization (125 g portions, each with 500 ml buffer using Polytron PT35), the cytosol and crude nuclear sediment are obtained on centrifugal separation (20 minutes at 9500 *g*). Tritiated estradiol is added to the cytosol at a final concentration of 10 n*M* followed by incubation for 60 minutes. Meanwhile, one-half of the sediment is rehomogenized with 3200 ml TK_{400}, pH 8.5 buffer, incubated 20 minutes, and collected by 20 minutes centrifugation at 9500 *g*, then washed with 3200 ml of 10 m*M* Tris buffer, pH 7.4, to reduce the salt concentration for the uptake. This washed, preextracted nuclear sediment is combined with the cytosol–estradiol mixture, gently rehomogenized (PT35 at setting 1), and incubated for 60 minutes at 25° with magnetic stirring. Following cooling, the mixture is centrifuged 20 minutes at 9500 *g* and the recovered sediment is washed twice (3200 ml 10 m*M* Tris buffer, pH 7.4) by resuspension and centrifugation. The nuclear extract is prepared by homogenization of the washed sediment with 3200 ml, TK_{400} buffer, pH 8.5, followed by incubation for 60 minutes and centrifugation for 60 minutes at 9500 *g*.

Step 2. Ammonium Sulfate Precipitation

To the nuclear extract (3200 ml), saturated ammonium sulfate (1370 ml) is added during 30 minutes to bring the final ammonium sulfate concentration to 30% of saturation. After stirring for an additional 30 minutes, the precipitate is collected by centrifugation for 25 minutes at 9500 *g* and resuspended in 320 ml of $T_{100}K_{400}$ buffer, pH 8.5, by gentle homog-

TABLE III
PURIFICATION OF UTERINE NUCLEAR RECEPTOR COMPLEX

	ml[a]	Total DPM[b]	Total protein (mg)	Specific activity[b] (DPM/mg protein)	Recovery % ^3H	Purification factor
A. Typical Purification—½ Nuclear Sediment						
Cytosol	6380	263×10^7 [c]	52,316	70,650[c]	100	—
Nuclear extract	3200	159×10^7	4,832	329,000	60.5	4.6
Ammonium sulfate precipitation	320	65.7×10^7	586	1,120,000	24.9	15.9
G-200 pool conc.	46	23.9×10^7	123.8	1,930,000	9.1	27.3
G-25 pool	70	23.8×10^7	111.6	2,130,000	9.0	30.2
DEAE pool conc.	29.4	9.9×10^7	13.23	7,495,000	6.2	106.1
Disc electrophoresis	—	—	—	>80,000,000[d]	—	>1000[d]
B. Modified Procedure—⅕ Nuclear Sediment						
Cytosol	2630	146×10^7	21,250	68,900[b]	100	—
Nuclear extract						
Low speed	1550	102×10^7	1,395	738,000	69.8	10.7
High speed	1550	96×10^7	698	1,380,000	65.6	20.0
Ammonium sulfate precipitation	261	47×10^7	112	4,180,000	32.1	60.7
G-200 pool conc.	74.5	10.8×10^7	11.9	9,070,000	7.4	131.6

[a] Corrected for material used for assay or for other reasons not carried on to subsequent steps.

[b] Specific activity based on tritiated estradiol, sp. act 57 Ci/mmole. On large scale preparations the specific activity is usually diluted 10-fold to reduce radiochemical decomposition in the highly purified product.

[c] Based on the total cytosol receptor found by sedimentation analysis. The total DPM added to the cytosol is generally about twice this amount.

[d] There is generally not enough protein obtained from a single preparation to determine protein concentration. By the amounts of staining impurities it is apparent that at least 90% of the protein stain occurs outside the tritium peak. If one estradiol is bound per receptor, total purity would correspond to a specific activity of 1.7×10^9 DPM/mg (for M. W. of 75,000).

enization. The resuspended precipitate is stirred for 60 minutes and clarified by centrifugation for 60 minutes at 78,000 g.

Step 3. Gel Filtration on Sephadex G-200

A 6800-ml (bed volume) column of Sephadex G-200 (K100/100 Column, Pharmacia Fine Chemicals, Inc.) is prepared and washed in $TK_{400}E$ buffer, pH 7.4. The reconstituted ammonium sulfate precipitate of the nuclear extract (310 ml) is applied by upward flow and eluted with the column buffer, flow rate 100–150 ml/hour at 15 cm pressure head. The elution profile indicates that although some of the bound tritium elutes in the void volume along with the major part of the contaminating protein, the major product elutes ($K_{av} = 0.13$) with smaller amounts of contaminating protein. Its elution position however suggests an unusually large Stokes radius for the receptor complex which in the same medium (400 mM KCl) shows a sedimentation coefficient of about 5 S. This proves to be of considerable advantage in the final stage of the purification (cf. Step 5).

Step 4. DEAE-Cellulose Fractionation

Since the nuclear receptor complex is not adsorbed on DEAE-cellulose in 0.4 M KCl at pH 7.4, the elution medium of the G-200 column, it is necessary to desalt the product from Step 3. In our experience, very poor recovery of the receptor complex occurs when the desalting is performed on dilute solutions of the receptor (such as seen with G-200 eluates). Therefore, the pooled G-200 included peak fractions are first concentrated about 25-fold by ultrafiltration using an XM-50 membrane (Amicon, Inc.) followed by desalting by passage through Sephadex G-25 in 10 mM Tris buffer, pH 7.4. This sample (80 mg protein) is applied to a 95-ml column of DEAE-cellulose (Whatman DE-52) packed in $TK_{30}E$ buffer. After washing the column with 300 ml $TK_{30}E$ buffer, pH 7.4, elution is performed with a 400-ml gradient of KCl (30–400 mM) in 10 mM Tris, 1 mM EDTA, pH 7.4 buffer. The peak of receptor complexed with tritiated estradiol elutes in the range of 0.2 to 0.3 M KCl somewhat ahead of, but not entirely separated from, the 280 nm O.D. peak. Although the purification obtained by the DEAE-cellulose chromatography is variable and generally low because the tritium peak is never entirely separated from the associated protein peak, it nonetheless provides the nuclear receptor complex in a form which can more readily move into the 7% acrylamide gels for final purification. Gel electrophoresis of this receptor complex, not purified by DEAE-cellulose chromatog-

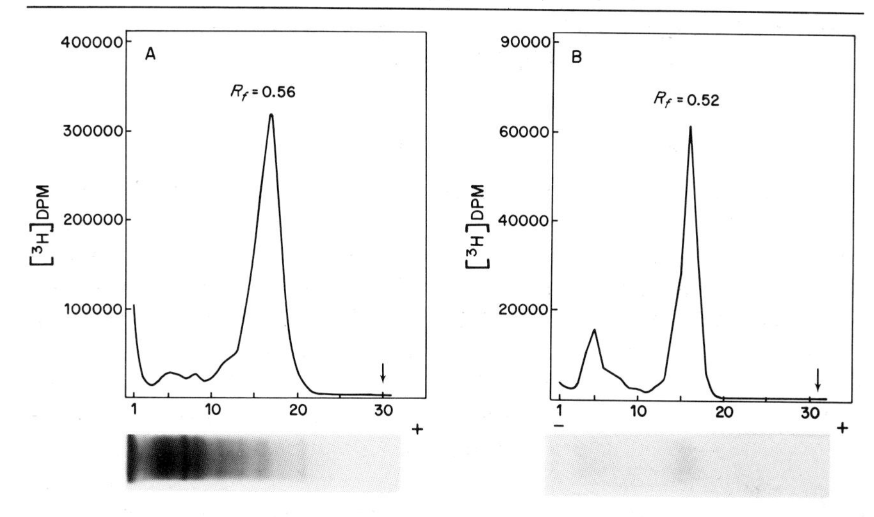

FIG. 2. Disc electrophoretic purification of nuclear 5 S complex. (A) Conditions as described for Fig. 1 using DEAE-cellulose purified nuclear 5 S complex (81 μg protein/gel). Gel at bottom stained with Coomassie blue since amido black stain does not detect any bands at position of tritium peak. (B) Rerun of a portion of the pooled tritium peak fractions of A under the same conditions. DPM determined after combustion of slices of one gel. Gel stained with Coomassie blue.

raphy, results in major portions of receptor complex which do not enter the separating gel.[16]

Step 5. Acrylamide Disc Gel Electrophoresis

For the reasons outlined,[13] we use the analytical size gels in a preparative fashion also for the final purification of the nuclear receptor. The DEAE-cellulose purified receptor complex (<200 μg) is made to 7% (w/v) with sucrose and layered on top of the stacking gel of 12 analytical acrylamide gels. The preparation, electrophoresis, slicing, and extraction of the gels is identical to the procedure outlined for the stable 4 S receptor purification. However, the 5 S receptor complex migrates with an $R_f \cong 0.5$, considerably faster than the protein-staining bands (Fig. 2). In fact, with the less sensitive amido black stain, no protein stain is evident in the region of the tritium peak. A very faintly staining band at the tritium position is picked up with the more sensitive Coomassie blue

[16] *Note added in proof:* Recently it has been found that by eliminating the stacking gel and conducting the electrophoresis in Tris-borate buffer, pH 9.2, the desalted, concentrated Sephadex G-200 pool can be purified directly by polyacrylamide gel electrophoresis.

stain. From the stain intensities, it is evident that more than 90% of the staining proteins are removed by this purification step. It appears probable that this high degree of purification results from contamination with proteins of very large molecular dimensions eluted from the G-200 (see Step 3).

Optional Modifications of Nuclear Receptor Procedure

Although the method for the medium scale purification described above gives quite reproducible yields and purification, several modifications may be of benefit (Table IIIB), especially when the procedure is conducted on a smaller scale. The first of these is the use of a smaller ratio of preextracted sediment to cytosol. The results shown in Table IIIB are from a preparation using all the cytosol from the 670 g of uterine mince homogenized with 2680 ml 10 m*M* Tris buffer but incubating with only one-fifth of the preextracted sediment from this amount of tissue. Although somewhat less receptor is taken up by this sediment, good yields of nuclear receptor are nonetheless accomplished by increasing the volume of extraction buffer (using 8 ml TK_{400} for the sediment of each gram of tissue, rather than 4 ml as described in the general procedure), resulting in an approximately twofold improvement in the purification at this point over that seen in Table IIIA.

The second improvement which is especially advantageous when working on a feasible scale is to further clarify the nuclear extract at high *g* force (1 hour at 78,000 *g*). This step removes up to one-half of the protein with only small losses of receptor complex. As can be seen in the table, the effectiveness of the gel filtration step with the more purified receptor is, if anything, somewhat better than with the less purified receptor. However, it has been found that the DEAE-cellulose chromatography step is still required with this more highly purified nuclear receptor complex since without it the complex does not readily migrate into acrylamide gels during disc electrophoresis.[16]

Cytosol 5 S Receptor Purification

As has been recently reported,[6] a partially purified 5 S form of the cytosol receptor can be obtained under appropriate conditions from the cytosol itself. Like the stable 4 S, this form of the receptor can be prepared even in the absence of estradiol. Although it is not yet known how this receptor (5 S in high salt) is obtained apparently from the native cytosol receptor (4 S in high salt), this 5 S receptor complex is similar in its uptake by nuclei to the 5 S complex of cytosol produced from the cytosol 4 S complex on warming with the hormone. For the preparation of this partially purified 5 S receptor it is preferable to homogenize the uterine tissue in buffer containing no EDTA to circumvent the possibility of any stable 4 S formation which can occur in cytosol from homogenates prepared with EDTA.

Step 1. Cytosol Preparation

Uterine mince (738 g) is homogenized with 10 m*M* Tris, pH 7.4 buffer (2950 ml), and the cytosol is obtained after low speed centrifugation (9500 *g*, 1 hour). Sufficient tritiated estradiol is added to the cytosol to provide a final concentration of 2×10^{-8} *M* E*2, followed by incubation for 1 hour in an ice bath.

Step 2. Ammonium Sulfate Precipitation

Saturated ammonium sulfate, 2°, pH 7.2 (1245 ml), is added to the cytosol during 30 minutes, bringing the final concentration to 30% of saturation. After an additional 30 minutes of stirring, the precipitate is collected by centrifugation (20 minutes, 9500 *g*). The precipitate is homogenized with $T_{100}K_{400}$, pH 7.4 buffer (290 ml), stirred for 1 hour to redissolve the precipitate, and the resulting suspension clarified by centrifugation (60 minutes, 78,000 *g*).

Step 3. Gel Filtration on Sephadex G-200

The reconstituted ammonium sulfate precipitate (270 ml) is applied to a Sephadex G-200 column (6800 ml) packed in $TK_{400}E$ buffer, pH 7.4, and eluted by upward flow with the same buffer. The elution profile shows the tritium peak eluting (K_{av} = .123) just after the peak of excluded proteins (V_0), indicating that like the behavior of the nuclear receptor, this cytosol–receptor complex elutes in a region of large Stokes radius unusual for a protein which sediments as 5 S on sedimentation analysis in the same medium (400 m*M* KCl). By the procedure described, a modest degree of purification is obtained (Table IV). We are currently pursuing further steps in the purification of this receptor.

Sedimentation Behavior of Purified Receptor Complexes

The sedimentation coefficient (4.5 S) of the stable 4 S form of the receptor in both high and low salt sucrose gradient analysis is essentially

TABLE IV
PURIFICATION OF CYTOSOL 5 S RECEPTOR COMPLEX

	ml[a]	Total DPM[b]	Total protein (mg)	Specific activity[b] (DPM/mg protein)	Recovery % ^3H	Purification factor
Cytosol	2900	25.8×10^8 [c]	27,550	93,700[c]	100	—
Ammonium sulfate precipitation	290	18.2×10^8	2,639	691,000	70.7	7.4
G-200 pool	1118	14.8×10^8	675	2,200,000	57.4	23.5

[a] Corrected for material used for assay or for other reasons not carried on to subsequent steps.

[b] Specific activity based on tritiated estradiol, sp. act 57 Ci/mmole. On large scale preparations the specific activity is usually diluted 10-fold to reduce radiochemical decomposition in the highly purified product.

[c] Based on the total cytosol receptor found by sedimentation analysis. The total DPM added to the cytosol is generally about twice this amount.

unchanged throughout the large degree of purification obtained, as described. It is also the same if characterized directly in the cytosol.[3]

The sedimentation behavior of the nuclear receptor, on the other hand, changes characteristically during its purification. The initial complex obtained on extraction from nuclei, as well as the complex purified by salt precipitation, sediments as 5 S complex on high salt sedimentation analysis but appears as 8 S complex with some aggregates on low salt analysis. However, after purification by gel filtration, the nuclear receptor sediments as a 5 S complex in either low salt or high salt sedimentation analysis. Therefore, it is likely that either a change has taken place in the nuclear receptor itself or, more probably, some nonreceptor proteins responsible for the aggregation to 8 S in low salt have been removed during this step. With subsequent steps in the purification this behavior is retained. While the degree of separation between sedimentation peaks of nuclear receptor (5 S) and albumin (4.6 S) decreases somewhat with the more highly purified receptor, the two can still be differentiated by sedimentation analysis.[17] Actual assignment of sedimentation coefficients must wait analysis with the analytical ultracentrifuge, work now in progress.

Molecular Parameters of Receptor Complexes

Although we have reported[2] the isoelectric point (by isoelectric focusing) of the stable 4 S receptor complex to be 6.4, we have awaited preparation of pure receptor complexes, which are now at hand, to obtain reliable physical parameters for the receptor complexes. However, as is apparent from comparison of Figs. 1 and 2, the purified stable 4 S and nuclear 5 S have retained characteristic differences in electrophoretic mobility as well as sedimentation behavior even after the high degree of purification of each.

Acknowledgments

Various parts of these investigations were supported by Grant 690-0109 from the Ford Foundation, Grant P-422 from the American Cancer Society, U.S. Public Health Service Grant CA-02897 from the National Cancer Institute, and Contract NIH-69-2108 from the National Institute of Child Health and Human Development. E.R.D is supported by Research Career Development Award HD-46,249 from the NICHD.

[17] E. V. Jensen, S. Mohla, T. Gorell, S. Tanaka, and E. R. DeSombre, *J. Steroid Biochem.* **3**, 445 (1972).

[32] Methods for the Purification of Androgen Receptors

By W. I. P. MAINWARING and R. IRVING

The purification of the receptors for androgenic steroid hormones, as for all classes of steroid–receptor complexes, is a demanding undertaking. Not only are the androgen receptors labile proteins present in only minute amounts in androgen-sensitive cells but end organ metabolism of the naturally secreted androgen, testosterone, can play a unique and integral part in the mechanism of action in certain tissues. Accordingly, the most appropriate or maximally bound androgen must be unequivocally established in preliminary experiments before the purification of a putative androgen receptor complex is undertaken. The active androgen is almost invariably 5α-dihydrotestosterone (17β-hydroxy-5α-androstan-3-one; termed simply dihydrotestosterone throughout) in the male accessory sexual glands[1] but nonmetabolized testosterone itself may be the active androgen in certain tissues such as mouse kidney[2] and rat levator ani muscle.[3] Using the accepted definition of a steroid receptor as a specific, high affinity steroid-binding protein but in the sense of tissue and steroid specificity, a method will be described for the purification of androgen receptors which was originally developed for the receptors in the rat ventral prostate gland. Recent studies in our laboratory have shown that the methods are equally applicable to other male accessory sexual glands and thus possibly applicable to androgen receptors in general. It should be borne in mind that until the precise function of the receptors is established such binding proteins may only be identified by their ability to bind tritiated steroids with high affinity and their ability to transfer labeled steroids into chromatin, *in vitro*.[4] Furthermore, after extensive purification, the labeled receptor complexes can only be routinely located by scintillation spectrometry rather than analysis for protein.

Materials

1. Preparative Media. 1 *M* Dithiothreitol is stored in small lots at —20° and added to media just prior to use. Tris base should be at least

[1] N. Bruchovsky and J. D. Wilson, *J. Biol. Chem.* **243,** 2012 (1968).
[2] L. P. Bullock, C. W. Bardin, and S. Ohno, *Biochem. Biophys. Res. Commun.* **44,** 1537 (1971).
[3] I. Jung and E.-E. Baulieu, *Nature (London), New Biol.* **237,** 24 (1972).
[4] W. I. P. Mainwaring and B. M. Peterken, *Biochem. J.* **125,** 285 (1971).

99.5% pure and glass-distilled water is used throughout. 1 *N* HCl should not contain $HgCl_2$ commonly added to commercially available volumetric standards but made up by dilution of concentrated HCl (85.5 ml of HCl to 1 liter of water). Medium A consists of 50 m*M* Tris-HCl buffer, pH 7.4, containing 0.25 m*M* EDTA (disodium salt) and 0.25 m*M* dithiothreitol; medium B additionally contains 10% (v/v) glycerol.

2. Ammonium Sulfate. This is either recrystallized from 5 m*M* EDTA or enzyme grade (Schwartz-Mann; low in metals) is equally suitable. A saturated solution (760 g/liter at 0°) is adjusted to pH 7.4 by the addition of 1 *N* NH_4OH.

3. Sephadex. Medium grade G-25 and G-50 gel beads (Pharmacia; 30 g) are added with magnetic stirring to 1 liter of water and placed on a boiling water bath for 3 hours (G-25) and 4.5 hours (G-50). After cooling, fines are decanted and the swollen beads are collected on a Büchner head, resuspended in either medium B or 5% (w/v) sucrose containing 0.2% NaN_3 as preservative, and stored in tightly closed polyethylene bottles in a refrigerator. An appropriate volume of gel is deaerated *in vacuo* prior to use. After 1 cm of settled matrix has collected in the column under gravity with the tap closed, the remaining gel is packed under maximum rates of flow.

4. DNA-Cellulose. Several methods are available for coupling DNA to an insoluble matrix,[5,6] but the procedure of Alberts[7] is recommended because it is a direct and highly reproducible procedure that requires no special equipment except a freeze dryer. A solution of highly polymerized calf thymus DNA (Sigma; 2.5–3.0 mg/ml) is prepared by stirring overnight in medium C (10 m*M* Tris-HCl buffer, pH 7.4, containing 1 m*M* EDTA). To each gram of Macherey-Nagel cellulose (type 2200ff) is added 3.0 ml of DNA, and after thorough mixing the mixture is allowed to stand for 3 hours at ambient temperature. DNA-cellulose is conveniently prepared in 10- or 20-g batches. After dilution with 4 volumes of medium C, the mixture is frozen as the thinnest possible film in a round-bottomed flask (minimum volume, 500 ml) in a solid CO_2–acetone freezing mixture and freeze dried for at least 24 hours. The DNA-cellulose is resuspended in 100 ml of medium C, washed three times on a Büchner head with this volume of medium, resuspended as 1 g original dry weight of cellulose per 3 ml of medium C, and stored in 3-ml lots at —20° for up to 3 months without deleterious change. Small samples of DNA-cellulose (approx. 0.1 g) are heated in 1.0 ml of 1 *N* $HClO_4$

[5] R. Litmann, *J. Biol. Chem.* **243**, 6222 (1968).

[6] M. R. Poonian, A. J. Schlabach, and A. Weissbach, *Biochemistry* **10**, 424 (1971).

[7] B. M. Alberts, F. J. Amodio, M. Jenkins, E. D. Gutmann, and F. L. Ferris, *Cold Spring Harbor Symp. Quant. Biol.* **33**, 289 (1968).

at 75° for 20 minutes, together with DNA standards and controls containing an identical weight of cellulose. DNA is determined in the acid-soluble fraction after centrifugation either by spectrometry at 260 nm or by colorimetry.[8] Each gram of DNA-cellulose contains 400–440 μg of adsorbed DNA.

Special Methods

1. Protein Determination. At early stages of purification, protein may be determined by the conventional Folin-Ciocalteau reagent[9] but more sensitive methods must be adopted with highly purified preparations. We recommend a fluorimetric procedure.[10] Bovine serum albumin serves as reference for both procedures.

2. Polyacrylamide Electrophoresis. This is conducted by the standard procedure[11] in 7.5% (w/v) acrylamide gels, but the maintenance of the receptor complexes necessitates the inclusion of 15 mM 2-mercaptoethanol and 10% glycerol in the analytical gels and the electrode buffers. The SH-protecting agent impedes gel formation at ambient temperatures but polymerization may be readily accomplished by setting the tubes in a 37° bath for 20 minutes. The stacking gels are prepared by the usual photochemical method.[11]

3. Isoelectric Focusing. In our experience, this procedure is often difficult to accomplish satisfactorily in conventional apparatus, volume 110 ml or larger, either because of excessive dilution of the labile complexes or local heating effects. However, focusing may be reproducibly performed in 14- or 24-ml columns made to the design of Osterman.[12] These may be readily made in the standard scientific workshop. A particular feature of the columns is the ease with which they may be cooled during focusing. Using a 14-ml column, the lower (anode) electrolyte is 1.0 ml of 0.1% (v/v) H_2SO_4 in 60% (w/v) sucrose, and this is overlaid with a linear 10 ml 10–50% sucrose gradient, containing 0.1% (v/v) of pH range 5–8 ampholytes (LKB) throughout. The gradient is most conveniently made with a mechanical gradient former or, if made manually, by a successive series of five 2.0-ml layers, decreasing in 10% sucrose per layer from 50% sucrose. Each layer contains 0.1% ampholytes. Sev-

[8] K. Burton, *Biochem. J.* **62**, 315 (1956).
[9] O. H. Lowry, N. J. Rosebrough, A. L. Farr, and R. J. Randall, *J. Biol. Chem.* **193**, 265 (1951).
[10] T. Hiraoka and D. Glick, *Anal. Biochem.* **5**, 497 (1963).
[11] L. Ornstein, *Ann. N.Y. Acad. Sci.* **121**, 321 (1964).
[12] L. Osterman, *Sci. Tools* **17**, 31 (1970).

eral columns may be run concomitantly from one high voltage power supply.

Application of the Methods of Purification

The usefulness of the methods will be illustrated by the partial purification of the cytoplasmic 8 S androgen receptor complex and the nuclear androgen receptor complex from rat ventral prostate gland, 36 hours after bilateral orchidectomy. The cytoplasmic receptor is distinguished by its sedimentation coefficient rather than other physicochemical properties since this was how it was originally described.[13] Successful application of the methods depends on the performance of all steps at temperatures as close to 0° as possible; all apparatus must be thoroughly chilled before use and the entire purification must mandatorily be performed in a cold room maintained at 0–2°. Both isolations to be described may be completed within 36 hours, including the performance of isoelectric focusing overnight.

Cytoplasmic 8 S Androgen Receptor Complex

Step a. Preparation of Tissue Extracts. Pooled prostate glands (4.0–4.5 g wet weight) from 18 animals are thoroughly chilled in ice-cold medium A, minced finely with curved scissors and homogenized in 18.0 ml of medium A, using a loose-fitting, motor-driven Potter-Elvehjem homogenizer. Excessive shearing, ensuing from the use of homogenizers with less than 1 mm clearance, leads to a marked reduction in the yield of receptor complex. After centrifugation at 100000 g for 1 hour at 2°, the clear supernatant is aspirated to within 5 mm of the sediment of particulate material and also avoiding the floating layer of lipids. After gentle mixing with one-tenth volume of glycerol in an ice-bath, [^{3}H]dihydrotestosterone (approx. 40 Ci/mmole; Amersham-Searle or New England Nuclear) is added to a concentration of 5 nM; the [^{3}H]steroid is added in 200 μl of 1,2-propanediol and after mixing the labeled extract is allowed to stand undisturbed for 2 hours in an ice-bath. Since other proteins (approx. 4 S or Stokes radius 40–46 Å) are also labeled under these conditions,[13] a small sample (0.25 or 0.50 ml) is chromatographed on a 2 × 20-cm column of Sephadex G-200, equilibrated with medium B, to assess the initial labeling of the 8 S (Stokes radius 80–84 Å) receptor–dihydrotestosterone complex.[13]

Step b. $(NH_4)_2SO_4$ Fractionation. The labeled 8 S complex is selectively precipitated together with the removal of the majority of free

[13] W. I. P. Mainwaring, *J. Endocrinol.* **45**, 531 (1969).

[^{3}H]dihydrotestosterone and the 4 S protein–dihydrotestosterone complexes.[4] With slow magnetic stirring in an ice-bath, one-half volume of saturated $(NH_4)_2SO_4$ is added dropwise over a 5-minute period and the extract is then left undisturbed for 10 minutes. After centrifugation at 5000 *g* for 5 minutes, the sediment is taken up in 4.0 ml of medium B.

Step c. DNA-Cellulose Chromatography. The 8 S receptor complex is extensively purified at this stage and any residual 4 S complexes are totally eliminated; the latter, like the complexes of [^{3}H]dihydrotestosterone with serum proteins, do not bind to DNA. The sample is applied to 2 g of DNA-cellulose (approx. 800 μg of DNA) in a column of 1-cm diameter and previously equilibrated with 25 ml of medium B. The columns are run under gravity alone without the application of external pressure. If the rate of flow declines, the upper surface of the absorbent should be gently stirred with a fine glass rod. Fractions of 1.0 ml are collected and sufficient medium B (usually 12–14 ml) is run through the column to elute nonretained radioactivity (20 μl of each fraction are counted). When background levels of tritium are reached, the column is eluted with medium B containing 0.5 *M* KCl. The labeled 8 S complex is essentially recovered in two fractions.

Step d. Desalting. This essential step prior to isoelectric focusing presents the principal hazard in the entire procedure and must be accomplished rapidly. There are always significant losses during this step. The 1 × 20-cm columns of Sephadex G-25 are calibrated with Dextran Blue 2000 (Pharmacia) and further equilibrated with 5% (w/v) sucrose. The sample is applied and elution continued with 5% sucrose. The 8 S complex is recovered precisely at the void volume of the columns, generally in two fractions, volume 1.0 ml each (20 μl samples from a series of 1.0-ml fractions approaching the void volume are counted). Conventional dialysis cannot be substituted for the recommended procedure.

Step e. Isoelectric Focusing. Further purification is accomplished by this procedure. While earlier steps are in progress, the focusing apparatus is cooled by the rapid circulation of ice-cold water. The labeled extract is layered over the preformed sucrose gradient, and after the application of an initial voltage of 300 V for 1 hour, focusing is conducted overnight at 800 V with the anode connected to the base of the apparatus. Circulation of ice-cold water is maintained throughout the entire procedure. Fractions (0.50 ml) are collected by way of the tap at the base of the apparatus, and after the measurement of the pH of individual fractions at 0–2°, using a glass electrode calibrated against potassium hydrogen phthalate, 20-μl samples are counted. The 8 S receptor complex has a pI of 5.8.

The result of a typical purification is presented in Table I. Major

TABLE I
PARTIAL PURIFICATION OF 8 S RECEPTOR–[^{3}H]DIHYDROTESTOSTERONE COMPLEX OF RAT PROSTATE GLAND

Stage of purification	Protein[a] (mg)	^{3}H Bound to 8 S receptor (dpm)	Specific radioactivity (dpm/mg)
Initial cytoplasmic fraction[b]	364	5.5×10^7	1.5×10^5
$(NH_4)_2SO_4$ precipitation	30.4	4.9×10^7	1.6×10^6
DNA-cellulose chromatography[c]	3.0; 2.96*	2.4×10^7	8×10^6
Desalting	2.9; 2.84*	7.4×10^6	2.6×10^6
Isoelectric focusing[d]	0.105*	5.8×10^6	5.3×10^7
Approximate purification: on recovery of protein, 3,500			
on specific radioactivity, 350			

[a] Protein measured only in pooled fractions, at each stage, by Folin or fluorimetric procedures (marked with an asterisk).
[b] Cytoplasmic fraction from 18 glands, labeled directly with 5 n*M* [^{3}H]dihydrotestosterone.
[c] Retained proteins, eluted only at high ionic strength.
[d] Proteins with pI of 5.8. Complex identified throughout by scintillation spectrometry.

losses are encountered at the desalting step. The degree of purification calculated on the specific radioactivity of the purified complex is not commensurate with that calculated on the overall recovery of protein. This may be attributed to the dissociation of bound [^{3}H]dihydrotestosterone from the 8 S receptor in the course of purification. A more elaborate isolation procedure may be conducted in which excess ligand, 1 n*M* [^{3}H]dihydrotestosterone, is added to all preparative media throughout. This procedure involves the running of many Sephadex G-25 columns to assess protein-bound [^{3}H]steroid in fractions collected during each purification step, but the degree of purification based on the specific radioactivity of the purified 8 S receptor complex is now compatible with that calculated on the recovery of protein.[13a]

In terms of sedimentation coefficient ($s_{20,w}$ 8.0), isoelectric point (pI 5.8), and the ability to transfer [^{3}H]dihydrotestosterone into chromatin in a reconstituted cell-free system,[4] the receptor maintains its fundamental properties after considerable purification. This is further illustrated by its constant electrophoretic ability in polyacrylamide gels after purification (Fig. 1). The identity of the specific 8 S receptor complex is confirmed by the use of the antiandrogen, 6α-bromo-17β-hydroxy-17α-

[13a] W. I. P. Mainwaring and R. Irving, *Biochem. J.* **134,** 113 (1973).

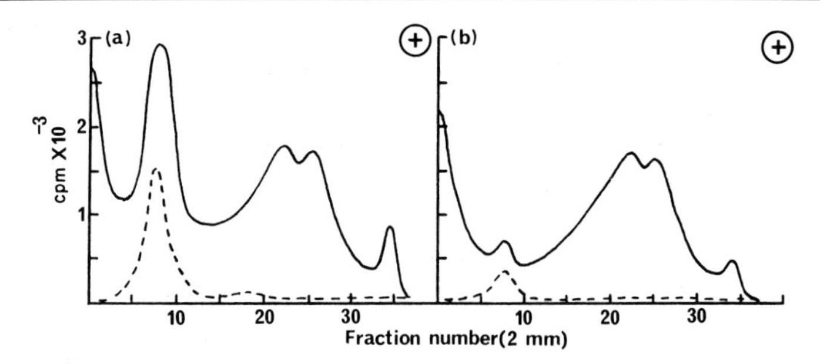

FIG. 1. The constant electrophoretic ability of the 8 S androgen receptor in polyacrylamide gels after extension. Samples of whole prostate cytoplasm or purified 8 S receptor after the isoelectric focusing step were analyzed in polyacrylamide gels. Slices of 0.2 mm were dispersed in KOH and counted. The initial cytoplasm was labeled with (a) n*M* [^{3}H]dihydrotestosterone alone or (b) in the additional presence of 1 μM nonradioactive antiandrogen, BOMT. In both figures, unfractionated cytoplasm, ——, purified 8 S receptor, - - -.

methyl-4-oxa-5α-androstan-3-one (BOMT: Hofmann-La Roche). This competes for the specific high affinity dihydrotestosterone-binding sites on the 8 S receptor[14] but does not displace dihydrotestosterone from the nonspecific, low affinity 4 S androgen-binding proteins.

Nuclear Receptor Complex

The purification of this complex may be described more briefly since many steps are common with the procedure described for the cytoplasmic 8 S receptor complex. Problems particular to the nuclear receptor complex, however, warrant additional explanation. First, the nuclear receptor complex cannot be labeled directly with [^{3}H]dihydrotestosterone to any appreciable extent but only after incubation of whole prostate tissue at 30°. Second, the presence of the basic histones in nuclear extracts creates artifacts through association with the relatively acidic nuclear receptor complex and must accordingly be removed at any early stage of the purification.

Step a. Isolation of Labeled Nuclei. Directly after removal from animals, prostate glands are immersed in Eagle's minimal medium at 30°. A coarse mince of tissue derived from 18 animals is incubated for 40 minutes at 30° in 25 ml of Eagle's medium containing 5 n*M* [^{3}H]dihydrotestosterone. After several rinses in 25 ml of tritium-free medium, the tissue is thoroughly cooled in an ice-bath and homogenized in 20 ml of ice-cold 0.25 *M* sucrose containing 3 m*M* $CaCl_2$, using a loose-fitting Pot-

[14] F. R. Mangan and W. I. P. Mainwaring, *Steroids* **20**, 331 (1972).

ter homogenizer. Impure nuclei, collected by sedimentation at 1000 *g* for 10 minutes, are evenly dispersed in 20 ml of 2.2 *M* sucrose and purified by sedimentation at 50000 *g* for 1 hour.

Step b. Extraction of Nuclear Receptor Complex. The nuclear receptor complex is solubilized and freed of nuclear components which otherwise interfere with subsequent purification procedures, notably isoelectric focusing. The labeled nuclei are stirred gently in an ice-bath with 10 ml of medium B containing 0.5 *M* KCl. After 15 minutes, the viscous extract is treated with 1.0 ml of 15 mg/ml dextran sulfate (Pharmacia) and centrifuged 15 minutes later at 10000 g for 10 minutes. The sediment is discarded.

Step c. Desalting. Somewhat larger columns of Sephadex G-50 (2.0×20 cm, equilibrated with medium B) are used since they efficiently separate the labeled nuclear receptor (recovered at the void volume) from residual histones of significantly lower molecular weight. The larger volume of desalted receptor, sometimes recovered in six fractions, volume 1.0 ml each, is not disadvantageous for the next step.

Step d. DNA-Cellulose Chromatography. The desalted receptor is adsorbed selectively onto this matrix (2 g of DNA-cellulose) and eluted in medium B containing 0.5 *M* KCl, as described.

Step e. Desalting. The smaller columns of Sephadex G-25 (1×20 cm), equilibrated with 5% sucrose, are used.

Step f. Isoelectric Focusing. This is conducted as described, but the nuclear receptor complex has a pI at 6.5. The nature of the nuclear components removed by the dextran sulfate has never been established but their presence creates heterodisperse peaks of bound radioactivity during isoelectric focusing.

The result of a typical purification is presented in Table II. A distinctive feature of this form of receptor is the lower degree of dissociation of [^{3}H]dihydrotestosterone during purification. The purified complex fully retains the ability to transfer [^{3}H]dihydrotestosterone into chromatin in a reconstituted, cell-free system.[4]

Concluding Remarks

In neither case have the receptor complexes been purified to a state approaching homogeneity. Several additional bands of protein are revealed by staining analytical polyacrylamide gels with Coomassie Brilliant Blue R, but receptor-bound radioactivity is associated with a detectable protein band. Residual ampholytes must be leached from the gels in cold 5% (w/v) trichloroacetic acid prior to staining to avoid the formation of artifacts.)

TABLE II
PARTIAL PURIFICATION OF NUCLEAR RECEPTOR–[^{3}H]DIHYDROTESTOSTERONE COMPLEX OF RAT PROSTATE GLAND

Stage of purification	Protein[a] (mg)	^{3}H Bound to receptor (dpm)	Specific radioactivity (dpm/mg)
Purified nuclei[b]	8.10	5.6×10^6	6.9×10^5
Extraction, dextran sulfate	6.12	3.9×10^6	6.4×10^5
Desalting Sephadex G-50	1.54	3.5×10^6	2.3×10^6
DNA-cellulose chromatography[c]	0.37; 0.370*	3.4×10^6	9.2×10^6
Desalting, Sephadex G-25	0.352*	3.2×10^6	9.1×10^6
Isoelectric focusing[d]	0.071*	3.0×10^6	4.3×10^7

[a] Protein measured only in pooled fractions at each stage, by Folin or fluorimetric procedures (marked with an asterisk).
[b] Nuclei recovered from 18 glands after incubation of whole tissue with 5 n*M* [^{3}H]dihydrotestosterone.
[c] Retained protein eluted only at high ionic strength.
[d] Proteins with pI of 6.5. Complex identified throughout by scintillation spectrometry.

The methods described are probably suitable for the partial purification of androgen receptors from most male accessory sexual glands without major modification in experimental procedure. In studies on entirely new systems, however, the specificity of the androgen-binding protein in terms of the most suitable [^{3}H]steroid for maximum labeling must be established prior to any purification attempt. The binding of the putative androgen receptor to DNA-cellulose should then be investigated as a matter of priority.

[33] Synthesis and Use of Affinity Labeling Steroids for Analysis of Macromolecular Steroid-Binding Sites

By JAMES C. WARREN, FERNANDO ARIAS, and FREDERICK SWEET

Introduction

Affinity labeling (site-directed irreversible binding) of a macromolecular steroid binding site requires synthesis of appropriate steroid derivatives bearing reagent groups capable of reacting with amino acid residues present at that site. Concentration of the reagent group at the binding

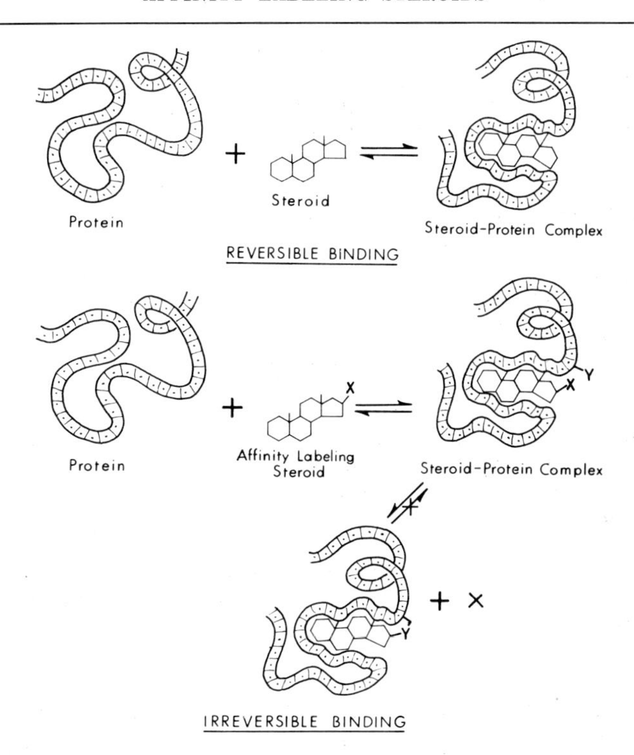

FIG. 1. Reversible vs. irreversible (affinity labeling) binding of a steroid to a protein (i.e., receptor or enzyme). Y represents an amino acid residue capable of forming a covalent bond with a steroid containing a reagent (alkylating) group represented by X.

site by the reversible binding step of the steroid moiety favors covalent bond formation at the binding site as compared to the protein molecule in general. The basic scheme is indicated in Fig. 1. Here reversible binding of steroid and protein is contrasted with the situation where the steroid bears a reagent group (X) capable of reacting with an amino acid residue (Y) at the binding site. Note that the covalent bond prevents dissociation of the steroid from the protein.

A typical sequence of events is to select a macromolecule for study, pick a steroid which has high affinity for the steroid binding site, and modify the steroid by adding a reagent group. The reagent groups are essentially small molecules capable of alkylating various amino acid residues in the protein. Many of these are delineated in an earlier volume in this series dealing with protein modification.[1] If the derivative forms

[1] Enzyme structure, Vol. 11 (1967).

a covalent bond with the macromolecule in question, one may then ascertain the existence of the primary reversible binding step. Ultimately, with strategically selected derivatives, it should be possible to delineate the topography of the steroid-binding site under study, permanently occupy such sites on receptors (with exclusion of natural steroids), and determine whether cytoplasmic steroid receptors play an obligatory and final role in the mechanism of steroid action. This chapter summarizes the information that can be gained by this technique and delineates synthesis and application of several model compounds.

Synthesis of Mercuri-Estradiol Derivatives

4-Mercuri-17β-estradiol (4MEβ)

Principle. The preparation of 4MEβ acetate is accomplished by a single step mercuration of 17β-estradiol in acetic acid with one equivalent of mercuric acetate (Fig. 2).

Reagents

17β-Estradiol, 1.0 g (3.7 mmoles) (Sterloids Inc., Pawling, New York)

Mercuric acetate, 1.17 g (3.7 mmoles) (Fisher Scientific Co., St. Louis, Missouri or Mallinckrodt Co., St. Louis, Missouri)

Reagent grade glacial acetic acid, 15 ml (Fisher Scientific Co., St. Louis, Missouri)

Procedure. 17β-Estradiol (1.0 g, 3.7 mmoles) is dissolved in hot glacial acetic acid (15 ml) with stirring. After all the steroid has dissolved, the solution is cooled to 35° and mercuric acetate (1.17 g, 3.7 mmoles) in acetic acid (15 ml) is added dropwise. After all the mercuric acetate has been added the reaction mixture is stirred at 35° for 9 hours. During this time the white crystalline precipitate which forms is collected at 3-hour intervals by filtration with vacuum. The three crops of crude prod-

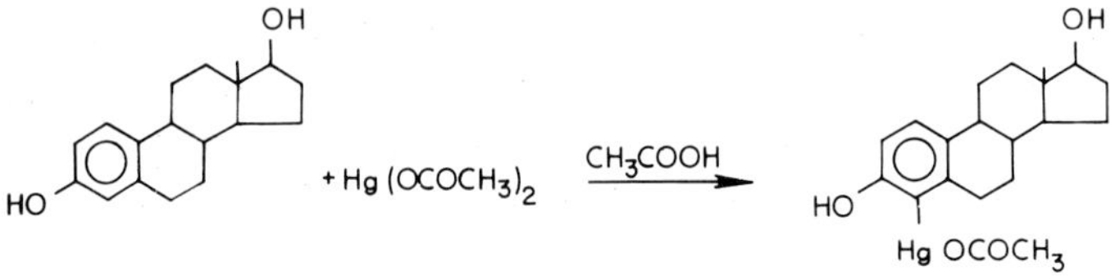

OH HO + $Hg(OCOCH_3)_2$ $\xrightarrow{CH_3COOH}$ OH HO $Hg\ OCOCH_3$

FIG. 2. Synthesis of 4MEβ by direct mercuration of 17β-estradiol.

uct thus obtained are combined, washed with a small volume of cold acetic acid, then dried overnight in a vacuum dessicator, then recrystallized two times from absolute ethanol (250 ml/g of product). The final product, after drying in a vacuum dessicator, weighs 0.74 g (38% yield). The purified 4MEβ acetate melts with decomposition at 235°.

Analysis of Purity and Characterization of 4MEβ Acetate. Thin-layer chromatography of the product is carried out with Eastman Silica Gel Sheets (No. 6060). The chromatogram is developed with methanol–water (8:2) and dried. Spraying with the Folin-Ciocalteu reagent[2] reveals a single purple spot with an R_f of 0.1. Paper chromatography of 200 μg of the product is carried out by using the systems of Zaffaroni[3] with methanol–formamide as the stationary phase. With chloroform as the mobile phase, the Folin-Ciocalteu reagent reveals a single spot (R_f 0.15). With chloroform–acetone (95:5) as the mobile phase, a single spot (R_f 0.3) is observed. By control paper chromatograms as little as 5.0 μg of product or 17β-estradiol are clearly detected by these methods.

A solution of 4MEβ acetate in 0.1 *M* sodium phosphate buffer, pH 7.5, exhibits an ultraviolet absorption maximum at 286 nm (ε 2,220). An ethanolic solution of 4MEβ acetate absorbs at 290 nm which compares with 280 nm for 17β-estradiol. In solutions of higher pH values, 4MEβ acetate displays a hypochromic shift in absorption maximum to 314 nm.

The infrared absorption spectra of 4MEβ acetate and 17β-estradiol are very similar. However, nuclear magnetic resonance spectra (NMR) of these two steroids possess distinct differences. Nuclear magnetic resonance spectroscopy of 4MEβ acetate may be carried out in CD_3COOD at 30° with tetramethylsilane as internal reference. Because of solubility limitations, it is necessary to do 400 scans of the relatively dilute 4MEβ solution and average out noise with a Varian C-1024 computer of average transients. The spectrum displays doublet peaks in the aromatic ring hydrogen region, centered at 6.56 ppm (tetramethylsilane) and 7.07 ppm (tetramethylsilane), each with $J = 9.4$ Hz. This provides unequivocal evidence that the two hydrogen atoms remaining on the A ring are *ortho* (i.e., that mercuration takes place at position 4). [Substitution of the mercury at position 2 would leave the unsubstituted hydrogens in the *para* position and one would predict a coupling constant (J) of 0 to 1 Hz.] Peaks representing the angular methyl hydrogen (0.68 ppm) and acetate hydrogen (1.93 ppm) are of similar magnitudes. The spectrum of 17β-estradiol reveals no acetate hydrogen peak.

[2] O. Folin and V. Ciocalteu, *J. Biol. Chem.* **73**, 627 (1927).

[3] A. Zaffaroni, *Recent Progr. Horm. Res.* **8**, 51 (1953).

4-Mercuri-17α-estradiol (4MEα) Acetate

Principle. The preparation of 4MEα acetate is virtually identical to that of 4MEβ acetate described above.

Procedure. 17α-Estradiol (1.0 g, 3.7 mmole) is dissolved in hot glacial acetic acid (15 ml), then the solution is cooled to 35°. Mercuric acetate (1.17 g, 3.7 mmole) in acetic acid (15 ml) is slowly added to the solution with stirring. Crystals accumulate within the first 12 hours, and after a total reaction time of 36 hours the precipitated product is collected by vacuum filtration, washed with a small quantity of cold acetic acid, and after drying under vacuum for 24 hours recrystallized 3 times from ethanol (250 ml/g of product). The pure product decomposes at 225°. The ultraviolet absorption maximum of 4MEα acetate in 0.1 M sodium phosphate buffer at 7.5 is 290 nm (ϵ 2200). This contrasts with 17α-estradiol which in the same solvent system possesses λ_{max} = 280 nm (ϵ 2000). The homogeneity of the preparation can be further evaluated with the thin-layer chromatographic system described above for 4MEβ acetate.

Synthesis of Bromoacetoxyprogesterone Derivatives

Two procedures have been employed for the synthesis of bromoacetates of the corresponding hydroxyprogesterone precursors, as represented by equations in Fig. 3. They shall be referred to here as Method A (bromoacetyl bromide–dimethylformamide) and Method B (bromoacetic acid–dicyclohexylcarbodiimide).

The conventional method for preparation of esters with bromoacylbromide in ether when applied to 16α-hydroxyprogesterone was found to give mostly 16-dehydroprogesterone. Formation of this product is due to the acid catalyzed dehydration of the β-hydroxy keto system, which is more rapid than acylation under these experimental conditions. Therefore, a nonacidic medium is required for the desired acylation of 16α-hydroxyprogesterone, and dimethylformamide (the solvent in Method A) provides the optimum reaction medium.

Preparation of radioactive bromoacetoxyprogesterone derivatives by Method A, wherein the tritium (or ^{14}C) label is located in the bromoacetoxy functional group, would require appropriately radiolabeled bromoacetyl bromide. This reagent is unavailable commercially and small-scale preparation of this rather unstable compound would be beyond the capacity of most research laboratories.

Bromoacetic acid containing tritium or ^{14}C is commercially available, and therefore Method B is most useful for preparing radioactive bromoacetoxyprogesterone derivatives.

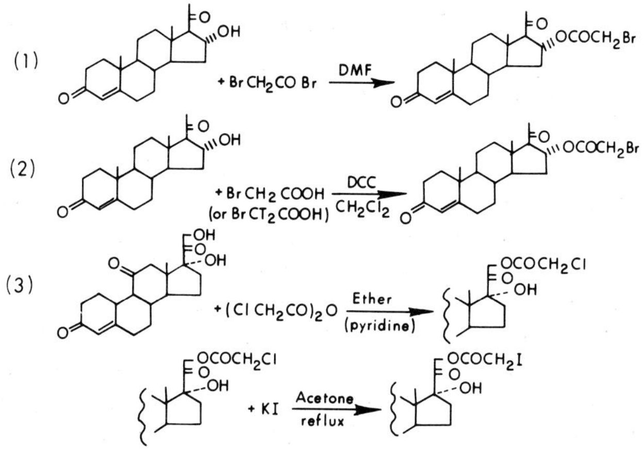

FIG. 3. Synthetic schemes for preparation of (1) bromoacetoxyprogesterone derivatives by Method A; (2) bromoacetoxyprogesterone derivatives by Method B; (3) cortisone 21-iodoacetate by halogen exchange. DMF = *N,N*-dimethylformamide; DCC = dicyclohexylcarbodiimide.

16α-Bromoacetoxyprogesterone by Method A

Principle. A solution of bromoacetyl bromide is added to the hydroxysteroid in DMF. After the appropriate reaction time the bromoacetoxyprogesterone is recovered by careful addition of water to the reaction mixture. After purification by column chromatography the product is obtained in crystalline form.

Reagents

Dimethylformamide (8.0 ml) freshly distilled from BaO into a dry reaction flask[4]

16α-Hydroxyprogesterone, 264 mg (0.8 mmole)[5]

Bromoacetyl bromide, 0.100 ml (1.15 mmoles)

[4] Commercial *N,N*-dimethylformamide (DMF) is allowed to stand overnight over anhydrous barium oxide, then distilled prior to use into predried reaction flasks. DMF tends to absorb atmospheric moisture so that use of freshly dried solvent is imperative. Water present in this solvent rapidly hydrolyzes the bromoacetyl bromide to bromoacetic acid, thus destroying this reagent.

[5] Although 16α-hydroxyprogesterone is commercially available its high cost ($7.00/mg at this printing) prompted us to prepare this compound in large quantity in a single step by reducing the inexpensive 16α,17α-epoxyprogesterone according to the methods of W. Cole and P. I. Julian [*J. Org. Chem.* **14,** 131 (1954)]. A simple method for preparing the highly *air sensitive* reducing agent, chromous acetate, is reported in the literature [D. H. R. Barton, N. K. Basu, R. H. Hesse, F. S. Morehouse, and M. M. Pechet. *J. Amer. Chem. Soc.* **88,** 3016 (1966)].

Procedure. To a stirred solution of 264 mg (0.8 mmole) of 16α-hydroxyprogesterone in 8.0 ml of *N,N*-dimethylformamide (DMF), 0.1 ml (1.15 mmoles) of freshly distilled bromoacetyl bromide is added[6] and kept at room temperature for 3 hours. The reaction mixture is cooled in an ice bath, and ice water (40 ml) is added dropwise with vigorous stirring. During addition of water a white precipitate accumulates.[7] After all the water has been added, stirring is continued at 0° for 30 minutes; then the crude product is filtered with vacuum and the residual solid is washed well with several portions of cold water to complete the removal of DMF. The dried solid product (about 150 mg) is analyzed with thin-layer chromatography.[8] If there is found to be a significant amount of starting material present this is removed by chromatographing the crude product on a column containing 10 g of silica gel G. A small amount of methyl red is added to serve as a marker. All the steroidal bromoacetates described here are eluted well before the dye while the hydroxysteroid migrates behind it, when reagent grade chloroform (which contains 0.5–1.0% ethanol) is used as the eluent. Fractions containing the product are pooled and removal of the solvent under reduced pressure leaves a crystalline residue of about 230 mg of pure product. Recrystallization of this material from ethanol–water gives 210 mg (58%) of needles, m.p. 120–122°; ultraviolet absorption maximum (ethanol): 240 nm (ϵ 16,400). Infrared spectrum[9] which is shown in Fig. 4 (KBr pellet of 1.0 mg of steroid in 85 mg anhydrous KBr) contains characteristic absorptions at 1730 cm^{-1} (O—C═O), 1705 cm^{-1} (C—20, C═O), 1660 cm^{-1} (C—3, C═O), 680 cm^{-1} (characteristic of the bromoacetate), specific rotation (chloroform) $[\alpha]_D^{25}$, +87°.

[6] Addition can be most conveniently accomplished with a 100-μl micropipet.

[7] Too rapid addition of water may cause formation of a gum. However, working the gum with water removes the entrapped DMF and provides the solid.

[8] Eastman 20 × 20-cm thin-layer chromatography sheets of silica gel G sheets containing fluorescent indicator are cut into 2.5 × 8-cm plates. For analysis 10 μl of a solution, containing 1–2 mg of product in 250 μl of chloroform, is applied to a plate and eluted with chloroform by ascending chromatography in an Eastman TLC chamber. Visualization of the dried chromatogram under ultraviolet light (260 nm) generally reveals bromoacetoxyprogesterones with approximate R_f = 0.6 and hydroxyprogesterones with R_f = 0.1 when reagent grade chloroform is used. Somewhat similar results are obtained with benzene–ether 1:1 as eluents. These systems can also be used to analyze the ring-bromo progesterone derivatives.

[9] Infrared spectra were obtained with a Beckman IR 18-A instrument wherein the scan time was 20 minutes. KBr pellets containing 1.0 mg of steroid in 85–95 mg of KBr were generally used to obtain infrared spectra.

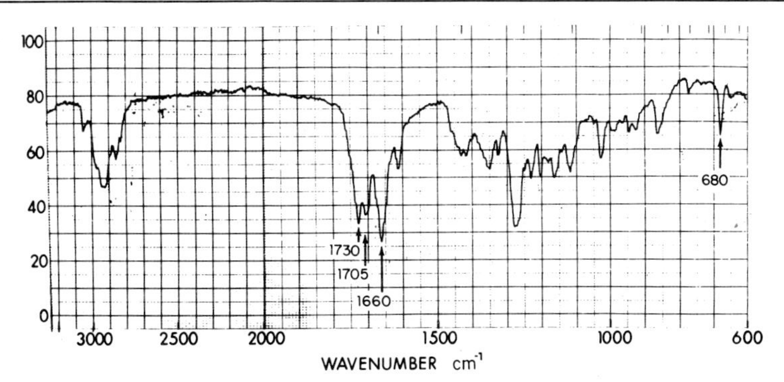

FIG. 4. Infrared absorption spectrum of 16α-bromoacetoxyprogesterone (1.0 mg of steroid in 80–95 mg KBr). Characteristic absorption maxima of bromoacetoxyprogesterone derivatives occur in the regions 1715–1735 cm^{-1} (O—C═O); 1690–1705 cm^{-1} (C-20, C═0); 1650–1665 cm^{-1} (C-3, C═0); 665–685 cm^{-1} (characteristic of the bromoacetate).

11α-Bromoacetoxyprogesterone by Method A

The synthesis of 11α-bromoacetoxyprogesterone according to Method A is carried out with 330 mg (1.0 mmole) of 11α-hydroxyprogesterone in 8 ml of DMF to which is added 0.100 ml of bromoacetyl bromide. Stirring is continued for 3 hours at room temperature, then the product is worked up as described above and the crude bromoacetate is purified by column chromatography (15 g silica gel G, elution with chloroform) to give, after two recrystallizations from diisopropyl ether, 285 mg (63%) of colorless crystals, m.p. 154–156°. Ultraviolet absorption maximum (ethanol) is 240 nm (ε 16,400); characteristic infrared absorption bands are similar to those obtained with 16α-bromoacetoxyprogesterone, shown in Fig. 4; specific rotation (chloroform) $[\alpha]_D^{25}$, +148°.

21-Bromoacetoxyprogesterone by Method A

Procedure. With the same quantities of hydroxysteroid, solvent, and reagents and under similar conditions to those described for the preparation of 11α-bromoacetoxyprogesterone, 330 mg of 21-hydroxyprogesterone affords 320 mg of pure (needles) 21-bromoacetoxyprogesterone after two recrystallizations from acetone–water (or methanol–water). The product melts at 105–106°; ultraviolet absorption maximum (ethanol) is 240 nm (ε 16,300); characteristic infrared absorption maxima are similar to those of 16α-bromoacetoxyprogesterone, shown in Fig. 4.

16α-Bromoacetoxyprogesterone by Method B

Principle. An excess of dicyclohexylcarbodiimide in dry methylene chloride is added to a chilled solution of a hydroxyprogesterone and bromoacetic acid in the same solvent. The progress of the reaction is followed by thin-layer chromatography,[8] and when the reaction is completed the crude bromoacetoxyprogesterone is purified by column chromatography. Method B is thus far the only procedure through which the radioactive [2-^{3}H]bromoacetoxyprogesterone derivatives are available.

Reagents

[2-^{3}H]Bromoacetic acid (100 mCi/mmole)
Solutions of dicyclohexylcarbodiimide (515 mg) and bromoacetic acid (215.5 mg) each in methylene chloride (10 ml, freshly distilled from anhydrous calcium chloride)
16α-Hydroxyprogesterone (33 mg, 0.1 mmole), pyridine (0.01 ml)

Procedure. To a stirred solution of 33 mg (0.1 mmole) of 16α-hydroxyprogesterone in 3 ml of anhydrous methylene chloride cooled to 0° is sequentially added 1.00 ml of a solution containing 25.15 mg of bromoacetic acid and 5 mCi of [2-^{3}H]bromoacetic acid, 1.00 ml of a solution containing 51.5 mg (0.2 mmole) of dicyclohexylcarbodiimide in 1.00 ml of methylene chloride, and 0.010 ml (0.12 mmole) of pyridine (previously dried over potassium hydroxide). The resulting solution is stirred at 0° for 1 hour then at room temperature for an additional hour. During the first 2 to 3 minutes of reaction a crystalline solid (dicyclohexylurea) precipitates and continues to accumulate as the reaction proceeds. After a total of 2 hours the reaction is complete (confirmed by thin-layer chromatography[8]). Then acetic acid (0.16 mmole) is added and stirring is continued at room temperature for 20 minutes. The reaction mixture is concentrated to dryness,[10] then 3-ml portions of acetone are added and evaporated from the crude product in order to remove the last traces of methylene chloride. The residue is mixed well with 3 ml of acetone and the resultant mixture is filtered. The solid residue (dicyclohexylurea) is washed with two 1-ml portions of acetone. Addition of 0.5 ml of water to the combined filtrates, chilled in an ice bath, precipitates an additional small amount of dicyclohexylurea which is removed by filtration. Further chilling of the filtrate and dropwise addition of 2.5 ml of water precipitates 40 mg of crystals (small needles) of product which after collection by filtration and recrystallization from acetone–water provides 33 mg

[10] For small scale (0.1 mmole) reactions the solvents are removed with a stream of nitrogen while for large scale (1.0 mmole) reactions solvents are flash evaporated under reduced pressure.

(73%) of pure 16α-bromoacetoxyprogesterone, m.p. 120–122°. This material is identical in all respects (mixed melting point, thin-layer chromatography, UV and IR spectra) with the product obtained above by Method A.

11α-Bromoacetoxyprogesterone by Method B

Procedure. By application of the above procedure, 1.00 g (3.03 mmoles) of 11α-hydroxyprogesterone in 30 ml of dry methylene chloride was treated with 0.84 g (6.04 mmoles) of bromoacetic acid and 1.25 g (6.04 mmoles) of dicyclohexylcarbodiimide in 20 ml of methylene chloride, and 0.50 ml of pyridine gives, after work up, 1.5 g of crude product. Two recrystallizations of this material from diisopropyl ether[11] give 1.25 mg of pure 11α-bromoacetoxyprogesterone, m.p. 155–157°. This product is identical in all respects to that obtained by Method A above.

6β-Bromoacetoxyprogesterone by Method B

Procedure. To 1.00 g (3.03 mmoles) of 6β-hydroxyprogesterone in 30 ml of anhydrous methylene chloride (freshly distilled from $CaCl_2$) stirred at 0° are added in sequence 0.84 g of bromoacetic acid and 1.25 g (6.04 mmoles) dicyclohexylcarbodiimide in 20 ml CH_2Cl_2. After 5 minutes of stirring, 0.50 ml of pyridine is added. Following reaction times of 1 hour at 0° then 1 hour at room temperature the product is isolated in crude form as described above. Recrystallization of the crude product with ethanol–water or butanol–hexane provides crystals with a melting point of 140–143° (softens at 110°), ultraviolet absorption maximum (ethanol) is 238 nm (ε 16,400); the infrared spectrum resembles those of the above steroidal esters (Fig. 4); $[\alpha]_D^{25} = +131°$.

Synthesis of Active Iodo Derivatives of Cortisone

Both 21-iododeoxycortisone and cortisone 21-iodoacetate have been shown to react rapidly with free nucleophilic amino acids.[12] Although the steroidal iodoacetate inactivates 20β-hydroxysteroid dehydrogenase by affinity labeling, the α-iodoketone, although a substrate for this enzyme, does not. The synthesis of 21-iododeoxycortisone has been previously described by Borrevang.[13]

[11] Crystallization of this compound is quite slow, often requiring 12 to 24 hours to attain completion.

[12] M. Ganguly and J. C. Warren, *J. Biol. Chem.* **246**, 3646 (1971).

[13] P. Borrevang, *Acta Chem. Scand.* **9**, 587 (1955).

There are two methods which may be employed for synthesis of cortisone 21-iodoacetate, represented by equations in Fig. 3.

The 21-chloroacetate, obtained by treating cortisone with chloroacetic anhydride, is converted to the corresponding 21-iodoacetate by the Finkelstein halogen exchange synthesis, with potassium iodide in boiling acetone. In order to obtain the radioactive cortisone 21-[2-^{3}H]iodoacetate a procedure is used similar to Method B described above for preparation of the progesterone bromoacetate derivatives.

Cortisone 21-Iodoacetate by Halogen Exchange Method

Principle. Cortisone 21-chloroacetate is prepared initially by reaction of cortisone with chloroacetic anhydride. The 21-chloroacetate product is treated with potassium iodide in boiling acetone to give cortisone 21-iodoacetate.

Procedure. STEP 1. During a 30-minute period, 3.6 g of chloroacetic anhydride in 5 ml of acetone are added dropwise to an ice-cold solution of 300 mg of cortisone in 60 ml of acetone containing 0.6 ml of pyridine. The reaction mixture is stirred at 0° for 1 hour and allowed to stand overnight at 4°, then concentrated to a volume of about 20 ml under N_2 and finally added dropwise to a vigorously stirred solution of ice water (400 ml). After 1 hour, the resulting precipitate is collected by filtration, washed thoroughly with cold water, dissolved in a minimum volume of acetone, and crystallized by slow addition of water. The dried product melts at 228–230°. The yield is 300 mg (80%). Ultraviolet absorption maximum (ethanol) is 238 nm (ϵ 15,500) and the specific rotation in chloroform is $[\alpha]_D^{23} = +196°$.

STEP 2. Cortisone 21-chloroacetate (200 mg) and KI (300 mg) are heated under reflux in 30 ml of acetone for 4 hours. Then the acetone is evaporated with N_2 to a residue of 5 ml which is added to 25 ml of an ice-cold 5% solution of $Na_2S_2O_3$ in water. The mixture is stirred for 15 minutes at 0° and the precipitate is collected by filtration. Recrystallization from acetone by addition of water, as above, yields 160 mg (66%), m.p. 180°. Ultraviolet absorption maximum (ethanol) of the product is 238 nm (ϵ 14,000) and the specific rotation in chloroform is $[\alpha]_D^{23} = +160°$.

Cortisone 21-[2-^{3}H]Iodoacetate by Method B

Principle. The preparation of cortisone 21-iodoacetate by Method B is the same as described above in the section on progesterone bromoacetates.

Procedure. [2-^{3}H]Iodoacetic acid (10 mCi) is mixed with unlabeled iodoacetic acid and the specific activity of the mixture is determined. A solution of 0.24 mmole of iodoacetic acid (22 mCi/mmole), 0.1 mmole of cortisone, and 0.3 mmole of pyridine in 4.5 ml of dry CH_2Cl_2 is cooled to 0° and reacted with 0.25 mmole of dicyclohexylcarbodiimide in 0.1 ml of cold CH_2Cl_2. Within 5 minutes, dicyclohexylurea precipitates. The suspension is stirred at 0° for 1 hour and at 23° for a second hour. Then, 0.025 ml of glacial acetic acid is added to this suspension, and it is stirred at 23° for 15 minutes. Dichloromethane is evaporated with a stream of N_2. The solid residue is stirred with 3 ml of acetone and filtered. The filtrate is poured into 50 ml of a stirred solution of 2% $Na_2S_2O_3$ in water at 0°. The suspension which forms is stirred for 15 minutes and then filtered. The collected precipitate is stirred with 5 ml of acetone, then filtered to remove any remaining dicyclohexylurea. The filtrate is concentrated to a small volume in a stream of N_2 and reprecipitated with cold water, washed with a small volume of cold methanol, and recrystallized from aqueous acetone. Yield 15 mg, m.p. 180° (dec.). This product migrates as a single spot on thin layer chromatograms[8] with a mobility identical to that of cortisone 21-iodoacetate prepared by the halogen exchange method.

Synthesis of Ring-Bromoprogesterone Derivatives

Although 2α-, 6α-, and 6β-bromoprogesterone[14,15] have been synthesized prior to our application of these compounds to affinity labeling, the earlier syntheses were all reexamined and optimized to provide maximal yields. Of these three ring-bromoprogesterone derivatives only 6α- and 6β-bromoprogesterone were found to significantly inactivate 20β-hydroxysteroid dehydrogenase.[16] Tritiated 6β-bromoprogesterone for affinity radiolabeling was prepared by small-scale bromination of [1,2-^{3}H]progesterone. Scaling down this reaction resulted in lower yields of product then obtained in large-scale reactions.

Synthesis of 6β-Bromo[1,2-^{3}H]progesterone

Principle. A solution of [1,2-^{3}H]progesterone and *N*-bromosuccinimide in carbon tetrachloride is heated. After working up the reaction mixture, homogeneous tritiated 6β-bromoprogesterone is obtained.

[14] G. R. Allen, Jr. and M. J. Weiss, *J. Amer. Chem. Soc.* **82,** 1909 (1960).
[15] F. Sondheimer, S. Kaufman, J. Romo, H. Martinen, and G. Rosenkranz, *J. Amer. Chem. Soc.* **75,** 4712 (1953).
[16] C. C. Chin and J. C. Warren, *Biochemistry* **11,** 2720 (1972).

Procedure. A solution of [1,2-^{3}H]progesterone (3.65 mCi) in 3 ml of anhydrous carbon tetrachloride is mixed with unlabeled progesterone (90 mg, 0.236 mmole; the specific activity of the final mixture was determined to be 16.1 Ci/mole). *N*-Bromosuccinimide (50 mg, 0.30 mmole) is added and the mixture heated under reflux for 1 hour. After cooling to room temperature, the reaction mixture is filtered and the filtrate is concentrated to dryness with a stream of nitrogen. The residue is triturated with hexane, filtered, and the solid crude product recrystallized twice from 3-ml portions of acetone: yield, 10.0 mg (9%); specific activity 8.75 Ci/mole; m.p. 147–150° (dec.); single spot on thin-layer chromatogram[8] (R_f value identical with unlabeled 6β-bromoprogesterone).

Reaction of Bromoacetoxyprogesterone Derivatives with Free Amino Acids

In order for steroidal derivatives to possess affinity labeling potential they must be capable of alkylating constitutional amino acid residues in steroid binding sites of proteins. L-Cysteine, L-histidine, L-methionine, and L-tryptophan are the more nucleophilic amino acids which are likely candidates for alkylation at steroid binding sites by affinity labeling steroids. The following experiments are designed to test the alkylating capacity of bromoacetoxyprogesterone derivatives.

Principle. A standard solution of a bromoacetoxyprogesterone in ethanol is added to a solution of an amino acid in 0.05 *M* potassium phosphate buffer at pH 7.0. Formation of the steroid–amino acid conjugate is monitored by thin-layer chromatography. The structure of the conjugate is deduced after characterization of the modified amino acid obtained by hydrolysis of the steroid–amino acid conjugate.

Procedure. A solution of 21-bromoacetoxyprogesterone (4.51 mg, 0.01 mmole) in 5 ml of ethanol is added to a stirred solution of L-cysteine (or other amino acid) (1.45 mg, 0.012 mmole) in 5 ml of 0.05 *M* potassium phosphate buffer at pH 7.0, at 25°. Aliquots (0.02 ml) of the reaction mixture are applied to 7.5 × 7.5-cm chromatographic plates of silica gel G (Eastman No. 6060) containing a fluorescent indicator. The stock steroid and amino acid solution are applied to the same chromatograms to serve as standards.[17] The reaction with cysteine is generally completed

[17] The thin-layer chromatograms are developed by ascending chromatography with butanol–acetic acid–water (12:3:5:, by volume) as eluent. The plates are visualized with ultraviolet light to locate steroid containing spots, then sprayed with ninhydrin reagent and heated at 100° for 2 minutes. Spots which are ultraviolet absorbing and also react with ninhydrin (to give a blue spot) are taken to be steroid–amino acid conjugates.

in 1 hour, others should be followed for 8–24 hours. After dilution with 10 ml of water the reaction mixture is extracted sequentially with 10 ml each of ether, ethyl acetate, and butanol. The extracts contain unreacted steroid, steroid *plus* steroid–amino acid conjugate, and the steroid–amino acid conjugate, respectively. No free cysteine or other amino acid is ordinarily detected in any of the extracts. The butanol extract is evaporated under reduced pressure and the residue is hydrolyzed by heating it for 1 hour in 3 ml of 0.1 *M* NaOH. After neutralization with 0.1 *M* HCl, thin-layer chromatography of the hydrolysate (in the case of cysteine) reveals two spots with mobilities identical with authentic 21-hydroxyprogesterone and *S*-carboxymethyl-L-cysteine. Analysis of the hydrolysate in an amino acid analyzer confirms the presence of *S*-carboxymethyl-L-cysteine. These results indicate that the steroid–amino acid conjugate is 21-*S*-L-cysteinylacetoxyprogesterone. Similar results are obtained with the other bromoacetoxyprogesterone derivatives when subjected to the above conditions.

Reaction of the Bromoprogesterones with Amino Acids

Procedure. Solutions (2.5×10^{-4} *M*) of each of the bromoprogesterones are incubated with (1.25×10^{-3} *M*) L-cysteine, L-methionine, L-histidine, L-lysine, L-tyrosine, or L-tryptophan in 0.1 *M* potassium phosphate buffer MeOH (50:50) (pH 7.0) at room temperature for 12 hours. The reaction mixtures are dried and the residues extracted with 1.0 ml of methanol. Control solution of each of the bromoprogesterones and amino acids (separately) are treated in a similar way. Any new compounds produced from coupling of the bromoprogesterones with amino acids can be detected by thin-layer chromatography as spots with different mobilities from the starting materials and which absorb ultraviolet light and react with ninhydrin reagent. Chromatograms are developed with butanol–acetic acid–water (4:1:2) to examine the mobility of amino acids and with chloroform–hexane (9:1) to examine the mobility of steroids.

Synthesis of Progesterone-6-*S*-L-cysteine

Principle. Authentic conjugates between steroids and amino acids, which are formed as a result of affinity labeling of macromolecules, must be prepared so that the amino acid which is alkylated at the steroid binding site can be identified. When 6β-bromoprogesterone is used to affinity label 20β-hydroxysteroid dehydrogenase a cysteine residue is alkylated at the active site.[16] This was verified by synthesis of progesterone-6-*S*-L-cysteine which possesses the same elution characteristics in amino acid analysis as the steroid–amino acid conjugate obtained from the 6 *N* HCl hydrolysis of the affinity radiolabeled enzyme.

Procedure. 6β-Bromoprogesterone (500 mg, 1.25 mmoles) is dissolved in 100 ml of 95% ethanol through which a flow of nitrogen was maintained for 30 minutes. L-Cysteine·HCl (160 mg, 1.0 mmole) was added over a period of 30 minutes and the pH value adjusted to 8.0 with 1.0 *N* NaOH. The solution is kept at room temperature for 3 hours under nitrogen with the pH maintained at 8.0. After cooling at 0–5° for 3 hours, the precipitated NaBr and NaCl is removed by filtration. The filtrate is concentrated to dryness in an evaporator. Progesterone-6-*S*-L-cysteine in the residue is extracted with warm absolute ethanol at 50–60°. Undissolved materials containing unreacted cysteine HCl, NaBr, and NaCl are filtered and the filtrate is adjusted to pH 7.0 with concentrated HCl. Additional precipitate of NaCl is removed and the filtrate is kept at 0–5° overnight. The yellow precipitate of progesterone-6-*S*-L-cysteine is collected with a Büchner funnel. The filtrate is condensed to one-half of its original volume and left at 0–5° overnight to give a second crop of product. The combined crude product is recrystallized from ethanol–ether: yield, 18%; m.p. 201–203° (dec); ultraviolet absorption maximum (methanol) 240 nm (ϵ 9300); infrared spectrum (KBr) 1715 cm^{-1}, (20 C=O), 1675 cm^{-1} (3 C=O), 1590 cm^{-1}, 1410 cm^{-1} (COO^-), 1527 cm^{-1}, 1465 cm^{-1} (NH_3^+), and no band at 2550 cm^{-1}, indicating absence of the SH group; R_f 0.75 on thin-layer chromatography with acetone–MeOH (80:20) as solvent. The product migrates as a single spot, absorbs ultraviolet light, and produces a blue spot with ninhydrin reagent spray.[17]

Applications of Affinity Labeling Steroids

4-Mercuri-17β-estradiol (4MEβ)

Biochemical applications of 4-mercuri-17β-estradiol (4MEβ) are enhanced by the ultraviolet spectral change produced on mercaptide formation (Fig. 5). This conveniently permits evaluation of the rates at which the compound forms mercaptides. When 4MEβ reacts with glutathione under physiological conditions, mercaptide formation is evidenced by a maximal increase in absorbance at 305 nm with $\Delta\epsilon$ for the mercaptide of 3.4×10 M^{-1} cm^{-1}. According to affinity labeling theory, the presence of natural 17β-estradiol should slow the rate of mercaptide formation by 4MEβ at a macromolecular estradiol binding site (as it competes during the initial reversible binding step) but have no effect when mercaptide formation is nonspecific. Table I shows that this is the case. When 4MEβ

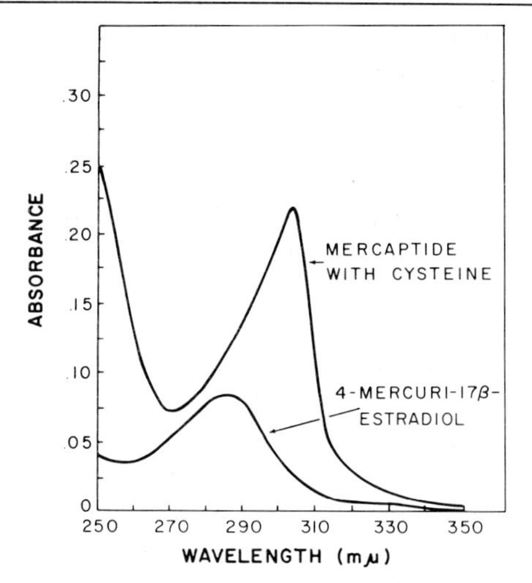

FIG. 5. Absorbance of 39 μM 4-mercuri-17β-estradiol and its mercaptide with 41 μM cysteine in 0.10 M sodium phosphate buffer, pH 7.5, 25°. From C. C. Chin and J. C. Warren, *J. Biol. Chem.* **243**, 5056 (1968).

reacts with glutathione or egg albumin, 33 μM 17β-estradiol has no effect on the rate of mercaptide formation as determined spectrally. Under similar conditions, the rates of mercaptide formation of 4MEβ with glutamate dehydrogenase and pyruvate kinase (which have allosteric binding sites for 17β-estradiol) are decreased in the presence of the natural steroid. These observations indicate (a) the presence of a cysteine residue at the

TABLE I[a]
EFFECTS OF STEROIDS ON RATE OF 4MEβ AND PCMB MERCAPTIDE FORMATION

Sulfhydryl donor	Reagent[b]	Steroid (33 μM)	Inhibition of mercaptide formation
1. Reduced glutathione	4MEβ	17β-Estradiol	No
2. Egg albumin	4MEβ	17β-Estradiol	No
3. Glutamate dehydrogenase	4MEβ	17β-Estradiol	Yes
4. Pyruvate kinase	4MEβ	17β-Estradiol	Yes
	4MEβ	Estriol	No
	4MEβ	Cortisol	No
	PCMB	17β-Estradiol	No

[a] From C. C. Chin and J. C. Warren, *J. Biol. Chem.* **243,** 5056 (1968).
[b] 4MEβ, 4-mercuri-17β-estradiol; PCMB, *p*-chloromercuribenzoate.

allosteric steroid binding sites of glutamate dehydrogenase and pyruvate kinase and (b) that it is with these residues that 4MEβ forms a mercaptide bond.

Study of Steroid–Enzyme Active Sites

The study of steroid binding sites of enzymes or receptor proteins with affinity labeling steroids is an extension of the general application of covalent binding substrate derivatives to the study of enzyme active sites. The elegant work of Schoellmann and Shaw,[18] in which L-(1-tosylamido-2-phenyl)ethylchloromethyl ketone was used to demonstrate the presence of histidine in the active site of chymotrypsin, is an illustrative example. In studying steroid specific enzymes, advantage is taken of the high affinity of the enzyme for modified steroid molecules which are capable of forming a covalent bond with amino acids at the active site. The introduction of radioactive labeling in the modified steroid molecule allows the identification of the modified amino acid residue. Variation in the position of the reagent group on the steroid molecule permits the identification of different amino acids relative to the molecule and eventual mapping of the active site. This approach has been successfully used in our laboratory to study the active site of the steroid specific enzyme 20β-hydroxysteroid dehydrogenase from *Streptomyces hydrogenans* with cortisone 21-iodoacetate,[12] 6α- and 6β-bromoprogesterone,[16] 16α-bromoacetoxyprogesterone,[19] 11α-bromoacetoxyprogesterone, and 6β-bromoacetoxyprogesterone[20] as the modified steroids. The use of these compounds permitted identification of the amino acids L-histidine, L-cysteine, and L-methionine at the enzyme active site and furthered our understanding of the interaction between the steroid and the macromolecule. The methods applied to obtain these results are as follows:

A. Assessment of the stability of the modified steroid
B. Reaction of affinity-labeling steroid with amino acids and model sulfhydryl compounds, e.g. [reduced glutathione: 5,5′-dithiobis(2-nitrobenzoic acid)]
C. Study of the characteristics of enzyme inactivation
D. Obtaining evidence that the affinity labeling steroid binds at the enzyme active site
E. Inactivation of the enzyme with radioactive affinity labeling steroid to obtain stoichiometry
F. Identification of the modified amino acid in the enzyme active site

[18] G. Schoellmann and E. Shaw, *Biochemistry* **2,** 252 (1963).
[19] F. Sweet, F. Arias, and J. C. Warren, *J. Biol. Chem.* **247,** 3424 (1972).
[20] F. Arias, F. Sweet, and J. C. Warren, *J. Biol. Chem.* **248,** 5641 (1973).

A. Assessment of Stability of the Modified Steroid

Procedure. Crystalline steroid (0.3 μmole of 0.2 ml of ethanol) is incubated at 25° in 5 ml of 0.05 M potassium phosphate buffer, at pH 7.0 and 8.0 (conditions under which the 20β-dehydrogenase is stable). At 1-hour intervals, aliquots of the incubation mixture are extracted with chloroform and the extracts are analyzed by thin-layer chromatography.[8]

Discussion. The appearance of hydrolytic products of the affinity labeling steroid is indicative of its instability, which complicates the interpretation of experiments with the enzyme. For example, cortisone 21-iodoacetate is hydrolyzed at pH 7.0 with liberation of iodoacetic acid, and 1,3-dicarboxymethylhistidine is obtained when this steroid is used to inactivate 20β-hydroxysteroid dehydrogenase.[12] This makes the mechanism of dicarboxymethylation of histidine difficult to interpret. The instability of the steroid raises the possibility that the second carboxymethylation of histidine may result from a reaction between a steroid–carboxymethylhistidine (initially formed in the course of the reaction) and the iodoacetic acid present in the incubation mixture due to the hydrolysis of the steroidal ester. The other derivatives, however, are stable under these conditions.

B. Reaction of Affinity Labeling Steroid with Amino Acids and Model Sulfhydryl Compounds

Principle. The kinetics of alkylation of an affinity labeling steroid can be evaluated with 2-nitro-5-mercaptobenzoic acid (—SH nucleophile) as a nucleophilic substrate. The aromatic thiol possesses an absorption maximum at 412 nm due to the auxichromic sulfhydryl anion. Alkylation of the sulfhydryl group, as represented in Fig. 6, suppresses the absorbance at 412 nm, so that the rate of decrease in absorbance can be used to determine the rate of alkylation.

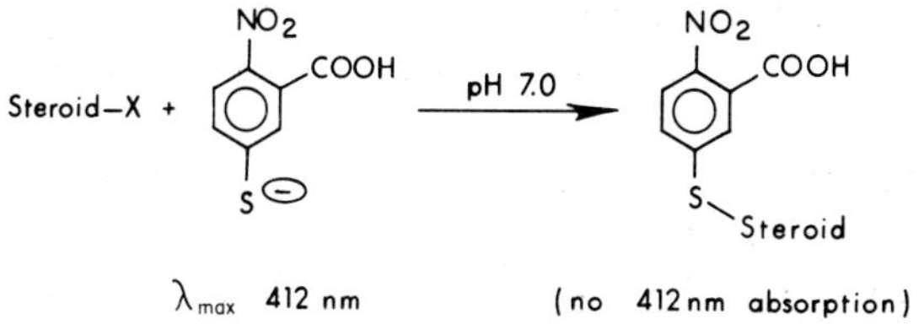

Steroid–X + [2-nitro-5-mercaptobenzoate anion: NO_2, COOH, $S^{\ominus}$] $\xrightarrow{\text{pH 7.0}}$ [NO_2, COOH, S–Steroid]

λ_{max} 412 nm (no 412 nm absorption)

FIG. 6. Reaction of affinity labeling steroid with 2-nitro-5-mercaptobenzoic acid to obtain kinetic rate of alkylation of sulfhydryl group.

Procedure. The incubation mixture contains 0.1 mole of 5,5′-dithiobis-(2-nitrobenzoic acid) and 0.1 μmole of reduced glutathione in 2.8 ml of 0.05 *M* phosphate buffer, pH 7.0. After formation of the yellow anion the affinity labeling steroid (0.05 mole in 0.2 ml of ethanol) is added and the decrease in absorbance at 412 nm with time is then recorded. Alternatively, reduced glutathione (0.1 μmole) is reacted with affinity labeling steroid (0.05 μmole) in 3 ml of 0.05 *M* potassium phosphate buffer, pH 7.0, under an atmosphere of nitrogen. At various intervals 0.2 ml of the reaction mixture is added to 2.8 ml of a solution containing 0.2 μmole of 5,5′-dithiobis(2-nitrobenzoic acid) and the amount of unreacted reduced glutathione is determined by the change in absorbance at 412 nm.

Discussion. Reaction rates of modified steroid with model sulfhydryl compound show distinct differences. For example, 21-iododeoxycortisone reacts faster than cortisone 21-iodoacetate (Fig. 7), and 6β-bromoproges-

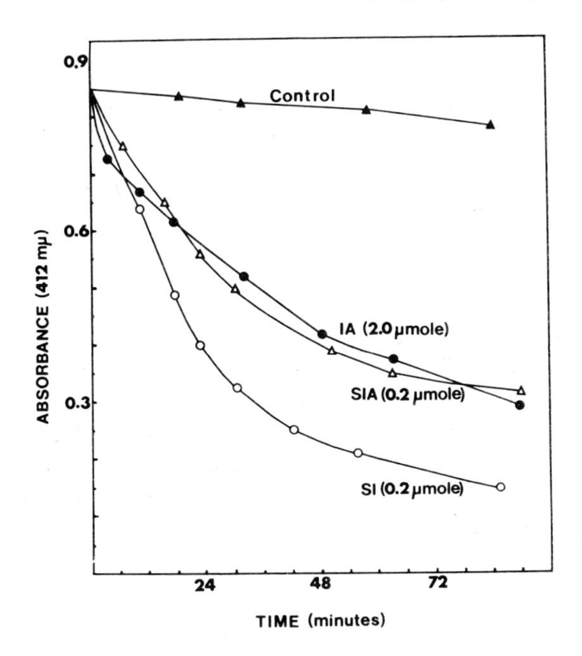

FIG. 7. Reaction of iodo compounds with colored anionic form of 5,5′-dithiobis(2-nitrobenzoic acid). Iodo compounds in the amounts shown were added separately in 0.20 ml of alcohol to the previously formed color of reduced glutathione and 5,5′-dithiobis(2-nitrobenzoic acid) (0.2 μmole each) in 2.80 ml of 0.05 *M* phosphate buffer, pH 7.0, 23°. The curves compare the control (▲) of 5,5′-dithiobis(2-nitrobenzoic acid) with solutions containing iodoacetic acid (●), cortisone 21-iodoacetate (△), and 21-iododeoxycortisone (○). From M. Ganguly and J. C. Warren, *J. Biol. Chem.* **246**, 3646 (1971).

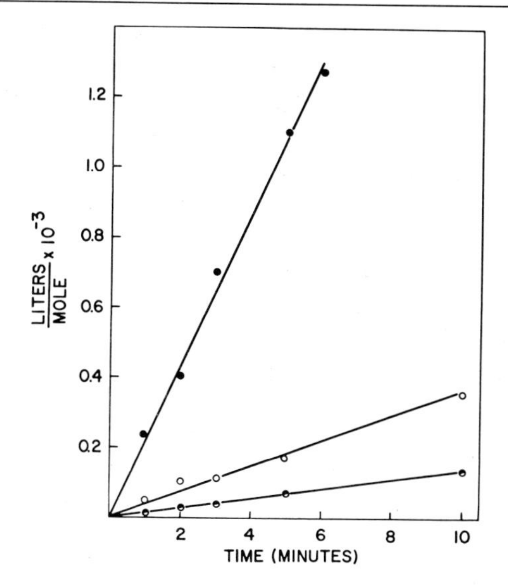

FIG. 8. Second-order plots for the reaction of 0.05 μmole of bromoprogesterones (A) with 0.05 μmole of reduced glutathione (B) in 0.05 *M* potassium phosphate buffer, pH 7.0, 25°, for times shown. Unreacted glutathione was determined by reaction with 5,5′-dithiobis(2-nitrobenzoic acid) at 412 mμ. Reaction $A + B \rightarrow x$ with $[A] = [B]$; the ordinate represents $[x]/[A]([A] - [x])$. (●)6β-Bromoprogesterone, (○)6α-bromoprogesterone, and (◓)2α-bromoprogesterone. From C. C. Chin and J. C. Warren, *Biochemistry* **11**, 2720 (1972).

terone reacts much more rapidly than 2α- and 6α-bromoprogesterone with reduced glutathione. The methods above can be used to establish rate constants (Fig. 8). There is a wide range of variation in the reactivity of affinity labeling steroids with amino acids *in vitro*. For example, bromoacetoxyprogesterones react readily with several amino acids while 4-mercuri-17β-estradiol and 6β-bromoprogesterone react only with cysteine at physiologic pH.

C. Study of the Characteristics of Enzyme Inactivation (Illustrated with 20β-Hydroxysteroid Dehydrogenase)

Principle. Enzyme activity is evaluated by determining the rate at which cofactor is consumed as a result of the reduction of progesterone which serves as the standard substrate with substrate and cofactor at concentrations near enzyme saturation. The slope of the tracing from the initial linear decrease in absorbance at 340 nm (due to oxidation of NADH) with time is used to estimate enzyme activity. Assays may be carried out at 25 ± 1°.

Procedure. A quantity of enzyme (generally 0.1–0.3 nmole) is incubated with 0.1 to 0.3 μmole of affinity labeling steroid. The incubation is conducted at 25°, in 5 ml of 0.05 *M* potassium phosphate buffer, pH 7.0, containing 5% ethanol. Enzyme activity is assayed before the addition of the affinity labeling steroid, immediately following addition, then at regular time intervals thereafter. Assays are carried out spectrophotometrically in 0.05 *M* potassium phosphate buffer, pH 6.5, at 25°, in a final volume of 3 ml, with 0.3 μmole of NADH as cofactor and 0.3 μmole of progesterone serving as the substrate. The assay is initiated by adding an aliquot of the enzyme-inhibitor incubate (containing 0.5–1.0 μg of enzyme) to the NADH–progesterone solution. The slope of the tracing from the initial linear decrease in absorbance at 340 nm with time is used as a measure of enzyme activity. A graph is constructed by plotting the percent of enzymatic activity in the enzyme-inhibitor mixture relative to control on a logarithmic scale along the ordinate and the time of incubation on a linear scale along the abscissa (e.g., Figs. 9 and 12). From this graph the $t_{1/2}$ of inactivation for a given affinity labeling steroid can be calculated. The irreversibility of the inactivation is assessed with enzyme-inhibitor solution, which has reached about 50% inactivation, by dialyzing the solution against 0.05 *M* potassium phosphate buffer, pH 7.0, for 12 hours. If the enzyme inactivation is irreversible, no change in the state of enzyme activity will be appreciated after dialysis. Irreversibility of the reaction can also be observed by adding a 100 molar excess (relative to steroid) of 2-mercaptoethanol to a 50% inactivated enzyme-inhibitor solution (see Fig. 10). The mercaptan combines with unreacted affinity labeling steroid preventing further inactivation, but this does not restore enzyme activity.

Discussion. We have observed a large variation in the $t_{1/2}$ of inactivation of 20β-hydroxysteroid dehydrogenase with affinity labeling progesterones. The slowest rate of inactivation has been observed for 11α-bromoacetoxyprogesterone with $t_{1/2}$ of 11 hours and the fastest for 6β-bromoacetoxyprogesterone with $t_{1/2}$ of 22 minutes. Figure 9 shows the rate of inactivation for 6α- and 6β-bromoprogesterone and also the absence of an inactivating effect on the enzyme with a bromine atom at the 2α-position in the steroid ring. Figure 10 illustrates the affect of 2-mercaptoethanol which prevents further alkylation but does not restore enzymatic activity by combining with unreacted steroid. These differences depend on presence or absence of a reactive amino acid residue, the nature of that residue, and K_s of the steroid.

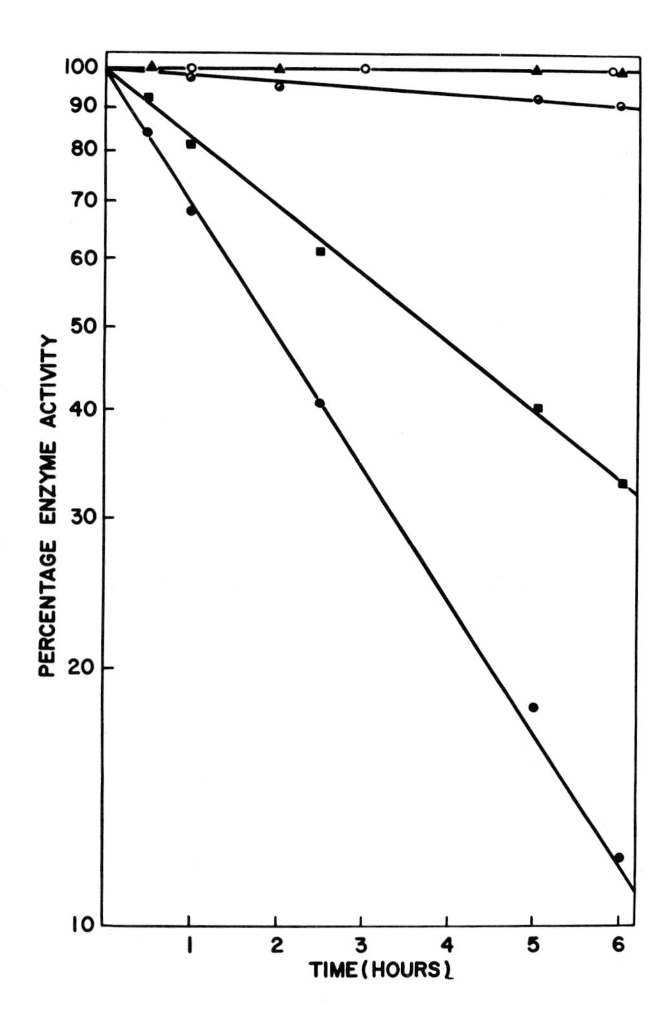

FIG. 9. Pseudo-first-order plots for the effect of progesterone and bromoprogesterones on the activity of 20β-hydroxysteroid dehydrogenase. Preincubations conducted with 50 μg of enzyme in 48 ml of 0.05 M potassium phosphate buffer, pH 7.0, 25°, to which was added at zero time 0.05 μmole of steroids in 2 ml of ethanol separately. At times indicated, 0.1 ml of this solution was assayed as described in the text using near saturated concentrations of substrate and cofactor. (▲)Progesterone, (○)2α-bromoprogesterone, (◒)6α-bromoprogesterone, (●)6β-bromoprogesterone, and (■) 0.05 μmole of 6β-bromoprogesterone and 0.10 μmole of progesterone. From C. C. Chin and J. C. Warren, *Biochemistry* **11**, 2720 (1972).

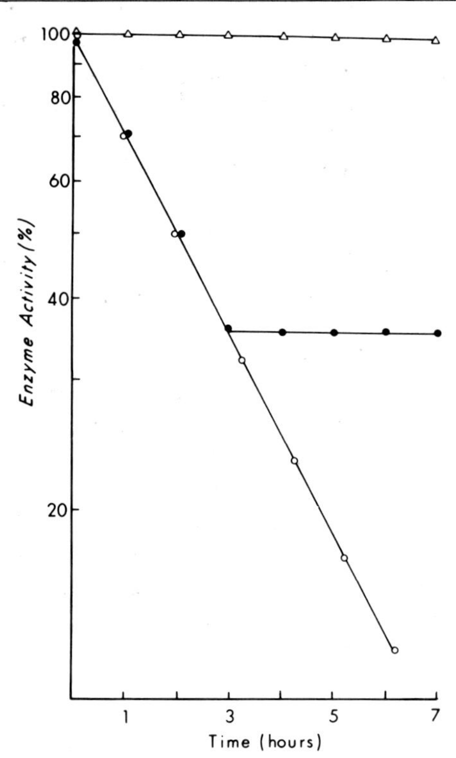

FIG. 10. Effect of 2-mercaptoethanol on enzyme inactivation. The enzyme (0.6 nmole) in 4.8 ml of 0.05 *M* phosphate buffer, pH 7.0, was incubated with 16α-bromoacetoxyprogesterone (0.3 μmole) in 0.2 ml of ethanol (○). Excess 2-mercapto-ethanol (30 μmoles) was added to the enzyme simultaneously with the steroid (△) or after 3 hours of incubation of the enzyme–steroid mixture (●). At 1-hour intervals, 0.1 ml of the incubation mixture was removed and used for assay of enzyme activity. The percentage of enzyme activity is plotted on a logarithmic scale along the *ordinate* and the time incubation is plotted on a linear scale along the *abscissa*. The values plotted are the means from at least three assays. From F. Sweet, F. Arias, and J. C. Warren, *J. Biol. Chem.* **247**, 3424 (1972).

D. Obtaining Evidence That the Affinity Labeling Steroid Binds at the Enzyme Active Site

Principle. Evaluate the affinity labeling steroid as a substrate; serving as substrate is proof that it goes to the active site. Inactivate the enzyme in the presence of a natural substrate, which competes for the binding site with the affinity labeling steroid. This is evidenced by a reduction in the rate of inactivation.

Procedure. To evaluate an affinity labeling steroid as a substrate,

0.5–1.0 μg of enzyme is dissolved in 0.1 ml of 0.05 *M* phosphate buffer, pH 7.0, and added to a cuvet containing (a) 2.5 ml of 0.05 *M* phosphate buffer, pH 6.5; (b) 0.3 μmole of NADH in 0.2 ml of 0.05 *M* phosphate buffer, pH 7.0, and (c) affinity labeling steroid in increasing concentrations (usually 0.1–0.3 μmole) in 0.2 ml of ethanol. After mixing the components the enzymatic activity is determined from the slope of the tracing produced by the initial decrease with time in the absorbance at 340 nm. The value for enzymatic activity (generally expressed as nmoles of NADH oxidized/minute/μg of enzyme) is plotted against the concentration of affinity labeling steroid present in the incubation mixture, with a Lineweaver-Burk double reciprocal plot (Fig. 11), to obtain the V_{max} and the K_m for each compound. To study the effect of cofactor or a natural substrate on the rate of enzyme inactivation 3.0 nmoles of enzyme are dissolved in 4.6 ml of 0.06 *M* phosphate buffer, pH 7.0, and 0.3–0.6 nmole of either cofactor of substrate; in 0.2 ml of buffer or ethanol, respectively, is added to the solution before the final addition of 0.3 μmole of affinity labeling steroid in 0.2 ml of ethanol. The rate of enzyme inactivation is recorded as described above and compared with parallel experiments wherein the enzyme is separately incubated with inhibitor, cofactor, substrate, and solvent.

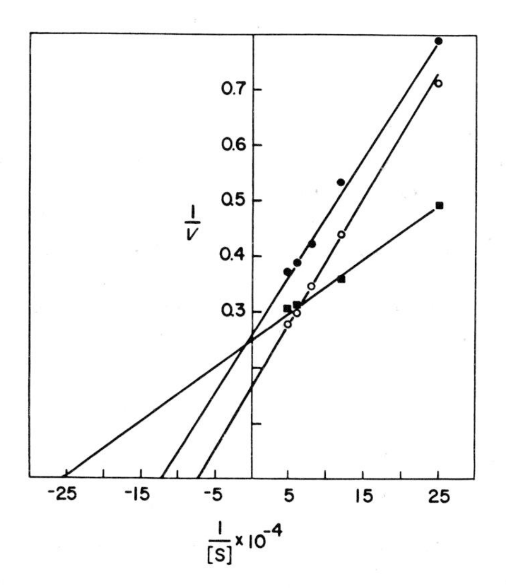

FIG. 11. Double-reciprocal plot of reduction of the bromoprogesterones by 20β-hydroxysteroid dehydrogenase. *V*, velocity in nmoles of NAD^+ generated per minute; [S], molar concentration of substrate; (●)6β-bromoprogesterone; (○)6α-bromoprogesterone; and (■)2α-bromoprogesterone. From C. C. Chin and J. C. Warren, *Biochemistry* **11**, 2720 (1972).

TABLE II
KINETIC DATA FOR REDUCTION OF STEROID DERIVATIVES BY 20β-HYDROXYSTEROID DEHYDROGENASE

Substrate	K_m Value ($\times 10^{-5}$ M)	V_{max} (nmoles/minute/μgE)	$t_{1/2}$ of 20β-HSDH inactivation (hours)	Ref.
Progesterone	0.395	11.58	—	a, b, c
Cortisone	5.10	12.52	—	a, b, c
6β-Acetoxyprogesterone	2.87	0.5	—	d
6β-Bromoacetoxyprogesterone	2.0	3.72	0.36	d
11α-Acetoxyprogesterone	2.78	9.00	—	d
11α-Chloroacetoxyprogesterone	1.02	7.70	—	d
11α-Bromoacetoxyprogesterone	2.5	7.10	5.0	d
16α-Acetoxyprogesterone	12.5	0.33	—	c
16α-Chloroacetoxyprogesterone	9.1	0.59	—	c
16α-Bromoacetoxyprogesterone	14.5	1.81	2.0	c
Cortisone 21-iodoacetate	10.0	10.0	3.5	a
21-Iododeoxy cortisone	14.0	20.0	—	a

[a] M. Ganguly and J. C. Warren, *J. Biol. Chem.* **246,** 3646 (1971).
[b] C. C. Chin and J. C. Warren, *Biochemistry* **11,** 2720 (1972).
[c] F. Sweet, F. Arias, and J. C. Warren, *J. Biol. Chem.* **247,** 3424 (1972).
[d] F. Sweet, F. Arias, and J. C. Warren, *J. Biol. Chem.* **248,** 5641 (1973).

Discussion. The K_m values obtained for each affinity labeling steroid can be taken only as an approximation of their actual K's values. Various compounds, e.g., 6β-bromoprogesterone and 6α-bromoprogesterone or 16α-bromoacetoxyprogesterone and 16α-chloroacetoxyprogesterone, exhibit similar K_m and V_{max} values (Table II). The fact that an affinity labeling steroid is a substrate for an enzyme does not in itself constitute proof that inactivation involves binding solely at the active site. However, if the inactivation process is significantly retarded by the presence of a natural substrate (Fig. 12), the conclusion necessarily follows that the natural substrate and the modified substrate compete for the enzyme active site.

E. Inactivation of the Enzyme with Radioactive Affinity Labeling Steroid to Obtain Stoichiometry

Principle. Incubations of enzyme and radioactive affinity labeling steroid are performed under identical conditions to those described in Section

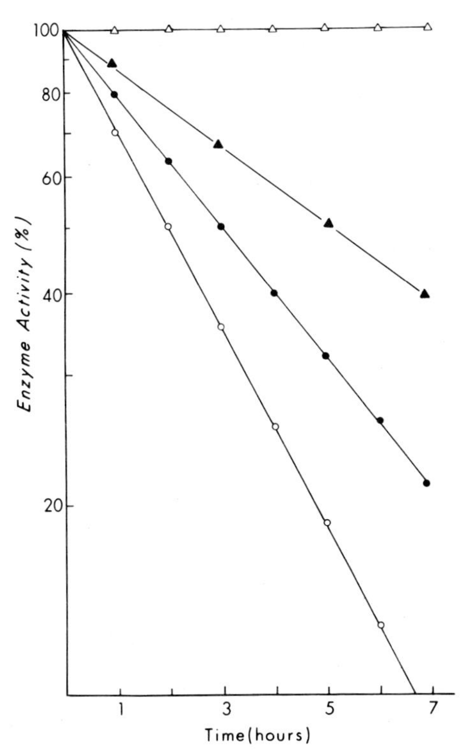

FIG. 12. Inactivation of 20β-hydroxysteroid dehydrogenase by 16α-bromoacetoxyprogesterone, and protection against enzyme inactivation by cofactor or substrate. Enzyme (3.0 nmoles) dissolved in 4.8 ml of 0.05 *M* phosphate buffer, pH 7.0, mixed with: 16α-bromoacetoxyprogesterone, (0.3 μmole) in 0.2 ml of ethanol (○); 16α-bromoacetoxyprogesterone (0.3 μmole) and progesterone (0.6 μmole) in 0.2 ml of ethanol (●); progesterone (0.3 μmole), or bromoacetic acid (0.3 μmole), or 16α-acetoxyprogesterone (0.3 μmole), or 16α-chloroacetoxyprogesterone (0.3 μmole) separately added in 0.2 ml of ethanol (△). Enzyme (3.0 nmoles) and NAD^+ (0.6 μmole) in 4.8 ml of 0.05 *M* phosphate buffer at pH 7.0 mixed with 16α-bromoacetoxyprogesterone (0.3 μmole) in 0.2 ml of ethanol (▲). At various time intervals, 0.1 ml of each mixture was assayed for enzyme activity. The percentage of enzyme activity is plotted on a logarithmic scale along the *ordinate* and the time of incubation is plotted on a linear scale along the *abscissa*. The values plotted are the means from at least four assays.

C above. The enzyme inactivation is quenched by addition of excess 2-mercaptoethanol, which almost instantaneously reacts with unreacted steroid. Elimination of the alkylating capacity of the steroid stops the covalent reaction with the enzyme. Small molecules are separated from the radiolabeled enzyme by exhaustive dialysis; protein is quantitated with radioactivity determined. Knowing the specific activity of the affinity labeling steroid, and the quantity of enzyme inactivated, the stoichiometry of enzyme inactivation can be calculated.

Procedure. The enzyme (0.1 μmole) is dissolved in 100 ml of 0.05 *M* phosphate buffer, pH 7.0. A 3.0-ml aliquot is removed to serve as a control and replaced with 3.0 ml of ethanol which contains 60 μmoles of radioactive affinity labeling steroid. Enzyme activity is determined at the same time intervals as in the enzyme inactivation experiments with nonradioactive steroid. Aliquots (30 ml) of 25 and 50% inactivated incubation mixtures are removed and 20 mg of 2-mercaptoethanol are added to each. The remainder of the incubation mixture (40 ml) is treated with 30 mg of 2-mercaptoethanol when 75–80% inactivation is obtained. Controls should retain the original enzyme activity throughout this experiment. The 25, 50, and 75% inactivated samples are separately dialyzed against distilled water with frequent changes of the dialysate. When the dialysate is found to contain radioactivity at the "background" level, dialysis is terminated and the radioactivity and protein concentration of the contents of the dialysis bag is determined. The results are used to calculate the stoichiometry of enzyme inactivation.

Discussion. Ideally the stoichiometry of inactivation shows that 1 mole (or in some cases 2 moles) of radioactive steroid is incorporated per mole of enzyme present in the incubation mixture (Fig. 13). When steroids with prolonged $t_{1/2}$ of inactivation are used one may obtain an artifactual stoichiometry after extensive inactivation. In such cases we have found nonspecific denaturation as shown by loss of activity in the enzyme control. Enzyme denaturation makes several amino acids available for binding to the modified steroid with increase in the number of radioactive groups incorporated in the macromolecule. This problem can be avoided by shortening time of inactivation.

F. Identification of the Modified Amino Acid in the Enzyme Active Site

Principle. Enzyme inactivation with radioactive steroid is conducted as outlined above. After dialysis, the enzyme samples are digested and the hydrolysates analyzed in an amino acid analyzer. The location of the radiolabeled, modified amino acid is determined by its elution profile. Authentic modified amino acid is then analyzed together with the radioactive enzyme hydrolysate to verify the identity of the labeled amino acid residue.

Procedure. Inactivated enzyme samples obtained at 25, 50, and 75% inactivation are concentrated to dryness under reduced pressure and at

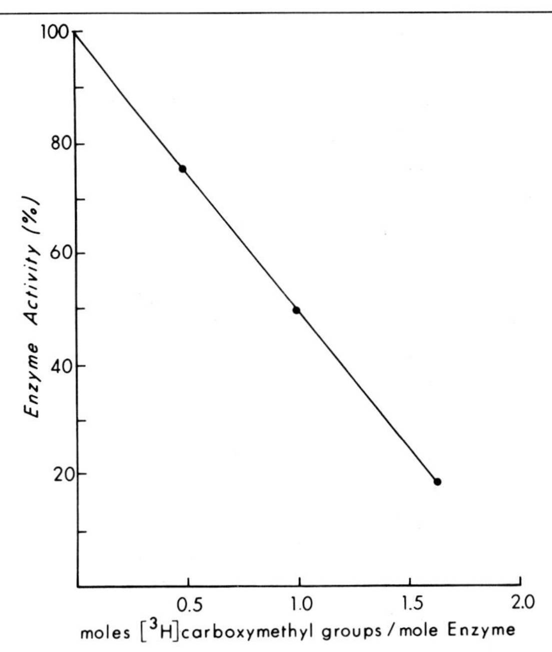

FIG. 13. Stoichiometry of incorporation of [^{3}H]carboxymethyl groups correlated with decrease in enzyme activity. The radiolabeled enzyme preparations (25, 50, and 100% inactivated) were exhaustively dialyzed against distilled water, concentrated to dryness, and the residues dissolved in phosphate buffer 0.05 *M*, pH 7.0. The protein content and radioactivity of the resulting solutions were quantitated. The percentage of enzyme activity is plotted along the *ordinate* and moles of [^{3}H]carboxymethyl groups incorporated per mole of enzyme plotted along the *abscissa*. From F. Sweet, F. Arias, and J. C. Warren, *J. Biol. Chem.* **247,** 3424 (1972).

low temperature.[21] The dry protein samples are digested with constant boiling 6 *N* HCl or with proteolytic enzymes. For acid hydrolysis, samples (1 mg protein) are heated at 110° for 18 to 24 hours in 6 *N* HCl (1 ml) in evacuated, sealed tubes. Then the hydrolysate is concentrated to dryness.[21] The residue is dissolved in 1 ml of 0.2 *M* sodium citrate buffer, pH 3.2, and subjected to amino acid analysis. These conditions provide sufficient final sample to obtain four amino acid analyses. Two methods have been used in our laboratory for enzymatic digestion of labeled enzymes. One method by Hill and Schmidt[22] uses papain, leucine aminopeptidase, and prolidase. More recently we have serially digested

[21] It is important to collect the condensate from this evaporation to quantitate the radioactivity which it may contain, especially after acid hydrolysis as some derivatives may break up into volatile products which appear in the condensate.

[22] R. L. Hill and W. R. Schmidt, *J. Biol. Chem.* **237,** 389 (1962).

the protein with protease type VI (Sigma Chemical Co., St. Louis, Missouri) and aminopeptidase M (Henley and Co., Inc., New York, New York). In the latter case 1 mg of dry inactivated enzyme is dissolved in 0.5 ml of 0.05 *M* potassium phosphate buffer, pH 7.0, to which 0.25 ml of a similar buffered solution containing 20 μg of protease VI are added, then the resultant mixture is incubated at 37° for 18 hours. A solution of buffer (0.25 ml) containing 500 milliunits of aminopeptidase M is added to the incubation which is continued an additional 24 hours. At the end of the incubation period, the mixture is concentrated to dryness (checking for evaporation losses), and the residue is dissolved in 1 ml of 0.2 *M* sodium citrate buffer, pH 3.25, for amino acid analysis. Amino acid analysis is performed according to a procedure by Spackman *et al.*[23] or by Bradshaw *et al.*[24] Analysis of radiolabeled hydrolysate is accomplished by collecting effluent fractions from the analyzer column which emerge after the ninhydrin reaction (at a rate of 1.5 ml/minute). Aliquots (0.5 ml) of each fraction are dissolved in 15 ml of scintillation fluid [0.05%, 2,5-diphenyloxazole, 0.01% 1,4-bis(5-phenyloxazolyl)benzene, and toluene–Triton X-100 (2:1)], and the radioactivity is quantitated in a liquid scintillation spectrometer. In the case of carboxymethyl derivatives, location of the radioactive peaks which emerge before aspartic acid is critical for accurate identification of the radiolabeled product (Fig. 14). Therefore, ^{14}C-labeled L-aspartic acid or synthetic L-carboxymethylcysteine is added as a marker to the sample prior to analysis. A recognizable elution profile of the radioactive products from the amino acid analysis does not in itself constitute an absolute proof of identity. Following tentative identification of the modified amino acid based on the time of appearance of the radioactive peak, it is necessary to carry out amino acid analysis of a sample to which has been added a suitable amount of the authentic material. The ninhydrin positive peak generated by the analyzer must exactly coincide with the radioactive peak obtained from the hydrolysate. In the event that the radioactive product obtained in the analysis is not ninhydrin positive a double-labeling technique, with the authentic material containing an isotope different from that of the hydrolysate, is used to obtain confirmatory identification. It is also desirable to obtain evidence of the identity of the radioactive product by cocrystallization. In this case the hydrolysate is mixed with pure authentic product and submitted to several recrystallizations from the appropriate solvent. A constant value for the specific activity of the mixture throughout several recrystallizations is excellent evidence of identity.

[23] D. H. Spackman, W. H. Stein, and S. Moore, *Anal. Chem.* **30**, 1190 (1958).

[24] R. R. Granberg, K. A. Walsh, and R. A. Bradshaw, *Anal. Biochem.* **30**, 454 (1969).

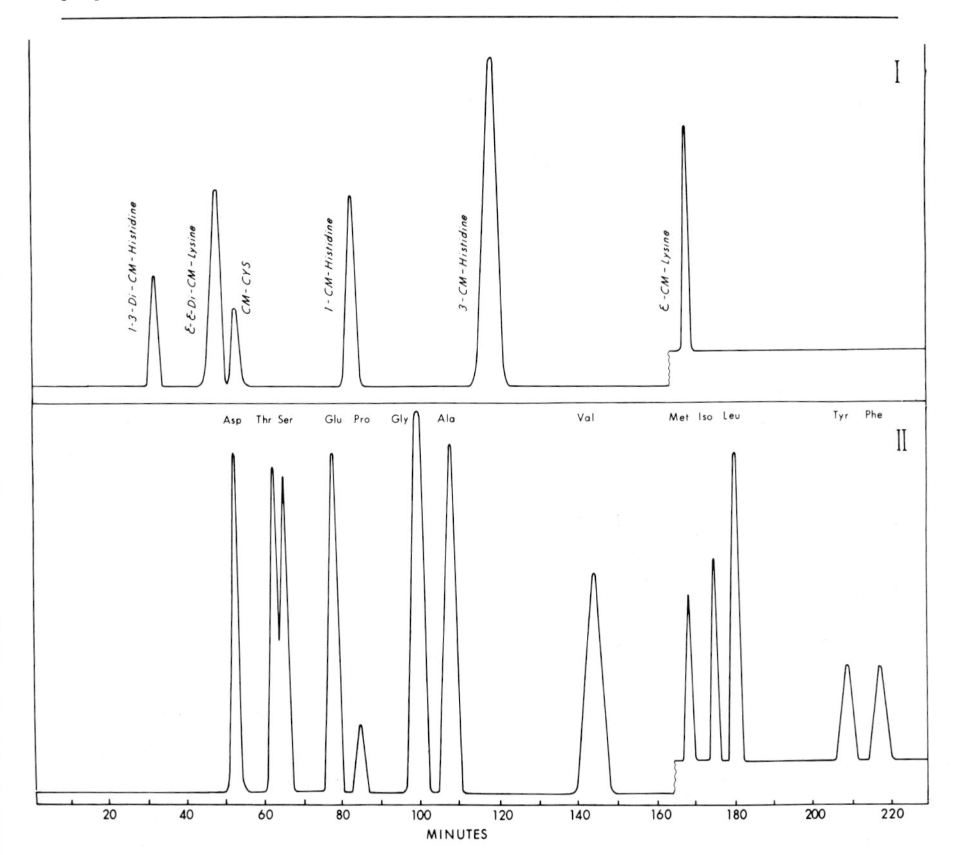

FIG. 14. Elution profile of carboxymethylated amino acids. Mixture containing 1,3-dicarboxymethylhistidine (1-3-Di-CM-Histidine), ϵ,ϵ-dicarboxymethyllysine (ϵ,ϵ-Di-CM-Lys), carboxymethylcysteine (CM-CYS), 1-carboxymethylhistidine (1-CM-Hist), 3-carboxymethylhistidine (3-CM-Hist), and ϵ-carboxymethyllysine (ϵ-CM-Lys) (I) compared with that of 6 N HCl hydrolysate of 20β-hydroxysteroid dehydrogenase (II). From F. Sweet, F. Arias, and J. C. Warren, *J. Biol. Chem.* **247,** 3424 (1972).

Discussion. It has been found that amino acid analysis is not very useful for identifying the labeled amino acid residue when the alkylating group is a bromine attached to the steroid ring. In this case a covalent bond is established between the steroid ring and the amino acid and is resistant to acid hydrolysis. The steroid amino acid conjugate is retained in the long column of the analyzer an appears only with the NaOH wash, which precludes positive identification. In this case it is necessary to use chromatographic techniques to identify the labeled amino acid residue (Fig. 15). With bromoacetoxysteroids, both acid and enzymatic hydroly-

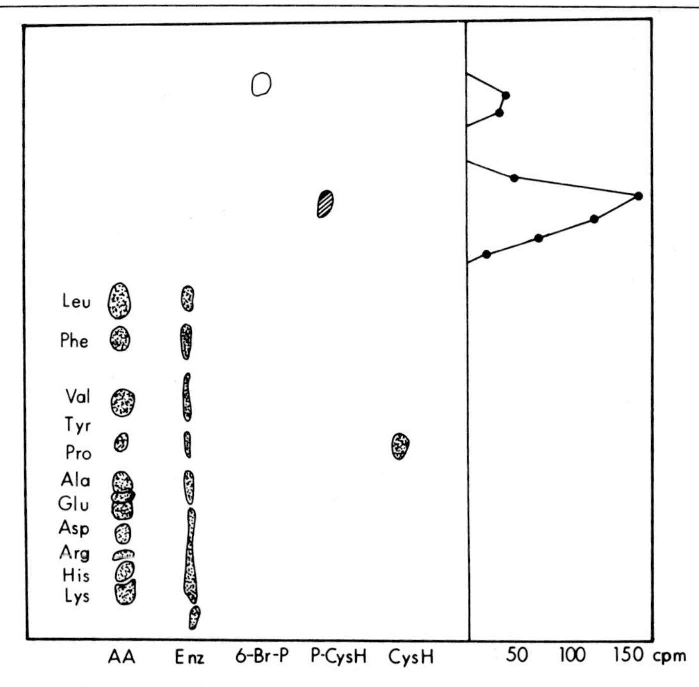

FIG. 15. Distribution of radioactivity on thin-layer chromatography of hydrolysate of 20β-hydroxysteroid dehydrogenase after treatment with tritiated 6β-bromoprogesterone. (●) Ninhydrin positive, (○) ultraviolet light absorbent, (◍) ninhydrin positive and ultraviolet light absorbent. AA, standard amino acids; Enz, enzyme hydrolysate; 6-Br-P, 6β-bromoprogesterone; P-CysH, progesterone-6-*S*-L-cysteine; CysH, L-cysteine. From C. C. Chin and J. C. Warren, *Biochemistry* **11**, 2720 (1972).

sis break the ester bond generating a carboxymethyl amino acid. In these cases a substantial amount of the radioactivity appears in the pre-aspartic region of the amino acid elution profile. Examples of pre-aspartic peaks are 1,3-dicarboxymethyl-L-histidine, glycolic acid, ε,ε-dicarboxymethyl-L-lysine, and carboxymethyl-L-cysteine. Products from carboxymethylation of L-histidine produce two peaks, in addition to the early peak due to 1,3-dicarboxymethylhistidine (Fig. 14). One of the peaks due to 1-carboxymethyl-L-histidine superimposes on L-proline while that due to 3-carboxymethyl-L-histidine appears between alanine and L-valine (Fig. 14). Carboxymethylation of L-lysine produces several pre-aspartic peaks corresponding to various products of mono- or dicarboxymethylation of the α or ε groups of the amino acid. However, identification can readily be made because of the distinct peak of ε-carboxymethyl-L-lysine, which appears in the elution profile superimposed on the methionine peak (Fig. 14). The carboxymethylation of L-cysteine yields a product charac-

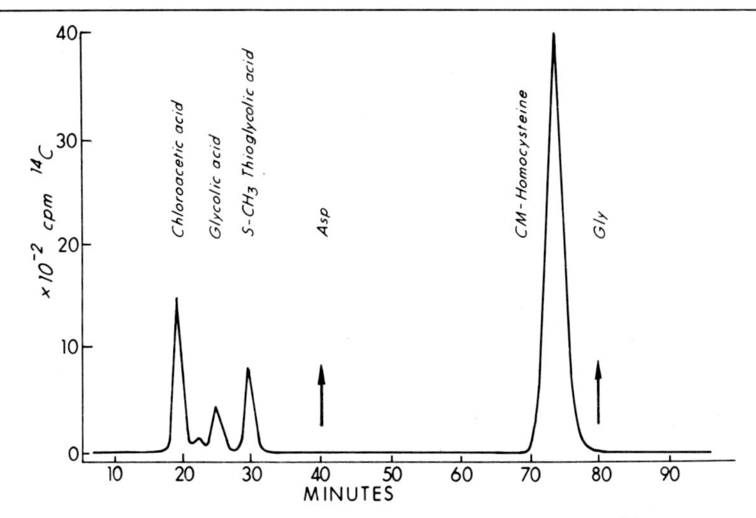

FIG. 16. Elution profile of 6 N HCl hydrolysis products of S-[^{14}C]carboxymethyl-methionine sulfonium iodide.

terized by its appearance just before the L-aspartic acid peak. Figure 14 shows the elution profile of several carboxymethyl derivatives of amino acids as they normally appear in amino acid analysis. The identification of glycolic acid as a radioactive product of enzyme labeling poses special problems since it could arise from carboxymethylation of either L-aspartic acid, L-glutamic acid, the terminal carboxyl group of the protein, or from S-carboxymethyl-L-methionine. In the last case, positive identification is possible because of the appearance of other characteristic products from the hydrolysis of carboxymethylmethionine sulfonium salt. In particular, carboxymethyl-L-homocysteine appears just after glutamic acid (Fig. 16). It is important to recognize that chloroacetic acid, thioglycolic acid, and to some extent glycolic acid, all products of carboxymethyl-L-methionine hydrolysis, are volatile compounds which may be partially or completely lost during evaporations carried out prior to amino acid analysis.

Biological Applications of Affinity Labeling Steroids

Estrogenic Activity of 4-Mercuri-17β-estradiol (4MEβ)

Procedure. For biological applications, direct application to the target organ is preferred. In the case of uterus, this is done by placing an otoscope speculum in the vagina which allows visualization of the cervix. A blunt (sharp end cut off), 16 or 18 guage spinal needle bent 30° about 1 cm from the end can then be inserted into the endometrial cavity of the anethetized animal. By abdominal palpition of the supine animal and with practice, one can determine the position of the needle (i.e., which horn and how far up). Application is made in 1.8% saline with volumes of 0.05 ml or so. The vehicle should also contain a small amount of alcohol sufficient to keep the steroid in solution.

Discussion. Direct intraluminal application of 4MEβ (or its mercaptide with 2-mercaptoethanol) induces morphological effects in the rat uterus typical of estrogenic stimulation, increases the activity of glucose-6-phosphate dehydrogenase and 6-phosphogluconate dehydrogenase and promotes increase in uterine weight and glycogen deposition.[25] These are typical estrogenic responses.[26] A dose-response analysis of 17β-estradiol after intraluminal application clearly indicates that the effects observed with mercury steroid are not the result of contamination with 17β-estradiol. The increase in glucose-6-phosphate dehydrogenase activity induced by 4MEβ or its mercaptide with 2-mercaptoethanol is not detected after the administration of either *p*-mercuribenzoate or its mercaptide. Direct intraluminal application of 4MEβ minimizes its reaction with sulfhydryl groups in blood. It was feared that the same property might result in nonspecific binding to membrane sulfhydryl groups even after intraluminal injection but this was certainly not entirely the case.

One might presume that 4MEβ (as a result of its capacity to form a pseudocovalent bond with the receptor) should not only show estrogenic activity but that its estrogenic activity might be prolonged as compared to 17β-estradiol. Studies on the levels of uterine glucose-6-phosphate dehydrogenase activity as a function of time following administration of 4MEβ indicate that this is, indeed, the case. Figure 17 shows enzyme activity in the uterus after administration of 4MEβ (or 4MEα) and 17β-estradiol. In both cases, 100 ng of compound are deposited in the uterine horn. After the time periods shown the uteri are removed, homogenized, and the supernatant assayed for glucose-6-phosphate dehydrogenase (G6PD) using methods of Barker and Warren.[27] When 17β-estradiol is administered, enzyme activity returns to base levels 140 hours after administration while in the case of 4MEβ activity is diminished only slightly from the maximum values seen at 20 hours. Similar effects are observed in terms of intrauterine glycogen content. These observations indicate that 4MEβ has a prolonged estrogenic activity which may result

[25] T. G. Muldoon and J. C. Warren, *J. Biol. Chem.* **244,** 5430 (1969).
[26] K. L. Barker, M. H. Nielson, and J. C. Warren, *Endocrinology* **79,** 1069 (1966).
[27] K. L. Barker and J. C. Warren, *Endocrinology* **78,** 1205 (1966).

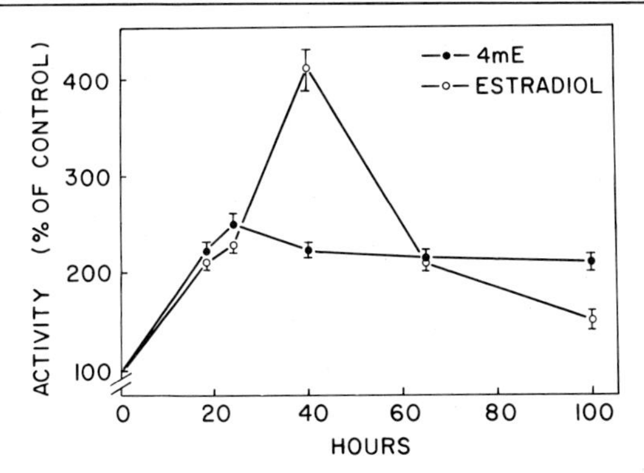

FIG. 17. Uterine glucose-6-P dehydrogenase activity at various intervals after a single intraluminal application of 0.1 μg of 17β-estradiol or 4-mercuri-17β-estradiol (4ME). Activity calculated as units per uterus for each group of four animals. Values shown are mean ± standard error expressed as percentage of controls for comparison. Rats weighed 200–220 g and were 3 weeks postovariectomy. From T. G. Muldoon and J. C. Warren, *J. Biol. Chem.* **244**, 5430 (1969).

from its pseudocovalent bond formation. Nevertheless, this compound can be slowly released from macromolecules. Thus, one cannot rule out the possibility that the prolonged activity may be due to its retention in the uterus by reaction with nonspecific sulfhydryl groups and subsequent continuous exchange onto the receptor macromolecule. At various time periods after administration of 4MEβ, injected uteri are homogenized and extracted. No free 17β-estradiol is detected, indicating that reductive cleavage *in vivo* of the carbon–mercury bond does not occur in uterine tissue.

Evidence that 4MEβ reacts with the cytoplasmic estradiol receptor is presented in Fig. 18. It can be seen that 4MEβ binds to a macromolecule which has a mobility in sucrose density gradient identical with the receptor protein to which 17β-estradiol binds. Furthermore, exposure of the supernatant samples to 17β-estradiol at 0° prior to the additon of 4MEβ prevents binding of the latter compound to the receptor. Finally, when fractions 8–18 are extracted with chloroform 98% of the 17β-estradiol is extracted while in the case of 4MEβ only 4% can be found in organic extract, suggesting that 4MEβ is bound to the receptor protein by a pseudocovalent bonding process. Thus, there is a distinct difference between the completely reversible binding of 17β-estradiol and that of 4MEβ.

The above methods can be used to demonstrate that 4-mercuri-17β-estradiol resembles 17β-estradiol in the production of several biological responses and also binds to the cytoplasmic receptor. 4MEβ mainly

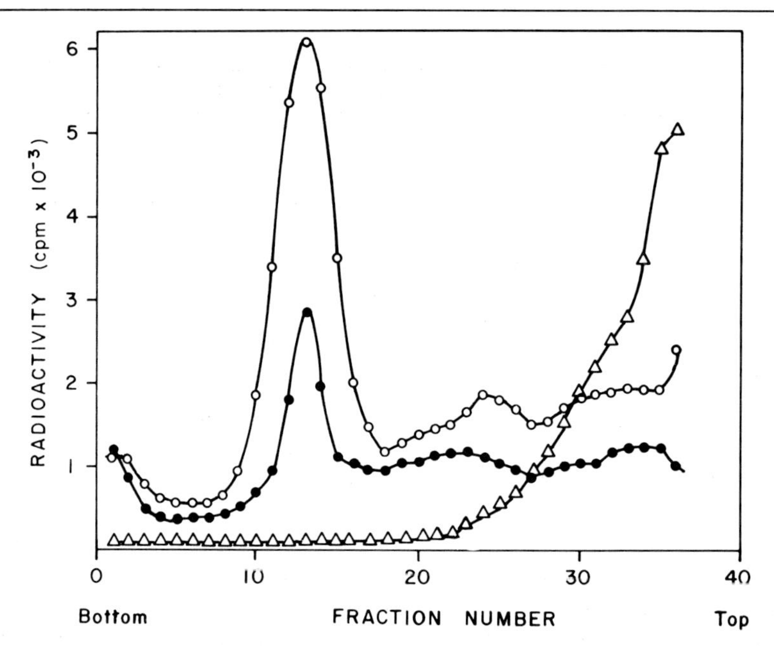

FIG. 18. Density gradient sedimentation patterns of radioactivity in uterine supernatant fractions after 15 minutes of incubation with 17β-estradiol-6,7-^{3}H (1.2×10^{-8} M) (○), 4-mercuri-17β-estradiol-6,7-^{3}H (1.3×10^{-8} M) (●), unlabeled 17β-estradiol (1.2×10^{-8} M) followed by 15 minutes of incubation with 4-mercuri-17β-estradiol-6,7-^{3}H (1.3×10^{-8} M) (△). Samples of 200 μl were layered onto 5 to 20% sucrose gradients and centrifuged for 13 hours at 220,000 g at 2°. From T. G. Muldoon and J. C. Warren, *J. Biol. Chem.* **244**, 5430 (1969).

differs from 17β-estradiol in its persistent biological activity and in its inability to be extracted with organic solvents, once it is bound to the cytoplasmic estrogen receptor.

Antiestrogenic Activity of 4-Mercuri-17α-estradiol (4MEα)

Discussion. While 4MEβ, on direct intrauterine application, possesses long-acting estrogenic activity, the epimeric 4-mercuri-17α-estradiol (4MEα) displays long-acting antiestrogenic activity. Although 17α-estradiol competes with 17β-estradiol at the receptor site, the 17α-epimer has little measurable estrogenic activity. Several reports have suggested that low estrogenicity and significant antiestrogenicity of compounds like 17α-estradiol are due to rapid clearance from the target organ.[28,29]

[28] L. Martin, *Steroids* **13,** 1 (1969).

[29] S. G. Korenman, *Steroids* **13,** 163 (1969).

TABLE III[a]
ESTROGENIC ACTIVITY OF 4-MERCURI-17α-ESTRADIOL (4MEα) AND 4-MERCURI-17β-ESTRADIOL (4MEβ)[b]

Exp.	Treatment	(ng/horn)	G6PD Activity (mU/horn)
1	Saline	—	98.0 ± 5.8
2	4MEβ	50	208.6 ± 8.9
3	4MEα	50	105.0 ± 5.9
4	4MEα	100	105.7 ± 3.6
5	4MEα	200	122.5 ± 7.7

[a] From R. W. Ellis and J. C. Warren, *Steroids* **17,** 331 (1971).
[b] Steroids given by uterine intraluminal application to Holtzman rats (200–220 gm) ovariectomized 4 weeks prior to use. Enzyme activities determined 24 hours later. Each value represents six animals; 5 is significantly greater than 1, $p < 0.02$.

When 4-mercuri-17α-estradiol (4MEα) is synthesized and structure and purity assured,[30] intraluminal application into the uterus of the castrate animal shows it to be a much less potent estrogen than 4MEβ (Table III). Following intraluminal application of 4MEβ (50 ng/horn), uterine G6PD activity is slightly more than doubled while 100 ng/horn of 4MEα does not cause a significant increase. Only with very high levels of 4MEα do G6PD levels increase somewhat slightly above base values.

Table IV shows that 4MEα possesses antiestrogenic activity. Intraluminal uterine application of 4MEα or 17α-estradiol prior to intrauterine application of 17β-estradiol significantly attenuates the elevation of uterine G6PD activity. 17α-Estradiol and 4MEα are almost equally potent in their antiestrogenic activity when given 2 hours prior to administration of 17β-estradiol. The persistent antiestrogenic activity of 4MEα as compared to that of 17α-estradiol is impressive. After 24 and 48 hours between administration of 4MEα (or 17α-estradiol) and 17β-estradiol, the antiestrogenic activity of 17α-estradiol falls while that of 4MEα persists. This effect of 4MEα after 48 hours is similar to that observed after 2 hours.

These observations show that 4MEα is retained in the target organ. Under these conditions, this steroid displays barely measurable estrogenicity but has considerable antiestrogenicity which is effected by its exclusion of 17β-estradiol from the receptor site.

The "Achilles heel" in the use of affinity labeling mercury steroids is the chemical nature of the mercaptide bond which they form. Rather

[30] R. W. Ellis and J. C. Warren, *Steroids* **17,** 331 (1971).

TABLE IV[a]
ANTIESTROGENIC ACTIVITY OF 4-MERCURI-17α-ESTRADIOL (4MEα) AND 17α-ESTRADIOL[b]

Exp.	First administration	Time (hours)	Second administration	G6PD Activity (mU/horn)
1	Vehicle	2	Saline	106.2 ± 10.1
2	Vehicle	2	17β-Estradiol	172.5 ± 15.6
3	17α-Estradiol	2	17β-Estradiol	125.9 ± 10.2
4	4MEα	2	17β-Estradiol	131.0 ± 10.8
5	17α-Estradiol	24	17β-Estradiol	155.0 ± 11.7
6	4MEα	24	17β-Estradiol	133.5 ± 2.8
7	17α-Estradiol	48	17β-Estradiol	177.2 ± 10.1
8	4MEα	48	17β-Estradiol	132.9 ± 5.8

[a] From R. W. Ellis and J. C. Warren, *Steroids* **17,** 331 (1971).

[b] Steroids given by uterine intraluminal application to Holtzman rats (200–220 gm) ovariectomized 4 weeks prior to use. Each value represents six animals. First administration was vehicle, 17α-estradiol (50 ng/horn) or 4MEα (100 ng/horn). After the times shown, saline or 17β-estradiol (4 ng/horn) applied to same horn. Enzyme activity in injected horn evaluated 24 hours after second administration; 2 is significantly greater (at least $p < 0.05$) than 3, 4, 6, and 8, but not 5 and 7. Further, 7 is significantly greater than 8 ($p < 0.01$).

than being covalent, it is a pseudocovalent one and there is the possibility of exchange of the steroid from one cysteinyl residue to another. Thus, long acting biological activity could result from retention of the 4ME compounds by nonspecific sulfhydryl groups in a tissue and subsequent transfer to receptor molecules. Indeed, distinct absence of transport of 4MEβ into the target cell nucleus[31] may be due to its early dissociation from cytoplasmic receptor. According to recent observations by Jensen, estradiol may activate the cytoplasmic receptor, then dissociate from the activated receptor while the latter goes on to stimulate RNA synthesis in uterine cell nuclei *in vitro*.[32] Affinity labeling steroids capable of forming stable covalent bonds are attractive because they cannot dissociate once such a bond is established. Their biological applications should be similar to those illustrated with the mercury steroids.

[31] T. G. Muldoon and J. C. Warren, *Biochemistry* **10,** 3780 (1971).

[32] E. V. Jensen, *Biochem. Biophys. Res. Commun.* **48,** 1601 (1972).

[34] Synthesis of Affinity Labels for Steroid-Receptor Proteins

By Howard E. Smith, Jon R. Neergaard, Elizabeth P. Burrows, Ross G. Hardison, and Roy G. Smith

A. Introduction

The general considerations underlying active-site-directed irreversible enzyme inhibition[1] (affinity labeling of enzymes[2]) have been applied to the study of steroid-binding proteins,[3] restricting attention to cytoplasmic androgen[4,5] and progestagen[6,7] receptors.

There are two requisites for the affinity labeling of a steroid receptor: The steroid must bind reversibly at the active site of the receptor, and the steroid must possess a functional group capable of forming a covalent bond with an amino acid residue either within (endo[1]) or adjacent to (exo[1]) the binding site.

The first requisite seems relatively easy to fulfill. The progesterone receptor of chick oviduct forms a highly stable ($k_d \simeq 10^{-10}$) complex with progesterone,[6,7] and competition studies show that other steroids form complexes somewhat less stable but with $k_d \leq 10^{-8}$.[8] Thus, in the absence of progesterone, the progesterone-binding protein should strongly bind other steroids.

The second requisite is more difficult. The amino acids in the polypeptide chain at the binding site are not known. Cysteine may be one since it has been shown for both a rat prostate 5α-dihydrotestosterone receptor[5] and the chick oviduct progesterone receptor[6] that sulfhydryl groups are involved either in binding of the steroid to the receptor or in maintenance of the active structure of the protein. We do assume that one or both of the oxygen functions of 5α-dihydrotestosterone (C-3 and C-17) and

[1] B. R. Baker, "Design of Active-Site-Directed Irreversible Enzyme Inhibition." Wiley, New York, 1967.

[2] L. Wofsy, H. Metzger, and S. J. Singer, *Biochemistry* **1**, 1031 (1962).

[3] U. Westphal, "Steroid-Protein Interaction." Springer-Verlag, Berlin and New York, 1971.

[4] S. Fang and S. Liao, *J. Biol. Chem.* **246**, 16 (1971).

[5] W. I. P. Mainwaring. *J. Endocrinol.* **45**, 531 (1969).

[6] M. R. Sherman, P. L. Corvol, and B. W. O'Malley, *J. Biol. Chem.* **245**, 6085 (1970).

[7] W. T. Schrader and B. W. O'Malley, *J. Biol. Chem.* **247**, 51 (1972).

[8] H. E. Smith, R. G. Smith, D. O. Toft, J. R. Neergaard, E. P. Burrows, and B. W. O'Malley, *J. Biol. Chem.*, in press.

of progesterone (C-3 and C-20) are involved in binding. A reactive substituent at a position near one of these oxygen functions would probably be closer to the polypeptide chain in the steroid–receptor complex than a more remotely positioned substituent and consequently would be more likely to undergo reaction.

Four general classes of chemical reactions that have been used in attempts at affinity labeling apply here: alkylation reactions, photochemical insertion reactions, disulfide bond formation, and mercaptide bond formation. An amino acid with a nucleophilic group such as cysteine, serine, threonine, tyrosine, tryptophan, histidine, lysine, arginine, and methionine may participate in an alkylation reaction with a steroid appropriately substituted with a leaving group such as a halogen atom.[9] Insertion reactions of carbenes generated by the photolysis of diazo steroids can take place with any amino acid.[2] Disulfide bond formation may occur between a sulfur-containing steroid derivative and cysteine.[10] Mercaptide bond formation involves the formation of a mercury–sulfur bond between a mercurated steroid and cysteine.[11,12]

Steroid reagents for these reactions may be divided into general classes according to the functional group present for the formation of a covalent bond with the receptor. A portion of the work in these laboratories has involved the preparation and use of halo and haloacetoxy derivatives of progesterone for alkylation reactions, but this discussion will be limited to the preparation and utility of diazo, sulfur-containing, and mercurated steroids for affinity labeling of androgen and progestagen receptor proteins.

In the procedures outlined below, we give details for the formation and purification of compounds of each of these types. Sufficient spectral data are included so that the identity and purity of subsequent preparations can be evaluated.

For nuclear magnetic resonance spectra (NMR), chemical shifts (δ) are reported in parts per million (ppm) downfield from tetramethylsilane as an internal standard and were determined as solutions in deuteriochloroform. Abbreviations used for these spectra are: s, singlet; bs, broad singlet; d, doublet; t, triplet; q, quartet; and m, multiplet. Infrared spectra (IR) were determined as potassium bromide disks. Ultraviolet spectra (UV) were determined as solutions in absolute ethanol. Silica

[9] C.-C. Chin and J. C. Warren, *Biochemistry* **11**, 2720 (1972).
[10] I. Field, *in* "Organic Chemistry of Sulfur" (S. Oae, ed.). Plenum, New York, in press.
[11] P. D. Boyer, *J. Amer. Chem. Soc.* **76**, 4331 (1954).
[12] C.-C. Chin and J. C. Warren, *J. Biol. Chem.* **243**, 5056 (1968).

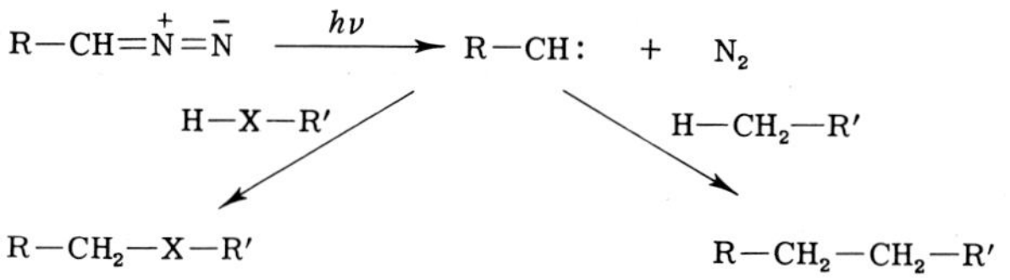

$$R-CH=\overset{+}{N}=\overset{-}{N} \xrightarrow{h\nu} R-CH: + N_2$$

$H-X-R'$ → $R-CH_2-X-R'$; $H-CH_2-R'$ → $R-CH_2-CH_2-R'$

FIG. 1

$$R-CH=\overset{+}{N}=\overset{-}{N}: \; (\text{H-S-Protein}) \longrightarrow R-CH_2-\overset{+}{N}\equiv N \; (\bar{S}\text{-Protein}) \longrightarrow R-CH_2-S\text{-Protein} + N_2$$

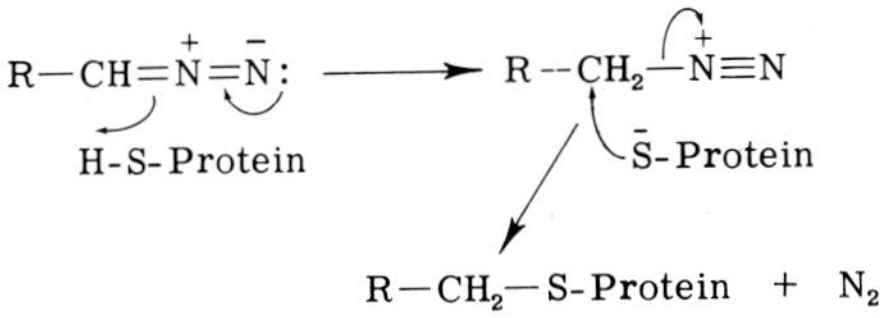

FIG. 2

gel was used for all thin-layer chromatograms (TLC). Magnesium sulfate was the drying agent for organic solutions.

B. Diazo Steroids

Diazo steroids are of potentially great utility for photochemical insertion reactions at a receptor active site. On irradiation, a carbene is generated which then can insert into a carbon–hydrogen bond or react with other groups such as an hydroxyl, an amino, or a thiol group (Fig. 1).[13] Diazo compounds may also alkylate a protein thiol group (Fig. 2) as shown by the reaction of azaserine with the reactive sulfhydryl group of 2-formamido-*N*-ribosylacetamide 5′-phosphate amidotransferase.[14]

Two types of diazo steroids are discussed here: α-diazo keto steroids, and ethyl diazomalonate and diazoacetate esters of hydroxy steroids. 2-Diazo-3-keto steroids are obtained by oximination of the 3-keto steroid with 2-octyl nitrite (*n*-butyl nitrite is unsatisfactory)[15] followed by treatment of the α-oximino ketone with chloramine in aqueous tetrahydrofuran[16] (THF) (Fig. 3). 21-Diazoprogesterone constitutes a special case as it is conveniently prepared from sodium 4-androsten-3-one-17β-carboxylate in two steps[17] (Fig. 4).

[13] W. Kirmse, "Carbene Chemistry," 1st ed. Academic Press, New York, 1964.
[14] T. C. French, I. B. Dawid, and J. M. Buchanan, *J. Biol. Chem.* **238,** 2186 (1963).
[15] H. E. Smith and A. A. Hicks, *J. Org. Chem.* **36,** 3659 (1971).
[16] M. P. Cava and B. R. Vogt, *J. Org. Chem.* **30,** 3775 (1965).
[17] A. L. Wilds and C. H. Shunk, *J. Amer. Chem. Soc.* **70,** 2427 (1948).

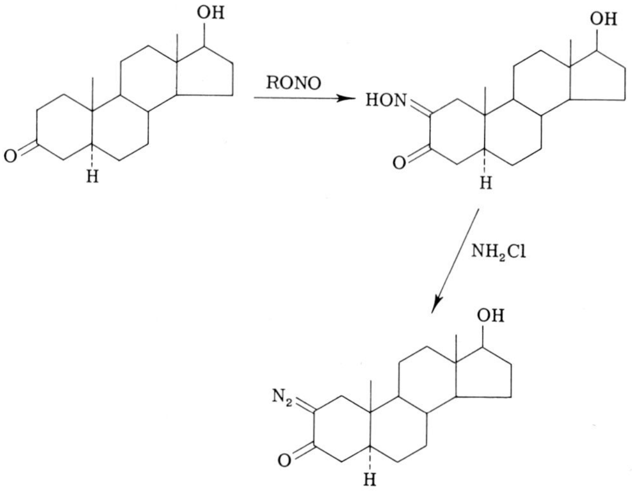

OH RONO HON OH NH$_2$Cl N$_2$ OH

FIG. 3

CO_2Na $(COCl)_2$ COCl CH_2N_2 N$_2$ O

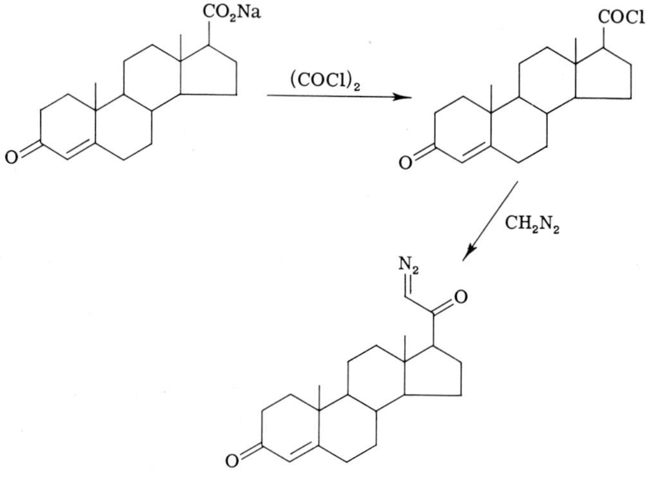

FIG. 4

Initially, ethyl diazomalonate esters of steroidal alcohols, rather than the respective diazoacetates, were the compounds of choice because of their simplicity of synthesis. Reaction of ethyl diazomalonyl chloride, prepared by reaction of phosgene with ethyl diazoacetate,[18] with an ap-

[18] H. Staudinger, J. Becker, and H. Hirzel, *Ber. Deut. Chem. Ges.* **49,** 1978 (1916); R. J. Vaughan and F. H. Westheimer, *Anal. Biochem.* **29,** 305 (1969).

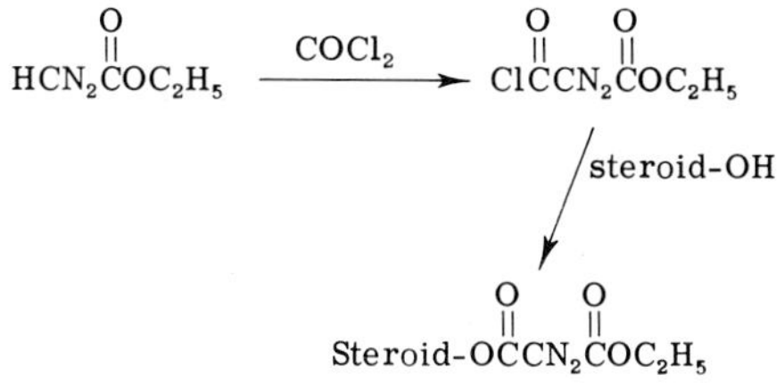

$$HCN_2COC_2H_5 \xrightarrow{COCl_2} ClCCN_2COC_2H_5 \xrightarrow{\text{steroid-OH}} \text{Steroid-OCCN}_2\text{COC}_2\text{H}_5$$

(each C of the carbonyl groups shown with =O)

FIG. 5

$$C_6H_5CH_2OCNHCH_2COH \xrightarrow{SOCl_2} \text{N-carboxyglycine anhydride} \xrightarrow{\text{steroid-OH/HCl}} \text{Steroid-OCCH}_2\text{NH}_3\text{Cl} \xrightarrow{NaNO_2} \text{Steroid-OCCHN}_2$$

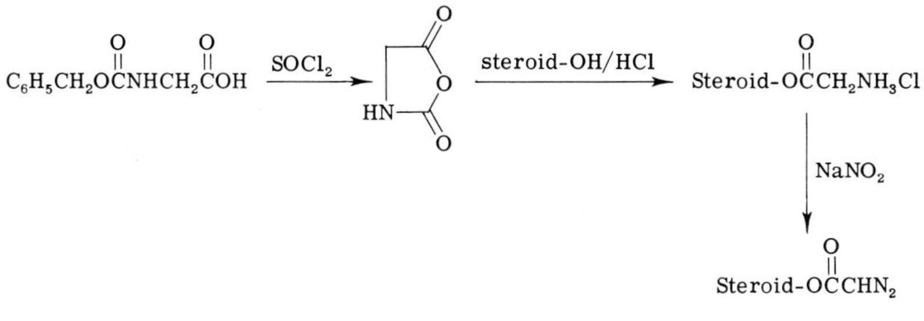

FIG. 6

propriate steroidal alcohol gives the ethyl diazomalonate ester in good yield (Fig. 5).

For better success in receptor protein-labeling experiments, however, the less readily prepared and less sterically hindered diazoacetates may be required. The latter are prepared as shown in Fig. 6. Treatment of *N*-carbobenzoxyglycine with thionyl chloride yields *N*-carboxyglycine anhydride,[19,20] which, on heating with the appropriate steroidal alcohol in ethyl acetate-containing dry hydrogen chloride, gives the steroid glycinate hydrochloride. Treatment of the latter with aqueous sodium nitrite gives the desired diazoacetate (Fig. 6).

2-Oximino-5α-androstan-17β-ol-3-one. This compound was prepared previously[15] in 12% yield from 5α-androstan-17β-ol-3-one and 2-octyl nitrite. It is advised that a procedure strictly analogous to that described below for the preparation of 17α-methyl-2-oximino-5α-androstan-17β-ol-3-one be used. 2-Oximino-5α-androstan-17β-ol-3-one has m.p. 265–268° dec; TLC R_f 0.3 (9:1 benzene–methanol); $[\alpha]^{26}_D$ +67° (*c* 0.243, absolute ethanol); IR 1620 (C=N), 1720 (C=O), 3150 (OH), and 3480 cm^{-1} (OH); UV_{max} 243 (ε 7300) and 340 nm (33) (shoulder).

2-Diazo-5α-androstan-17β-ol-3-one. To a stirred, ice-cooled mixture of 2-oximino-5α-androstan-17β-ol-3-one (302 mg, 0.951 mmole) in 10%

[19] M. Bergmann and L. Zervas, *Ber. Deut. Chem. Ges.* **65,** 1192 (1932).
[20] Y. Go and H. Tani, *Bull. Chem. Soc. Jap.* **14,** 510 (1939).

aqueous sodium hydroxide (3 ml), water (1.5 ml), and tetrahydrofuran (15 ml) was added, in one portion, concentrated ammonium hydroxide (3 ml), and then followed by 5.25% aqueous sodium hypochlorite (Clorox) (7.5 ml) dropwise during 10 minutes. The mixture was stirred 45 minutes at 0°, then allowed to come to room temperature and stirred 2 hours longer. The tetrahydrofuran layer was separated and extracted with half-saturated sodium chloride until the aqueous layer was neutral (3 portions). After drying, removal of the tetrahydrofuran gave a yellow, crystalline residue (253 mg, 84%) of 2-diazo-5α-androstan-17β-ol-3-one, homogeneous on TLC, R_f 0.5 (9:1 benzene–methanol): $[\alpha]^{25}{}_D + 37°$ (c 1.04, chloroform); NMR δ 0.76 (s, 3, C-18 H), 0.95 (s, 3, C-19 H), and 3.68 ppm (t, 1, J = 8 Hz, C-17 H); IR 1620 (C=O), 2100 (C=N=N), and 3460 cm^{-1} (OH); UV_{max} 262 (ϵ 4700) (shoulder) and 291 nm (8400). It had no definite melting point but gradually decomposed with gas evolution when heated above 170°. An analytical sample was recrystallized from chloroform–acetone.[21]

17α-Methyl-2-oximino-5α-androstan-17β-ol-3-one. Under nitrogen, potassium (about 100 mg) was dissolved in *tert*-butyl alcohol (15 ml), and to the stirred solution was added in one portion 17α-methyl-5α-androstan-17β-ol-3-one (304 mg, 1.00 mmole) followed by a solution of 2-octyl nitrite[22] (162 mg, 1.02 mmoles) in *tert*-butyl alcohol (10 ml) dropwise during 10 minutes. The mixture was stirred for 5 hours at room temperature, then diluted with water (125 ml) and extracted with methylene chloride (50 ml). The aqueous layer was separated, acidified with 5% hydrochloric acid, and extracted with methylene chloride. The methylene chloride layer was washed with water, dried, and evaporated to a white, crystalline residue (86 mg, 26%), homogeneous on TLC, R_f 0.3 (9:1 benzene–methanol): $[\alpha]^{25}{}_D + 84°$ (c 2.05, pyridine); IR 1605 (C=N), 1710 (C=O), 3180 (OH), and 3440 cm^{-1} (OH); UV_{max} 228 (ϵ 5100) (shoulder), 243 (6700), and 330 nm (60) (shoulder). An analytical sample was recrystallized from methanol and had m.p. 256–260° dec.[21]

2-Diazo-17α-methyl-5α-androstan-17β-ol-3-one. The procedure outlined above for the preparation of 2-diazo-5α-androstan-17β-ol-3-one was used with 17α-methyl-2-oximino-5α-androstan-17β-ol-3-one (175 mg, 0.525 mmole), 10% sodium hydroxide (2 ml), water (1 ml), concentrated ammonium hydroxide (2 ml), and Clorox (5 ml). The mixture was stirred

[21] This previously unreported compound, on combustion analysis for carbon, hydrogen, and nitrogen, showed elemental composition in agreement with the assigned structure.

[22] M. Pezold and R. L. Shriner, *J. Amer. Chem. Soc.* **54**, 4707 (1932). Prepared by a modification of the procedure described for n-butyl nitrite [W. A. Noyes, *Org. Syn. Collect. Vol.* **2**, 108 (1943)], and distilled, b.p. 54–55° (10 mm).

45 minutes at 0°, then allowed to come to room temperature and stirred 5 hours longer. Work-up as before afforded a yellow semicrystalline solid (63 mg) which on one recrystallization from methanol gave pure crystalline 2-diazo-17α-methyl-5α-androstan-17β-ol-3-one (42 mg, 24%): TLC R_f 0.5 (9:1 benzene–methanol; $[\alpha]^{25}_D$ +47° (c 1.1, absolute ethanol); NMR δ 0.88 (s, 3, C-18 H), 0.95 (s, 3, C-19 H), and 1.22 ppm (s, 3, C-17 CH_3); IR 1610 (C=O), 2100 (C=N=N), and 3420 cm^{-1} (OH); UV_{max} 262 (ε 4700) (shoulder) and 291 nm (8600). Its behavior on melting was similar to that of 2-diazo-5α-androstan-17β-ol-3-one.[21]

21-Diazoprogesterone. A solution of 17β-carboxy-4-androsten-3-one (1.00 g; 3.16 mmoles) in 0.1 *N* sodium hydroxide (38 ml) was lyophilized, and the resulting pale yellow powder was dried under reduced pressure at 100° for 8 hours. A stirred solution of this salt in dry benzene (10 ml) was cooled to 0° and pyridine (3 drops) was added followed by excess oxalyl chloride (4 ml) dropwise. The mixture was allowed to warm slowly to 15°, and when no further evolution of gas occurred, the solvent was evaporated under reduced pressure. A solution of the residual acid chloride in benzene (20 ml) was filtered, diluted with ether (20 ml), and added dropwise to a stirred ethereal solution of excess diazomethane kept at −15°. The mixture was kept at −15° for 30 minutes and at 0° for another 30 minutes, and the solvents were evaporated under reduced pressure. Trituration of the residue with ether followed by filtration yielded crude crystalline 21-diazoprogesterone (459 mg, 43%). It was dissolved in hot acetone, and the solution was filtered and cooled to give pure 21-diazoprogesterone: m.p. 174–176° (lit[17] 177–178° dec); TLC R_f 0.7 (9:1 benzene–methanol); NMR δ 0.95, 1.21 (2 s, 6, C-18 and C-19 H), 5.22 (s, 1, C-21 H), and 5.76 ppm (s, 1, C-4 H); IR 1635 (C-20 C=O), 1660 (C-3 C=O), 2180 (C=N=N), and 3160 cm^{-1} (C-21 C-H stretch); UV_{max} 244 (ε 23,000) and 273 nm (7800) (shoulder).

Ethyl Diazomalonyl Chloride. Phosgene was bubbled in benzene (43.5 ml) until the total volume of the solution was 50 ml. This solution was cooled in an ice bath, and ethyl diazoacetate (22.1 g, 0.194 mole) in benzene (65 ml) was added dropwise with stirring. The mixture was allowed to come to room temperature and then was stored for 6 hours. Most of the benzene was removed under reduced pressure, and distillation of the residue gave ethyl diazomalonyl chloride (9.87 g, 58%) as a pale yellow oil: b.p. 49–54° (0.07 mm) [lit[23] b.p. 35° (0.03 mm)].

Testosterone Ethyl Diazomalonate. Ethyl diazomalonyl chloride (0.35 g, 2.0 mmoles) followed by ether (1.0 ml) was added with swirling to a solution of testosterone (288 mg, 1.00 mmole) in dry pyridine (1 ml). After 1 hour, the mixture was diluted with ether (20 ml), washed with

[23] D. J. Brunswick and B. S. Cooperman, *Proc. Nat. Acad. Sci. U.S.* **68**, 1801 (1971).

dilute hydrochloric acid and with water, and dried. The solvent was removed under reduced pressure and the residue was dissolved in benzene and chromatographed on silica gel. Elution with 4:1 benzene–ether gave the diazo ester (0.31 g, 72%) as a syrup, pure by TLC (1:1 benzene–ether). This syrup was crystallized with difficulty from ethanol–water to give pure testosterone ethyl diazomalonate as white needles: m.p. 80–82°; NMR δ 0.86 (s, 3, C-18 H), 1.20 (s, 3, C-19 H), 1.32 (t, 3, $J = 7$ Hz, OCH_2CH_3), 4.29 (q, 2, $J = 7$ Hz, OCH_2CH_3), 4.73 (m, 1, C-17 H), and 5.74 ppm (s, 1, C-4 H); IR 2145 cm^{-1} (C=N=N); UV_{max} 242 nm (ϵ 25,400). Recrystallization from methanol–water gave the analytical sample, m.p. 81–83°.[21]

6β-Hydroxyprogesterone Ethyl Diazomalonate. Ethyl diazomalonyl chloride (1.06 g, 6.00 mmoles) was added with swirling to a solution of 6β-hydroxyprogesterone (198 mg, 0.599 mmole) in dry pyridine (3.0 ml), and the reaction mixture was kept at room temperature for 16 hours. Work-up as described for testosterone ethyl diazomalonate gave a syrup (0.61 g) which was chromatographed on silica gel. Elution with 9:1 benzene–ether gave 6β-hydroxyprogesterone ethyl diazomalonate (235 mg, 84%) as a gum which solidified and was pure by TLC (1:1 benzene–ether): NMR δ 0.71 (s, 3, C-18 H), 1.31 (t, 3, $J = 7$ Hz, OCH_2CH_3), 1.32 (s, 3, C-19 H), 2.15 (s, 3, C-21 H), 4.32 (q, 2, $J = 7$ Hz, OCH_2CH_3), 5.65 (m, 1, C-6 H), and 6.02 ppm (s, 1, C-4 H); IR 2120 cm^{-1} (C=N=N). A benzene solution of the diazo ester was chromatographed a second time on silica gel. Elution with 9:1 benzene-ether gave a syrup. Lyophilization of a benzene solution of this syrup gave the analytical sample as an amorphous, white powder.[21]

11α-Hydroxyprogesterone Ethyl Diazomalonate. Ethyl diazomalonyl chloride (1.77 g, 10.0 mmoles) was added dropwise with stirring to 11α-hydroxyprogesterone (662 mg, 2.00 mmoles) in pyridine (3.6 ml). The mixture was allowed to stand overnight. Work-up was similar to that described for testosterone ethyl diazomalonate except that methylene chloride was the organic solvent instead of ether. The crude product was chromatographed on silica gel. Elution with 8:1 benzene–ether gave the diazo ester (861 mg, 92%) which was recrystallized from ethyl acetate–hexane to give pure 11α-hydroxyprogesterone ethyl diazomalonate (402 mg, 43%): m.p. 155–156° dec; $[\alpha]^{24}_D$ +50.7° (c 3.02, dioxane); IR 2120 cm^{-1} (C=N=N).[21] The NMR spectrum was similar to that of 6β-hydroxyprogesterone ethyl diazomalonate except for the positions of the C-11 and C-4 protons at 5.52 and 5.84 ppm, respectively.

Deoxycorticosterone Ethyl Diazomalonate. Ethyl diazomalonyl chloride (0.35 g, 2.0 mmoles) was added to a swirled solution of deoxycorticosterone (330 mg, 1.00 mmole) in pyridine (2.0 ml), and the mixture was kept 8 hours at room temperature. Work-up as described for testosterone ethyl diazomalonate gave the crude product (0.31 g) as a syrup which was chromatographed on silica gel. Elution with 85:15 benzene–ether gave deoxycorticosterone ethyl diazomalonate (0.27 g, 58%), pure by TLC (1:1 benzene–ether), as a gum. Lyophilization of a benzene solution of this gum gave the analytical sample as an amorphous, light tan powder.[21] Its NMR spectrum was similar to that of 11α-hydroxyprogesterone ethyl diazomalonate except for the absence of the C-11 proton multiplet and the appearance of the C-21 protons as 2 doublets (J = 18 Hz) at 4.69 and 4.95 ppm.

Testosterone Diazoacetate. *N*-Carboxyglycine anhydride[19,20] (101 mg, 1.00 mmole) was added to a solution of testosterone (288 mg, 1.00 mmole) in ethyl acetate (2 ml) containing dry hydrogen chloride. The mixture was stirred 15 minutes at 60° and then kept 24 hours at room temperature. The precipitate was collected by filtration and was washed three times with hot ethyl acetate. This crude testosterone glycinate hydrochloride (229 mg) was dissolved in water (2 ml), methylene chloride (4 ml) was added, and the mixture was cooled (—1°) in an acetone–dry ice bath. Sodium nitrite (70 mg, 1.01 mmoles) was added to the stirred mixture, the bath temperature was lowered to —9°, and 5% sulfuric acid (2 ml) was added dropwise. After 10 minutes, the mixture was allowed to come to room temperature and was stirred for an additional 30 minutes. The methylene chloride layer was separated, washed with 5% sodium bicarbonate and with water, and dried. Removal of the solvent and crystallization from methanol gave testosterone diazoacetate (27 mg, 8% based on testosterone) as pale yellow prisms: m.p. 163–165°; NMR δ 0.65 (s, 3, C-18 H), 1.23 (s, 3, C-19 H), 4.86 (m, 1, C-17 H), 4.87 (s, 1, $N_2CHC{=}O$), and 5.93 ppm (s, 1, C-4 H); IR 2100 cm^{-1} ($C{=}N{=}N$).[21]

C. Sulfur-Containing Steroids

For the preparation of sulfur-containing derivatives of testosterone such as steroidal sulfenyl thiolcarbonates[24] (steroid—S—S—CO_2R) steroidal sulfenyl thiocyanates[25] (steroid—S—SCN), and steroidal alkyl or aryl disulfides[26] (steroid—S—S—R), which might react with cysteine to form a disulfide bond at the active site of a receptor protein,[10] the synthesis of 4-androstene-17β-thiol-3-one was investigated in some detail.

[24] S. J. Brois, J. F. Pilot, and H. W. Barnum, *J. Amer. Chem. Soc.* **92,** 7629 (1970).
[25] R. G. Hiskey and W. P. Tucker, *J. Amer. Chem. Soc.* **84,** 4789 (1962).
[26] E. E. Reid, "Organic Chemistry of Bivalent Sulfur," Vol. III. Chem. Publ. Co., New York, 1960.

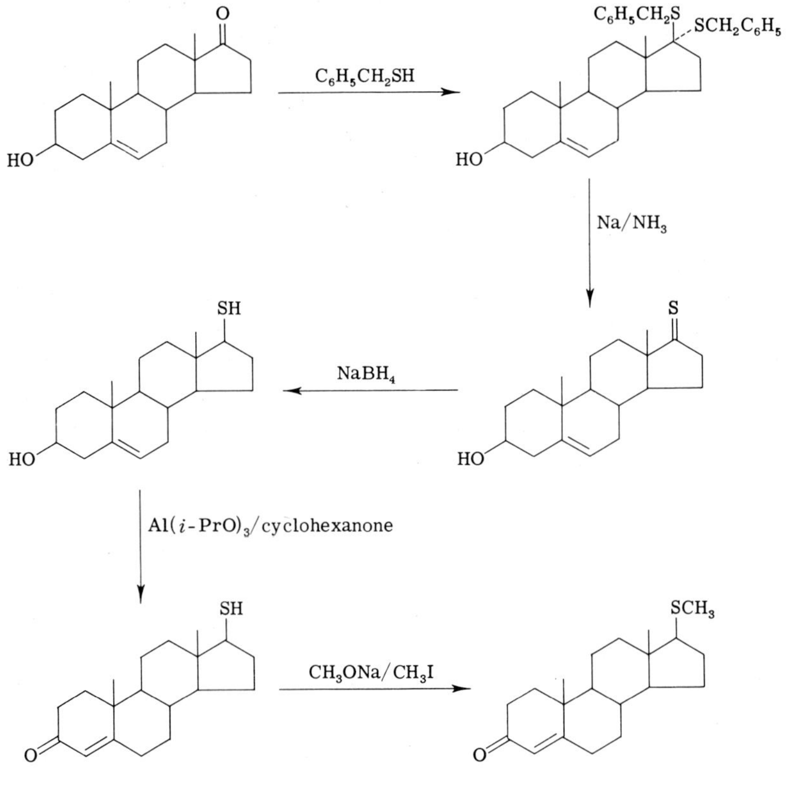

O
$C_6H_5CH_2SH$
$C_6H_5CH_2S$ $SCH_2C_6H_5$
HO
HO
Na/NH_3
SH
S
$NaBH_4$
HO
HO
$Al(i\text{-}PrO)_3$/cyclohexanone
SH
SCH_3
CH_3ONa/CH_3I
O
O

FIG. 7

The synthetic scheme was outlined earlier[27,28] (Fig. 7), but it lacked many experimental details. As shown in Fig. 7, a sulfur atom was substituted for the carbonyl oxygen atom of 5-androsten-3β-ol-17-one by reduction of its dimercaptal derivative. Reduction of 5-androsten-3β-ol-17-thione gave 5-androstene-17β-thiol-3β-ol, which was oxidized to the desired thiol analog of testosterone. Treatment of the sodium salt of the thiol with methyl iodide gave 17β-methylthio-4-androsten-3-one.

5-Androsten-3β-ol-17-one Dibenzyl Thioketal. A solution of 5-androsten-3β-ol-17-one (10.0 g, 34.7 mmoles), benzyl mercaptan (10.0 ml, 85.2 mmoles), and *p*-toluenesulfonic acid (1.0 g) in glacial acetic acid (50 ml) was allowed to stand at room temperature for two days. The precipitate which formed was collected and washed several times with water

[27] R. M. Dodson and P. B. Sollman, U.S. Patent 2,763,669 (1956); *Chem. Abstr.* **51,** 5134a (1957).
[28] C. Djerassi and D. Herbst, *J. Org. Chem.* **26,** 4675 (1961).

and then was boiled 2 hours in 1 *N* ethanolic potassium hydroxide (500 ml). The mixture was poured into ice water (about 400 ml) and acidified with glacial acetic acid. The precipitate was collected and recrystallized from 95% ethanol to give 5-androsten-3β-ol-17-one dibenzyl thioketal (12.4 g, 69%) as white needles: m.p. 184–186° (lit[29] 184–186°); NMR δ 1.05 (s, 3, C-19 H), 1.11 (s, 3, C-18 H), 3.62 (m, 1, C-3 H), 4.06 (s, 4, SCH_2), 5.52 (m, 1, C-6 H), and 7.52 ppm (m, 10, aromatic H); IR 700 cm^{-1} (phenyl out of plane bending).

5-Androsten-3β-ol-17-thione. 5-Androsten-3β-ol-17-one dibenzyl thioketal (5.19 g, 10.0 mmoles) was added to a stirred solution of sodium (2.6 g, 0.11 g-atom) in twice-distilled liquid ammonia (about 300 ml) and anhydrous ether (100 ml). After 0.5 hours additional sodium (2.6 g, 0.11 mole) was added, and the deep blue reaction mixture was stirred for an additional 4 hours. Additional anhydrous ether (100 ml) was then added, the excess sodium was destroyed by careful addition of absolute ethanol, and the ammonia was allowed to evaporate, leaving a clear, colorless solution. Ice was added, the mixture was stirred 0.5 hour, and then extracted with ether. The aqueous layer was acidified with glacial acetic acid and extracted with chloroform. The chloroform layer was dried and evaporated to give crude 5-androsten-3β-ol-17-thione (2.75 g, 90%) as an orange solid with no IR absorption at 700 cm^{-1}.

This material was usually used without purification. Chromatography on silica gel (eluted with 9:1 benzene–ether) gave material homogeneous to TLC (9:1 benzene–ethyl acetate) with an unchanged IR spectrum.

5-Androstene-17β-thiol-3β-ol. Excess sodium borohydride was added with stirring to a solution of 5-androsten-3β-ol-17-thione (2.75 g, 9.03 mmoles) in dry methanol (100 ml) until a clear, colorless solution was obtained. The mixture was acidified with glacial acetic acid and reduced to near dryness under reduced pressure. Ice and water were added, and the mixture was extracted with methylene chloride. The methylene chloride solution was dried, and the methylene chloride removed to give crude 5-androstene-17β-thiol-3β-ol (2.42 g, 87%), which was chromatographed on silica gel. Elution with 9:1 benzene–ether gave the thiol (1.95 g, 70%) as a white amorphous solid, pure by TLC (9:1 benzene–ethyl acetate). Crystallization from hexane afforded the thiol as white needles: m.p. 169–171° (lit[27] 174.5–175.5°); NMR δ 0.75 (s, 3, C-18 H), 1.05 (s, 3, C-19 H), 3.62 (m, 1, C-3 H), and 5.49 ppm (m, 1, C-6 H); IR 2540 (SH, weak) and 3250 cm^{-1} (OH).

4-Androstene-17β-thiol-3-one. A mixture of 5-androstene-17β-thiol-3β-ol (500 mg, 1.63 mmoles), aluminum isoproxide (500 mg, 2.45

[29] R. H. Levin and J. L. Thompson, *J. Amer. Chem. Soc.* **70**, 3140 (1948).

mmoles), cyclohexanone (5 ml), and toluene (10 ml) was boiled for 16 hours. The cooled mixture was then acidified with glacial acetic acid, filtered through Celite, and the Celite was washed with benzene. The solvents were removed from the combined filtrate and washings under reduced pressure yielding a pale yellow liquid (4.75 g) which was chromatographed on silica gel. Elution with benzene gave 2-cyclohexenyl cyclohexanone (~2.22 g).[30] Elution with 9:1 benzene–ether gave 4-androstene-17β-thiol-3-one (466 mg, 94%) as a syrup, pure by TLC (9:1 benzene–ethyl acetate), which was rechromatographed on silica gel to give 323 mg (65%) of syrup which crystallized: m.p. 99–109° (lit[27] 118–119°); IR 2520 (SH, weak) and 1665 cm^{-1} (conjugated C=O).

In a similar preparation using 400 mg of 5-androsten-3β-ol-17β-thiol, the reaction was worked up by addition of a solution of potassium sodium tartrate (4.0 g) in water (25 ml). The mixture was stirred for 1 hour and then steam-distilled to remove toluene and cyclohexanone. The aqueous residue was extracted with methylene chloride and the extract was evaporated under reduced pressure to a syrupy residue (400 mg) which was chromatographed on silica gel. Elution with 9:1 benzene–ether afforded pure 4-androstene-17β-thiol-3-one (0.21 g, 53%) which crystallized from methanol–water as white needles: m.p. 110–113°.

17β-Methylthio-4-androsten-3-one. Sodium (90 mg, 3.9 mg-atoms) dissolved in dry methanol (50 ml) was added with swirling to 4-androsten-17β-thiol-3-one (690 mg, 2.27 mmoles), and the mixture was heated on a steam plate for 10 minutes. Excess methyl iodide (about 2 ml) was then added, and the mixture was heated for an additional 10 minutes. It was then poured on ice, acidified with glacial acetic acid, and extracted with methylene chloride. The extract was dried, the solvent was removed under reduced pressure, and the residue (0.66 g) was chromatographed on silica gel. Elution with 9:1 benzene–ether gave crystalline 17β-methylthio-4-androsten-3-one (319 mg, 44%), pure by TLC (9:1 benzene–ethyl acetate). Recrystallization from ethanol followed by sublimation at 95° (0.02 mm) afforded the compound as white needles: m.p. 128–130° (lit[27] m.p. 131.0–132.5°); NMR δ 0.81 (s, 3, C-18 H), 1.20 (s, 3, C-19 H), 2.12 (s, 3, SCH_3), and 5.73 ppm (s, 1, C-4 H).

D. Mercurated Steroids

A previous report of affinity labeling of an estradiol receptor protein using 4-acetoxymercuri-estradiol[12] suggested the synthesis of mercurated

[30] R. A. Abramovitch and A. R. Vinutha, *J. Chem. Soc. C* p. 2104 (1969).

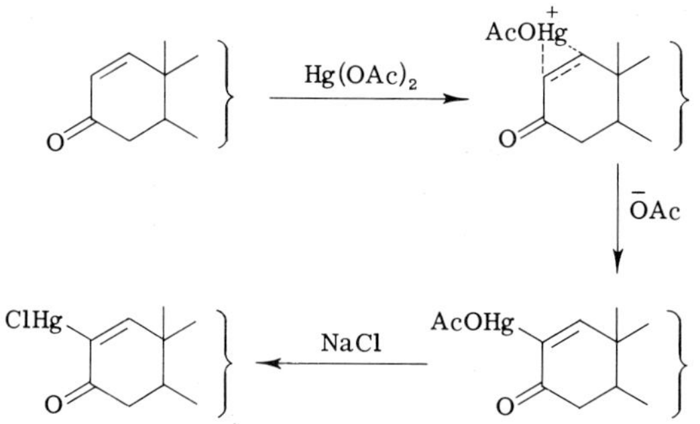

AcOHg
Hg(OAc)$_2$
ŌAc
ClHg
NaCl
AcOHg

FIG. 8

steroids, in expectation that these might form mercaptide bonds with cysteine residues of androgen and progestagen receptors. The aim was to introduce mercury into the testosterone or progesterone molecule adjacent to the C-3 carbonyl group.[31] In contrast to a number of acyclic model compounds, α,β-unsaturated 3-keto steroids do not react with mercuric acetate in methanol at room temperature. 3-Keto steroids with a readily abstractable allylic proton, such as progesterone, appear to be oxidized on heating of the steroid in methanol and in acetic acid with mercuric acetate but no well-defined products were isolated. 3-Keto steroids without this structural feature and with a C-1 double bond react in boiling acetic acid with addition of acetoxymercuri occurring at the C-1 double bond. The proton at C-2 is abstracted, and the 2-acetoxymercuri-1-en-3-one is formed (Fig. 8). For ease of purification, the acetate ion was replaced with a chloride ion. Steroidal 1,4-dien-3-ones and 1,4,6-trien-3-ones also react similarly and more readily, the latter being the most reactive.

It is possible to follow the conversion of the steroidal substrate to the 2-acetoxymercuri derivative using NMR spectroscopy. On mercuration, the doublet at 7.1 ppm (J = 10 Hz) assigned to the C-1 proton in the substrate spectrum collapses to a singlet and is shifted to a slightly lower field.

The mercurated steroids are extremely difficult to purify since even after purification they tend to form gums. They also appear to retain solvents tenaciously as indicated by the NMR spectra of purified compounds which showed additional protons characteristic of the solvent used. For these reasons the yield of pure mercurated compound in some

[31] R. G. Smith, H. E. Ensley, and H. E. Smith, *J. Org. Chem.* **37**, 4430 (1972).

cases was low although conversion to the mercurated steroid was high.

2-Chloromercuri-5α-androst-1-ene-3,17-dione. 5α-Androst-1-ene-3,17-dione (200 mg, 0.698 mmole) and mercuric acetate (1.00 g, 3.14 mmoles) in glacial acetic acid (5.0 ml) were boiled for 24 hours. After cooling the solution was poured into a stirred saturated solution of sodium chloride (50 ml). The precipitated gum was taken up in chloroform and the solution was washed with water until the wash water was neutral. Evaporation of the dried chloroform solution left a gum which on trituration with ether gave a white solid. Four recrystallizations from chloroform–ether gave pure 2-chloromercuri-5α-androst-1-ene-3,17-dione as white microcrystals (40 mg, 11%): m.p. 180° dec; NMR δ 7.25 ppm (s, 1, C-1 H); IR 1640 (conjugated C=O) and 1730 cm^{-1} (C=O).[31] This compound retained chloroform so tenaciously that it was never obtained completely free of chloroform.[31]

2-Chloromercuri-1,4-androstadiene-3,17-dione. 1,4-Androsta-1,4-diene-3,17-dione (2.00 g, 7.03 mmoles) and mercuric acetate (10.0 g, 31.4 mmoles) were boiled in glacial acetic acid (125 ml) for 6 hours. After cooling, the mixture was poured into a saturated sodium chloride solution. The resulting yellow precipitate was taken up in chloroform. Careful addition of hexane to this solution precipitated the mercurated steroid. Reprecipitation on cooling from 95% ethanol gave 2-chloromercuri-1,4-androstadiene-3,17-dione (0.16 g, 4%) as a white amorphous solid: m.p. 282–283° dec; NMR δ 6.20 (s, 1, C-4 H), and 7.17 ppm (s, 1, C-1 H); IR 1640 (conjugated C=O) and 1725 cm^{-1} C=O).[31]

2-Chloromercuri-1,4,6-androstatriene-3,17-dione. 1,4,6-Androstatriene-3,17-dione (5.00 g, 17.7 mmoles) and mercuric acetate (13.0 g, 40.8 mmoles) were boiled in glacial acetic acid for 30 minutes. Dilution of the reaction mixture with water (250 ml) containing sodium chloride (15 g) precipitated a solid which was washed with water and extracted with hot acetone (400 ml). This solution was reduced in volume to 75 ml, and dilution with hexane (100 ml) precipitated a white solid. Crystallization of this solid from acetone gave 2-chloromercuri-1,4,6-androstatriene-3,17-dione (0.867 g, 9.5%) as white needles: m.p. 280–285° dec; $[\alpha]^{25}_D$ +27° (*c* 0.84, chloroform); NMR δ 6.05–6.45 (m, 3, C-4, C-6, and C-7 H), 7.28 (s, 1, C-1 H), and 7.28 ppm (d, about 0.2, J = 280 Hz, C-1 H coupled to ^{199}Hg).[31]

17α-Methyl-1,4,6-androstatrien-17β-ol-3-one. A solution of 17α-methyl-5-androstene-3β,17β-diol (6.1 g, 20 mmoles) and 2,3-dichloro-5,6-dicyanobenzoquinone (DDQ) (13.9 g, 61 mmoles) in dry dioxane (300 ml) was boiled under nitrogen for 16.5 hours. The mixture was allowed to cool, filtered to remove the hydroquinone, and evaporated at reduced pressure to yield a brown gum. The gum was dissolved in ethyl acetate and chromatographed on alumina (neutral, Brockmann activity I) (160 g). Elution with ethyl acetate gave the crude, crystalline product. Recrystallization from ethyl acetate gave pure 17α-methyl-1,4,6-androstatrien-17β-ol-3-one (1.2 g, 20%): m.p. 136–138° (lit[32] m.p. 139–140°); NMR δ 5.92–6.36 (m, 4, C-2, C-4, C-6, and C-7 H) and 7.08 ppm (d, 1, J = 10 Hz, C-1 H).

2-Chloromercuri-17α-methyl-1,4,6-androstatrien-17β-ol-3-one. A solution of 217α-methyl-1,4,6-androstatrien-17β-ol-3-one (0.813 g, 2.72 mmoles) and mercuric acetate (4.00 g, 12.6 mmoles) in glacial acetic acid was boiled for 15 minutes. After cooling, the solution was poured into a stirred, saturated sodium chloride solution (250 ml). The precipitated crude product was washed with water and recrystallized twice from 95% ethanol to give 2-chloromercuri-17α-methyl-1,4,6-androstatrien-17β-ol-3-one (0.370 g, 25%) as pale yellow microcrystals: m.p. 180–181° dec (lit[33] 155–162°); NMR δ 6.0–6.3 (m, 3, C-4, C-6, and C-7 H) and 7.29 ppm (s, 1, C-1 H); IR 1585 (C=C) and 1612 cm^{-1} (C=O).[31]

1,4,6-Pregnatriene-3,20-dione. Using the procedure as described for the preparation of 17α-methyl-1,4,6-androstatrien-17β-ol-3-one, 5-pregnen-3β-ol-20-one was dehydrogenated with DDQ to give 1,4,6-pregnatriene-3,20-dione (28%), eluted from an alumina (neutral, Brockmann activity I) column with ethyl acetate, and recrystallized from ethyl acetate: m.p. 148–149° (lit[34] m.p. 150–152°); NMR δ 5.91–6.37 (m, 4, C-2, C-4, C-6, and C-7 H) and 7.08 ppm (d, 1, J = 10 Hz, C-1 H).

2-Chloromercuri-1,4,6-pregnatriene-3,20-dione. 1,4,6-Pregnatriene-3,20-dione (4.40 g, 14.2 mmoles) and mercuric acetate (22.0 g, 69.0 mmoles) in glacial acetic acid (40 ml) were boiled for 30 minutes. Dilution of the cooled reaction mixture with saturated sodium chloride precipitated a solid, which was washed with water and then extracted into chloroform and dried. Evaporation of the chloroform left a gum which on heating in 95% ethanol and then cooling gave solid 2-chloromercuri-1,4,6-pregnatriene-3,20-dione (2.5 g, 32%): m.p. (135° softens) 150° dec. An analytical sample was obtained by heating a suspension of the solid in ethanol for a few minutes, allowing the mixture to cool to 50°, decanting the ethanolic solution from a gummy residue, and allowing crystallization to proceed at 0° for 24 hours. Two additional recrystallizations in this manner gave 2-chloromercuri-1,4,6-pregnatriene-3,20-dione (0.510 g, 7%) as off-white microcrystals: m.p. (137° shrinks) 145–150° dec;

[32] G. O. Weston, D. Burn, D. N. Kirk, and V. Petrow, British Patent 854,343 (1960); *Chem. Abstr.* **55**, 18813f (1961).
[33] M. Kocor and M. Gumulka, *Tetrahedron Lett.* p. 3067 (1969).
[34] S. K. Pradhan and H. J. Ringold, *J. Org. Chem.* **29**, 601 (1964).

$[\alpha]^{25}_{D}$ +58° (*c* 1.0, chloroform); NMR δ 5.94–6.36 (m, 3, C-4, C-6, and C-7 H) and 7.28 ppm (s, 1, C-1 H).[31]

Acknowledgments

The Agency for International Development, under Contract AID/csd 2491, administered by the Population Council, supported this work. The Center for Population Research and Studies in Reproductive Biology at Vanderbilt University is supported by National Institutes of Health Grant HD-05797.

Section VI

Steroid Hormone Effects on Biochemical Processes

[35] Methods for Assessing Kinetics of Hormone Effects on Energy and Transport Mechanisms in Cells in Suspension

By ALLAN MUNCK[1] and LAWRENCE ZYSKOWSKI

Introduction

The effects of glucocorticoids on glucose transport in thymus cells in suspension are among the most rapid observed with any steroid hormones. To determine the rates of appearance and disappearance of these effects we have developed the short (2–5 minutes) assays, referred to as "pulse" methods, that will be described here. We have also been concerned with the problem of dealing with limited amounts of material, particularly in studying hormone effects on tissues from fetal and neonatal rats, and the last section in this chapter gives a simple procedure for incubating small volumes of cell or tissue suspensions. With appropriate modifications, some or all these methods should be applicable to the study of hormone effects on metabolic and transport processes in other preparations.

The system for which the methods were designed is a suspension of rat thymus cells in Krebs-Ringer bicarbonate buffer equilibrated with an atmosphere of 95% O_2–5% CO_2. Cell concentrations can be expressed as the cytocrit V (milliliters of packed cells per milliliter of cell suspension),[1a] determined by standard microhematocrit procedures; quantities of cell-associated substances such as glucose 6-phosphate are given per milliliter of packed cells.

Glucose Pulse: Formation of Glucose 6-Phosphate[1a,2]

Multiple aliquots of cell suspension (0.5 ml or more; V = 0.2–0.35) are incubated without substrate, and with or without hormone. At the time the magnitude of the hormone effect is to be determined, glucose is added to each aliquot as a 2% solution in buffer, to give a final glucose concentration of 5.5 m*M*. Five minutes later the incubations are stopped by adding 1 ml of ice-cold 12% perchloric acid per milliliter of cell suspension. The mixtures are kept cold, centrifuged, and the supernatants

[1] Supported in part by Research Grant AM-3535 and Research Career Development Award AM-K3-16.

[1a] A. Munck, *J. Biol. Chem.* **243**, 1039 (1968).

[2] K. M. Mosher, D. A. Young, and A. Munck, *J. Biol. Chem.* **246**, 654 (1971).

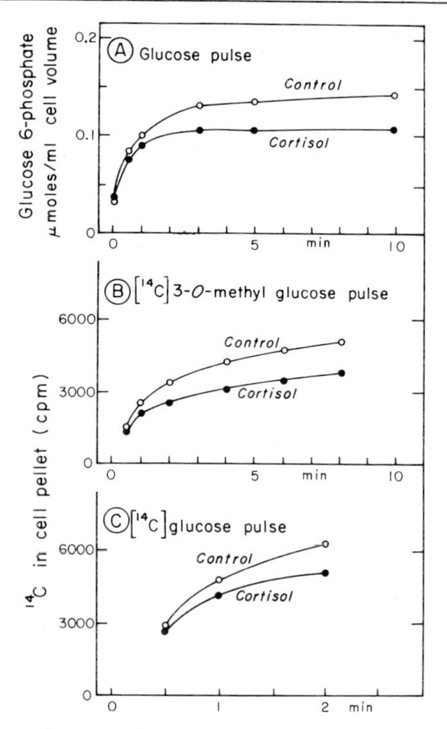

FIG. 1. Time course of accumulation of glucose 6-phosphate or ^{14}C in rat thymus cells at 37° following addition of various sugars. The cortisol concentration used was 10^{-6} *M*. (A) Cells were first incubated for 20 minutes at 37° without substrate, and with or without cortisol. At zero time, glucose was added to give a concentration of 5.5 m*M*. The values for 0 minute are from samples taken just before glucose addition. Each point is the mean of three incubations.[2] Standard errors for each point are smaller than the size of the symbols used. (B) Cells were first incubated for 60 minutes at 37° without substrate, and with or without cortisol. At zero time [^{14}C]3-*O*-methyl glucose (spec. act. 18.3 Ci/mole) was added to give a concentration of about 10 n*M*. Each point is the mean of two incubations, from each of which triplicate samples were taken. (C) Cells were first incubated for 25 minutes at 37° without substrate, and with or without cortisol. At zero time uniformly labeled [^{14}C]D-glucose (spec. act. 183 Ci/mole) was added to give a concentration of about 1 n*M*. Each point is the mean of four incubations, from each of which triplicate samples were taken.

neutralized with 60% KOH and assayed for glucose 6-phosphate by standard enzymatic procedures.[3]

Figure 1A shows the time course of the increase in glucose 6-phosphate levels in cortisol-treated and control thymus cells following such a glucose

[3] H.-U. Bergmeyer, ed., "Methods of Enzymatic Analysis." Academic Press, New York, 1965.

pulse at 37°. Low initial levels such as those given for 0 minute are consistently found in cells prepared and incubated without substrate; they presumably reflect the lack of significant rates of glycogenolysis and gluconeogenesis in these cells, and the dependence of glucose 6-phosphate levels on externally supplied glucose. The final levels of glucose 6-phosphate are close to those observed *in vivo*. These steady-state levels, clearly reduced in the cells exposed to cortisol, are reached by 5 minutes, and persist almost unchanged for at least 30 minutes. Beyond 5 minutes the length of the pulse at 37° is therefore not critical. A glucose pulse can also be used at 20° to detect a cortisol effect, but does not reach steady-state levels of glucose 6-phosphate.

This assay has been used to study the time course of development of the cortisol effect and the influence on this effect of temperature, inhibitors of RNA and protein synthesis, and hormone competitors.[1a,2,4]

[^{14}C]3-*O*-Methyl Glucose Pulse

In lymphoid cells as in many other cells, 3-*O*-methyl glucose is transported—without subsequent metabolism—by a transport system shared partly with glucose. Transport of both 3-*O*-methyl glucose and glucose is inhibited by cortisol.

A 2-minute assay for a cortisol effect on the amount of [^{14}C]3-*O*-methyl glucose taken up by lymphoid cells can be carried out as follows. Aliquots of a cell suspension (V = 0.05–0.3) in Erlenmeyer flasks are incubated at 37° with or without hormone. At the time the hormone effect is to be determined, [^{14}C]3-*O*-methyl glucose in trace amounts is added to the cells. One-hundred seconds after the addition of labeled sugar to a flask, triplicate 20-μl aliquots of the suspension are removed with an Eppendorf microliter pipet. Each aliquot is deposited inside a Beckman Microfuge tube, on the wall near the top. The lower half of the tube already contains 200 μl ice-cold buffer. At 120 seconds, centrifugation is started and continued for 30 seconds, sufficient time for the cells to be sedimented into a firm pellet after being rapidly washed by passage through the cold buffer.

The supernatant is immediately shaken off or removed by suction, and the tubes are placed upside down at room temperature on paper towels. Subsequently the cell pellets are transferred to counting vials by cutting off the tips and proceeding as described in Chapter 22 [Munck and Wira]. Replicate measurements rarely differ by more than 3%.

Figure 1B shows the time course of the increase of [^{14}C]3-*O*-methyl glucose in cortisol-treated and control cells. For a given time of exposure

[4] C. Hallahan, D. A. Young, and A. Munck, *J. Biol. Chem.* **248**, 2922 (1973).

of cells to cortisol, the magnitude of the cortisol effect obtained in this assay is consistently smaller than that observed in either the preceding or the following assays with glucose pulses; it is relatively large in Fig. 1B because of the longer cortisol treatment. If excess (20 m*M*) unlabeled 3-*O*-methyl glucose is added to the cells 10 minutes before the pulse, the ^{14}C found associated with the cells is 10–15% of that after 2 minutes without unlabeled sugar, and changes little with time.

[^{14}C]Glucose Pulse

The 2-minute assay just described for [^{14}C]3-*O*-methyl glucose may be carried out using [^{14}C]glucose, in which case a larger cortisol effect is obtained. Two modifications are introduced: (a) The 200 μl cold buffer in the Beckman Microfuge tubes contain 5.5 m*M* glucose; (b) after removal of supernatant from the cell pellets, to stop further metabolism in the cells the tubes are cooled immediately in a dry ice–acetone mixture for 10 seconds, and then stored upside-down in a freezer until the pellets are to be transferred to counting vials and counted. Replicates by this method rarely differ by more than 6%.

The time course of ^{14}C accumulation in control and cortisol-treated cells is shown in Fig. 1C. As with 3-*O*-methyl glucose, if excess (10 m*M*) unlabeled glucose is added to the cells, only 10–15% as much ^{14}C is found in the cells as after 2 minutes without unlabeled glucose.

Comparison of the Pulse Methods

With all three methods described, the cortisol effect in thymus cells incubated with 10^{-6} *M* cortisol at 37° appears abruptly between 15 and 20 minutes after exposure to hormone. The reproducibility of the cortisol effect measured by any of these methods is about the same, the smaller effect obtained with the 3-*O*-methyl glucose pulse being compensated for by somewhat less scatter. Precision of timing is considerably less important for the first method than for the two methods using radioactivity. The latter two methods, however, are probably more widely applicable than the former and have the great advantage of requiring scintillation counting rather than a tedious enzymatic assay for their final values.

Small-Scale Incubation Procedure

Volumes of cell suspensions in the 40–200 μl range can be incubated with efficient stirring in the plastic tubes made for the Eppendorf Microcentrifuge 3200, arranged as shown in Fig. 2. After the suspension is

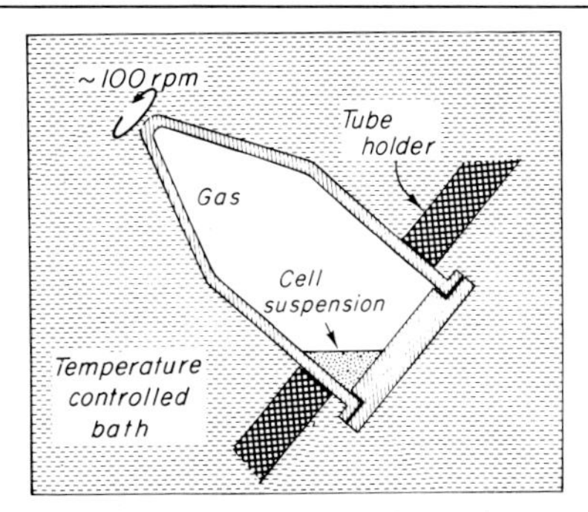

FIG. 2. Tube arrangement for small-scale incubation.

placed in a tube, gassed if necessary, the tube is capped and shaken or swung so as to drive the suspension to the capped end. The whole tube is then immersed, cap down, at an angle of roughly 45° in a thermostatted bath and rotated at about 100 rpm. The suspension rolls in the bottom of the trough formed by the cap and inner wall of the tube, thereby being stirred. Volumes of suspension less than 40 μl may become fixed to the tube by surface tension and stirred inadequately. At the end of an incubation the suspensions can be centrifuged without transfer in the Eppendorf Microcentrifuge 3200.

A simple holder for almost any number of tubes can be made by cutting holes of appropriate size with a cork-borer in a sheet of porous polyethylene 1/8 in. thick (Bel-Art Products). If a hole is made in the center of the holder, when it is loaded with tubes the holder can be slipped onto a motor spindle, immersed in a bath, and rotated at the required rate.

With this method we have scaled down by a factor of 25—from 1 ml to 40 μl—our earlier procedure[5] for measuring hormone effects on glucose uptake, with little if any loss in accuracy.

[5] J. Kattwinkel and A. Munck, *Endocrinology* **79**, 387 (1966).

[36] Techniques for the Study of Steroid Effects on Membraneous ($Na^+ + K^+$)—ATPase

By PETER LETH JØRGENSEN

Introduction

The identity established between the sodium pump and the enzyme activity, ($Na^+ + K^+$)-ATPase,[1] offers an opportunity to study the hormonal regulation of the sodium pump by applying principles used in the examination of the regulation of other enzyme systems. Studies of this kind require reliable methods for determination of the amount of enzyme in tissues and for examination of the properties of the enzyme. For ($Na^+ + K^+$)-ATPase these analyses are hampered by technical difficulties which are explained mainly by the firm association of the enzyme with the plasma membrane of the cells and by the complex and yet unresolved kinetics of this enzyme system.

The present techniques represent an attempt to solve some of these difficulties. They were developed during a study of the *in vivo* effects of steroid hormones on ($Na^+ + K^+$)-ATPase in kidney,[2-4] but may be adopted to work on other tissues. A summary of other studies of steroid hormone effects on ($Na^+ + K^+$)-ATPase is given in Table I with reference to alternative techniques. It has not been possible to demonstrate direct effects of steroid hormones *in vitro* on this enzyme activity.

Determination of the Total Activity of ($Na^+ + K^+$)-ATPase in a Tissue

During homogenization of tissues, the plasma membranes of the cells are broken into fragments of varying sizes. This explains that ($Na^+ + K^+$)-ATPase is found in all sediments after a conventional differential centrifugation. The specific activity of ($Na^+ + K^+$)-ATPase in a single subcellular fraction, e.g., a microsomal fraction, is therefore not a safe measure of the amount of enzyme in the tissue. Analysis of the activity in whole homogenates or in a particulate fraction containing all ($Na^+ + K^+$)-ATPase activity of the homogenate should be preferred.

Another consequence of the association of ($Na^+ + K^+$)-ATPase with the plasma membranes is that a major part, about 80%, of the activity

[1] J. C. Skou, *Physiol. Rev.* **45**, 596 (1965).
[2] P. L. Jørgensen, *Biochim. Biophys. Acta* **151**, 212 (1968).
[3] P. L. Jørgensen, *Biochim. Biophys. Acta* **192**, 326 (1969).
[4] P. L. Jørgensen, *J. Steroid Biochem.* **3**, 181 (1972).

TABLE I
STUDIES OF STEROID HORMONE EFFECTS ON $(Na^+ + K^+)$-ATPASE *in Vivo*

Hormone	Tissue	Preparation	Reference
Corticosterone Hydrocortisone	Rat kidney	Homogenate, microsomal fraction	[a,b]
Aldosterone (adrenalectomy)	Rat kidney	Quantitative histochemistry	[c]
Aldosterone Dexamethasone Cortisone	Rat kidney	Plasma membranes	[d]
Testosterone Dihydrotestosterone	Rat ventral prostate	Microsomal fraction	[e,f]

[a] C. F. Chignell and E. Titus, *J. Biol. Chem.* **241,** 5083 (1966).
[b] E. D. Hendler, J. Toretti, L. Kupor, and F. H. Epstein, *Amer. J. Physiol.* **222,** 754 (1972).
[c] U. Schmidt and U. C. Dubach, *Eur. J. Clin. Invest.* **1,** 307 (1971).
[d] N. G. De Santo, H. Ebel, and K. Hierholzer, *Pfluegers Arch.* **324,** 26 (1971).
[e] K. Ahmed and H. G. Williams-Ashman, *Biochem. J.* **113,** 829 (1969).
[f] W. E. Farnsworth, *Biochim. Biophys. Acta* **150,** 446 (1968).

in fresh homogenates or subcellular fractions is latent. The latency can be explained by limited access of substrate or activators to the inside of plasma membrane vesicles formed during homogenization. Activation of latent enzyme activity has been observed after treatment with concentrated salts,[5] storage in urea,[6] freezing and thawing,[7] and after treatment with detergent.[8] In the present procedure, latent enzyme activity is demasked by incubation with detergent.

The content of $(NA^+ + K^+)$-ATPase per cell may vary greatly from one cell type to the other within the same organ. The tissue is therefore dissected to obtain the highest possible degree of uniformity of cell population in the preparation.

Tissue Preparation

Principle. The procedure is adapted to small amounts of tissue (25–300 mg). After dissection and homogenization, the particulate fraction is prepared by centrifugation at high speed. The particulate fraction contains all subcellular particles and thus all ATPase activity in the homogenate.

[5] T. Nakao, Y. Tashima, K. Bagano, and M. Nakao, *Biochem. Biophys. Res. Commun.* **19,** 755 (1965).
[6] R. L. Post and A. K. Sen, Vol. 10, p. 762.
[7] O. J. Møller, *J. Exp. Cell Res.* **68,** 347 (1971).
[8] P. L. Jørgensen and J. C. Skou, *Biochim. Biophys. Acta* **233,** 366 (1971).

The soluble phase, removed by centrifugation, is devoid of ATPase activity and contains about 40% of the protein in the homogenate. Besides, ions and metabolites which may influence the enzyme activity in an uncontrollable way are removed.

Procedure. All operations are carried out at 0–4°. The organ is removed immediately after killing and bleeding the animal and is stored in ice-cold 0.25 *M* sucrose, 30 m*M* histidine, adjusted to pH 7.2 (20°) with HCl (sucrose–histidine). Dissection is done with a scalpel on a block of frozen sucrose–histidine or on a cooled metal plate covered with filter paper moistened with sucrose–histidine. The desired portion of the tissue is dissected free and vessels and connective tissue framework are removed. The tissue is placed in tared vials with 5 ml ice-cold sucrose–histidine and the exact weight is determined. The tissue is homogenized with five strokes in an ice-cooled glass homogenizer with a tight-fitting Teflon pestle operated at 1000 rpm. The homogenate is centrifuged for 45 minutes at 60,000 rpm ($\omega^2 t = 1.2 \times 10^{11}$). The sediment is resuspended in the original volume with a glass rod and is centrifuged again for 45 minutes at 60,000 rpm. The sediment is resuspended with four strokes in a Teflon-glass homogenizer to a protein concentration of 2–4 mg $\times$ ml^{-1}. For kidney tissue, the amount of protein recovered in the particulate fraction is about 6 mg protein/100 mg tissue.

Incubation with Detergent and Enzyme Assay

Principle. The activation of ($Na^+ + K^+$)–ATPase by detergents is most likely due to opening of vesicular membrane structures, resulting in free access of substrate and activators to their respective sites on the membrane. Maximum activation is obtained at the critical micelle concentration of the detergents and depends on the incubation time, the temperature, pH, and the concentration of protein and detergent in the incubation medium.[8] Incubation with detergent is done prior to the enzyme assay and only a small volume, 25 μl, of the incubation medium is transferred to test tubes with 1 ml of enzyme assay medium. By this dilution, the concentration of detergent in the assay medium is 3–4 fold lower than the concentrations which affect the enzyme reaction. The amount of particulate fraction added to the medium for incubation with detergent is adjusted so that the amount of enzyme in 25 μl of this medium when added to 1 ml of assay medium will cleave 5–10% of the total amount of ATP present (3 μmoles) in 5 or 10 minutes. The concentration of deoxycholate (DOC) necessary for maximum activation of ($Na^+ + K^+$)–ATPase at a given protein concentration is determined for each tissue and for each regulatory condition examined.

Procedure. Portions of the particulate fraction each containing the calculated amount of protein are incubated in a series of test tubes with 2 m*M* EDTA, 50 m*M* imidazole (pH 7.0 at 20°), and DOC in concentrations increasing stepwise from 0 to 1 mg DOC/ml. After incubation on a water bath at 20° for 30 minutes, 25 μl of this medium is transferred to test tubes for enzyme assay. The test tubes are equilibrated on a water bath at 37° for 5 minutes prior to addition of enzyme and contain 1 ml of a solution of 3 m*M* ATP (Tris salt), 3 m*M* $MgCl_2$, 130 m*M* NaCl, 20 m*M* KCl, 30 m*M* histidine (pH 7.5 at 37°), and where appropriate 1 m*M* ouabain. After 5 or 10 minutes at 37° the reaction is stopped with 100 μl 50% TCA, and inorganic phosphate is measured.[9] The $(Na^+ + K^+)$–ATPase activity is the difference in activity with and without ouabain. It is controlled that the ATP hydrolysis both in presence and absence of ouabain is a linear function of time and of enzyme concentration.

A plot of the $(Na^+ + K^+)$–ATPase activity versus the concentration of DOC gives a biphasic curve with a narrow optimum.[8] From this curve the highest concentration of DOC which gives maximum activation of $(Na^+ + K^+)$–ATPase is chosen and used in the routine incubations of the particulate fraction.

Protein Analysis. Protein is determined by the method of Lowry *et al.*[10] or by the Kjeldahl method[11] after precipitation and wash with trichloroacetic acid at 0–4°.

Expression of Results. Enzyme activity is expressed in μmoles P_i min^{-1} per milligram protein or per milligram tissue and mean values ± SE are calculated for groups of at least five animals. The significance of differences between means can be evaluated by Student's *t*-test. The biological variation of the activity of $(Na^+ + K^+)$–ATPase is usually moderate. For kidney tissue, the standard deviation is 5–8% of the mean specific activity in the particulate fractions from the outer cortex and the outer medulla of five normal rats.[3]

From Specific Activity to Amount of Enzyme

It is usually assumed that the specific activity of $(Na^+ + K^+)$–ATPase measured under proper conditions of assay is proportional to the amount of enzyme. This is only the case if the molecular activity, i.e., the transfer of Pi per minute per enzyme molecule at optimal substrate and activator

[9] C. H. Fiske and Y. Subbarow, *J. Biol. Chem.* **66,** 375 (1925).

[10] O. H. Lowry, N. I. Rosebrough, A. L. Farr, and R. I. Randall, *J. Biol. Chem.* **193,** 265 (1951).

[11] R. Ballentine, Vol. 3, p. 984.

concentrations,[12] is constant. In fact, measurements of the molecular activity of $(Na^+ + K^+)$-ATPase have given values from 1,000 to 16,000 Pi min^{-1}[8,13-15] depending on the nature of the preparation and on the method used for determination of the number of enzyme sites. If changes in specific activity of $(Na^+ + K^+)$-ATPase are to be related to changes in amount of enzyme, it is therefore advisable to measure the number of enzyme sites in each of the regulatory conditions examined. Further, the properties of the enzyme should be analyzed to ensure that the enzyme assay is conducted under optimal conditions.

Purification of $(Na^+ + K^+)$-ATPase. Active site determinations and examination of the properties of the enzyme are more reliable and accurate if done on preparations of relatively high specific activity of $(Na^+ + K^+)$-ATPase. Following a procedure described before,[16] a tenfold purification of the enzyme can be achieved by a single isopycnic-zonal centrifugation of the particulate fraction after incubation with detergent.

The Properties of the Enzyme

The kinetic properties of $(Na^+ + K^+)$-ATPase are studied intensively[17] and only a few parameters are so well defined that they can be used in examination of the hormonal regulation of the enzyme.

ATP and Mg^{2+}. As determined in ATP-binding studies, the enzyme–ATP dissociation constant is 0.1–0.7 μM, depending on the K^+ concentration.[14] The concentration of ATP for half-maximal activity at a Mg/ATP ratio of 1:1 is about 0.2 mM. At concentrations higher than 0.5–2 mM, the initial velocity is independent of the ATP concentration.[4]

Na^+ and K^+. To find the optimum concentrations of Na^+ and K^+ the ratio between Na^+ and K^+ in the test tube is varied at a constant ionic strength. A characteristic curve is obtained from which the relative affinities of the enzyme for K^+ and Na^+ can be calculated from the K/Na and the Na/K ratios for half-maximum activation.[17]

pH Optimum. The pH optimum is about 7.5 with a relatively broad optimum.[2]

Ouabain. The rate and the extent of the binding of ouabain to the enzyme is highly dependent on the composition of the medium.[18] The

[12] Enzyme nomenclature, *Compr. Biochem.* **13**, 9 (1965).
[13] H. Baden, R. L. Post, and G. H. Bond, *Biochim. Biophys. Acta* **150**, 41 (1968).
[14] J. G. Nørby and J. Jensen, *Biochim. Biophys. Acta* **233**, 104 (1971).
[15] J. Kyte, *J. Biol. Chem.* **247**, 7634 (1972).
[16] P. L. Jørgensen, Vol. 32.
[17] J. C. Skou, *Bioenergetics* **4**, 14 (1972).
[18] H. Matsui and A. Schwartz, *Biochim. Biophys. Acta* **151**, 655 (1968).

ouabain–enzyme dissociation constant can be determined by [^{3}H]ouabain-binding studies and have values of 10–100 n*M*.[19] The sensitivity of the enzyme to ouabain varies. The enzyme in crab nerve[20] and in rat kidney[2] is, for example, relatively insensitive to ouabain.

Active Site Determinations. Estimates of the number of enzyme molecules may be obtained by measurement of the number of binding sites for [^{3}H]ouabain,[18,19] or ATP,[14] or of the number of sites labeled with ^{32}P from [^{32}P]ATP in the presence of Mg^{2+} and Na^+.[13] Procedures for measurement of the ouabain-binding capacity[16] and phosphorylation[21] were described before. ATP binding can be determined by the equilibrium dialysis technique[14] or by a centrifugation technique.[22] If it is inferred that a change in the amount of enzyme has occurred, the measurements should show that the number of sites (nmoles/mg protein) has changed in parallel with the specific activity of the enzyme (μmoles Pi/minutes/mg protein), in other words, that the molecular activity (Pi/min) is unchanged.

Immunochemical techniques are valuable for measuring the relative content of enzyme proteins. Antibodies to ($Na^+ + K^+$)–ATPase inhibit the enzyme reaction and the active transport of sodium and potassium in red cells,[23] but they have not yet been used in measurements of the amount of enzyme.

[19] O. Hansen, *Biochim. Biophys. Acta* **233,** 122 (1971).
[20] J. C. Skou, *Biochim. Biophys. Acta* **42,** 6 (1960).
[21] R. L. Post and A. K. Sen, Vol. 10, p. 773.
[22] C. Hegyvary and R. L. Post, *J. Biol. Chem.* **246,** 5234 (1971).
[23] P. L. Jørgensen, O. Hansen, I. M. Glynn, and J. D. Cavieres, *Biochim. Biophys. Acta* **291,** 795 (1973).

[37] Methods for Assessing the Effects of Aldosterone on Sodium Transport in Toad Bladder

By Geoffrey W. G. Sharp

Aldosterone and other adrenocortical steroids with mineralocorticoid activity stimulate sodium reabsorption by kidney (the distal convoluted portion of the nephron), intestines, sweat glands and salivary glands. Because of the difficulty of studying sodium transport and the effect of mineralocorticoids directly in these tissues, simpler sodium transporting tissues such as frog skin and toad bladder have been used as model systems. The ease with which sodium transport can be measured in such

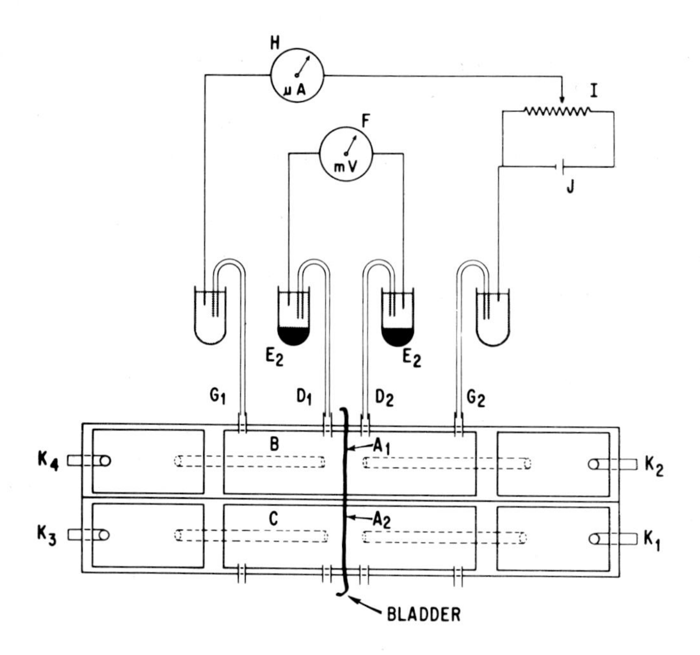

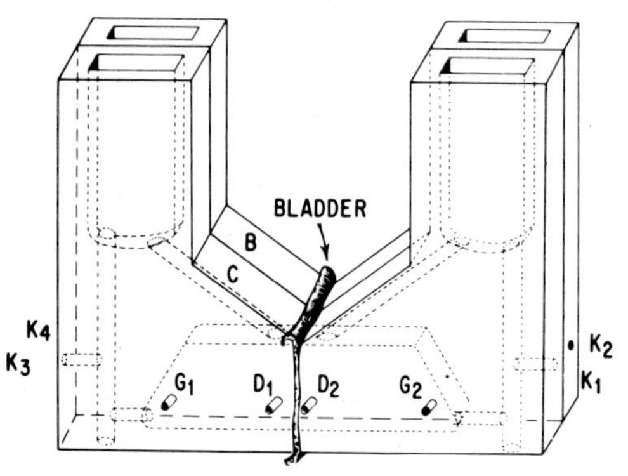

FIG. 1. Double-chamber apparatus used to detect the stimulation of sodium transport by aldosterone. In the upper figure, a horizontal section through the chamber as viewed from above. The bladder membrane is shown mounted as quarter bladders (A_1 and A_2) in the two parts (B and C) of the double chamber. Sodium Ringer's solution bathing the two opposite surfaces of the bladders is mixed and oxygenated by an air-bubble lift entering at K_1, K_2, K_3, K_4. Bridges made of 0.11 *M* potassium chloride in 2% agar in polyethylene tubing (D_1 and D_2) connect the mucosal and serosal bathing media with calomel electrodes (E_1 and E_2) and the potentiometer (F). Two more bridges (G_1 and G_2) connect the Ringer's solution in the chambers with the external circuit. This comprises a microammeter (H), voltage divider (I), and power supply (J). The potential difference across the membrane may be reduced to zero by the applied EMF in the external circuit and the current recorded. Only one of the two identical circuits is shown, that connected to chamber B. In the lower figure, the double chamber is shown in side views exposing the fluid compartments, air inlet, and bridge leads to chamber C.

epithelial membranes is due to the development of the short-circuit current technique by Ussing and Zerahn.[1] The toad bladder is a bilobed saclike tissue and it has advantages to the investigator in that it may be studied *in vitro* as a large membrane and because the transporting cells comprise a single cell layer. Thus the asymmetry of the transepithelial sodium transport system may be explored directly.

The Short-Circuit Technique for Measurement of Sodium Transport

When the toad bladder is clamped as a membrane between two halves of a glass or lucite chamber and both sides bathed in oxygenated Ringer's solution, an electrical potential difference can be measured across the bladder. This is achieved by means of potassium chloride–agar bridges placed in the Ringer's solution close to the bladder and connected via saturated KCl–calomel electrodes to a potentiometer. The potential measured is due to the active transport of ions by the bladder and can be reduced to zero by the addition of metabolic poisons. Two more potassium chloride–agar bridges are placed in the solutions bathing the two sides of the bladder (at some distance from it) and connected through silver electrodes in potassium chloride solution to a battery source, variable resistance, and an ammeter. The short-circuit is then complete (see Fig. 1). By the application of current to the external circuit the potential difference across the bladder can be reduced to zero. Under these short-circuited conditions of zero potential and with Ringer's solution of equal composition on the two sides, all chemical and electrical driving forces for the net movement of ions are removed. It is found, however, that an electrical current still flows in the external circuit and can be measured on the microammeter. This then is due to ion movements generated actively by the tissue. By the use of radioisotopes of sodium it has been shown that the short-circuit current is due entirely to the movement of sodium ion from the mucosal side of the tissue to the serosal side[2] (i.e., lumen to body). Because of this equality of active sodium transport and short-circuit current, the former can be accurately and conveniently measured. It is important that the equality of short-circuit current and active sodium transport be tested for each set of conditions used experimentally. Changed circumstances may uncover other ion transport systems. Care should be taken that the toads used are identified as to species and source because of differences in the behavior of different species and subspecies of toads.

[1] H. H. Ussing and K. Zerahn, *Acta Physiol. Scand.* **23,** 110 (1951).

[2] A. Leaf, J. Anderson, and L. B. Page, *J. Gen. Physiol.* **41,** 657 (1958).

Effects of Aldosterone

The first demonstration of an effect of aldosterone on sodium transport in the toad bladder was by Crabbe in 1961.[3] Using the short-circuit current technique of Ussing and Zerahn, an increased rate of sodium transport was observed some 4 hours after the addition of aldosterone to the serosal bathing medium. The toad bladder, however, presents problems to the study of hormone action because of the variability of sodium transport rate after being mounted in Ringers' solution *in vitro*. Sodium transport in the bladder may increase or decrease with time after removal from the toad, or it may even follow complex wave forms. In the study of aldosterone effects, which are expressed over several hours, adequate controls are required. Thus refinements of the short-circuit technique have been developed for the study of this hormone.

Double-Chamber Techniques.[4] Using this method, toad hemi-bladders are mounted in double chambers so that short-circuit current is monitored in adjacent portions of tissue (see Fig. 1). Spontaneous variations in transport are similar and parallel in the two pieces of tissue. This means that whether the rate of sodium transport rises or falls with time, the stimulation of sodium transport by aldosterone will be superimposed on the control rate of transport and can be easily detected. Other considerations are important. For instance, in the study of adrenocorticosteroid action in the usual laboratory animals, adrenalectomy is required to eliminate the effect of endogenous steroids. In the toad, complete removal of the equivalent inter-renal glands is not easily achieved and endogenous steroid hormones are present in circulating blood. As aldosterone is the major steroid hormone secreted by the toad inter-renal glands the production of this hormone must be minimized. This can be achieved by keeping the toads partially immersed in 0.6% saline solution for a period of 2–5 days. Stress-induced steroid secretion is also minimized by keeping the toads under quiet conditions and pithing them rapidly prior to use. Thus, the effects of exogenous aldosterone are more easily seen. As with other physiological functions in amphibia, sensitivity to aldosterone is subject to marked seasonal variation.

Prolonged Incubation of Bladders in Ringer's Solution. Porter and Edelman[5] took advantage of the ability of toad bladder to survive and to transport sodium ions over long periods of time, even in the absence of metabolic substrate in the bathing medium. Thus by prolonged incuba-

[3] J. Crabbe, *J. Clin. Invest.* **40**, 2103 (1961).
[4] G. W. G. Sharp and A. Leaf, *Nature (London)* **202**, 1185 (1964).
[5] G. A. Porter and I. S. Edelman, *J. Clin. Invest.* **43**, 611 (1964).

tion the effects of endogenous factors controlling sodium transport, particularly the effects of endogenous steroids, are dissipated. The bladders can be made dependent on a supply of exogenous substrate, and because of the decreased and relatively constant rate of sodium transport, the effects of agents which are stimulatory can be seen against a low background of control transport. The method they developed is as follows.

Toads are kept under constant conditions in the laboratory before use and the hemi-bladders removed after double-pithing. The bladders are rinsed in Ringer's solution (Na^+, 114; K^-, 3.5; Cl^-, 120.4; Ca^{2+}, 5.4; HCO_3^-, 2.5 mEq/liter, osmolality 228 mOsm/liter) containing 2×10^{-5} *M* methacholine chloride to contract the bladders. The contracted bladders are mounted between the chambers of the short-circuit current apparatus and supported on both sides by nylon mesh. The methacholine is removed by rinsing the chamber with Ringer's solution. Short-circuit current is then measured intermittently as required. For the study of aldosterone effects, the hemi-bladders are incubated overnight. Any bladders which do not maintain a potential difference of 5 mV or greater are discarded. After the overnight incubation, one hemi-bladder is designated test and one control and the overnight Ringer's solution replaced with fresh solution containing 5.5 m*M* glucose. All additions must be isosmotic. Two hours later aldosterone is added to the test hemi-bladders and the diluent (ethanol or methanol) added to the control bladder. Under these conditions the bladder gives a reproducible increase in sodium transport, the response being characterized by a latent period of approximately 60–90 minutes and a steady linear increase in sodium transport over the next 5 hours. The response can be quantitated and expressed as a ratio by dividing the current at any particular time after addition of aldosterone by the current at the time of addition of aldosterone.

Other Methods

Variations of these two techniques (double chamber and prolonged incubation) have been used and include differences in the substrates used, different concentrations, preincubation conditions, and timing. These variations are relatively unimportant and are used to meet the needs of individual investigations. For example, bladders can be preincubated overnight at 4° rather than at room temperature. This depletes the bladders of endogenous steroids with minimal depletion of energy-supplying substrates. The double chamber has also been combined with the overnight incubation technique to obtain the advantages of both systems.

Characteristics of the System

Aldosterone stimulates sodium transport after a latent period of 60–90 minutes. The effect can be detected at concentrations of aldosterone between 10^{-9} and 10^{-7} M. In fresh tissue the stimulation is maximal at 3–4 hours after addition of hormone, while in depleted tissue the rise may continue for 6 hours or more. Other steroids affect the transport in toad bladder. Their order of potency as stimulators of sodium transport approximates that expected from their known pharmacological properties in mammalian systems. Aldosterone, deoxycorticosterone, and 9α-fluorocortisol are powerful stimulators of sodium transport. Cortisol, corticosterone, and prednisone are less effective. Caution should be exercised in the interpretation of results with toad bladder because occasional differences in steroid specificity do exist. Dexamethasone provides a good illustration of this—thought to be almost inactive on sodium transport in mammalian systems, dexamethasone has a powerful stimulatory action on sodium transport in toad bladder. Spirolactones, as in mammals, act as structural competitive antagonists of aldosterone. Inhibition is also seen with inhibitors of RNA and protein synthesis such as actinomycin D, puromycin, and cycloheximide. This led to the development of the theory that aldosterone stimulates sodium transport by synthesis of specific proteins involved in the transport system.

[38] Methods for Assessing Estrogen Effects on New Uterine Protein Synthesis *In Vitro*[1]

By BENITA S. KATZENELLENBOGEN and JACK GORSKI

Estrogens stimulate the growth and maturation of the female reproductive system. Involved in this growth are increased net syntheses of RNA and protein. The sequence of these new synthetic events following estrogen interaction with a "target" or hormone-responsive tissue has been particularly well characterized in rat uterine tissue. Utilization of a combination of *in vitro* and *in vivo* techniques has enabled a detailed analysis of the time course of appearance and nature of these RNA and protein species. Whereas net increases in RNA and protein are not seen until 2 to 4 hours after estrogen administration, synthesis of a specific uterine protein (called "induced protein" or IP) is detectable by 40 min-

[1] This work was supported by Grant HD6726 from the National Institutes of Health and Ford Foundation Grant 700-0333.

utes[1a,2] and increased IP synthesizing capacity (presumably mRNA) can be detected within 10 minutes.[3,4]

The methods described here are those we have used to monitor the effects of estrogen on protein synthesis in rat uterine tissue *in vitro* during the first few hours of estrogen stimulation. These methods could also prove useful in analyzing hormonal effects on protein synthesis in other hormone-responsive tissues.

Method of Estradiol Administration *in Vivo* and *in Vitro*

Stock Solutions

1 mg estradiol-17β/1 ml absolute ethanol[5]
0.15 M NaCl
10^{-3} M estradiol-17β in absolute ethanol

The procedures described here utilized immature female rats, 22–25 days old (Holtzman). When estrogen is to be administered *in vivo,* 100 μl of 1 mg/ml estradiol stock solution is mixed with 9.9 ml of hot (to insure that the steroid is fully dissolved) 0.15 M NaCl. The mixture is allowed to cool slightly and then 0.5 ml of the warm mixture, containing 5 μg of estradiol and 1% ethanol, is injected intraperitoneally into each experimental animal. Control animals receive an equal volume of ethanol plus saline alone. At designated time intervals after injection, the animals are decapitated, and their uteri are quickly excised, stripped of all surrounding fatty tissue, and processed as indicated in the next section. Each rat is processed separately to minimize the time between killing and placing the uterus into the incubation medium.

When estradiol is to be administered *in vitro,* uteri from untreated animals are collected as described above and immediately transferred to unmodified Eagle's HeLa medium[6] (Difco) containing designated concentrations of estradiol-17β or vehicle control. Steroids are added in ethanol solution to give a final concentration of under 2% ethanol (see below). Incubations typically contain 3 uteri/2 ml or medium and are conducted in silicone rubber stoppered 10-ml Erlenmeyer flasks maintained in a shaking ($\sim$60 cycles/minute) water bath at 37° under an atmosphere of 95% O_2 and 5% CO_2.

[1a] A. Notides and J. Gorski, *Proc. Nat. Acad. Sci. U.S.* **56,** 230 (1966).
[2] A. Barnea and J. Gorski, *Biochemistry* **9,** 1899 (1970).
[3] A. DeAngelo and J. Gorski, *Proc. Nat. Acad. Sci. U.S.* **66,** 693 (1970).
[4] B. S. Katzenellenbogen and J. Gorski, *J. Biol. Chem.* **247,** 1299 (1972).
[5] 17β-Estradiol is 1,3,5(10)-estratrien-3,17β-diol.
[6] H. Eagle, *Science* **122,** 501 (1955).

Double-Isotope Labeling of Uterine Proteins and Preparation of Uterine Soluble Proteins

After indicated intervals of exposure to hormone or vehicle control either *in vitro* or *in vivo,* (1) uteri are thoroughly rinsed with HeLa medium and transferred to flasks containing fresh HeLa medium (again at 3 uteri/2 ml) and radioactive amino acid with or without actinomycin D (see below), or (2) additions of radioactive amino acid with or without actinomycin D are made directly to the original flask. Uterine proteins are labeled with 20 μCi/ml [4,5-^{3}H]L-leucine (2.0 Ci/mmole in 10^{-2} *M* HCl) or 5 μCi/ml [^{14}C]L-leucine (316 Ci/mole in 10^{-2} *M* HCl) for 2 hours at 37°, with or without 30 μg/ml actinomycin D, in Eagle's HeLa medium under an atmosphere or 95% O_2 and 5% CO_2. Incorporation of radioactive leucine into IP and into general protein is linear over this time period. This concentration of actinomycin D has been shown to nearly completely block overall uterine RNA synthesis *in vitro*[7] and to completely block induction of IP synthesizing capacity *in vitro.*[8] Because the ^{14}C amino acids are added in small volumes of 10^{-2} *M* HCl, ^{3}H incubations are also equalized in both HCl and volume. This amount of HCl does not detectably change the pH of the HeLa medium, and, whether included or omitted, has no effect on IP induction.

Estrogen-treated uteri are incubated with one isotope and the controls with the second. At the end of the incorporation period, the uteri of both groups are rinsed thoroughly with three changes of ice-cold 0.05% Na_2EDTA and homogenized separately or together in this EDTA solution (6 uteri/1.2 ml) using a motor-driven Duall glass homogenizer. Homogenates are centrifuged for 50 minutes at 27,000 *g* at 4° in a Sorvall RC2B centrifuge, and the resulting supernatant fraction is frozen until use. This *in vitro* incubation procedure is outlined in Fig. 1.

Optimal Conditions for Induction of IP Synthesis and for General Uterine Protein Synthesis *in Vitro*

Estradiol Concentration. For induction of IP synthesis *in vivo,* typically 5 μg of esthadiol-17β per immature rat has been injected. However, injection of 0.2 μg of estradiol (i.e., a 25-fold lower concentration) gives full induction, whereas 0.05 μg per animal gives significantly reduced induction.

For *in vitro* induction, 2–3 × 10^{-8} *M* estradiol gives maximal induction, and 2–3 × 10^{-9} *M* estradiol gives 50% of maximal induction. The dose-response curve for induction *in vitro* is shown in Fig. 2. Dose-re-

[7] A. B. DeAngelo, Ph.D. Thesis, University of Illinois, Urbana (1970).
[8] B. S. Katzenellenbogen, unpublished data (1970).

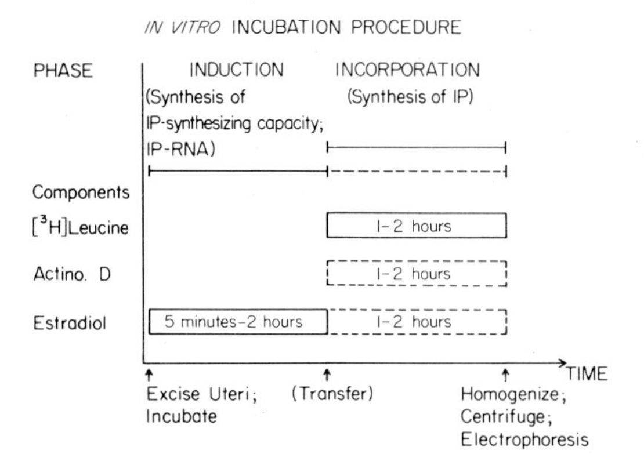

FIG. 1. General scheme for the *in vitro* incubation of rat uteri and the induction of IP synthesis. Omission of actinomycin D allows induction to continue through incorporation phase (dashed line). In some cases, exposure to estradiol is continued throughout the incorporation phase. See text for further details.

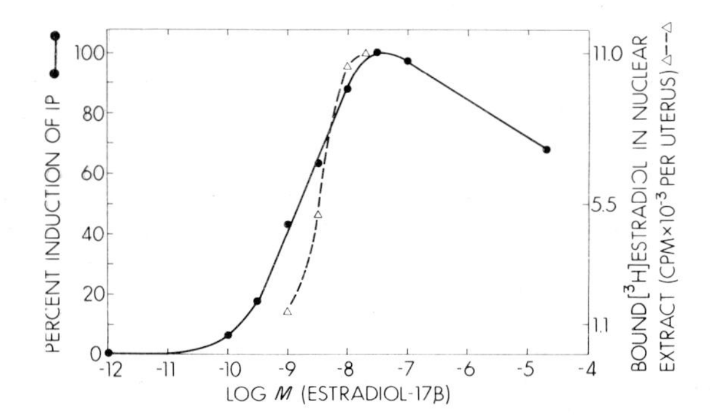

FIG. 2. Effect of *in vitro* estradiol-17β concentration on the rate of IP synthesis. Uteri (five per group) excised from untreated rats were incubated in 2.0 ml of Eagle's HeLa medium containing various concentrations of estradiol-17β (or ethanol for controls) for 60 minutes at 37°. At this time, [³H]leucine plus actinomycin D were added to experimental flasks and [¹⁴C]leucine plus actinomycin D were added to control flasks, and amino acid incorporation was allowed to proceed for 2 hours at 37°. Uterine soluble proteins were subjected to electrophoresis on polyacrylamide gels, and IP synthesis was quantitated as described in the text. Results are expressed relative to IP induction with 3×10^{-8} *M* estradiol being 100%. Values for bound [³H]17β-estradiol in immature rat uterine nuclear extracts after a 60-minute incubation with various concentrations of [³H]estradiol in Eagle's HeLa medium at 37° are from G. Giannopoulos and J. Gorski, *J. Biol. Chem.* **246,** 2524 (1971). [Reproduced from Katzenellenbogen and Gorski.[4]]

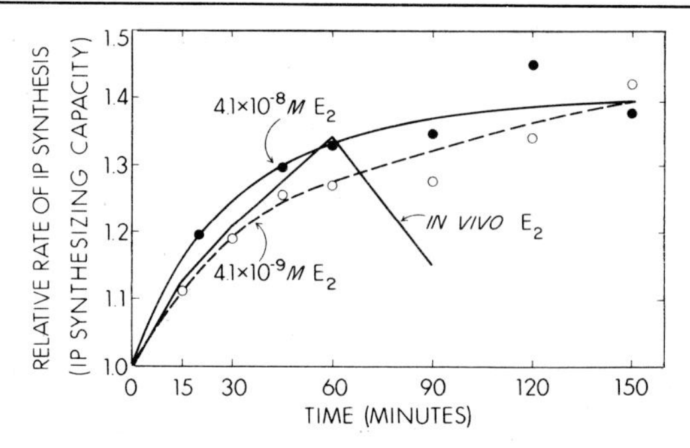

FIG. 3. Time course of the relative rate of IP synthesis at two different estradiol-17β concentrations *in vitro*. Uteri (three per group) excised from untreated animals were incubated in 0.9 ml of Eagle's HeLa medium containing either 4.1×10^{-8} *M* or 4.1×10^{-9} *M* estradiol (experimentals) or ethanol (controls) at 37° for time intervals up to 150 minutes. At the end of each time period, experimental uteri were allowed to incorporate [^{3}H]leucine and control uteri were allowed to incorporate [^{14}C]leucine into protein in the presence of actinomycin D for 2 hours at 37°. Control and estrogen-treated uteri were homogenized together, and the supernatant fraction of centrifuged homogenates was separated by polyacrylamide gel electrophoresis. From analysis of such gels, the relative rate of IP synthesis (experimental/control) was determined as described in the text. The solid line indicates the time course of the relative rate of IP synthesis following different periods of *in vivo* estradiol (5 μg/rat) and is from data reported by DeAngelo and Gorski.[3] [Reproduced from Katzenellenbogen and Gorski.[4]]

sponse curves with similar shapes but different midpoints are obtained with other estrogens such as estrone and estriol.[9]

Length of Induction Period (Exposure to Estradiol). The relative rate of IP synthesis or accumulation of IP synthesizing capacity is maximal after exposure for 1 hour to high concentrations of estradiol either *in vitro* (4×10^{-8} *M*) or *in vivo* (5 μg). At lower estradiol concentrations *in vitro* (e.g., 4×10^{-9} *M*) the rate of accumulation of IP synthesizing capacity is slower although the same maximal response is eventually obtained (Fig. 3).

Temperature. Optimal induction is obtained at 37°. If the 1-hour exposure to estradiol *in vitro* is conducted at 30°, only 40–50% of the maximal (37°) IP induction is obtained; likewise, incorporation of amino acids into IP relative to control proteins at 30° (after estradiol exposure *in vitro* at 37°) is only 66% the incorporation at 37°.

pH. While a thorough study of IP induction as a function of pH has not been done, the incubation pH of 7.4 that is routinely used appears to be optimal for rat uterine tissue. Although minor increases in pH do

[9] T. S. Ruh, B. S. Katzenellenbogen, J. A. Katzenellenbogen, and J. Gorski, *Endocrinology* **92**, 125 (1973).

not affect the response, marked decreases in pH are deleterious. Incubation in Eagle's HeLa medium at pH 6.4 gives only 15% of maximal IP induction while overall amino acid incorporation is reduced to one-third of the pH 7.4 level.

Ethanol Concentration. The concentration of ethanol (from the estradiol solution) in the *in vitro* incubations should be kept below 2%. Ethanol concentrations up to 2% do not affect IP induction or overall protein synthesis. However, incubations of uteri in media containing 5% ethanol show no IP induction; furthermore, overall amino acid incorporation is less than one-half of normal, implying that 5% ethanol has an adverse effect on the tissue as a whole. Estradiol concentrations below 10^{-6} *M* are quite soluble in low (under 1%) concentrations of ethanol.

Rate of Tissue to Incubation Medium. Ideally, uteri should be incubated in a sufficiently large volume of medium so that depletion of estradiol from the incubation by tissue uptake does not significantly alter the concentration of estrogen in the medium. Above 1×10^{-9} *M* estradiol, IP responses of equal magnitudes are obtained with incubations of either 2 uteri/ml of medium or 2 uteri/32 ml of medium. Below 1×10^{-9} *M* estradiol, the magnitude of the IP response in small volume incubations appears to be somewhat less than that in large volume incubations. It should be pointed out that each immature rat uterus contains 1 to 2 pmoles of specific estrogen-binding sites. A 10^{-9} solution contains only 1 pmole estrogen/ml.

Analysis of Uterine Soluble Proteins

The following discussion relates to characterization of newly synthesized uterine soluble proteins prepared as described in the first section. Soluble proteins (~0.5–1.0 mg/immature uterus) account for 20–40% of the total uterine protein, and likewise these newly synthesized soluble proteins represent the same fraction of the total newly synthesized uterine proteins (up to at least 6 hours after estradiol).

Polyacrylamide Gel Electrophoresis

Solutions

I. 6% Cyanogum-41 gelling agent (Fisher)[10] in TBE buffer
II. 100 mg ammonium persulfate/ml water (freshly prepared)
III. TBE buffer: 0.066 *M* Tris-(hydroxymethyl)aminomethane, 0.02 *M* boric acid, 0.003 *M* Na_2EDTA, pH 8.6 at 25°

[10] Cyanogum-41 gelling agent is a 95:5 mixture of acrylamide–methylene bisacrylamide.

IV. 13% Ficoll (Pharmacia) containing a few grains of Bromophenol Blue (Canalco)

Procedure. To prepare 6% polyacrylamide gels in TBE buffer, 20.0 ml of solution I is mixed with 0.35 ml of solution II, and then 45 μl of *N,N,N′,N′*-tetramethylethylenediamine (TMEDA; Eastman or Bio-Rad) is added. After the solution is mixed, aliquots (5 ml) are dispensed to each of four gel tubes (9×0.7 cm), carefully layered with 2–3 mm of buffer, and allowed to polymerize. (As the gels polymerize they turn opaque. Polymerization should be complete within 30 minutes.) Uterine soluble proteins (supernatant fractions) 200–300 μl, equivalent to 1–2 uteri/gel, from estrogen-treated and/or control uteri are mixed with 100 μl of solution IV (to increase sample density and provide a tracking dye marker), applied to the top of the gels, and carefully layered with buffer. The electrophoresis, from cathode to anode, is conducted at room temperature at 1 mA/tube for 1 hour and then at 2 mA/tube for the remainder of the run (~5 hours).

At the termination of a run, gels are removed from the tubes by forcing ice water between the gel and the tube wall, using a syringe fitted with a long needle. The position of the dye band and the length of the gel are noted. Gels to be stained are treated with 1% Amido Schwarz in 7% acetic acid for 3–16 hours and are then electrophoretically destained.[11] Staining reveals three faint bands migrating faster than the albumin band. The position of the newly synthesized radioactive IP corresponds with the middle one of these three stained protein bands.

Determination of Radioactivity in Gels. Gels are allowed to remain several hours in tubes containing 7.5% acetic acid and are then freed of radioactive amino acids by electrophoresing in 7.5% acetic acid for 45 minutes in an electrophoretic destainer. Gels are then frozen over dry ice and sectioned into 1–2 mm discs using a multiple razor blade slicer. The proteins in the gel slices are eluted by incubating the discs overnight at room temperature in scintillation vials containing 0.8 ml of 1% sodium dodecyl sulfate in water, with agitation on a shaker.[12] After being thoroughly mixed with 15 ml of toluene-based scintillation fluid (0.03% dimethyl POPOP {1,4-bis-[2-(4-methyl-5-phenyloxazolyl)]-benzene}) and 0.5% PPO (2,5-diphenyloxazole) (New England Nuclear) that is 10% in Biosolv-BBS3 (Beckman), samples are counted. Radioactivity determinations on duplicate samples with gels processed using the hydro-

[11] Gels stained with the more sensitive Coomassie Brilliant Blue dye (0.25% in methanol–acetic acid–water, 5:1:5) destain very poorly (with 7.5% acetic acid, 5% methanol); hence, Amido Schwarz is routinely used for staining proteins in this gel system.

[12] K. Weber and M. Osborn, *J. Biol. Chem.* **244**, 4406 (1969).

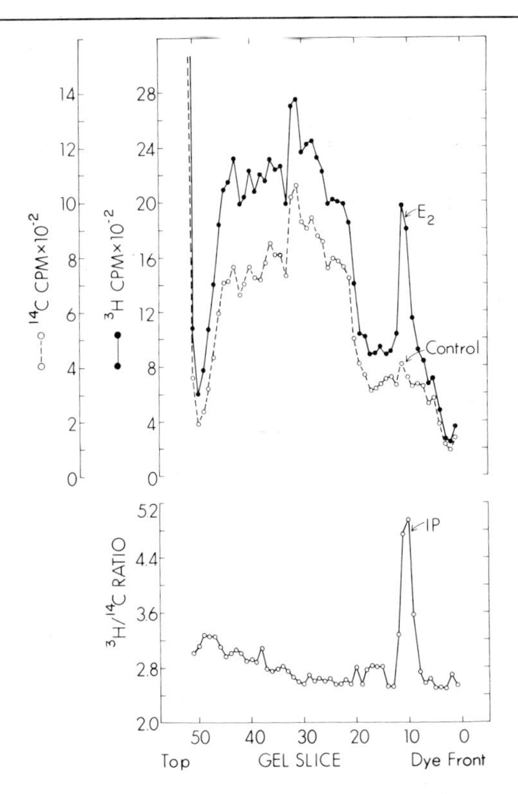

FIG. 4. Electrophoretic distribution on polyacrylamide gels of uterine soluble proteins synthesized *in vitro* following a 1-hour *in vitro* incubation with 3.7×10^{-8} *M* estradiol-17β. Uteri (three per group) excised from untreated animals were first incubated with either 3.7×10^{-8} *M* estradiol or ethanol only (1%) in 2.0 ml of Eagle's HeLa medium for 60 minutes at 37° and then allowed to incorporate labeled leucine (^{3}H for estradiol-treated and ^{14}C for control uteri) into protein for 2 hours at 37° in the presence of 28 μg/ml of actinomycin D. Control and estradiol-treated uteri were homogenized together, and the supernatant fraction of centrifuged homogenates was separated by polyacrylamide gel electrophoresis. The radioactivity and ^{3}H:^{14}C ratio in each gel slice were determined. [Reproduced from Katzenellenbogen and Gorski.[4]]

gen peroxide and Nuclear Chicago Solubilizer method[13] give similar radioactivity profiles and count recoveries. However, as the peroxide method is considered by us to be more laborious and more liable to variability, the dodecyl sulfate-count elution method is used routinely. The results of a typical polyacrylamide gel electrophoretic separation of newly synthesized uterine soluble proteins is seen in Fig. 4.

[13] P. V. Tishler and C. J. Epstein, *Anal. Biochem.* **22,** 89 (1968).

Starch Gel Electrophoresis

Solutions

I. 14% hydrolyzed starch (Connaught) in TBE buffer
II. TBE buffer: 0.033 *M* Tris-(hydroxymethyl)aminomethane, 0.01 *M* boric acid, 0.0015 *M* Na_2EDTA, pH 8.6 at 25°

Procedure. Starch gels (14%) are made in TBE buffer. Hydrolyzed starch is dissolved in buffer at 90°, then deaerated by evacuation. The starch can then be poured onto molds for gel blocks[1] or drawn into glass tubes such as used for the polyacrylamide gels.[2] After the gel solidifies, the plates or tubes are covered with Saran Wrap (Dow) and stored at 4° until use within 24 hours.

Fifty or one-hundred microliters of the uterine soluble fraction are placed in each sample slot of the starch gel block. The sample slots are covered with melted vaseline, and vertical electrophoresis is carried out at 270 mV at 4° for 10–12 hours. At the completion of the electrophoretic run, the gel is removed from the plate and sliced lengthwise into two. One-half of the gel is stained with 1% Amido Schwarz in 7% acetic acid for at least 1 hour. The excess dye, as well as unincorporated labeled amino acid, is then removed in an electrophoretic destainer. The electrophoretic patterns usually extend 10 cm from the origin, and the band containing IP migrates approximately 1½ cm in front of the serum albumin band.

In order to measure the radioactivity associated with the soluble protein bands, the other half of the starch gel is sectioned into 1.8-mm slices by means of a multiple razor blade slicer. Each starch gel slice is placed in a scintillation vial and dissolved in 0.5 ml of 98% formic acid at room temperature for 4–5 hours after which the formic acid is evaporated by a draft of air in a hood. Hydroxide of Hyamine (0.5 ml; Packard) is added to each vial; the vial is capped and allowed to stand for 12–24 hours at 45° to digest the residue. Toluene-based scintillation fluid (10 ml) is added, and the radioactivity in each vial determined. Starch gels in tubes (10×0.8 cm) are treated similarly to acrylamide gels. Samples mixed with Ficoll (see above) are layered over the gel. Buffer is layered over the sample and electrophoresis carried out at 1.5 mA/tube for 3 hours.

Quantitation of IP Synthesis

Procedure. The relative rate of IP synthesis is determined by analysis of double-labeled experimental and control proteins that have undergone coelectrophoresis on polyacrylamide or starch gels. Either one of two methods is used. In the first method, which is the one routinely used, the ^3H:^{14}C ratio (or estrogen to control ratio) determined for each gel slice is plotted, and the increase in area under the ratio curve in the IP region is determined. The control area (A_c) plus estrogen-increased area (A_e) divided by control area, $(A_c + A_e)/A_c$, is used as an estimate of relative rate of IP synthesis. In the second method, the ratio of cpm in the IP band (average of the peak slices) to the cpm in several gel slices above and several slices below the IP band is calculated for both ^3H and ^{14}C. The ratio so obtained for ^3H (estrogen) is divided by the ratio for ^{14}C (control) within a gel, and the value obtained is used as an estimate of relative rate of IP synthesis.

Data calculated by both methods give similar relative values within a series of gels, although absolute values calculated by the two methods are obviously different. Such analytical procedures are necessary, as we have no way at present of determining the absolute amount of IP.

Sodium Dodecyl Sulfate (SDS)—Polyacrylamide Gel Electrophoresis[12]

Solutions

I. 22.2% acrylamide–0.6% methylene bisacrylamide in water
II. 15 mg ammonium persulfate/ml water (freshly prepared)
III. 0.2 *M* sodium phosphate, 0.2% SDS, pH 7.2

Reservoir Buffer

0.1 *M* sodium phosphate, 0.1% SDS, pH 7.2

Sample Buffer

0.25% SDS, 0.01 *M* sodium phosphate, pH 7.2, 1.5% β-mercaptoethanol, 10% (v/v) glycerol, 0.002% Bromophenol Blue (Canalco)

Preparation of SDS Gels. To prepare 10% polyacrylamide–0.1% SDS, pH 7.2 gels, 9.0 ml of solution I is mixed with 1.0 ml of solution II, 10.0 ml of solution III, and 15 μl of TMEDA. This provides enough polyacrylamide for four 9 × 0.7-cm gels. Gels are poured, layered with 2–3 mm of buffer, and polymerization is complete in about 45 minutes.

Preparation of IP Sample and Protein Molecular Weight Standards. Following polyacrylamide gel electrophoresis in TBE buffer the two peak IP gel slices (from gels which had not been fixed in acetic acid) are cut from the gel and mashed into small pieces.[14] The proteins in these gel

[14] Similar treatment of gel slices immediately adjacent to the two peak IP slices indicates that they may contain small amounts of IP and that this IP migrates as a single band, coincident with that of the peak IP material.

fragments are eluted by incubation with 0.8 ml of sample buffer containing 1% SDS for 40 hours on a rotary shaker at room temperature. This procedure gives over 95% elution of counts. Six hundred microliters of this eluate are then heated to 60–65° for 10 minutes prior to direct application to an SDS gel. Protein molecular weight standards (0.5 mg each) are mixed with 1.0 ml of Sample Buffer, heated at 60–65° for 10 minutes, and 100 μl (50 μg) are applied per gel.

Running and Processing of SDS Gels. Electrophoresis is conducted at room temperature, migration from cathode to anode, at ~35 V with a constant current of 8 mA/gel, and is allowed to proceed for ~15 hours (until the bromophenol blue tracker dye is just above the bottom of the gel). The length of the gel and the dye position are noted, and the gels are stained for proteins for at least 3 hours with 0.25% Coomassie Brilliant Blue in methanol:acetic acid:water (5:1:5) and destained with 7.5% acetic acid, 5% methanol. Radioactive gels are frozen and sliced into discs; counts from each slice are eluted in 0.8 ml of 1% SDS as described above under "Polyacrylamide Gel Electrophoresis." Recovery of IP counts on the SDS gel is approximately 100%.

The molecular weight of IP can be estimated by comparison with the protein molecular weight standards run simultaneously in parallel gels and with internal radioactive protein markers run together with IP in the same gel. IP is well resolved as a single protein of molecular weight ~42,000 when an initial purification step (such as polyacrylamide gel electrophoresis) is first done to separate the bulk of the other proteins from the IP. Induced protein is not resolved in SDS gel runs of unpurified whole uterine supernatant.

Use of RNA and Protein Synthesis Inhibitors to Study Effects of Estrogen on IP Induction and New RNA and Protein Synthesis

Inhibitor Treatment of Animals. For studies in which RNA or protein synthesis inhibitors are used *in vivo*, actinomycin D (4 mg/kg or 8 mg/kg; Merck, Sharp and Dohme) is injected 15 to 30 minutes before estrogen administration; puromycin (200 mg/kg; Nutritional Biochemicals) or cycloheximide (4 mg/kg; Calbiochem) is administered 30 minutes prior to the hormone. All inhibitors are dissolved in 0.5–1.0 ml of 0.15 *M* NaCl and administered intraperitoneally.

Inhibitor Treatment of Uteri in Vitro. For *in vitro* studies, uteri are incubated with 0.3 μg/ml (low dose) or 20–30 μg/ml (high dose) actinomycin D in Eagle's HeLa medium for designated periods of time prior to estradiol exposure.

Pretreatment for 15 or 30 minutes *in vivo* with the RNA synthesis inhibitor actinomycin D blocks cytidine incorporation into RNA by 85–90% and effectively blocks the estradiol-induced synthesis of IP. These RNA synthesis inhibitors only slightly reduce (less than 20%) the rate of labeled amino acid incorporation into total soluble protein, whereas IP synthesis is depressed to 23% ± 2% after pretreatment with 4 mg/kg actinomycin D and is further depressed to 10% ± 4% at the higher (8 mg/kg) actinomycin D dose.[3] Likewise, 30-minute pretreatment with 20 or 30 μg/ml actinomycin D *in vitro* completely blocks subsequent IP induction by estrogen, although a lower (0.3 μg/ml) concentration of actinomycin D *in vitro,* which presumably blocks ribosomal RNA synthesis, is without effect on IP induction.[15]

It should be noted that exposure of uteri to 20 μg/ml actinomycin D for periods of up to 6 hours *in vitro* does not appear to cause any serious metabolic impairment in the uteri. Whereas one finds increases in glucose metabolism, protein synthesis, total lipid synthesis, and perchloric acid-soluble radioactivity in the absence of inhibitor, the presence of actinomycin D maintains these metabolic parameters at, or slightly above, the zero time level.[7]

Although IP induction is blocked by actinomycin D, production of IP synthesizing capacity is not dependent on prior protein synthesis.[16] Thirty-minute pretreatment of uteri *in vivo* (prior to a 30-minute *in vivo* estrogen stimulation) with levels of puromycin (250 mg/kg) or cycloheximide (4 mg/kg) which reduce total protein synthesis to less than 5% of control values is without effect on the estrogen-stimulated induction of IP synthesizing capacity.[3]

Determination of the Time Course of Accumulation of IP Synthesizing Capacity and IP. Since the production of IP synthesizing capacity is actinomycin D sensitive, it is possible to obtain an estimate of the time course of accumulation of IP synthesizing capacity by incubating uteri with estrogen for various periods of time prior to the addition of actinomycin D (to stop production of IP synthesizing capacity) and labeled amino acid. The relative rate of IP synthesis can be determined from analysis, by polyacrylamide gel electrophoresis, of experimental and control uterine soluble proteins after different intervals of estradiol exposure and RNA synthesis (Fig. 3).

The time course of synthesis of IP has been measured *in vivo* after estrogen stimulation *in vivo* and direct intraluminal injection of radioactive amino acid.[2]

[15] B. S. Katzenellenbogen, unpublished data (1971).

[16] It should be noted that studies based on use of inhibitors (actinomycin D) constitute indirect evidence of RNA involvement.

[39] A Molecular Bioassay for Progesterone and Related Compounds

By STANLEY R. GLASSER

Bioassay techniques are no longer of primary importance in the determination of the content of progesterone or other progestins in any given material. Far more accurate and direct physical and chemical methods are now available. However, after physical, chemical and immunological methods have been used to extract, purify, and characterize natural and/or synthetic compounds similar to progesterone, it is the bioassay which must be used to estimate if the compound in question has activity *in vivo*. It is the bioassay which is used to evaluate the nature and degree a suspected progestin can compete with, replace, or even supercede the natural progesterone.

In comparison with the bioassay of other hormones, only a few of the methods developed over the years are suitable for routine use. Historically, interest in progesterone has been sporadic and not aggressive. By the time of the advent of progestins in contraceptive therapy stimulated concerted efforts to determine progestin concentrations, emphasis was already on physical and chemical technology. The limitation to the use of many bioassay methods is that they require, particularly for screening purposes, rather large amounts of material which is often available in small quantities in order to obtain useful results. Other methods are too involved and some even more uneconomical of time and material.

Bioassays for progestational compounds may be grouped on the basis of either the target organ or the effect produced. The more practical and useful assays include the following:

(1) Assays based on changes in the cytology or the biochemical pattern of the glandular endometrium

- (a) Corner-Allen: a comparison of glandular to stromal portions of the uterine mucosa of spayed rabbits following treatment. This assay is based on planimetric measurement of histological sections.[1]
- (b) Clauberg: a modification of the Corner-Allen utilizing immature rabbits. These animals are primed for 5–8 days with an estrogen prior to treatment (5 days) with a test substance. This assay also depends on planimetric measurement of the mucosal com-

[1] G. W. Corner and W. M. Allen, *Amer. J. Physiol.* **83**, 326–339 (1929).

ponents and the calculation of their ratios.[2] A widely used modification was introduced by McPhail.[3]

(c) McGinty: a further modification of the Clauberg-McPhail genre but somewhat sparing of test compound. The test substance is introduced directly into the uterine lumen. The final comparison, like those above, is based on a semiquantitative evaluation of histological changes.[4]

(d) Carbonic Anhydrase Test: this assay is based on the reported correlation between the activity of the enzyme and progesterone-stimulated proliferation of rabbit endometrium.[5] The assay utilized the "Clauberg rabbit" as a source of enzyme. Numerous modifications, including a manometric analysis of enzyme activity,[6] confirm the original finding[7] but no correlation can be developed between carbonic anhydrase activity and deciduoma formation in rodents.

(2) Assays based on changes in uterine stromal histology

(a) Hooker-Forbes: this assay uses ovariectomized (at least 2 weeks prior to use) CH mice into which not more than 0.6 μg of test material is injected directly into a ligated segment of uterine horn. Cytological analysis of the transformation of uterine stromal nuclei is made 48 hours later.[8] This test is very sensitive but tedious; it requires training and accuracy and its reliability has always been questioned.

(b) Decidual Cell Response: this assay is based on the precipitous growth response of the uterine stroma of a properly hormonally prepared rodent to chemical or physical stimulation. Originally described by Loeb[9] the assay has often been modified.[10] The techniques used are rapid, simple to master, and test animals are comparatively inexpensive. The endpoints used are histological or gravimetric. The latter offers a rational basis for quantitation.

[2] C. Clauberg, *Zentralbl. Gynaekol.* **54**, 2757–2770 (1930).

[3] M. K. McPhail, *J. Physiol.* (*London*) **83**, 145–156 (1934).

[4] D. A. McGinty, L. P. Anderson, and H. B. McCullough, *Endocrinology* **24**, 829–832 (1939).

[5] C. Lutwak-Mann, *J. Endocrinol.* **13**, 26–38 (1955).

[6] T. Miyake, *in* "Methods in Hormone Research," 1st ed. (R. I. Dorfman, ed.), pp. 127–178. Academic Press, New York, 1962.

[7] C. Lutwak-Mann and H. Laser, *Nature* (*London*) **173**, 268 (1954).

[8] C. W. Hooker and T. R. Forbes, *Endocrinology* **41**, 158–159 (1947).

[9] L. Loeb, *Proc. Soc. Exp. Biol. Med.* **5**, 102 (1908).

[10] V. J. DeFeo, *in* "Cellular Biology of the Uterus" (R. M. Wynn, ed.), pp. 192–290. Appleton, New York, 1967.

The correlation with rabbit assays is very good and this assay has been increasingly employed in the biological screening and evaluation of progestins. The purpose of this assay will be to describe the utilitarian aspects of this assay and to introduce a new endpoint, biochemical in nature, which shortens the assay, spares the test compound, increases accuracy and specificity, and gives promise of further productive modification.

An immediate word of caution is required before consideration of any of the assays based on alterations of endometrial or stromal histology or biochemistry. None of these tests provides reliable information on the most singular and important function of progesterone, i.e., the maintenance of normal pregnancy. This is the ultimate criterion. Our experience has indicated that there are many compounds which are very active in producing progestational changes in any of the bioassays cited above but cannot maintain a normal pregnancy in a castrate animal. Additionally there are compounds that not only are unable to maintain pregnancy but may interfere with the conduct of gestation in intact animals either by terminating the pregnancy or producing developmental abnormalities particularly of the urogenital system. Pregnancy maintenance tests are spuriously simple to perform but require skill and experience to interpret.

The decidual cell response (DCR) of the uterine stroma is a biological assay which provides indirect evidence for the existence of a functional corpus luteum and has been extensively utilized as an analog for ovoimplantation. Of primary importance is its utility as an assay that provides evidence for the presence of hormones, associated with the corpus luteum. Thus, the DCR has been used to assay progesterone[11] as well as deoxycorticosterone,[12] pregnenolone,[13] as well as 19-norsteroids.[14] Alternately, interference with the proper development of the progestational state has served to evaluate the putative antiprogestational action of drugs,[15,16] gonadotrophins,[17] natural and synthetic steroids,[18] and radiation.[19] The physiology of DCR permits extension of the assay to a variety of physio-

[11] E. B. Astwood, *J. Endocrinol.* **1**, 49–55 (1939).

[12] G. Masson, *Proc. Soc. Exp. Biol. Med.* **54**, 196–197 (1943).

[13] M. R. Cohen and I. F. Stein, *Amer. J. Obstet. Gynecol.* **40**, 713–724 (1940).

[14] G. Pincus, M. C. Chang, M. X. Zarrow, E. S. F. Hafez, and A. Merrill, *Endocrinology* **59**, 695–707 (1956).

[15] M. C. Shelesnyak and A. Barnea, *Acta Endocrinol.* (*Copenhagen*) **43**, 469–476 (1963).

[16] M. J. K. Harper, *J. Reprod. Fert.* **7**, 211–220 (1964).

[17] J. M. Yochim, D. R. Hiestermna, and J. E. Keever, *Endocrinology* **77**, 508–519 (1965).

[18] G. M. Stone and C. W. Emmens, *J. Endocrinol.* **29**, 147–157 (1964).

[19] S. R. Glasser and B. P. Grant, *Radiat. Res.* **25**, 191 (1965).

logical, neuroendocrine, and behavioral phenomena which might elicit luteotrophin materials (suckling; pituitary, hypothalamic secretions; neurotransmitters, etc.).[20-22]

Decidual cell response has been produced in guinea pigs, rabbits, dogs, monkeys, mice, hamsters, and rats.[10] The degree of response differs among different species but the response itself is always associated with those areas involved in early placentation. Decidualization is a naturally occurring phenomenon in all the species cited including man. It occurs during normal gestation and can be produced in pseudopregnancy. Experimental conditions allow the response to be maximized, involving the entire endometrium, so that it can serve as quantifiable assay. For purposes of assay the test animal should either be immature and/or castrated or hypophysectomized, thereby obviating any contribution of endogenous secretions and increasing the exclusivity and sensitivity of the assay.

The DCR is a hormonally dependent self-regulatory growth system. The life cycle of the DCR is characterized by four phases: (a) preinductive, (b) period of induction, (c) postinductive, and (d) regression (Fig. 1). The preinductive period is analogous to preimplantation in normal gestation. During this time the condition of uterine response or sensitivity is developed by ovarian hormones. The period of induction follows the requisite preparation. In normal reproductive biology the stimulus to decidualization derives from the presence of the blastocyst. Again, for purposes of the assay, ovariectomy and/or immaturity deletes any embryonic contribution. The endometrium is stimulated by any one of a variety of physical (electrical stimulation, crushing, threading, air embolus, etc.) or chemical (histamine, Hanks salt solution, croton oil, etc.) means.[10] The postinductive period is marked by rapid growth. Uterine weight increases 75–100% per day for the first three postinductive days. There are concomitant increases in water, protein, glycogen, and DNA. There is a specific increase in RNA.[23] From the fourth through the seventh postinductive day the growth rate may increase at the rate of 150% per day, depending on the hormonal regimen, before attaining a plateau (postinductive days 8–9). The period of regression then ensues, in spite of continued hormonal support, so that base line values may be measured by days 16–18.

[20] L. E. Brumley and V. J. DeFeo, *Endocrinology* **75**, 883–892 (1964).

[21] M. B. Nikitovich-Winer, *Endocrinology* **77**, 658–666 (1965).

[22] J. A. Coppola, R. G. Leonard, W. Lippman, J. W. Perrine, and I. Ringler, *Endocrinology* **77**, 485–490 (1965).

[23] S. R. Glasser, *in* "Reproductive Biology" (H. A. Balin and S. R. Glasser, eds.), pp. 776–833. Excerpta Med. Found., Amsterdam, 1972.

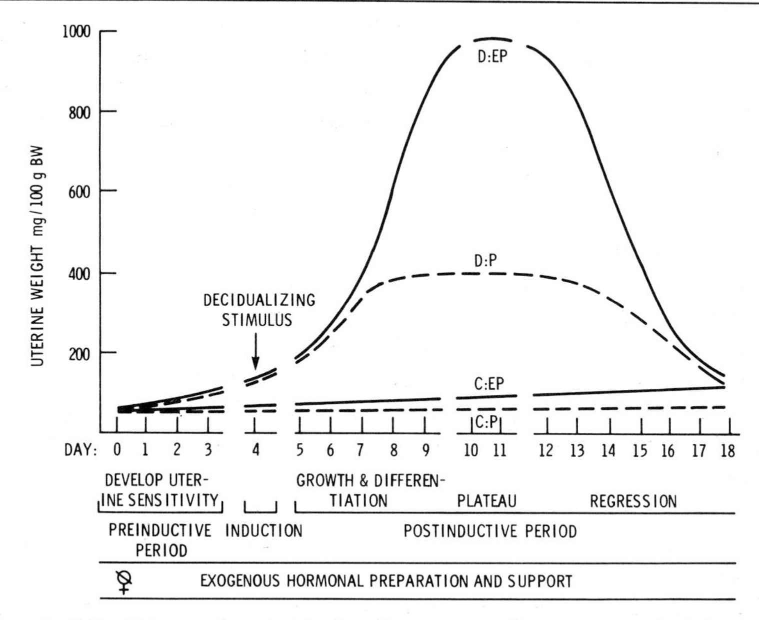

FIG. 1. Life history of a decidual cell response. C, nontraumatized horn; D, traumatized horn; EP, ovariectomized animal supported on 1 μg estrone and 2 mg progesterone per day; P, ovariectomized animal supported on 2 mg progesterone per day. Decidualizing stimulus (needle-scratch) given on day 4. Note lack of significant weight differences (D horn) between EP and P for the first 72 hours. After growth plateaus (D: EP, days 9–11) regression proceeds in spite of continued hormonal support.

The Gravimetric Assay

Animals. Either immature (21-day-old) female rats or regularly cycling females bilaterally ovariectomized the morning following ovulation (day 0) can be used. Validation of those physiological conditions incident to ovulation is provided by either artificial or natural cervical stimulation. Mice may also be used. There is increasing interest in the hamster because that animal does not appear to require estrogen to elicit implantation phenomena and has a very favorable endometrial myometrial ratio.

Hormones. Estrone, 1 μg in oil, and progesterone, 2 mg in oil, are most frequently employed in preparing the animal. Test compounds should be prepared in the same vehicle. Daily injections are given subcutaneously in a volume adjusted so that the daily dose is given in 0.1 ml vehicle.

Preinductive Period. Animals should be randomly assigned to appropriate control and experimental groups. We always include a vehicle (no hormone) control. For the standard assay there should be a group receiving the optimal estrone : progesterone (1:2000 μg) dose as well as appropriate test groups (E:X). The number of experimental groups depends on the availability of the test compounds and the number of dose levels to be tested.

An alternative, routinely used in this laboratory, is to inject progesterone alone during the preinductive period. It has been shown[23] that sensitivity to DCR can be developed with this hormone alone (2 mg/animal/day). The use of a test compound suspected of possessing progesterone-like activity, without the concomitant injection of estrogen, is a more rigorous test of these substances. The specific role of estrogen in the preinductive regimen is not clear. In most cases, estrogen will facilitate the action of a progestational compound, but there is the distinct possibility that the estrogen, if its relative and absolute concentrations are not proper, may inhibit and occlude the action of the unknown.

If the day of castration is day "0" the animals are injected on that day through day 3 of the preinductive period or a total of 4 injections.

Induction. The sensitivity of the uterus to DCR is maximal on day 4. Sensitivity is a transient phenomenon even on day 4, and for maximal consistency the inductive stimulus should be applied between 1000 and 1200 hours.

For purposes of the assay the surest method of eliciting DCR is to traumatize the endometrium of one horn of the bicornuate uterus with a specially designed knife or barbed needle.[10] The literature is replete with discussions of the capacity and conditions for other types of physical or chemical stimuli to elicit DCR. We have successfully used the intraluminal instillation of Hank's salt solution in a variety of situations including animals supported on progesterone alone. However, needle traumatization is a stimulus of such magnitude that if there is any response to be elicted from the uterus it will be provoked by scratching the uterus with a needle along the entire length of its antimesometrial wall.

Traumatization is accomplished by exteriorizing a uterine horn through either a dorsolumbar incision (ovarian end) or a ventral incision (cervical end). The needle is passed into and along the length of the uterine lumen with its tip against the antimesometrial wall. Tension is applied by raising the needle to stretch the uterus and then pulling the needle out, with continued tension, scratching the entire length of the wall. The puncture wound may be sealed by cautery. The contralateral horn is left untraumatized and serves as an internal control. All procedures are performed under light surgical anesthesia using clean technique. The animal is withdrawn from the ether and given the first postinductive hormone injection.

Postinductive Period. The length of time this period is continued is determined by the investigator. The most commonly used periods of assay are either 72 (day 7) or 120 hours (day 9) after eliciting the DCR. The animals are injected daily and are killed 24 hours after the last injection (day 6 or 8).

After the animal is killed the biocornuate uterus is removed, stripped of fat, the cervix is removed, and the experimental and control horns are separated and weighed. Weight comparisons may be made on an absolute or relative (mg uterus/100 gm body weight) basis. Prototype data are presented in Fig. 1.

It is recommended, for each new compound to be tested, that the investigator check the histology and perhaps the biochemical composition of the "decidualized" tissue. There have been, in our experience, a number of putative progestational compounds which did not elicit DCR but were uterotropic. The elevated weight of the control horn should alert the investigator but conclusive validation is provided by the failure to note decidual cells in histological sections of the traumatized horns.

The Biochemical Assay

The primary reason for introducing this modification was the specificity of its endpoint, i.e., the synthesis of RNA. It is also sparing of test compound. Minimally it offers, at least, the precision and accuracy of the gravimetric assay. The types of animals and the hormones are also the same.

Routinely we use 2 mg progesterone/animal/day (or test compound) through the preinductive period. On day 4, one horn of the uterus is traumatized with a needle to evoke DCR and the puncture wound is sealed by cautery. A fifth injection is given.

At either 4 or 24 hours after traumatization both the control and experimental uterine horns are exteriorized while the animal is under light ether anesthesia. Into each horn 25 μCi [^{3}H]cytidine is instilled in a volume of Hank's solution not greater than 0.1 ml or less than 0.05 ml. This fills the entire luminal volume and does not create pressure sufficient to open the cervix and spill the isotope. Each investigator should determine the amount of isotope to be used in his laboratory. The cytidine is injected through a 27- or 30-g needle. A 0.5-ml hypodermic syringe is particularly convenient to use. The shaft of the needle is held in position by a fine forceps lightly, but firmly, applied around the uterus below the puncture point. After injection the needle is withdrawn while continued pressure is applied by forceps. With the forceps still in place the puncture point is sealed by cautery, the muscle-skin incision closed and the animal withdrawn from the ether. Both horns of the uterus can be injected in a 60-second period.

Sixty minutes after instillation of the isotope the animal is killed. The uterus is removed, cleaned, and the experimental (traumatized) horn separated from the control. Each horn is weighed and placed in a tube of cold 50% saline. Reproducibility is enhanced if there is at least 50 mg wet tissue in each sample.

Samples are then homogenized in a Polytron Pt-10 (half-speed, 15–30 seconds). Prior to homogenization the saline is discarded and tissue disruption is done in 0.01 *M* Na acetate buffer, pH 5.2, with 0.01 *M* disodium EDTA and 0.05% Bentonite added (2.5 ml buffer/100 mg wet tissue; all operations are done in the cold). The entire fraction is saved. Remove duplicate aliquots of 0.5 ml. Precipitate with 3.0 ml cold 10% TCA. Spin at 1000 *g* for 1 minute. Wash successively with intermediate spins (1000 *g* for 1 minute) twice with cold 5% TCA, once with 95% ethanol (an ethanol–ether wash, 1:3, is optional) and once with ether. Centrifuge 2500 *g* for 5 minutes and air dry. Add 50 μl distilled water, 1.0 ml of NCS or similar reagent, and allow to stand overnight. Add 10 ml of an appropriate liquid scintillation cocktail, mix, and count. RNA is determined by analysis with the orcinol reaction.[23] Corrected counts and RNA concentrations are used to derive values for specific activity (cpm [^{3}H]cytidine/μg RNA). Comparison of specific activity data, control vs. experimental, is indicative of progestational activity and the data are expressed as the percentage increase in incorporation over the control (Fig. 2).

The specificity of this bioassay is seen in the comparisons offered in the table. The rankings provided by the biochemical parameter are more consistant with the ultimate bioassay, i.e., pregnancy maintenance tests, than are the rankings derived from the gravimetric DCR assay or the competition assay in which the test compound competes with [^{3}H]progesterone for a site on the specific cytosol receptor macromolecule.

Use of the biochemical endpoint for DCR offers additional advantages. As it is presently run, this modification of the assay decreases the amount of test compound required by 37.5%. With this saving it is possible to test increased numbers of compounds which can be synthesized only in limited amounts. Two further variants of the assay are being studied which should conserve even greater amounts of test steroid. The first is an augmentation type of assay in which all but the last and possibly the next to last injection are with progesterone. The test compound is used only for the final injections. The second modification is conserving not only of material but time. In this variation, castrate animals are

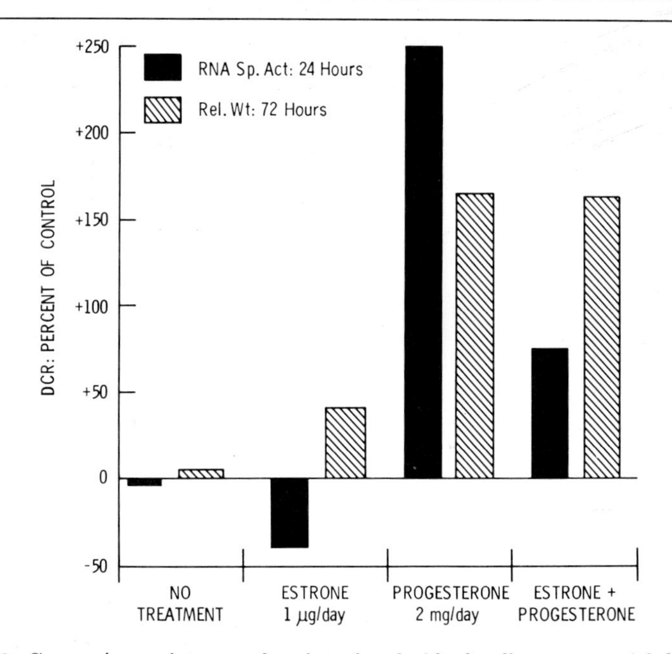

FIG. 2. Comparison of two end points for decidual cell response. Adult cycling females were castrated (day 0) and received either no treatment, 1 μg estrone, 2 mg progesterone, or 1 μg estrone + 2 mg progesterone subcutaneously on days 1, 2, and 3. One horn of the uterus was traumatized on day 4 and the hormonal regimen continued. Biochemical assay: The incorporation (60 minutes) of [^{3}H]cytidine into RNA was measured 24 hours after induction of DCR. Gravimetric assay: Cornua weights taken 72 hours (day 7) after DCR evoked. There are no differences between progesterone and estrone + progesterone groups for the first 24 hours. Thereafter the absolute weight of deciduoma of estrogen + progesterone animals is greater but relative weights actually decrease. Note that the 40% uterotropic response to estrone is not reflected biochemically and that there is significantly more [^{3}H]cytidine incorporation in the progesterone animals although weight increases in that group and in estrone + progesterone rats are comparable.

injected intraluminally and only once, with the steroid being studied. The specific incorporation of [^{3}H]cytidine into the test and control (vehicle only) horns is assayed at either 4 or 24 hours. The amounts used in this assay range from 100 to 200 μg and the animals can be used within 4 days of castration. Both modifications are yielding data which demonstrate their promise as methods to measure progestational activity.

At the present time the development of new chemical agents to be used in population control requires that a number of different species of assays be used in screening successful test compounds. This too is demonstrated in the table. A compound which vigorously and successfully com-

Comparison of Three Assay Modes Used to Screen for New Progestational Agents

	Specific activity [^{3}H]cytidine/ μg RNA	DCR weight increase	% Inhibition binding of [^{3}H]-progesterone to receptor
Progesterone	100	100	100
19-Norprogesterone	98	193	77
20 α-Hydroxyprogesterone	0	24	90
5α-Pregnone-3,20-dione	0	5	83
5β-Pregnone-3,20-dione	0	3	70
17α-Hydroxyprogesterone	20	6	88
Deoxycorticosterone	0	26	94
Testosterone	0	16	65
17-Ethynyl-19-nortestosterone	5	0	93

petes with progesterone for a site on the progesterone receptor yet has no biological activity (DCR or pregnancy maintenance) could serve to effectively block the biochemical action of progesterone and thus interfere with the establishment or maintenance of pregnancy. The identification of developmental compounds requires both the competition and the DCR assays. However, the definitive assay in the study of agents which regulate reproduction is the biochemical variant of the DCR for it is evident that the number of compounds that compete with progesterone for the receptor site are more legion than the ones with biological activity. The final exclusion can only be made with a sensitive, reproducible, quantitative bioassay.

[40] Reduced Nicotinamide Adenine Dinucleotide Phosphate: Δ^4-3-Ketosteroid 5α-Oxidoreductase (Rat Ventral Prostate)*

By RONALD J. MOORE and JEAN D. WILSON

Testosterone[1] + NADPH + H^+ → Dihydrotestosterone + $NADP^+$

Assay Method

Principle. The 5α-reduction of radioactive Δ^4-3-ketosteroids that do not contain substitutions on carbon-11 of the steroid molecule is followed in the presence of an excess of NADPH by measuring the appearance of the radioactive 5α-reduced metabolites utilizing a rapid thin-layer chromatographic separation. This assay has been developed because the low activity of the enzyme in most tissues precludes the measurement of NADPH disappearance spectrophotometrically.

Reagents and Materials.

Buffer, potassium phosphate, 0.1 *M*, pH 6.6

Stock substrate solution, [1,2-^{3}H]testosterone, available from several radiochemical sources at activities of approximately 50 Ci/mmole and about 95% purity, is diluted with benzene to a final activity of 0.1 mCi/ml and a concentration of approximately 2 μM. Tween 40 (1:50 w/v in ethanol) is added to provide 0.5 mg Tween/μg radiolabeled testosterone. The solution, stored at 4°, is stable for several months

* Supported by Grant AM 03892 from the National Institutes of Health.

[1] The trivial names used are testosterone, 17β-hydroxyandrost-4-en-3-one; dihydrotestosterone, 17β-hydroxy-5α-androstan-3-one; androstandiol, 5α-androstan-3α,17β-diol and 5α-androstan-3β,17β-diol; epitestosterone, 17α-hydroxyandrost-4-en-3-one; 17α-hydroxyprogesterone, 17α-hydroxypregn-4-ene-3,20-dione; progesterone, pregn-4-ene-3,20-dione; deoxycorticosterone, 21-hydroxypregn-4-ene-3,20-dione; cortexolone, 17α,21-dihydroxy-pregn-4-ene, 3,20-dione; androstenedione, androst-4-ene,3-17-dione; androstenediol, androst-4-ene,3β,17β-diol; corticosterone, 11β,21-dihydroxy-pregn-4-ene-3, 20-dione; cortisol, 11β,17α,21-trihydroxypregn-4-ene-3,20-dione; cortisone, 17α,21-dihydroxypregn-4-ene,3,11,20-trione.

Substrate–coenzyme solution, a volume of stock substrate solution in benzene is evaporated to dryness under a gentle stream of air. NADPH and buffer are added to provide a solution containing 1×10^{-7} *M* testosterone (approximately 5 μCi/ml) and 1×10^{-3} *M* NADPH

Enzyme, a suitable dilution of enzyme in buffer containing 1–40 units/ml

Chloroform:methanol (2:1 v/v)

Reference steroid solution, each of three reference steroids, dihydrotestosterone, testosterone, and androstandiol, is dissolved in chloroform at a concentration of 0.25 mg/ml and stored at room temperature

Silica gel thin-layer chromatography plate, precoated plastic sheets (20×20 cm) of silica gel without gypsum and with a layer thickness of 0.25 mm (Polygram Sil G-HY) from the Machery-Nagel Co., Düren, West Germany (distributed by Brinkmann Instruments, Inc.)

Chloroform, spectroquality grade

Methanol

Anisaldehyde reagent, 100 ml glacial acetic acid
2 ml concentrated H_2SO_4
1 ml *p*-anisaldehyde

Liquid scintillation fluor, 0.4% 2,5-diphenyloxazole in toluene:methanol (10:1 v/v)

Procedure.[2,3] The assay, described below for a final volume of 1.0 ml, can be scaled down to a semimicro assay of 0.1 ml final volume. To 0.5 ml enzyme suitably diluted in buffer is added 0.5 ml substrate–coenzyme solution to provide 5×10^{-8} *M* [1,2-^{3}H]testosterone and 5×10^{-4} *M* NADPH. The assay mixture is incubated 1 hour at 25° in 16×150-mm culture tubes. Five volumes (5.0 ml) of chloroform:methanol (2:1 v/v) are added, and the mixture is shaken or vortexed vigorously for 15 seconds. The emulsion is broken by low speed centrifugation (5 minutes at 500 *g*), and the upper phase and precipitated interphase protein are removed by aspiration and discarded. Approximately 0.3 ml of the lower phase (containing about 3×10^5 dpm) is transferred to a 6×50-mm disposable culture tube and evaporated to dryness. The residue is dissolved in 20 μl reference steroid solution and applied to a silica gel plate. Chromatographic separation of the steroids is accomplished by two ascents of the solvent front (origin to solvent front = 15 cm) in the system chloroform:methanol (98:2.5) at 25°. The developed plates are

[2] D. W. Frederiksen and J. D. Wilson, *J. Biol. Chem.* **246,** 2584 (1971).
[3] R. J. Moore and J. D. Wilson, *J. Biol. Chem.* **247,** 958 (1972).

sprayed with anisaldehyde reagent and heated at 100° for 15 minutes (or longer).

Within each lane of the chromatogram the zones corresponding to the solvent front, dihydrotestosterone, testosterone, androstandiol, and the origin are marked, and the entire lane is cut into five appropriate fragments with scissors. Each fragment is transferred to a liquid scintillation vial and assayed for ^{3}H following the addition of 10 ml fluor.

The total ^{3}H radioactivity recovered per lane is employed to calculate the percentages of ^{3}H radioactivity cochromatographic with each of the reference steroids. A substrate blank (substrate minus enzyme) correction for slight impurities in or degradation of the substrate is subtracted from the appropriate metabolite in each enzyme assay and is generally 1% or less for both androstandiol and dihydrotestosterone. Under these conditions, the rate of enzyme reaction is calculated as follows: picomoles testosterone reduced per hour equals the fraction of ^{3}H recovered in the dihydrotestosterone and androstandiol areas times the initial testosterone in the assay (50 pmoles). Androstandiol is measured in addition to dihydrotestosterone because the primary reaction product may serve as substrate for secondary reactions, principally 3-keto reduction to androstandiol by the 3-hydroxysteroid dehydrogenase system present in many tissues. Although these two reaction products (dihydrotestosterone and androstandiol) account for virtually all the 5α-reduced metabolites recovered from homogenates containing NADPH, the entire chromatographic lane is counted routinely to be certain that other metabolites might not form under some assay conditions.

A tissue slice procedure utilizing 10–50 mg of tissue slices incubated in Krebs-Ringer phosphate buffer containing [1,2-^{3}H]testosterone has also been utilized for the assay of the enzyme.[4-7] The slice procedure gives results that are on an average about 25% higher than in homogenates.[5]

Definition of Unit and Specific Activity. One unit is the amount of enzyme that catalyzes the formation of 1 picomole of 5α-reduced steroid per hour at 25°. Specific activity is defined as units of activity per milligram of protein.

Preparation of Enzyme (All Operations at 0–4°)

Homogenization. Freshly excised rat ventral prostates are cleaned from the adjacent connective tissues, blotted, weighed, and minced with

[4] R. E. Gloyna and J. D. Wilson, *J. Clin. Endocrinol. Metab.* **29**, 970 (1969).
[5] R. E. Gloyna, P. K. Siiteri, and J. D. Wilson, *J. Clin. Invest.* **49**, 1746 (1970).
[6] J. D. Wilson and I. Lasnitzki, *Endocrinology* **89**, 659 (1971).
[7] J. L. Goldstein and J. D. Wilson, *J. Clin. Invest.* **51**, 1647 (1972).

a razor blade. Twenty-five percent w/v homogenates are prepared (1 g wet weight minced prostate + 3 volumes 0.88 *M* sucrose–1.5 m*M* $CaCl_2$) in a Dounce homogenizer using 30 strokes of a loose pestle (clearance = 0.15 mm). The homogenate is filtered over 8 layers of surgical gauze and rehomogenized using 20 strokes of a tighter pestle (clearance = 0.09 mm). The filtered and rehomogenized preparation is used directly for subcellular fractionation and membrane isolation as described below.

Membrane Isolation[3]. Although the enzyme has not been solubilized, it has been possible to prepare nuclear membranes and endoplasmic reticulum with 90- and 4-fold enrichment respectively over the starting homogenate.[3]

Nuclear Membranes. STEP 1. PREPARATION OF PURIFIED NUCLEI IN BULK. The homogenate is centrifuged at 800 *g* for 10 minutes. The pellet is rehomogenized in the Dounce homogenizer in 2.0 *M* sucrose–0.5 m*M* $CaCl_2$ with a volume equal to 2.5 ml/g of starting prostate. Aliquots of this suspension (20 ml) are layered over 10 ml of 2.2 *M* sucrose–0.5 m*M* $CaCl_2$, and the tubes are centrifuged at 56,000 *g* in a Spinco SW 25.1 rotor for 60 minutes. The supernatant is decanted, and the pellets are resuspended in 1 ml of 0.88 *M* sucrose–1.5 m*M* $CaCl_2$/g of starting prostate. The yield averages 7×10^7 nuclei/g of prostate, and, on the basis of light microscopy and marker enzyme analyses, the nuclei appear to be about 96% pure. The DNA in this preparation averages about 0.6 mg/g of prostate, approximately 60% of the total prostate DNA (1.03 mg/g). The nuclei may be stored at −20°.

STEP 2. PREPARATION OF MEMBRANE EXTRACTS. For preparation of membrane extracts, frozen nuclei are thawed at 4° and centrifuged at 100,000 *g* for 30 minutes. The nuclear pellet is resuspended in 0.05 *M* potassium phosphate buffer, pH 6.6, containing 10^{-4} *M* EDTA and 10 m*M* dithiothreitol (KED buffer) at a protein concentration of approximately 2 mg/ml by triturating the suspension sequentially through an 18-gauge and then a 20-gauge needle. The suspension is again centrifuged at 100,000 *g* for 30 minutes, and trituration is repeated at a protein concentration of approximately 10 mg/ml. Aliquots (2 ml) in cellulose nitrate tubes are immersed in ice and subjected to sonic disruption with a Bronwill Biosonik 1 ultrasonic generator for four 5-second periods with 15-second intervals between sonic oscillations for temperature equilibration.

The density (*d*) of this disrupted suspension is then adjusted to 1.42 by adding an equal volume of d 1.84 cesium chloride in KED buffer (62.4%, w/w) in 10 equal increments with careful stirring between each addition. When the density is raised in this manner and the concentration of dithiothreitol is kept at 10 mM, gel formation is minimized. Alternatively, gel formation can be prevented by preincubation of the extract with DNase prior to the addition of the cesium chloride (500 μg DNase/ml) of sonicated extract incubated 30 minutes at 25°.

STEP 3. DENSITY GRADIENT CENTRIFUGATION IN CESIUM CHLORIDE. The cesium chloride extract is transferred to a cellulose nitrate tube (1 × 3 in.) that has previously been boiled in 1 mM EDTA. The sample is overlaid with a 28-ml continuous density gradient (d 1.05 to 1.40) of cesium chloride. This gradient is prepared in a two chamber mixing device filled with 15.5 ml of 5.9% cesium chloride in KED buffer (d 1.05) and 12.5 ml of 38.8% cesium chloride in KED buffer (d 1.40). The tubes are centrifuged in a Spinco SW 25.1 rotor for 1 hour at 56,000 g. At the end of the run, the tubes are punctured through the bottom with an 18-gauge needle, and 50-drop fractions are collected. Under these conditions enzyme activity from nuclei is recovered in a turbid band at approximately d 1.23 to 1.27.

Endoplasmic Reticulum. The supernatant remaining from the first 800 g centrifugation is centrifuged at 12,000 g for 10 minutes. The resulting supernatant is then sedimented at 105,000 g for 1 hour. The final supernatant is discarded, and the pellet is resuspended in a volume of 0.88 M sucrose–1.5 mM $CaCl_2$ equivalent to 1 ml/g of starting material. The suspended pellets are stored at −20° for subsequent study. The washing procedure, dilution, sonic disruption, addition of cesium chloride, and density gradient centrifugation are performed as described for nuclei. The enzyme activity from endoplasmic reticulum is found in a turbid band of d 1.18 to 1.22. Furthermore, when extracts of endoplasmic reticulum and nuclei are mixed, no hybridization of the membranes is seen. The purification and recovery of enzyme from membranes of nuclei and endoplasmic reticulum are illustrated in the table.

Properties

Distribution and Intracellular Localization. The enzyme exhibits distinctive tissue localization in that most activity is found in the organs of accessory reproduction[3-10] and in liver.[3,11-13] Wide variations exist in

[8] J. D. Wilson, N. Bruchovsky, and J. N. Chatfield, *Proc. Int. Congr. Endocrinol., 3rd, 1968* Int. Congr. Ser. No. 184, p. 17 (1969).
[9] J. D. Wilson and R. E. Gloyna, *Recent Progr. Horm. Res.* **26**, 309 (1970).
[10] M. D. Morgan and J. D. Wilson, *J. Biol. Chem.* **245**, 3781 (1970).
[11] J. S. McGuire, Jr. and G. M. Tomkins, *J. Biol. Chem.* **234**, 791 (1959).
[12] J. S. McGuire, Jr. and G. M. Tomkins, *J. Biol. Chem.* **235**, 1634 (1960).
[13] J. S. McGuire, Jr., V. W. Hollis, Jr., and G. M. Tomkins, *J. Biol. Chem.* **235**, 3112 (1960).

PURIFICATION AND RECOVERY OF 5α-REDUCTASE IN MEMBRANES OF NUCLEI AND ENDOPLASMIC RETICULUM[3]

Fraction	Protein (mg)	Enzyme: Recovery (E.U.[a])	Enzyme: Specific activity (E.U.[a] mg protein^{-1})	Enzyme: Enrichment	Enzyme: Yield (%)
Whole homogenate	1670	4100	2.5	1	100.0
Nuclei					
800 *g* pellet	328	3120	9.5	4	76.0
2.2 *M* sucrose pellet	43.0	1530	35.5	14	37.2
CsCl extract	29.0	908	31.3	13	22.1
Membranes (*d* 1.23–1.27)	1.7	386	224	90	9.4
Endoplasmic reticulum					
105,000 *g* pellet	39.8	146	3.7	2	3.6
CsCl extract	15.0	172	11.4	5	4.2
Membranes (*d* 1.18–1.22)	3.4	30	8.8	4	0.7

[a] E.U., enzyme units.

the relative activities between different tissues and among species for a given tissue. For the adult male rat, however, the following generalization concerning activities may be made for the accessory sex tissue: prostate > epididymis > seminal vesicle > external genitalia and preputial gland.[4] During embryonic development the enzyme can be demonstrated in urogenital sinus and urogenital tubercle prior to the differentiation of these tissues but appears in the Wolffian and Müllerian duct derivatives only after initiation of differentiation.[6]

Both in mammalian tissues[3-18] and in bacteria[19,20] the enzyme is membrane bound. In studies of rat ventral prostate, over half of the enzyme activity is located in nuclei,[2,3,9,18] even under circumstances in which enzyme markers are used to demonstrate that the nuclei contain only about 4% contamination with mitochondria and endoplasmic reticulum,[3] and this nuclear activity is associated exclusively with the nuclear envelope.[3] The remainder of the enzyme in rat prostate is located in the cytoplasmic membranes, principally endoplasmic reticulum. A major por-

[14] J. Chamberlain, N. Jagarinec, and P. Ofner, *Biochem. J.* **99,** 610 (1966).
[15] I. Björkhem, *Eur. J. Biochem.* **8,** 345 (1969).
[16] A. B. Roy, *Biochimie* **53,** 1031 (1971).
[17] W. Voigt, E. P. Fernandez, and S. L. Hsia, *J. Biol. Chem.* **245,** 5594 (1970).
[18] J. Shimazaki, T. Horaguchi, Y. Ohki, and K. Shida, *Endocrinol. Jap.* **18,** 179 (1971).
[19] H. R. Levy and P. Talalay, *J. Biol. Chem.* **234,** 2014 (1959).
[20] S. J. Davidson and P. Talalay, *J. Biol. Chem.* **241,** 906 (1966).

tion of the enzyme is also found in the nucleus in dog prostate;[5] in other organs of accessory reproduction in the male rat including epididymis, seminal vesicle, and preputial gland;[9] and in hen oviduct.[10] In rat liver, which contains about 500 times as much total 5α-reductase activity as the prostate, only about 1% is found in nuclei, the vast bulk of activity being recovered in the endoplasmic reticulum.[3]

pH Optimum. The 5α-reductase activity of nuclei and of cytoplasmic membranes from rat ventral prostate has an apparent pH optimum of 6.6 ± 0.3.[2]

Stability. The enzyme activity in all preparations deteriorates rapidly at 37° but appears to be stable at 25° during short-term incubations (1 hour). No activity is lost when stored at —20° for as long as 6 months, and about half the activity is lost when the homogenates are stored in ice at 4° for 2 days.

Inhibitors. The enzyme is strikingly inhibited by certain divalent cations. Cu^{2+} and Zn^{2+} at concentrations of 1 mM cause almost complete inhibition. Ca^{2+}, Mn^{2+}, and Cd^{2+} cause less than 20% inhibition at this concentration, whereas Mg^{2+} and Ba^{2+} have no effect.[2] The activity does not appear to be affected by chelating agents or by sulfhydryl agents, and no evidence of product inhibition has been obtained.[2]

Kinetics. Utilizing a variety of preparations of whole homogenates and of subcellular fractions, the apparent K_m values for testosterone when measured at saturating NADPH levels (5×10^{-4} M) in several laboratories are about 10^{-6} M.[2,16,17] With different assay conditions and homogenates from a different strain of rats, Shimazaki *et al.* have reported a K_m of 3.2×10^{-5} M for testosterone.[18] At an NADPH concentration of 5×10^{-4} M or greater it was not possible to saturate the enzyme at concentrations of testosterone between 3.5×10^{-8} and 7.4×10^{-7} M, the latter concentration being at least 20 times the upper limit of the physiological concentration of testosterone in the circulation and approaching the upper limits of solubility of the substrate in aqueous–protein mixtures.[2] Although it is possible to solubilize larger amounts of testosterone with the aid of additional organic solvents or detergents, such solvents almost invariably inhibit enzyme activity.

Therefore, the concentration of testosterone that is utilized for routine assays (5×10^{-8} M) has been chosen arbitrarily because it approximates the upper limits of hormone available in the circulation of adult male animals of most species. Since this concentration provides first-order kinetics with respect to testosterone, valid assays are performed only under conditions in which rates are proportional to enzyme concentration and to time. For most enzyme measurements the linearity of the assay declines if the substrate concentration falls below 60% of the starting level. On an average, linear conditions are obtained with 0.1–5 mg of protein in endoplasmic reticulum and whole homogenate preparations and 0.01–0.5 mg of protein in nuclei.

Substrate Specificity. Of the substrates studied, four—progesterone, 20α-hydroxypregn-4-ene-3-one, epitestosterone, and 17α-hydroxyprogesterone—are all better substrates than was testosterone (apparent K_m values that varied from $0.2–0.4 \times 10^{-6}$ M). Three substrates—deoxycorticosterone, cortexolone, and androstenedione—are all approximately as active as testosterone itself (apparent K_m values of $0.9–1.3 \times 10^{-6}$ M). Androstenediol is a weak substrate, and three steroids—corticosterone, cortisol, and cortisone—are inactive over the concentration range studied.[1] Furthermore, when the ability of nonradioactive steroids to inhibit competitively the 5α-reduction of testosterone is studied, the K_i values in most instances are similar to the K_m.[2] It is concluded that the rat ventral prostate probably contains a single 5α-reductase enzyme and that this enzyme has broad specificity for Δ^4-3-ketosteroids that do not contain substitution on carbon-11 of the steroid molecule.

Mechanism of the 5α-Reduction. On the basis of isotopic-labeling studies performed in preparations of rat liver, it is clear that the enzyme exhibits an absolute requirement for NADPH, which provides the hydride ion for carbon-5, whereas a proton from the aqueous medium is attached to carbon-4.[21] Furthermore, the rate of incorporation of the 4β hydrogen of the nicotinamide ring of NADPH into the 5 position of Δ^4-3-ketosteroids is 5 times as great as that of the 4α-hydrogen.[15] All attempts to reverse the reaction utilizing mammalian enzymes have been unsuccessful, because although the first step in reversal (enolization) occurs, the energy barrier for hydride abstraction is extremely high.[21,22]

Regulation by Hormones. The enzyme activity in rat ventral prostate is under potent regulation by androgenic hormones.[23–25] Testosterone propionate treatment of the castrated rat results in a sevenfold increase in the total activity of the prostate enzyme above the control value and causes a similar increase in the specific activity of the enzyme both in whole homogenate and in the nucleus. On the basis of time sequence studies following testosterone propionate administration and of studies

[21] J. S. McGuire, Jr. and G. M. Tomkins, *Fed. Proc., Fed. Amer. Soc. Exp. Biol.* **19,** 29 (1960).

[22] D. C. Wilton and H. J. Ringold, *Proc. Int. Congr. Endocrinol., 3rd.* Int. Congr. Ser. No. 157, p. 105 (1968).

[23] J. Shimazaki, I. Matsushita, N. Furuya, H. Yamanaka, and K. Shida, *Endocrinol. Jap.* **16,** 453 (1969).

[24] J. Shimazaki, Y. Ohki, M. Matsuoka, M. Tanaka, and K. Shida, *Endocrinol. Jap.* **19,** 69 (1972).

[25] R. J. Moore and J. D. Wilson, *Endocrinology* **93,** 581, 1973.

comparing the effects of 11 different androgens on prostate growth and enzyme content, the enzyme activity correlates with the growth response of the prostate to androgen treatment. Thyroid, adrenal, pituitary, and estrogenic hormones do not appear to play a major role in the mediation of this androgenic control.[25]

[41] Biological Rhythmicity Influencing Hormonally Inducible Events

By STANLEY R. GLASSER

Introduction

The necessity for controlling the time factor in studies of enzyme regulation derives from an increasing number of investigations reporting significant circadian changes in the activity of several hepatic enzymes that metabolize amino acids. If the product of a specifically directed protein synthesis demonstrates cyclic behavior, it follows that the biochemical steps that precede that enzyme activity (enzyme synthesis and synthesis of specific RNA species) could also be expected to exhibit endogenous rhythmicity. Circadian rhythms have in fact been noted for DNA synthesis,[1] RNA synthesis,[2] Mitotic index,[3] and alterations in the nature of endoplasmic reticulum[4] as well as for a number of liver enzymes[5-9] and secretory products of endocrine glands.[6,7,10] In seeking to investigate cause and effect relationships it would seem necessary to delineate if reported changes were vicarious or programmatically directed. In the latter case a particular study could then be designed to maximize results and avoid confusion between coincidence and correlation.

The internal variations in cellular activity that have been cited are biological processes prone to synchronization and direction by external correlates. The integration of these signals is manifest in the operation

[1] B. Barbiroli and V. R. Potter, *Science* **172,** 738 (1971).

[2] C. P. Barnum, C. D. Jardetzky, and F. Halberg, *Amer. J. Physiol.* **195,** 301 (1958).

[3] J. J. Jaffee, *Anat. Rec.* **120,** 935 (1954).

[4] A. Chedid and V. Nair, *Science* **175,** 176 (1972).

[5] I. B. Black and J. Axelrod, *in* "Biochemical Actions of Hormones" (G. Litwack, ed.), vol. 1, p. 135. Academic Press, New York, 1970.

[6] M. I. Rapoport, R. D. Feigen, J. Burton, and W. R. Beisel, *Science* **153,** 1642 (1966).

[7] M. Civen, R. Ulrich, B. M. Trimmer, and C. B. Brown, *Science* **157,** 1563 (1967).

[8] F. M. Radzealowski and J. Bosquet, *J. Pharmacol. Exp. Ther.* **163,** 229 (1968).

[9] A. Colas, D. Gregonis, and N. Moir, *Endocrinology* **84,** 165 (1969).

[10] G. E. Shambaugh, D. A. Warner, and W. R. Beisel, *Endocrinology* **81,** 881 (1967).

of biological clocks which set the phasing and rhythm of cellular and holobiological events.

Kinds of Clocks

Homeostatic. The earliest work with biorhythm dealt with genesis of cycles. Because interference with endocrine function was an effective tool the idea developed that all clocks functioned to enable an organism to achieve homeostosis. Components of a feed-back system were found to be operational in endocrinology (hypothalamus, pituitary, and target). They sustained rhythmicity (estrus cycle) but accounted for the limited accuracy of homeostatic clocks (menses, 28 days; range 17–37).

Central Clocks. Further research, particularly with 24–hour periods, disclosed a second clock which kept almost perfect time independent of almost every internal or external change. Thus this clock which influences many functions cannot be understood in terms of homeostasis or feed-back mechanisms.

Compared with the homeostatic clock the central clock is extremely consistent and very accurate. It is apparently situated in the hypothalamus, which gives it the advantage of modulating many autonomic functions and holobiological responses without servomechanisms. The central clock is probably the more primitive of the two since 24-hour periodicity has been established in unicellular organisms.

Peripheral Clocks. These operate outside the CNS and are independent of nervous and homeostatic controls. They are more difficult to identify (apparent in intermittent hydrarthroses and peripheral agranulocytosis) but are also accurate and constant.

Clock Function (Evolution)

Biological clocks are not chronometers but serve as signal markers. They mark the beginning of each phase or its maxima or minima.

At the present time there is only one clock of utility in biomedical investigation, the 24-hour clock. These clocks had evolutionary value for primatively they signal optimum conditions for mating, hunting, and avoidance. The coordination of activities, throughout the 24-hour period, with the autonomic nervous system and hypothalamic centers is probably the matrix underlying all measureable periodic responses. However, in spite of the universality of clocks, synchronization of rhythms is absent or erratic in early developmental stages.

It is suggested that the *in utero* experience of viviparity imposes a prenatal synchronizing effect which is absent in species such as birds. However, hatching as well as parturition initiates synchronization so that certain endogenous rhythms become more evident and consistent with advancing maturity.

A primary proof of central (vs. cellular) control of diurnal phenomena is the consistent synchronization of the response pattern under all conditions. However, certain diurnal rhythms are more directly modulated by exogenous factors than by endogenous diurnal programs and are more prone to rapid desynchronization or phase shifts. Certain responses in man (blood pressure) are strictly under exogenous control. There is in fact much less central control in higher organisms than was anticipated. Synchronization, with or without a central modulating step, is achieved most frequently by a *Zeitgeber* operating through the eyes and/or the nervous system.

Zeitgebers or synchronizers are external agents whose rhythmicity may determine the duration, shape, and level of the exogenous response. Their primary characteristic is timing. For the reason that they give time cues or clues, entrain or phase responses, they have been given the name Zeitgeber.

Physiologically dominant synchronizers are light, temperature, and regimens, i.e., feeding patterns. Weak synchronizers, discernible when dominant ones are withdrawn or disappear, are more diffuse and subtle and include climate and weather. The shape of the response curve is less instructive than the frequency and timing (entrainment) of the modulator. In analyzing the potential influence of biorhythmicity on a specific response, particular attention should be paid to (1) the strength of the synchronizer, (2) the strength of the endogenous rhythm, and (3) the type and degree of coupling between (1) and (2).

Synchronizer information may be in the nature of an impulse, duration at any one level of synchronizer intensity (proportional *Zeitgeber*), or the rate of change of signal strength (differential *Zeitgeber*). The purpose of the synchronizer is to inform the organism about the period and phase of universal time. Caution must be exercised in interpreting the response to the synchronizing signal because the response or spectrum of responses of the organism may be diffuse and increasingly nonspecific.

Evidence of Clocks

Investigations particularly germane to the issue of endorhythmic control of cell function include the studies of Potter *et al.*[11-13] who reported

[11] V. R. Potter, R. A. Gebert, H. C. Pitot, C. Peirano, C. Lamar, Jr., S. Lesher, and H. P. Morris, *Cancer Res.* **36**, 1547 (1966).

[12] V. R. Potter, R. A. Gebert, and H. C. Pitot, *Advan. Enzyme Regul.* **4**, 247 (1966).

[13] V. R. Potter, M. Watanabe, J. E. Becker, and H. C. Pitot, *Advan. Enzyme Regul.* **5**, 303 (1967).

the detection of circadian changes in the activity levels of several liver enzymes involved in the metabolism of amino acids. Rapoport *et al.*[6] reported the inducibility of mouse liver tryptophan pyrrolase by corticoid hormones. In establishing the relationship between the circadian rhythms of plasma corticoids and liver tryptophan pyrrolase activity they showed that adrenalectomy abolished the rhythmicity of enzyme activity. This proved that adrenal cortical hormones were inducers and endogenous control, via those hormones, was central.

The basic problems confronting the investigator with an interest in the phenomenon of enzyme induction were outlined by Civen *et al.*[7] In posing the question as to the reality of inductive changes, e.g., were they merely reflections of the rhythm of a central clock or were there truly inducible activities altered by a daily rhythmic pattern, certain guidelines were provided for arriving at differential proofs. They postulated that enzymes exhibiting marked diurnalism should be inducible in a short time period, whereas an enzyme with no rhythmic patterns would not be inducible in the same time span.

Potter[14] measured the rate of DNA synthesis in normal rats and in animals during liver regeneration. All rats had been adapted to a controlled feeding schedule. Two different phenomena proved operable in the regulation of DNA synthesis. The first, related to hepatectomy, was independent of the time of day. The second was the presence of a constant diurnal variation following the stimulus of the controlled feeding schedule. Wurtman *et al.*[15] confirmed and extended these principles in studies of the daily rhythm of hepatic tyrosine transaminase activity. Deprivation of dietary protein extinguished daily rhythm which proved to be modulated by the presence of dietary amino acids, specifically tryptophan.[16] The rat normally consumes significant quantities of food, i.e., dietary protein, during the hours which precede the daily rise in tyrosine transaminase activity.

Feeding schedules are often synchronized by light vs. dark phasing of the day. Dark-dependent natural feeding proved operable in the diurnal variations in the activities of nucleolar[1] and nucleoplasmic[11] DNA-dependent RNA polymerases of rat liver noted by Glasser and Spelsberg.[17] Both polymerases displayed biphasic shifts in activity (Fig. 1A, B). The patterns of the individual shifts were not synchronous. Polymerase I peaked between 0100–1100 hours with its lowest activity be-

[14] V. R. Potter, *Science* **172,** 738 (1971).

[15] R. J. Wurtman, W. J. Shoemaker, and F. Larin, *Proc. Nat. Acad. Sci. U.S.* **59,** 800 (1968).

[16] R. J. Wurtman, *Advan. Enzyme Regul.* **7,** 57 (1969).

[17] S. R. Glasser and T. C. Spelsberg, *Biochem. Biophys. Res. Commun.* **47,** 951 (1972).

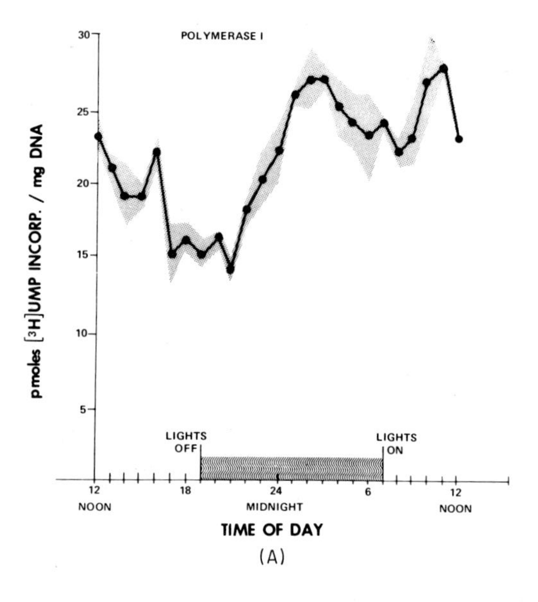

(A)

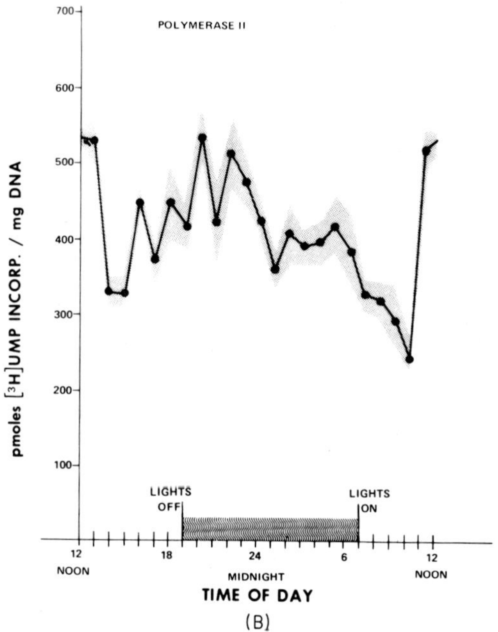

(B)

FIG. 1. (A) Fluctuations in the activity of the nucleolar DNA-dependent RNA polymerase (I) of rat liver during a 24-hour period. (B) Fluctuation in the activity of the nucleoplasmic DNA-dependent RNA polymerase (II) of rat liver during a 24-hour period. SEM is represented by the shaded area around each point. These rats consumed 75% of their daily food intake during the period of 2000 to 0100 hours. They were adapted to the light schedule over a 3-week period. From 1900 through 0600 hours rats were killed in total darkness.

tween 1700–2000 hours, whereas polymerase II peaked between 2400–0500 hours with a low point at 1000 hours. These authors were able to correlate the diurnal variation in a number of biochemical processes, occurring in the rat liver, which are related or dependent on protein synthesis, to light-dark and feeding cycles cited by Potter *et al.*[11] and Wurtman *et al.*[15] The parameters cited include DNA-dependent RNA polymerase I and II, DNA synthesis, mitoses, appearance of smooth endoplasmic reticulum, plasma corticosterone titers, and the activity of tyrosine transaminases. The table summarizes pertinent hormonally inducible biorhythmic phenomena.[1–4,6,7,10–13,17–32]

The correlated rhythmicity of these events would seem to require reevaluation and redesign of studies dealing with enzyme activity. Too often we can not conclude with certainty if the rise in enzyme activity reported as an inducible action is a fortuitous observation actually due to the daily rhythmicity in enzyme activity and synthesis. Even when proper controls are included it is possible that animals assayed during periods of peak enzyme activity may not be as "responsive" as animals used during the trough periods of enzyme activity.

Standardization

How can the careful investigator cope with the factors introduced by either circadian or even free-running rhythms? Rarely, if ever, is it possible to delete the various external factors or their special sense organ receptors. An operational alternative, that is economically feasible, is to

[18] J. Le Magnen and S. Tallon, *J. Phys. (Paris)* **58**, 323 (1966).

[19] J. J. Fabrikant, *Johns Hopkins Med. J.* **120**, 137 (1967).

[20] A. B. Novikoff and V. R. Potter, *J. Biol. Chem.* **173**, 223 (1948).

[21] E. D. Whittle and V. R. Potter, *J. Nutr.* **95**, 238 (1968).

[22] A. Fleck, J. Shepard, and H. M. Munro, *Science* **150**, 628 (1965).

[23] A. W. Pronczuk, Q. R. Rogers, and H. M. Munro, *J. Nutr.* **100**, 1249 (1970).

[24] A. J. Clifford, J. A. Ruimallo, B. S. Baliga, H. N. Munro, and P. R. Brown, *Biochim. Biophys. Acta* **227**, 443 (1972).

[25] A. V. Le Bouton and S. D. Handler, *FEBS Lett.* **10**, 78 (1970).

[26] C. N. Clark, D. J. Naismith, and H. N. Munro, *Biochim. Biophys. Acta* **23**, 587 (1957).

[27] W. L. Steinhart, *Biochim. Biophys. Acta* **228**, 301 (1971).

[28] R. J. Wurtman and J. Axelrod, *Proc. Nat. Acad. Sci. U.S.* **57**, 1594 (1967).

[29] E. C. C. Lin and W. Knox, *Biochim. Biophys. Acta* **26**, 85 (1957).

[30] C. A. Nichol and F. Rosen, *in* "Actions of Hormones in Molecular Processes" (G. Litwack and D. Kritschevsky, eds.), Wiley, New York, pp. 234–256.

[31] B. Barbiroli, M. S. Moruzzi, M. G. Monti, and B. Tadolini, *Biochem. Biophys. Res. Commun.* **54**, 62 (1973).

[32] S. Hayashi, Y. Aramaki, and T. Noguchi, *Biochem. Biophys. Res. Commun.* **46**, 795 (1972).

Biorhythmic Phenomena That Are Hormonally Inducible

Biological event	Reference
Stimulation of incorporation of DNA precursors and DNA synthesis	1–3, 18–20
Stimulation of incorporation of RNA precursors and RNA synthesis	2, 19, 21–24
Nucleotide metabolism	22–24
Polysome abundance	22–24
Alterations in endoplasmic reticulum	4
Protein synthesis	25, 26
Phospholipid synthesis	2
Increase in mitotic activity	2, 3, 19, 20
Increase in template activity of chromatin	27
Alteration in secretion of adrenal cortical hormones	6, 7, 10, 28–30
Stimulation of enzyme activation	
Tyrosine transaminase	7, 10–13, 28
Tryptophan pyrrolase	6, 11–13
RNA polymerase I	17–31
RNA polymerase II	17–31
Ornithine decarboxylase	32
Glutamine phosphoribosyl pyrophosphate amidotransferase	24

guarantee standardized animal or tissue sources. Standardization of laboratory conditions can be provided if the dominant synchronizers (temperature, humidity, food, and particularly light) are kept rigorously constant.

Animal sources. Many investigators, particularly biochemists, are studiously neglectful of the original reservoir of their material. Animals of any species should be purchased only from reputable dealers who adhere to National Institutes of Health regulations and will freely share any type of husbandry information with you. The animals should be disease-free and weight and/or age data should be sent with the order. It is equally important that the on-site facility for the care of animals be as vigorously standardized as is possible. Anything less than that leads to spurious data and is false economy. Be particularly wary of local suppliers who cannot provide these guarantees. Cage size, the number of animals per cage, temperature, humidity, and light must be maintained at recommended standards. Any deviation not only introduces oscillating input to biorhythmic control but introduces stress factors.

Investigators who do not deal with their own animals but obtain tissues from various suppliers are not immune from this type of problem.

Uniformity as to sex, size, time, and method of killing are minimum requirements. There is an advantage in dealing with a large supplier who collects large tissue pools which would minimize variability, but in cases of special orders the supplier should be provided with as extensive a list of specifications as the investigator can suggest.

On-site animal facilities should be automatically light and temperature regulated. Animal rooms should be free of windows and out of the traffic pattern. Food supplies should be generous and the food free of contaminating additives (diethylstilbestrol, antibiotics, etc.). Procedures should be routine and traffic restricted to minimize the "serial effect." This is a variable produced by repeated entries into the animal colony. This perturbation steadily alters the measurement of the desired parameter. Finally, for those studies obviously modulated by *Zeitgebers* and endogenous rhythmicity it would seem essential that pilot experiments be performed in order to determine the optimal set of external correlates necessary to produce the most unequivocal data.

In spite of a long history the principles of biological rhythm have only recently been integrated into experimental design and not yet with great regularity. We have learned to dampen the oscillations of the biological clock by strengthening the input of dominant synchronizers. This is a measure of our illiteracy for we do not know the nature of the clock, where it is located, or even if the very things we study, i.e., nucleic acids, proteins, the nucleus, etc., are the clock itself.

General References

E. Bünning, "The Physiological Clock: Endogeneous Diurnal Rhythms and Biological Chronometry." Academic Press, New York, 1964.

L. Ewing, ed., "Biology of Reproduction," Vol. 4, No. 4, pp. 239–357, 1971.

A. Sollberger, "Biological Rhythm Research." Elsevier, Amsterdam, 1965.

H. von Mayrsbach, ed., "The Cellular Aspects of Biorhythms." Springer-Verlag, Berlin and New York, 1967.

Section VII

Isolation of Biologically Active Metabolites of Steroid and Thyroid Hormones

[42] A Universal Chromatographic System for the Separation of Steroid Hormones and Their Metabolites

By Pentti K. Siiteri

Introduction

No other class of naturally occurring substances exhibits such complexity during their biosynthesis and metabolism as do the steroid hormones. Although only five or six kinds of hormonal activity are found among these, a variety of steroids which possess these activities are found in most biological systems. While a great body of knowledge has accumulated concerning this subject, it is still of considerable interest as exemplified by the recent observations which indicate that metabolites of testosterone appear to be the active hormonal agents in certain target tissues.[1] The success of such studies lies in the adequacy of methods used for separation and identification of closely related steroidal compounds. Chromatographic techniques utilizing paper and adsorbent materials found some of their earliest and most elegant applications in the steroid field.[2] While newer methods such as thin-layer[3] and gas-liquid chromatography[4] are currently in vogue because of their rapidity of operation, other methods which possess distinct advantages are available. A gradient elution liquid–liquid partition chromatographic system which has been applied to a wide variety of steroid separations is described herein. The advantages of this method lie in its simplicity, large capacity, excellent resolution, and universal applicability. For example, the basic method or slight modifications thereof have been used for such diverse problems as the isolation of estrogen metabolites from crude urinary extracts,[5] separation of radioactive intermediates obtained during tissue incubations,[6] and purification of plasma steroids prior to radioimmunoassay.[7]

Materials and Methods

Diatomaceous earth obtained from the Johns-Manville Company (Celite 545) is purified in large batches by allowing to stand in 6 *N* HCl

[1] J. D. Wilson, *N. Engl. J. Med.* **287**, 1284 (1972).
[2] I. E. Bush, "The Chromatography of Steroids." Macmillan, New York, 1961.
[3] B. P. Lisboa, Vol. 15, p. 3 (1969).
[4] H. H. Wotiz and S. J. Clark, Vol. 15, p. 158 (1969).
[5] P. K. Siiteri, *Steroids* **2**, 687 (1963).
[6] J. D. Wilson and P. K. Siiteri, *Endocrinology* **92**, 1182 (1973).
[7] G. E. Abraham, D. Tulchinsky, and S. G. Korenman, *Biochem. Med.* **3**, 365 (1970).

for 24 hours followed by exhaustive washing with water, distilled water, methyl alcohol, and ether and finally drying in air. For use in separation of steroids in plasma extracts prior to assay, the Celite is further purified by heating in a muffle furnace at 400°. The composite chromatogram illustrating the separation of C_{18}, C_{19}, and C_{21} steroids shown in Fig. 1 was obtained by chromatography using a 1.5×43-cm column. Celite (20 g) is thoroughly mixed with 15 ml of ethylene glycol in a polyethylene bag. The mixture is then tightly packed into the column in 12 equal portions with the aid of a tamper constructed from a wooden dowel and a tight fitting cork. Samples are dissolved in 1 ml of ethylene glycol and 0.5 ml of isooctane and mixed with 2 g of Celite and then packed on top of the column in the same way. Elution is carried out in four steps as shown in Fig. 1. The first consists of three 10-ml washes followed by 100 ml of isooctane (Eluant I). A linear gradient consisting of 200 ml of isooctane and 200 ml of isooctane–ethyl acetate (7:3) is then used

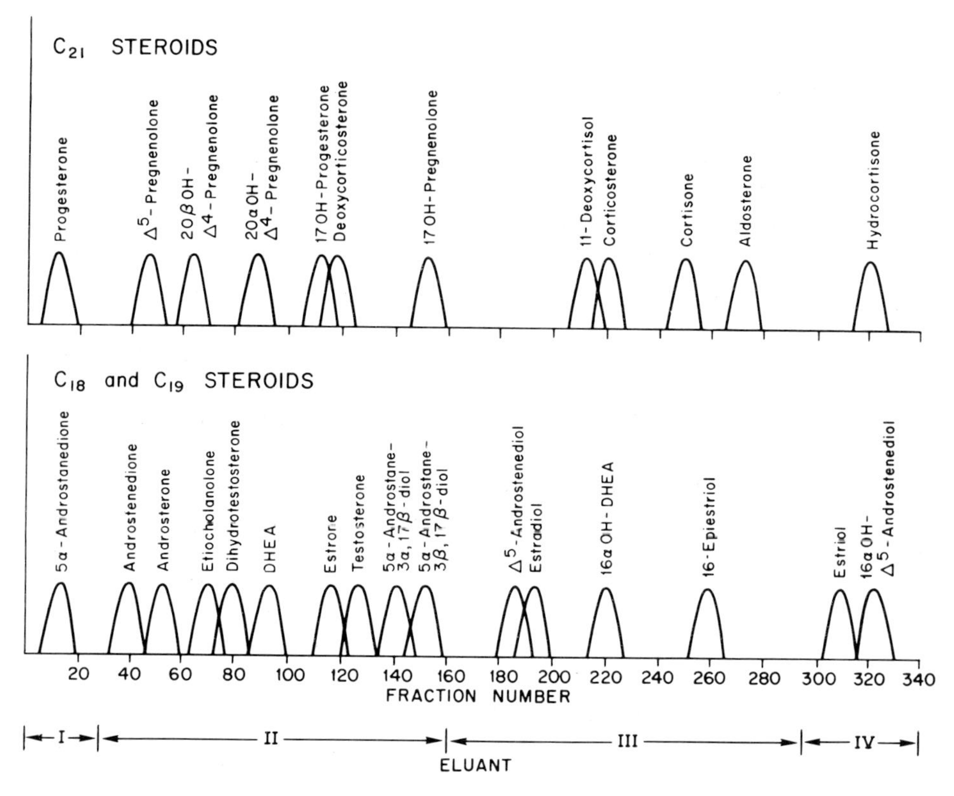

FIG. 1. Elution pattern of representative steroids from a Celite chromatographic column. See text for details.

(Eluant II). More polar metabolites are eluted using a linear gradient consisting of 200 ml of isooctane–ethyl acetate (7:3) and 200 ml of ethyl acetate (Eluant III). Finally, in order to elute very polar steroids such as estriol or hydrocortisone, pure ethyl acetate (100 ml) is passed through the column. Adequate flow rates are achieved by the application of air pressure to the gradient vessels. Experience has shown that resolution is not adversely affected at flow rates up to 300 ml/hour; however, extreme care is required when establishing the gradients in order to obtain reproducible chromatograms. Three-milliliter fractions are collected and aliquots of alternate fractions are assayed for radioactivity or by chemical methods. A double gradient is used in order to provide optimal resolution in the region of polarity in which the androgens and their metabolites appear, i.e., fractions 40 to 160. The gradient device consists of two identical cylinders previously described by Siiteri[5] and is commercially available from the Buchler Instrument Division of the Nuclear of Chicago Company, Fort Lee, New Jersey, and also from the Glenco Glass Company, Houston, Texas.

Discussion

As can be seen in Fig. 1, adequate resolution of many steroids of biological interest is obtained in this system. However, certain separations such as the pregnenolone and androstenedione pair are not achieved, but these are readily separated by subsequent thin-layer chromatography. On the other hand, this is one of the few chromatographic systems which effectively separates 17α-hydroxyprogesterone from testosterone or the 20 reduced isomers of progesterone, 20α- and 20β-hydroxy-4-pregnen-3-one. Of course in extremely complex mixtures of steroids such as extracts of pregnancy urine many other estrogen and progesterone metabolites will appear in the chromatogram. Nevertheless, once the behavior of the compounds of interest have been determined the system serves as a powerful tool in the identification of a particular metabolite. Since the peak tube for each steroid does not vary by more than ± 2 fractions when carried out by an experienced individual, the position of emergence and the shape of the peak are useful criteria for tentative identifications. Confirmation of the identity of the metabolite is then readily achieved by other methods such as thin-layer chromatography or by crystallization to constant specific activity with carrier steroid.

A disadvantage of this system is the stripping of ethylene glycol from the column as the polarity of the solvent increases. This becomes serious after approximately 200 fractions and requires separation of the ethylene glycol by extraction with water or the addition of methanol. Thus a homogenous solution can be obtained prior to removing aliquots for assay. Ethylene glycol can be removed simply by passing the pooled fractions through a bed of Celite. The ethylene glycol absorbs to the Celite, and the steroids are eluted with the original solvent. However, this is a minor problem in comparison with the system's advantages.

The first application of this basic system described the separation of 12 urinary estrogens in phenolic extracts of β-glucuronidase-treated 5-day collections of late pregnancy urine.[5] Such extracts are extremely crude, weighing 3–4 g and are not amenable to other chromatographic treatments without prior purification or partial fractionation by large scale columns or by counter-current distribution. Indeed, the original method has been modified by omitting the phenolic separation step in order to avoid losses of alkali sensitive compounds such as the 2-hydroxy- and 16,17α-ketol estrogens. No loss of resolution of estrogens has been noted despite the presence of large quantities of the whole spectrum of progesterone metabolites which are also well resolved. The original column procedure[5] differed from the one described here in that a 30-g column and a single gradient (900 ml isooctane/700 ml ethyl acetate) were used. Further improvements in the system include a reduction in the volume of solvents used (450 and 350 ml, respectively) and the use of higher flow rates. Accordingly, the time required for the chromatogram has been greatly reduced from 18 to 3 hours without significant loss in resolution.

The system as described herein recently has been used to investigate the biosynthesis of androgens by fetal rabbit testes.[6] Following incubation with either tritium-labeled progesterone or pregnenolone, tissue slices were extracted with methanol to which was added appropriate amounts of ^{14}C-labeled intermediates and products expected. These not only allowed positive identification of products formed but also served as internal recovery standards. Accurate yield of products could be calculated from the $^{3}H/^{14}C$ ratios observed following further purification by thin-layer chromatography and/or cocrystallization with carrier steroids. However, with the exception of androstenedione formation from pregnenolone, it was found that equally satisfactory results could be obtained simply by expressing the amount of tritium radioactivity associated with any peak on the chromatogram as a fraction of the total radioactivity and multiplying that value times the amount of substrate used. Thus, the analysis of such experiments was greatly simplified. These findings emphasize another advantage of this form of chromatography, i.e., essentially complete recovery of steroids is obtained.

A most useful modification of this chromatographic method has been the development of microcolumns for purification of plasma steroids prior to radioimmunoassay as described by Abraham *et al.*[7] These columns consist of disposable pipettes containing 0.5 g Celite and ethylene glycol, and elution is carried out in a stepwise manner. While considerable resolution is lost, separations are obtained extremely rapidly and are sufficient to eliminate potentially interfering steroids. The combination of this simple chromatographic method with assay methods having great inherent specificity (antibodies, enzymes, proteins, etc.) appears to be adaptable to the assay of any steroid. Most importantly the universally troublesome problem of nonspecific interference is virtually eliminated.

[43] Isolation of Progesterone Metabolites

By CHARLES A. STROTT

I. Introduction

Without the introduction of additional oxygen functions, there are essentially 20 possible metabolites of progesterone. (Reduction of the 3-ketone group without prior reduction of the Δ^4 double bond, though it has been reported, is generally at such a slow rate that, for all practical purposes, it can be disregarded.) A general scheme illustrating the *reductive* pathways in progesterone metabolism is shown in Fig. 1. While metabolism of progesterone has been studied in a variety of species utilizing a variety of tissues, most of these investigations were concerned primarily with total metabolism: that is, obtaining a general catalytic picture. In addition to such an approach, this section will contain a description of methods for looking at progesterone metabolism in relation to *target* tissue progesterone receptor activity.[1]

II. Principle

In part, the prevailing hypothesis on the mechanism of action of progesterone involves: (1) uptake of the progesterone molecule by the target cell; (2) binding to a cytoplasmic receptor molecule; (3) transport of the steroid–receptor complex into the nucleus; (4) binding of the steroid–receptor complex to chromatin acceptor sites; (5) alteration of the transcriptional machinery and, ultimately, stimulation of protein synthesis. The exact role of metabolic transformation of progesterone in target

[1] B. W. O'Malley, M. R. Sherman, and D. O. Toft, *Proc. Nat. Acad. Sci. U.S.* **67**, 501 (1970).

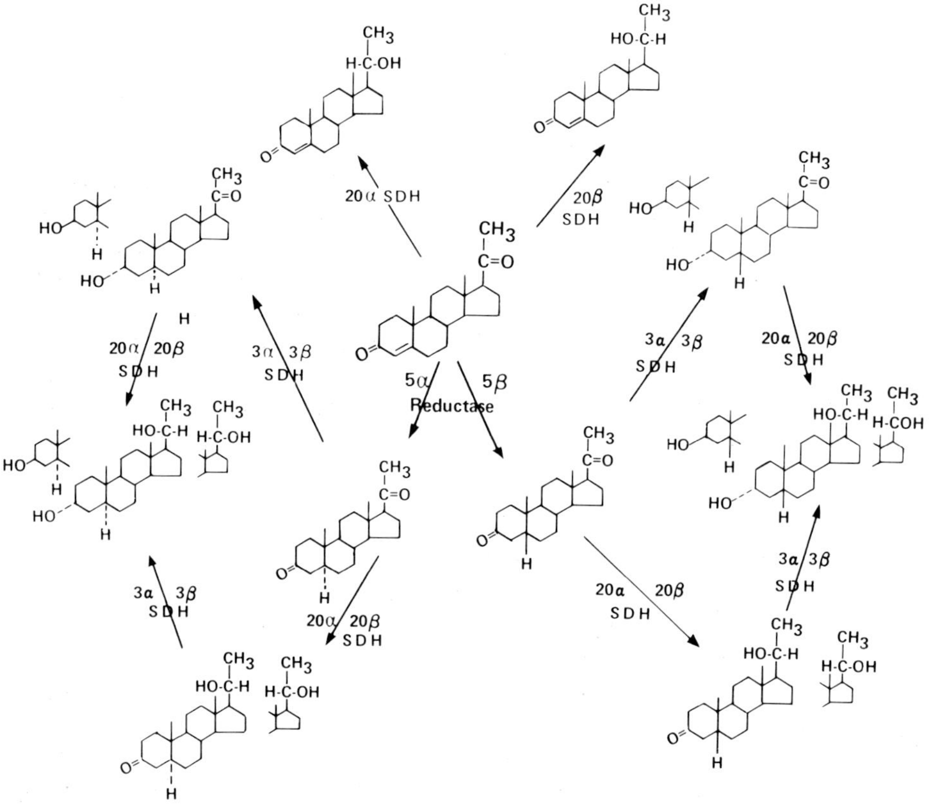

FIG. 1. General scheme illustrating reductive pathways in progesterone metabolism. Reductase, Δ^4-3-ketosteroid 5α- or 5β-oxidoreductase; SDH, steroid dehydrogenase.

tissue remains largely speculative at present. Thus, in an attempt to gain some insight into this problem, methods will be described for examining the metabolism of progesterone in relation to cytoplasmic and nuclear receptor activity. It is not unreasonable to suppose that the progesterone molecule which attaches to the receptor will be metabolized differently than the molecule which does not become receptor-bound. Therefore, a method will be described for examining this difference. For illustrative purposes the tissue employed is the oviduct magnum and shell gland obtained from estrogen-primed immature chicks.[2,3]

[2] B. W. O'Malley, W. L. McGuire, P. O. Kohler, and S. G. Korenman, *Recent Progr. Horm. Res.* **25,** 105 (1969).

[3] D. O. Toft and B. W. O'Malley, *Endocrinology* **90,** 1041 (1972).

III. Methods

A. Homogenization of Tissue

This procedure is performed in a 0–4° cold room. The oviduct tissue is finely chopped with a razor blade and placed in a Corex tube containing a medium of 0.25 *M* sucrose, 3 m*M* $MgCl_2$, and 50 m*M* phosphate (pH 7.4); approximately 1 g tissue/4 ml of medium. The tissue is homogenized using a polytron (PT-10, Brinkmann Instruments) with a rheostat setting of 5. The homogenizing tube is maintained in a salt ice bath and the tissue homogenized using 5 bursts (5 seconds each) while the tube is constantly rotated in the ice bath. The crude homogenate is filtered through cheese cloth and centrifuged at 700 *g* for 10 minutes in a refrigerated centrifuge.

B. Preparation of Subcellular Fractions

Nuclei are isolated by resuspending the 700 *g* pellet in 0.25 *M* sucrose, 0.4 m*M* $MgCl_2$, layering over 0.34 *M* sucrose, 0.4 m*M* $MgCl_2$, and centrifuging at 700 *g* for 10 minutes. The resultant pellet is resuspended in 2.2 *M* sucrose and centrifuged at 70,000 *g* for 1 hour at a temperature setting of 2°.

Mitochondria are isolated by centrifuging the 700 *g* supernatant fraction at 10,000 *g* for 30 minutes in a refrigerated centrifuge. The supernatant is decanted and the pellet resuspended in 0.25 *M* sucrose, 1 m*M* EGTA [ethyleneglycol-bis(aminoethyl) tetraacetate], 50 m*M* phosphate (pH 7.4) and centrifuged at 30,000 *g* for 10 minutes. The supernatant fraction is discarded and the light, buffy layer on top of the pellet is removed using a disposable pipet. The pellet is resuspended and the procedure repeated two more times.

Microsomes and *cytosol* are prepared by centrifuging the 10,000 *g* supernatant at 30,000 *g* to remove mitochondrial fragments; the 30,000 *g* supernatant is centrifuged at 200,000 *g* for 1 hour, yielding the microsomal pellet and the soluble cell fraction.

C. Isolation of Receptor Proteins

1. Sucrose Density Gradient Analysis

Cytosol which is prepared in 0.25 *M* sucrose (8%) is diluted 1:1 with 50 m*M* phosphate buffer to reduce the sucrose concentration to less than 5%. Nuclear protein is extracted by resuspending the purified nuclear pellet in 0.4 *M* KCl, 50 m*M* phosphate (pH 7.4), allowing the mixture to stand for 15 minutes and centrifuging at 12,000 *g* for 10 minutes.

Two- or three-tenths milliliter aliquots of either cytosol or nuclear protein preparations are applied to 4.8 ml sucrose gradients (5–20% sucrose in Tris-EDTA buffer) in the presence or absence of 0.3 *M* KCl and centrifuged at 200,000 *g* for 16 hours. Three-tenths molar KCl in sucrose gradients causes the proteins which bind progesterone (or a metabolite) to migrate in a single 4 S peak; whereas in the absence of KCl, an 8 S peak is found. In a *low-salt* gradient the 4 S peak (which is usually small to absent) is considered to contain some specific but largely nonspecific binding, while the 8 S peak contains only specific binding.[4] Following centrifugation the gradients are fractionated and eluates collected on an ISCO gradient analyzer (model D). Approximately 1/10 volume (25 μl) of each gradient fraction is removed for measurement of radioactivity to determine the gradient profile. Fractions (generally 4) from the top of the gradient (free steroids) and from the radioactive peaks (bound steroids) are pooled and the steroids extracted.

2. Ammonium Sulfate Precipitation

This procedure is a modification of a method used for purifying the progesterone receptor from chick oviduct cytosol.[5] A sufficient volume of a 100% $(NH_4)_2SO_4$ solution is added dropwise with constant stirring to the cytosol and nuclear protein preparations to achieve a 35% final concentration. The precipitate is centrifuged at 10,000 *g* for 20 minutes and the supernatant fraction decanted for extraction of steroids. The pellet is washed with a 50% solution of $(NH_4)_2SO_4$ to remove trapped nonprecipitable material. The final pellet is redissolved in water and the steroids extracted. This procedure precipitates approximately 1–2% of the total protein and 75–80% of the total receptor activity. The supernatant fraction which contains 98–99% of the total protein contains very little bound steroid when examined by density gradient analysis. Thus, this is a convenient method for separating receptor-bound steroids from the free, nonbound fraction. In addition, any loosely bound steroids (so-called nonspecific binding) will remain in the supernatant.

D. Extraction of Steroids

All pellets are resuspended in 1 ml water; nuclear pellets are allowed to stand at room temperature for 30 minutes to permit swelling and rup-

[4] M. R. Sherman, P. L. Corvol, and B. W. O'Malley, *J. Biol. Chem.* **245,** 6085 (1970).

[5] W. T. Schrader and B. W. O'Malley, *J. Biol. Chem.* **247,** 51 (1972).

ture. Samples are extracted 3 times with 3 volumes of ethyl acetate, the solvents pooled and evaporated in a 40° water bath under dry nitrogen gas. The lipid residues are dissolved in 70% methanol; the tubes are kept at −16° for 16–18 hours after which they are centrifuged and the supernatant fractions decanted. The pellets are rinsed with cold 70% methanol, and the rinsings added to the supernatants. The solutions are reduced in volume and the steroids extracted with ethyl acetate. This procedure results in 90–95% of the total radioactivity being extracted from each cell fraction. In the receptor studies, the defatting procedure can be eliminated.

E. Isolation of Steroids

The steroid extracts are transferred to precoated thin-layer plates (silica gel, F254, E. Merck A. G., Darmstadt, Germany) along with authentic steroid standards in parallel lanes as markers. The plates are developed in the solvent system benzene:ethyl acetate (8:2) unsaturated. The particular brand of thin-layer plate used with this solvent system effects an adequate separation of 5α- and 5β-pregnane-3,20-dione from each other (1% cross-contamination when checked with [^{3}H]5α-pregnane-3,20-dione and [^{14}C]5β-pregnane-3,20-dione) as well as from progesterone and further reduced monols and diols. This system also gives a fairly good separation of the various monols from the various diols. An illustration of this system is depicted in Fig. 2.

Following chromatographic development, the plates are fractionated according to the markers used; the silica gel is loosened with a razor blade and sucked into a disposable pipet plugged with glass wool. The pipets are placed in specially constructed racks and the pipet tips inserted into conical tubes. The pipets are filled to the top with acetone (approx. 2 ml) and the steroids eluted. Following elution, the acetone is evaporated and the residues redissolved in 1 ml of methanol.

The methanolic solutions are utilized as follows: (1) 200 μl are transferred to counting vials, the methanol evaporated, 5 ml of scintillation fluid added, and the radioactivity measured. The radioactivity in this aliquot is used to determine the total radioactivity present in each sample. (2) 400 μl of the methanolic solution are transferred to glass-stoppered tubes and the solvent evaporated. The residues are acetylated by adding 50 μl pyridine and 50 μl of acetic anhydride; the tubes are stoppered tightly, vortexed briefly, and placed in the dark for 16 hours at room temperature. Following the acetylation reaction, 1 ml of methanol is added to the tubes and the solution evaporated. The residues are transferred to thin-layer plates along with the appropriate steroid acetate

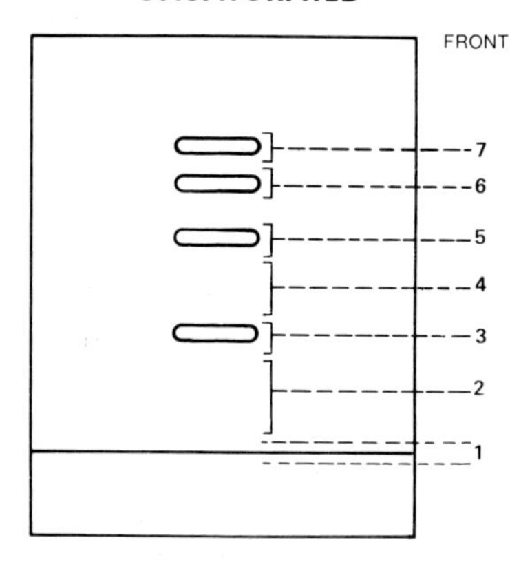

FIG. 2. An illustration of the thin-layer chromatography (TLC) system used in isolating progesterone metabolites. Seven fractions eluted according to the scheme illustrated contained essentially the following metabolites:

Fractions

(1) polar metabolites remaining at the origin

(2) 6β-, 14α-, 15α-, 16α-, and 19-hydroxyprogesterone
5β-pregnane-3α,20β-diol
5β-pregnane-3α,20α-diol

(3) 20α- and 20β-hydroxy-4-pregnen-3-one (*)
5β-pregnane-3β,20β-diol
5β-pregnane-3β,20α-diol
5α-pregnane-3β,20β-diol
5α-pregnane-3α,20α-diol
5α-pregnane-3β,20α-diol
5α-pregnane-3α,20β-diol

(4) 20α-hydroxy-5α-pregnan-3-one
20β-hydroxy-5α-pregnan-3-one
20α-hydroxy-5β-pregnan-3-one
20β-hydroxy-5β-pregnan-3-one
3α-hydroxy-5β-pregnan-20-one
3β-hydroxy-5α-pregnan-20-one

(5) progesterone (*)
3β-hydroxy-5β-pregnan-20-one
3α-hydroxy-5α-pregnan-20-one

(6) 5β-pregnane-3,20-dione (*)

(7) 5α-pregnane-3,20-dione (*)

* indicates marker steroids.

markers and the chromatograms developed in the benzene:ethyl acetate (8:2), unsaturated system. This system effects an adequate separation of the ring A-reduced 3,20-diacetates, 20-monoacetates, and 3-monoacetates from each other, as well as from the acetates of 20-dihydroprogesterone and various hydroxylated progesterone derivatives. An illustration of this system is shown in Fig. 3. (3) The remaining 400 μl of the methanolic solution are transferred to glass-stoppered tubes and the solvent evaporated. The residues are oxidized by adding 3 drops of a freshly prepared solution containing 0.4% chromium trioxide in 90% glacial acetic acid. The tubes are capped and allowed to stand for 30 minutes. Following this, 2 ml of water are added and the steroids extracted 3 times with 3 ml of ethyl acetate. The steroid extracts are transferred to thin-layer plates and developed in the benzene:ethyl acetate (8:2) system using progesterone and 5α- and 5β-pregnane-3,20-dione as markers. The chro-

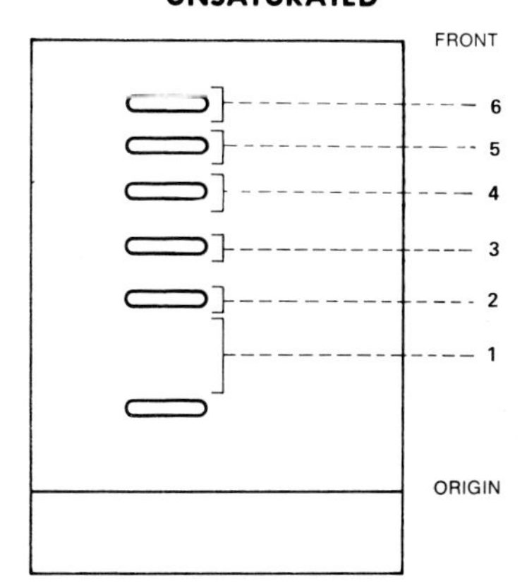

FIG. 3. An illustration of the thin-layer chromatography system (TLC) used in isolating acetates of progesterone metabolites. Six fractions eluted according to the scheme depicted contained essentially the following metabolite-acetates. The first spot indicated above the origin is 20β-hydroxy-4-pregnen-3-one which is used as a marker.

Fractions

(1) 6β-, 14α-, 15α-, 16α-, and 19-hydroxyprogesterone acetates
(2) progesterone
(3) 20α/20β-hydroxy-4-pregnen-3-one acetates
(4) 20α/20β-hydroxy-5α/5β-pregnan-3-one acetates
(5) 3α/3β-hydroxy-5α/5β-pregnan-20-one acetates
(6) 5α/5β-pregnane-3α/3β, 20α/20β-diol diacetates

matograms of the acetylated and oxidized samples are fractionated and eluted with acetone into counting vials as described above. The solvent is evaporated, 5 ml scintillation fluid added, and the radioactivity measured. The scintillation fluid is 4% (v/v) Spectrofluor or Liquifluor in toluene.

F. Digitonin Precipitation

To evaluate relative amounts of 3α and 3β reduction of the 3-ketonic group of progesterone, digitonin precipitation of 3β-hydroxy metabolites is performed.[6] After the initial TLC (*vide supra*), an aliquot of the 3-mono and diol fractions is dissolved in 1 ml ethanol containing 0.5 mg carrier steroid. Four milliliters of a solution of 1% digitonin in 50% ethanol are added; the mixture is incubated at room temperature for 16–24 hours. The resultant precipitate is centrifuged; an aliquot of the supernatant is transferred to a counting vial, the solvent evaporated, scintillation fluid added, and the radioactivity measured. The precipitate is washed with cold ethanol and redissolved in pyridine; water is added and the steroids extracted with ethyl acetate. The ethyl acetate is transferred to counting vials, the solvent evaporated, and the radioactivity measured. Knowing the total amount of radioactivity in the monol and diol fractions, a percent of the total material containing a 3β-hydroxyl (pellet) and 3α-hydroxyl (supernatant) group can be determined.

G. Identification of Metabolites by Isotope Dilution

Following elution of the various fractions from either the first or the second TLC after derivative formation, carrier steroids can be added to the appropriate samples and the steroids recrystallized in various solvent systems.

Positive identity is indicated when the specific activity (dpm/mg) becomes constant. An estimate of the percentage of the total radioactivity in the sample can be determined from the theoretical specific activity or the specific activity of the original mother liquor.

H. Data Analysis

Perhaps the most convenient way to analyze the data is to do so in a relative fashion. Thus, the total radioactivity is determined after the first TLC (*vide supra*); then, each fraction of progesterone metabolites,

[6] R. M. Haslam and W. Klyne, *Biochem. J.* **55**, 340 (1953).

as well as unreacted progesterone, is reported as a percent of the total recoverable radioactivity. Examination of the data can be broken down into eight categories:

1. Unreacted progesterone is determined following acetylation and a second TLC to remove contaminating 3-monols (Fig. 2).
2. 5α- and 5β-pregnane-3,20-dione can be determined after the first TLC (Fig. 2).
3. Total 3α/3β-hydroxy-5α/5β-pregnan-20-one is determined following acetylation and a second TLC (Fig. 3).
4. Relative amounts of 3α- and 3β-hydroxy-5α- and 5β-pregnan-20-one can be determined after the first TLC and subsequent digitonin precipitation.
5. Total 20α/20β-hydroxy-5α/5β-pregnan-3-one is determined following acetylation and a second TLC (Fig. 3).
6. Total 5α/5β-pregnane-3α/3β,20α/20β-diol is determined following acetylation and a second TLC (Fig. 3).
7. Total 5α and 5β reduction is determined following oxidation and a second TLC. Ring A-reduced monols and diols are converted to either 5α- or 5β-pregnane-3,20-dione (Fig. 2). This determination is exclusive of any material remaining at the origin of the chromatogram.
8. Total unknown polar material which remains at the origin of the chromatogram is determined after the first TLC. This material will contain compounds more polar than the ring A-reduced diols (Fig. 2). Thus, this material is composed of metabolites containing more than two oxygen functions. It should be noted that in some systems (e.g., the chick oviduct) this material can reach major proportions.

IV. *In Vitro* Experimental Procedures

A. Isolation of Steroids from Subcellular Compartments

Intact tissue (500–750 mg) is placed in a flask containing 2 ml 0.25 *M* sucrose, 50 m*M* phosphate (pH 7.4), and 2 μCi [^{3}H]progesterone (4×10^{-11} *M*). Incubations are carried out in air at 37° using a Dubnoff incubator with the shaker speed set at 100 strokes/minute. After incubating for the desired period of time, the tissue is removed from the flask, rinsed twice in ice-cold saline, and homogenized in fresh buffer. Subcellular fractions are prepared from the homogenates and the steroids extracted. If tissue from more than one animal is to be used, it may be necessary to subdivide and pool all the tissue fragments. The desired amount of tissue is then randomly removed from the pool for incubation. This technique tends to minimize any substantial difference between animals.

B. Progesterone Metabolites Specifically Bound, Nonspecifically Bound and Unbound, and Excreted from Target Tissue

1. Metabolism of Progesterone That Is Receptor-Bound

In this study, tissue is placed in medium containing 2 μCi [^{3}H]progesterone (4×10^{-11} *M*) at 0° for 10 minutes; after this time the tissue is removed from the medium, washed twice with cold saline, transferred to a flask containing fresh medium without [^{3}H]progesterone, and incubated at 37°. Following incubation, the tissue is removed and the medium saved for steroid analysis. The tissue is homogenized and nuclear protein and cytosol preparations subjected to either sucrose-density-gradient analysis or $(NH_4)_2SO_4$ precipitation. Studies have indicated that initially essentially all the radioactivity taken up by the tissue is bound to the receptor. The data obtained from this kind of study can be analyzed in terms of the pattern of progesterone metabolites: (1) bound to the specific progesterone receptor, (2) associated with nonspecific macromolecules, (3) unbound or free, and (4) released into the incubation medium.

2. Metabolism of Progesterone That Does Not Associate with the Receptor

This study is done exactly as described in the preceding section except that a 1000-fold excess of nonradioactive progesterone is added to the incubation medium with labeled progesterone giving a total progesterone concentration ("hot" plus "cold") of 4×10^{-8} *M*. An alternative is to add the nonradioactive progesterone first (preincubate) and then add the labeled steroid. In this situation essentially the same amount of radioactivity will enter the tissue as when there is no nonradioactive progesterone present. There will, however, be little uptake of radioactivity by the receptor since its capacity to bind progesterone will be exceeded by the excess "cold" steroid. Thus, in the experiment in Section IV,B,1, the metabolism of progesterone that *is* initially bound to the receptor can be examined, while the experiment in this section will indicate the metabolism of progesterone that *does not* bind to the receptor.

[44] Isolation of Cortisol Metabolites

By H. Leon Bradlow and David K. Fukushima

I. Isolation from Aqueous Media

A. From Urine (Endogenous or Labeled Compounds)

Cortisol[1] metabolites are present in the urine primarily in the conjugated form with β-glucuronic acid. In the guinea pig[1a] and some other species the metabolites are excreted unconjugated. Ethyl acetate is the solvent of choice to extract free compounds. Methylene chloride, chloroform, or ethyl ether are less suitable because they do not extract the more polar compounds such as 6-hydroxycortisol.

The cleavage of the glycosidic linkage of the conjugated metabolites is best accomplished with β-glucuronidase. Bacterial, molluscan, and mammalian preparations have been employed for this purpose. Although all of these will give good hydrolysis, the latter is to be preferred, because the extracts are less bulky and less liable to artifact formation.[2] The urine should be adjusted to pH 5, buffered with 10% v/v of 2 *M*, pH 5, acetate buffer and 300 units/ml of β-glucuronidase added. Following incubation for 3–5 days at 37°, the urine is extracted continuously with ethyl ether for 48 hours. The longer incubation period is required for complete hydrolysis of the glycosidic bond. However, over 90% hydrolysis is achieved in the shorter time. The incubation time can also be shortened by carrying out the incubation at 48°, especially with the snail and bacterial enzyme preparations. The extract is washed with 2 *M* NaCl to remove acidic material and then washed with saturated NaCl solution until neutral. Two additional separatory funnels containing ethyl

[1] Cortisol = 11β,17,21-trihydroxypregn-4-ene-3,20 dione; cortisone = 17,21-dihydroxypregn-4-ene-3,11,20 trione; 6-hydroxycortisol = 6β,11β,17,21-tetrahydroxypregn-4-ene-3,20 dione; THF = 3α,11β,17,21-tetrahydroxypregnane-20-one; ATHF = 3α,11β,17,21-tetrahydroxy-5α-pregnan-20-one; THE = 3α,17,21-trihydroxypregnane-11,20-dione; ATHE = 3α,17,21-trihydroxy-5α-pregnane-11,20-dione; cortols = 3α, 11β,17,20ζ,21-pentahydroxypentane; cortolones = 3α,17,20ζ,21-tetrahydroxypregnan-11-one; Reichstein's U = 17,20ζ,21-trihydroxypregn-4-ene-3,11-diones; Reichstein's E = 11β,17,20ζ,21-tetrahydroxypregn-4-en-3-ones, 11-oxy-C-19 compounds = 3α-hydroxy-5β-androstane-11,17-dione,3α,11β-dihydroxy-5α-androstan-17-one and 3α,11β-dihydroxy-5β-androstan-11-one.

[1a] S. Burstein and R. I. Dorfman, *J. Biol. Chem.* **206,** 607 (1954).

[2] H. L. Bradlow, *in* "Chemical and Biological Aspects of Steroid Conjugates" (S. Bernstein and S. Solomon, eds.). Springer-Verlag, Berlin and New York, 1971.

acetate are used as backwashes during the washing procedure in order to minimize loss of the more water-soluble metabolites. The extracts are combined and dried over anhydrous sodium sulfate. The solution is concentrated *in vacuo* and the residue weighed. Aliquots are taken for measurement of 17 ketosteroids (17 KS), 17-hydroxycorticosteroids (17-OH CS), and other assays as desired. Tracer cortisol studies showed that this extract, called the "glucosiduronate fraction," from most patients contained 50–60% of the cortisol metabolites initially present in the urine. In some severely ill patients the recovery in this fraction may be as low as 30%. The hydrolyzed metabolites could also be hand-extracted from the urine by ethyl acetate but the extracts contain much more nonsteroidal and pigmented substances. Solvents, such as chloroform, methylene chloride, and ethyl ether, are not satisfactory since polar compounds, such as THF, cortols, and cortolones, are not readily extracted.

If the metabolites in their conjugate forms are desired, they can be readily obtained from the urine by adsorption onto Amberlite XAD-2 and elution with methanol.[3] The extract is concentrated *in vacuo*. Approximately 90% of the original weight can be removed in this way. Tracer studies indicate that 90–96% of the metabolites, both free and conjugated, are recovered in this manner (a 40-fold reduction in volume can be achieved).

B. Isolation from Incubation Media

Cortisol metabolites can be isolated from the aqueous incubation media by adsorption onto Amberlite XAD-2, as described above. This serves to separate the steroid metabolites from small molecules (glucose, salts, urea, buffers, etc.) in the media. Recovery is usually greater than 95%. Alternatively the media could be extracted with ethyl acetate or chloroform to obtain unconjugated compounds.

C. Isolation from Tissues

Compounds present in tissues or cell preparations can be extracted with acetone or ethyl acetate. If necessary, hydrolysis can be carried out as described for urine.

II. Separation of Individual Metabolites

Although separations can be achieved by thin-layer chromatography, the separation of individual metabolites is best carried out by chroma-

[3] H. L. Bradlow, *Steroids* **11**, 265 (1968).

tography[4] on 18 × 118-cm sheets of Whatman No. 1 paper slit to give two 5-cm-wide side strips for reference standards and dye marker (isatin) and a 12-cm center strip for the extracts to be analyzed. After application of the samples the paper chromatogram is developed in the system benzene–methanol–ethyl acetate–water (1:1:0.1:1) without preequilibration until the dye marker reaches near the bottom of the paper or about 16 hours at 27°. The papers are then removed from the tanks, dried, and the side strips sprayed with appropriate reagents (blue tetrazolium for α-ketols, Zimmermann reagent for 17 ketosteroids, 1 M periodic acid followed by Zimmermann reagent for glycerol side chains, or 5% phosphomolybdic acid in ethanol as a universal reagent).[5] If the samples to be analyzed are radioactive, 5-cm-wide center strips are used and after development scanned for radioactivity. The radioactive peaks are identified by comparison with the standards on the side strips. The radioactive peak areas are cut up and eluted with ethanol and then quantitated. For nonradioactive samples a narrow strip (10% of the applied extract) is cut from the center and sprayed with the appropriate reagent to detect the various metabolites. The areas are cut, eluted, and quantitated by appropriate assays.

One finds the following on the 118-cm paper chromatogram in descending order from the origin: 6β-hydroxycortisol, the cortols, the cortolones plus the epimers of Reichstein's substance E, THF plus the epimers of Reichstein's substance U, ATHF, THE, cortisol plus ATHE, and, at the bottom of the paper, cortisone. The 11-oxygenated C-19 compounds are present in the effluent.

In many laboratories the 4-ft cylindrical glass tanks necessary for the 118-cm paper chromatogram are not available. Then, one must use the usual 2-ft glass tanks and 55-cm long paper. The chromatogram is developed however in a slower system, benzene–methanol–water (1:1:1) for a shorter time. The separation is not as good between the compounds by this procedure but adequate separations are achieved.

Since the extraction procedure described above gives a poor recovery of 6β-hydroxycortisol, a separate procedure (*vide infra*) should be employed for this polar metabolite. The 20-hydroxy epimers of cortols and cortolones can be separated by paper chromatography in a borate buffer system [isooctane–methanol–toluene–$M/10$ borate buffer, pH 9.2, (7:6:35:20)].[6] Using this system the epimers can be separated with the 20β-epimer running ahead of the 20α-epimer. Complete separation re-

[4] D. K. Fukushima, H. L. Bradlow, L. Hellman, and T. F. Gallagher, *J. Clin. Endocrinol. Metab.* **28**, 1618 (1968).

[5] I. E. Bush, *Chromatogr. Steroids, Int. Ser. Monogr. Pure Appl. Biol.* **2**, 373 (1961).

[6] J. J. Schneider and M. L. Lewbart, *Tetrahedron* **20**, 943 (1964).

quires about 40 hours. The β-epimers of cortolone and Reichstein's substance E run together ahead of the corresponding α-epimers. Separation of the cortolones from the corresponding Reichstein's substance E epimers can be achieved in one of two ways: (a) derivatization with pyridine and acetic anhydride to give the acetates which can be separated by thin-layer chromatography on silica gel G in the system cyclohexane–ethyl acetate (1:1); (b) oxidation with a 2% solution of periodic acid in water–methanol–acetic acid (1:1:1) followed by extraction with chloroform and acetylation of the extract. Thin-layer chromatography on silica gel G of this extract in the system ethyl acetate–cyclohexane (7:3) separates 3α-acetoxy-11β-hydroxy-5β-androstan-17-one derived from the cortolones and 11β-hydroxy-Δ^4-androstene-3,17-dione derived from the epimers of Reichstein's E. In some patients, measurable amounts of Reichstein's substance U is excreted. Its mobility on paper is similar to that of THF from which it can be readily separated by acetylation followed by thin-layer chromatography in the system ethyl acetate–cyclohexane (3:7). ATHF and THE can be further purified in the same manner if desired. Cortisol and ATHE can be separated as acetates in the same manner.

The 11-oxygenated C-19 metabolites which are in the effluent of the first paper chromatograms can be further separated by chromatography in the system toluene–isooctane–methanol–water (5:2:4:1) for 16 hours at 27° on 118-cm long Whatman No. 1 paper. The reference standards as well as the 10% portion cut from the center strip containing the endogenous metabolites are stained with the Zimmerman reagent to locate and identify the metabolites. The areas are then cut, eluted, and quantitated by the Zimmermann reaction.

6β-Hydroxycortisol is best isolated by the method of Burstein.[7] The aqueous sample, e.g., urine, is divided into 10 portions each of which is extracted sequentially with 5 volumes of ethyl acetate in each of 4 funnels. The ethyl acetate extracts are then combined, washed with 1% 1 *N* sodium hydroxide solution containing 15% sodium sulfate followed by brine, dried, and concentrated *in vacuo*. The extract is then subjected to paper chromatography or partition chromatography on Celite in the system benzene–ethyl acetate–methanol–water (1.5:0.5:1:1). The extract may also be acetylated and purified in the system ethyl acetate:chloroform on a thin layer of silica gel G. Quantitation is achieved with the Porter-Silber of blue tetrazolium reagents or preferably by the double isotope derivative method; addition of [^{3}H]6β-hydroxycortisol, acetyla-

[7] S. Burstein, H. L. Kimball, E. L. Klaiber, and M. Gut, *J. Clin. Endocrinol. Metab.* **30**, 491 (1967).

tion with standardized [^{14}C]acetic anhydride, and recrystallization to constant isotope ratio of the carrier added diacetate. In the studies following administration of [^{14}C]cortisol, radioactive recovery of [^{14}C]6β-hydroxycortisol can be achieved either by addition of [^{3}H]6β-hydroxycortisol to the urine and purification to constant isotope ratio or by addition of 15 to 20 mg of unlabeled 6β-hydroxycortisol to the extract and recrystallization to constant specific activity.

Although paper chromatography is preferred by most experimenters because of its speed and relative ease, column chromatographic systems which will also separate these metabolites have been described. The use of resin columns to separate these metabolites have been reported.[8] Although fairly good separations have been achieved the use of columns is more tedious.

[8] T. Seki and K. Matsumoto, *J. Chromatogr.* **27**, 423 (1967).

[45] Isolation and Synthesis of the Major Metabolites of Aldosterone and 18-Hydroxycorticosterone[1]

By Stanley Ulick and Leyla C. Ramirez

Aldosterone and 18-hydroxycorticosterone are conveniently considered together since both steroids are secreted by the glomerulosa zone of the mammalian adrenal cortex in response to the same stimulating factors.[2] The 18-hydroxysteroid is normally secreted at a 2–3 times greater rate[3] except in an inborn error of aldosterone biosynthesis[4] due to a defect in the terminal oxidase mechanism[5] in which there is marked overproduction of 18-hydroxycorticosterone relative to aldosterone.

[1] Nomenclature: aldosterone = 11β,21-dihydroxy-3,18,20-trioxo-4-pregnene; androstenedione = 4-androstene-3,17-dione; cortisol = 11β,17α,21-trihydroxy-4-pregnene-3,20-dione; 11-dehydrocorticosterone = 21-hydroxy-4-pregnene-3,11,20-trione; 18-hydroxycorticosterone = 11β,18-21-trihydroxy-4-pregnene-3,20-dione; 18-hydroxy-11-dehydrocorticosterone = 18,21-dihydroxy-4-pregnene-3,11,20-trione; 3α,5β-tetrahydroaldosterone = 3α,11β,21-trihydroxy-18,20-dioxo-5β-pregnane; tetrahydrocortisone = 3α,17α,21-trihydroxy-11,20-diketo-5β-pregnane.

[2] S. Ulick, G. Nicolis, and K. K. Vetter, *in* "Symposium on Aldosterone," (E. E. Baulieu nad P. Robel, eds.), p. 3. Blackwell, Oxford, 1964.

[3] S. Ulick and K. K. Vetter, *J. Clin. Endocrinol. Metab.* **25,** 1015 (1965).

[4] S. Ulick, E. Gautier, K. K. Vetter, J. R. Markello, S. Yaffe, and C. U. Lowe, *J. Clin. Endocrinol. Metab.* **24,** 669 (1964).

[5] S. Ulick, *Proc. Int. Congr. Endovrinol. 4th, 1972* Int. Congr. Ser. No. **256** (in press).

The metabolism of aldosterone *in vivo* has been studied best in man. With few exceptions,[6,7] little is known about metabolism in other species. The major metabolites in human urine are a tetrahydro derivative conjugated with glucuronic acid at C-3 and a pH 1 hydrolyzable conjugate which is probably an aldosterone C-18 glucuronoside.[8-10] Procedures for the isolation of aldosterone from this conjugate in urine and subsequent measurement have been extensively reviewed.[11,12] Several minor urinary metabolites have also been isolated following the administration of large amounts of aldosterone to human subjects,[13-15] but their production from endogenous aldosterone at normal rates of secretion has not been reported.

The 3α-hydroxy-5β-isomer of tetrahydroaldosterone is the most abundant metabolite of aldosterone in human urine, accounting for some 30–40% of the secreted hormone.[16] Much smaller amounts of the $3\alpha,5\alpha$- and the $3\beta,5\beta$-tetrahydro isomers have also been reported.[13] The major metabolite of 18-hydroxycorticosterone is 3α,18,21-trihydroxy-5β-pregnane-11,20-dione.[16] These tetrahydro metabolites of aldosterone and 18-hydroxycorticosterone can be measured as tri- and diacetates, respectively, but are more conveniently converted to stable etiolactones and quantitated by means of the labeled 3-monoacetate derivatives. For the determination of secretory rates, labeled aldosterone and 18-hydroxycorticosterone are injected into the patient and labeled metabolites isolated from the subsequent 24- to 48-hour urine collection and their specific activities measured. For excretory rates, the labeled metabolites are added to the urine before isolation. Two procedures are available.[3] In one, urine is treated with glucuronidase, and the α-ketol form isolated and converted to the etiolactone. The other procedure accomplishes simultaneous hydrolysis and oxidative cleavage by adding periodic acid directly to urine. The simplicity of the latter procedure is somewhat offset by the cruder extract which it yields, although a chemical purification procedure consist-

[6] S. Ulick and E. Feinholtz, *J. Clin. Invest.* **47**, 2523 (1968).
[7] S. Ulick, H. C. Rose, and L. C. Ramirez, *Steroids* **16**, 183 (1970).
[8] R. H. Underwood and J. F. Tait, *J. Clin. Endocrinol. Metab.* **24**, 1110 (1964).
[9] J. R. Pasqualini, F. Uhrich, and M. F. Jayle, *Biochim. Biophys. Acta* **104**, 515 (1965).
[10] R. H. Underwood and N. L. Frye, *Steroids* **20**, 515 (1972).
[11] J. P. Coghlan and J. R. Blair-West, *in* "Hormones in Blood" (C. H. Gray and A. L. Bacharach, eds.), Vol. 2, p. 391. Academic Press, New York, 1967.
[12] A. H. Brodie and J. F. Tait, *Methods Horm. Res.* 323 (1968).
[13] W. G. Kelly, L. Bandi, and S. Lieberman, *Biochemistry* **1**, 792 (1962).
[14] W. G. Kelly, L. Bandi, and S. Lieberman, *Biochemistry* **2**, 1243 (1963).
[15] W. G. Kelly, L. Bandi, and S. Lieberman, *Biochemistry* **2**, 1249 (1963).
[16] S. Ulick and K. K. Vetter, *J. Biol. Chem.* **237**, 3364 (1964).

ing of saponification and relactonization (described below) is available for its purification.

A. Isolation and Measurement of 3α,5β-Tetrahydroaldosterone and 3α,18,21-Trihydroxy-5β-pregnane-11,20-dione

a. General Procedures. Descending paper chromatography was carried out in the usual manner. Systems containing formamide or glycols as the stationary phase were applied by dipping the paper in a 25% solution of the nonvolatile phase in acetone. Alumina column chromatography was carried out on a 1-cm-diameter column containing 10 g of absorbent. A low-activity grade of neutral alumina suitable for the chromatography of alkali-sensitive lactone acetates was prepared by treating Harshaw alumina catalyst with ethyl acetate as described.[3]

b. Glucuronidase Hydrolysis. Either one-half or an entire 24-hour urine specimen is incubated with approximately 500 units mammalian glucuronidase/ml at pH 5 for 2 days at 45° and extracted with ethyl acetate. The extract is washed with 1.0 *N* sodium hydroxide and water. The neutral extract is chromatographed on a 15-cm-wide paper for 24 hours in the ethylene dichloride/formamide system. A narrow chromatogram strip stained with blue tetrazolium serves to locate tetrahydrocortisol and tetrahydrocortisone, normally present in the extract. Radioactive scanning reveals the position of the major zone for elution, migrating at 0.75 the rate of tetrahydrocortisone, which consists of a mixture of 3α,5β-tetrahydroaldosterone and 3α,18,21-trihydroxy-5β-pregnane-11,20-dione. This mixture dissolved in 5 ml methanol is oxidized with 5 ml periodic acid reagent (0.1 *M* containing 2 ml pyridine/100 ml) to yield a mixture of etiolactones which are separated as described below.

c. Oxidative Cleavage of Urinary Metabolites in Situ. Tritium-labeled internal standards are added to urine either as their α-ketol or etiolactone derivative. One-tenth volume 2.0 *M* periodic acid is added and the specimen allowed to stand at room temperature in the dark for 18 hours. A test for excess periodate[17] is carried out on aliquots at 1 hour and again at 18 hours and more periodic acid is added if needed. After oxidation, the urine is extracted with ethyl acetate or methylene chloride and the extract is washed with 1 *N* sodium hydroxide and water. Extracts containing iodine are also washed with a saturated solution of sodium bisulfite.

Saponification and relactonization. The neutral extract obtained after periodate oxidation of urine is evaporated, dissolved in 30 ml 95%

[17] An aliquot was diluted with water, saturated with sodium bicarbonate and a crystal of sodium iodide added. The evolution of iodine indicated excess periodate.

ethanol, and 15 ml 1.0 *N* sodium hydroxide is added. The mixture is warmed on the steam bath for 30 minutes or allowed to stand at room temperature overnight and the alcohol removed on a flash evaporator. The alkaline solution is washed twice with 10-ml portions of methylene chloride to remove nonsaponificable substances, acidified with 6 *N* hydrochloric acid, and extracted with five 20-ml portions of methylene chloride. The extract is washed with water, dried, and evaporated. The residue is dissolved in 40 ml of a saturated solution of *p*-toluenesulfonic acid in ethylene dichloride and heated under reflux for 1 hour. Lactonization can also be effected by allowing this solution to stand at room temperature for 18 hours. Acidic material is removed by washing the ethylene dichloride solution with two 10-ml portions of 0.1 *N* sodium hydroxide and with water. A purified lactone fraction remains after evaporation of the solvent. A small amount of acidic material represents nonlactonizable, 17-isoetioacid formed during saponification. Although toluenesulfonic acid is required for lactonization of the etioacid of tetrahydroaldosterone, the etioacids of the 18-hydroxy metabolite as well as other 18-hydroxyetioacids lactonize spontaneously on acidification with HCl. The reaction with *p*-toluenesulfonic acid in ethylene dichloride can therefore be eliminated if only 18-hydroxy metabolites are to be isolated.

d. Separation of Etiolactones and Conversion to Monoacetates. The mixture of etiolactones obtained by either hydrolytic procedure is chromatographed on methylcyclohexane–toluene (1:1)/formamide for 18 hours. The etiolactone of the 18-hydroxycorticosterone metabolite (3α,18-dihydroxy-11-keto-5β-etianic acid lactone) and of tetrahydroaldosterone (3α,11β-dihydroxy-18-oxo-5β-etianic acid lactone) migrate, respectively, at 0.5 and 0.7 the rate of 11β-hydroxyandrostenedione. Both etiolactones are eluted together and acetylated with [^{14}C]acetic anhydride in the usual manner and chromatographed on methylcyclohexane/formamide for 18 hours. Occasionally, this chromatogram was repeated when the radioscan indicated that a large amount of background ^{14}C radioactivity was present. The etiolactone monoacetates of the 18-hydroxy and of 18-oxo derivatives migrate, respectively, at 0.6 and 1.0 times the rate of androstenedione. Each monoacetate is eluted and chromatographed separately on neutral alumina. Each column is developed with 25 ml petroleum ether–benzene (4:1). The aldosterone metabolite is eluted with 5-ml fractions of petroleum ether–benzene (3:7) and the 18-hydroxycorticosterone metabolite eluted with petroleum ether–benzene (1:4). Some adjustment of the benzene content of the eluting solvent may be necessary to provide a hold-back volume of at least 50 ml for the etiolactone monoacetates. Specific activity is determined in the usual manner from the ^{3}H/^{14}C ratio of the fraction containing the ^{14}C peak provided the ratio of this fraction does not deviate by more than 10% from that of the two adjacent fractions.[3]

B. Syntheses

1. *3α,5β-Tetrahydroaldosterone*

The 3α,5β-isomer is more readily prepared by enzymatic reduction than by catalytic hydrogenation. Rat liver contains steroid 5α-reductases in the microsomal fraction and soluble 5β-reductase in the 100,000 *g* supernatant.[18] When aldosterone is reduced by unfractionated rat liver homogenates, the 3β,5α-tetrahydro isomer predominates.[19] A microsomal supernatant fraction can be used to obtain the 3α,5β isomer[20] but it is simpler to use an aqueous extract of a crude liver acetone powder. We prefer a rabbit liver preparation which gives a somewhat purer product and is readily available. A commercial source (Pel-Freeze Co., Rogers, Arkansas) of adult rabbit liver acetone powder, stored in the laboratory at —20° without desiccant, has retained its activity for more than 5 years.

a. Enzymatic Reduction. In a typical enzymatic synthesis, carried out in a cold room at 10°, 20 g rabbit liver acetone powder is suspended by gentle stirring in 200 ml 50 m*M* Tris buffer (pH 7.4) containing 10 m*M* $MgCl_2$. The suspension is centrifuged at 5000 rpm for 10 minutes and filtered through coarse fluted filter paper (Reeve Angel No. 802). To the clear filtrate is added 20 mg *d*-aldosterone (containing 1,2-^{3}H-labeled steroid to facilitate localization on chromatograms) in 1.5 ml 95% ethanol and 20 mg chemically reduced NADPH. The flask is stoppered and the mixture incubated on a Dubnoff-type shaker for 1 hour at 37°. An equal volume of 95% ethanol is added, the precipitated protein filtered, washed with ethanol, and the filtrate evaporated on a rotary evaporator. The residue is partitioned between 50 ml water and 100 ml methylene chloride and the aqueous phase is separated and extracted with two more portions of methylene chloride. The combined organic solvent extract is washed with 0.2 *N* NaOH and water, dried over sodium sulfate, and evaporated.

b. Isolation of 3α,5β-Tetrahydroaldosterone. The predominant tetrahydro isomer in the extract is the 3α-hydroxy-5β form. Small amounts (generally less than 5%) of the 3β-hydroxy-5β isomer may also be present. The formation of insoluble digitonides by 3β-hydroxy steroids is not

[18] G. M. Tomkins, see Vol. V, p. 499 (1962).
[19] H. Kohler, R. H. Hesse, and M. M. Pechet, *J. Biol. Chem.* **239**, 4117 (1964).
[20] G. L. Nicolis, H. H. Wotiz, and J. L. Gabrilove, *Endocrinology* **28,** 547 (1968).

very helpful for the separation of 3β,5β from 3α,5β isomers because the digitonide of the former precipitates only at relatively high steroid concentrations.[7] Digitonin can be used, however, to separate the 3β,5α isomer by precipitation. The 3β,5β-tetrahydro isomer, if present, can be isolated as described under "Other Tetrahydroaldosterone Isomers."

The incubation extract is chromatographed first on ethylene dichloride/formamide to the edge to remove unreacted aldosterone, if present, and next on ethyl acetate–toluene (7:3)/formamide for 8 hours to separate the 3α- and 3β-hydroxy-5β isomers which migrate at 0.5 and 0.6 times the rate of cortisol, respectively. Eluted steroids are dissolved in methylene chloride and washed with water to remove formamide. The yield of the 3α,5β-tetrahydro isomer after chromatograpy is approximately 9.0 mg. Since the isomers of tetrahydroaldosterone are difficult to crystallize and their infrared spectra differ only slightly from one another, they are best characterized by means of their etiolactone and etiolactone acetate derivatives.

c. Etiolactone and Etiolactone Acetate Derivatives. If the etiolactone of tetrahydroaldosterone is the desired derivative, the crude extract of the liver enzyme incubation can be oxidized directly with the methanol–periodic acid reagent and the etiolactone isolated chromatographically as described below. Alternatively, aldosterone etiolactone can be reduced enzymatically as described for aldosterone but with somewhat lower yields. However, the scale of the enzymatic reaction is limited somewhat by the low solubility of the etiolactone in ethanol. In addition, the pH (7.4) of enzymatic reduction may yield tetrahydro products partly in the etioacid form. Lactonization can be readily effected, however, by treating the incubation extract with *p*-toluenesulfonic acid in ethylene dichloride as described above.

3α,5β-Tetrahydroaldosterone in methanol is treated with an equal volume of periodic acid pyridine reagent at room temperature for 2–18 hours, the mixture extracted with methylene chloride, the extract washed with 0.1 *N* NaOH and water, dried, and evaporated. The etiolactone derivative is isolated from the residue by chromatography on methylcyclohexane–toluene (1:1)/formamide and located for elution by radioactive scanning. Chromatographic mobilities relative to other isomers are shown in the table. Further purification is carried out by chromatography on a column of neutral alumina developed with benzene and eluted with 2% ethanol in benzene. Recrystallization from methanol of recombined peak fractions gives needles melting at 254–255° for the 3α,5β isomer. Comparison of its infrared spectrum with published spectra[7] of the etiolactone isomers provides definitive identification.

The tetrahydroetiolactone isomeric acetates are not as readily sepa-

PREPARATION OF ISOMERIC TETRAHYDROALDOSTERONE ETIOLACTONES BY CATALYTIC HYDROGENATION

Tetrahydro isomer	Chromatographic mobility[a]	Yield (mg)
3β-Hydroxy,5α	0.69	4.6
3α-Hydroxy,5β	0.77	1.1
3β-Hydroxy,5β	0.95	4.1

[a] Relative to 11β-hydroxyandrostenedione in methylcyclohexane–toluene (1:1)/formamide run for 24 hours. The relative mobility of aldosterone etiolactone is 1.1.

rated chromatographically as the 3-hydroxy derivatives, but show distinctive infrared spectra. Acetylation of the etiolactone with acetic anhydride–pyridine (2:1) at room temperature yields the 3-monoacetate which can be isolated by chromatography on methylcyclohexane/formamide, a system in which both the 3α- and 3β-acetoxy-5β-etiolactones migrate at almost the same rate as androstenedione. Further purification on an alumina column developed with petroleum ether–benzene (4:1), elution with benzene, and crystallization of peak fractions gives a melting point of 279–281° for the 3α-acetoxy-5β isomer whose infrared spectrum can be compared with published spectra[7] of the isomeric etiolactone monoacetates.

Etiolactones and their monoacetates are stable during gas-liquid chromatography at 250° and can be collected directly from the chromatogram effluent in a state of purity suitable for infrared spectroscopic identification.[7,20]

2. 3α,18,21-Trihydroxy-5β-pregnane-11,20-dione

This metabolite of 18-hydroxycorticosterone can be prepared by enzymatic reduction of the corresponding Δ^4,3-ketone, 18-hydroxy-11-dehydrocorticosterone, which, in turn, can be prepared either by biosynthesis from corticosterone or by chemical oxidation of 18-hydroxycorticosterone.

a. Biosynthesis of 18-Hydroxy-11-dehydrocorticosterone. Rabbit adrenal tissue synthesizes the 18-hydroxy derivatives of both corticosterone and 11-dehydrocorticosterone from corticosterone. Freshly removed or fresh-iced or frozen rabbit adrenals (Pel-Freeze Co., Rogers, Arkansas) could be used but the latter two were less active. Three pairs of adrenals (1.0 g) are minced with scissors in 20 ml of mammalian Tris Ringers buffer[16] (pH 7.4) and incubated with 0.1–1.0 mCi [1,2-^{3}H]corticosterone and 10 mg chemically reduced NADPH at 37° for 2 hours. The incubation mixture is extracted with methylene chloride, the extract washed with water, dried, and chromatographed to the edge on ethylene dichloride–toluene (1:1)/ethylene glycol. Radioactive scanning revealed unreacted corticosterone, 11-dehydrocorticosterone, and their 18-hydroxy derivatives in 20–30% and 5–10% yield, respectively. The 18-hydroxy derivatives of corticosterone and 11-dehydrocorticosterone migrate at 0.25 and 1.0, respectively, the rate of aldosterone. Their identity is confirmed by oxidation of aliquots with periodic acid reagent. Chromatography of the neutral methylene chloride extract containing the etiolactone derivatives on the toluene/formamide system served to identify the etiolactone of 18-hydroxycorticosterone[21] at 1.1 the rate of corticosterone and the etiolactone of 18-hydroxy-11-dehydrocorticosterone and aldosterone at 0.9 and 1.3, respectively, the rate of 11β-hydroxyandrostenedione. Conversion of the 11β-hydroxy to the 11-ketoetiolactone has been described.[21] 18-Hydroxy-11-dehydrocorticosterone is rechromatographed in the ethyl acetate–toluene (7:3)/formamide system in which it migrates at 1.25 the rate of aldosterone, and 18-hydroxycorticosterone is rechromatographed on ethylene dichloride/ethylene glycol in which it migrates at 0.25 the rate of aldosterone. The ratio of 11β-hydroxy to 11-keto product in the rabbit adrenal system depends in part on the availability of NADPH. The biosynthesis of aldosterone is very low in this tissue.

b. Oxidation of 18-Hydroxycorticosterone. Labeled 18-hydroxycorticosterone prepared biosynthetically as described above with rabbit adrenal tissue or in a bullfrog adrenal system[22] can serve as an alternative source of the 11-keto derivative. [1,2-^{3}H]18-Hydroxycorticosterone is acetylated with 0.5 ml dry pyridine and 0.1 ml acetic anhydride at room temperature for 18 hours. Excess reagents are removed under vacuum and the residue treated directly with 10 mg chromic oxide in 1 ml pyridine for 2 hours at room temperature. Most of the pyridine is removed by repeated addition and evaporation of benzene. The residue is dissolved in methylene chloride, washed with 1 *N* HCl and water, dried, and evaporated. To remove the C-21 acetate, the residue is dissolved in 1.0 ml 95% ethanol, nitrogen is bubbled into the test tube fitted with a ground glass stopper, and 0.1 ml 10% tetramethylammonium hydroxide is added and the stoppered mixture allowed to stand for 30 minutes. The reaction is stopped with 1 *M* acetic acid, extracted with methylene chloride, washed with water, and dried. Chromatography of the product on ethylene dichloride–toluene (1:1)/ethylene glycol reveals more than 50% conversion to 18-hydroxy-11-dehydrocorticosterone. After elution, the identity

[21] S. Ulick and K. Kusch, *J. Amer. Chem. Soc.* **82,** 6421 (1960).
[22] G. L. Nicolis and S. Ulick, *Endocrinology* **76,** 514 (1964).

of the product is confirmed by oxidation of an aliquot in methanol with an equal volume of 0.1 *M* periodic acid reagent. The neutral diketoetiolactone oxidation product (18-hydroxy-3,11-diketo-4-etienic acid lactone)[21] is identified by its running rate in the toluene/formamide system (0.9 of 11β-hydroxyandrostenedione).

c. Enzymatic Reduction. Tritium-labeled 18-hydroxy-11-dehydrocorticosterone is incubated with the reductase system derived from 1 g rabbit liver acetone powder as described above. The reduction product is isolated from the neutral extract of the incubation mixture by chromatography on ethylene dichloride/formamide for 18 hours in which it migrates at 0.8 the rate of cortisol. Identity of the eluted metabolite is confirmed by periodic acid oxidation to 3α,18-dihydroxy-11-ketoetianic acid lactone which migrates at 0.5 the rate of 11β-hydroxyandrostenedione in the methylcyclohexane–toluene (1:1)/formamide system. An independent chemical synthesis of this etiolactone from tetrahydroaldosterone has also been described.[16]

3. Other Tetrahydroaldosterone Isomers

All four possible ring A reduction products are known. Rat liver homogenates or microsomal rat liver 5α-reductase yields the 3β,5α isomer predominantly[19] but the 3α,5α form can also be obtained if female rat livers are used.[13] The 3β,5β isomer is the major metabolite of aldosterone in the bullfrog[7] and is formed to a small extent (5–10%) by the same soluble 5β-steroid reductase system of rat or rabbit liver[16] which yields, as its major product, the 3α,5β form. The etiolactone forms of the 3β,5α- and 3β,5β-tetrahydro isomers are conveniently prepared by catalytic hydrogenation over platinum oxide of aldosterone etiolactone. The synthesis of 3β-hydroxy-5β-tetrahydroaldosterone by hydrogenation of aldosterone in the presence of palladium catalyst has also been described.[23]

a. Catalytic Hydrogenation. In a typical synthesis, 20 mg aldosterone etiolactone containing tracer 1,2-^{3}H-labeled steroid is dissolved in 10 ml glacial acetic acid and stirred with 10 mg platinum oxide catalyst for 18 hours at room temperature under 1 atm of hydrogen. The reaction mixture is filtered, evaporated, and the mixture of tetrahydro isomers separated by chromatography in methylcyclohexane–toluene (1:1)/formamide as shown in the table. For the isolation of each of these isomers in pure form, chromatography in the same paper system is repeated and final purification carried out on a column of neutral alumina developed

[23] Y. Lederman, R. Szpigielman, M. Bendcovsky, J. Herling, and M. Harnik, *Anal. Biochem.* **51**, 193 (1973).

with benzene and eluted with 2% ethanol in benzene. Etiolactones in peak fractions, monitored by radioactivity, are combined, crystallized, and identified by infrared spectroscopy[7] of the 3-hydroxy and 3-monoacetate derivatives. Further proof of structure can be obtained by oxidation of etiolactones to 3-keto derivatives with chromic oxide. The sign of the Cotton effect at 310 nm in the optical rotary dispersion spectrum of the 3-keto derivative confirms the configuration at C-5.[7,16]

[46] Preparation and Biological Evaluation of Active Metabolites of Vitamin D_3[1]

By M. F. Holick and H. F. DeLuca

Introduction

The demonstration, isolation, and eventual identification of the first biologically active metabolite of vitamin D_3, 25-hydroxyvitamin D_3 (25-OH-D_3),[1a] provided the impetus to search for other metabolites of vitamin D which possess biological activity. As of this writing there are four metabolites of vitamin D_3 which have been demonstrated to possess biological activity. They have been isolated in pure form and their structures assigned (Fig. 1).

The liver and the kidney appear to be the major sites for the hydroxylation of vitamin D_3. The vitamin is first hydroxylated in the liver on C-25 and then travels to the kidney to be hydroxylated either on C-1 or on C-24, depending on the physiological state of the animal. The site for the 26-hydroxylation of 25-OH-D_3 is not known. Many of the methods described here are related to the preparation, isolation, purification, and bioassay of vitamin D metabolites more polar than 25-OH-D_3. In a previous volume of "Methods in Enzymology"[2] the chemical synthesis of 25-OH-D_3, its isolation from biological sources, and many of the common bioassay methods for vitamin D active compounds have been described. These methods include antirachitic activity tests in rats and chicks, intestinal calcium transport activity in rats and chicks, bone calcium mobilization in rats, chemical methods of analysis, and methods of vitamin D chromatography. These will not be duplicated here and

[1] Supported in part by a grant from the USPHS No. AM-15512 and the Harry Steenbock Research Fund of the Wisconsin Alumni Research Foundation.

[1a] J. W. Blunt, H. F. DeLuca, and H. K. Schnoes, *Biochemistry* 7, 3317 (1968).

[2] H. F. DeLuca and J. W. Blunt, Vol. 18, p. 709 (1971).

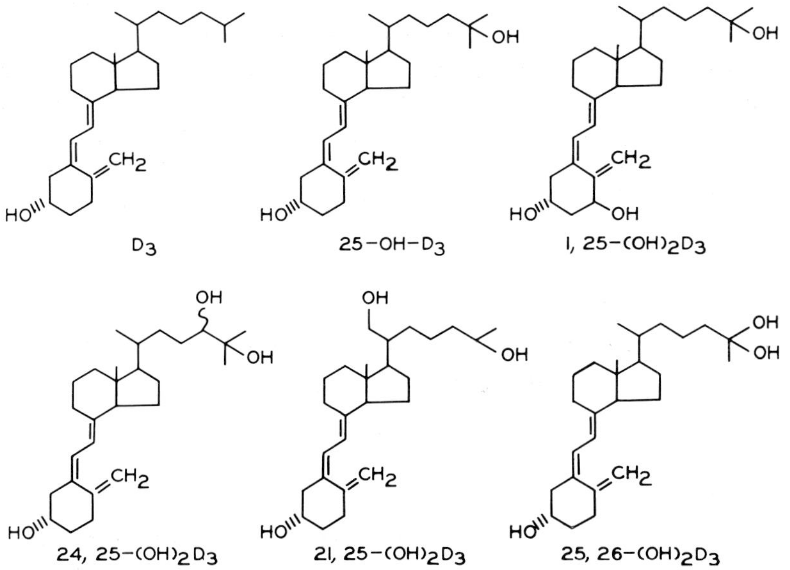

FIG. 1. Structures of vitamin D_3 and its metabolites. Abbreviations: D_3, vitamin D_3; 25-OH-D_3, 25-hydroxyvitamin D_3; 1,25-$(OH)_2D_3$, 1,25-dihydroxyvitamin D_3; 24,25-$(OH)_2D_3$, 24,25-dihydroxyvitamin D_3; 21,25-$(OH)_2D_3$, 21,25-dihydroxyvitamin D_3; 25,26-$(OH)_2D_3$, 25,26-dihydroxyvitamin D_3.

instead additional methods which have been developed since that time will be reported.

Preparation of 24,25-Dihydroxyvitamin D_3[3,4] and 25,26-Dihydroxyvitamin D_3[5] from Hog Plasma

Eight hogs, approximately 250 lb each, are given 6.25 mg of vitamin D_3 (Vita-Plus Corp., Madison, Wisc.) daily in their normal rations. After 28 days the pigs are slaughtered and their blood collected (approximately 2–3 liters/hog). The blood is immediately mixed with 0.1 volumes of 0.1 *M* sodium oxalate to prevent clotting and the plasma harvested by means of a DeLaval blood separator. The plasma is saturated to the level of

[3] T. Suda, H. F. DeLuca, H. K. Schnoes, G. Ponchon, Y. Tanaka, and M. F. Holick, *Biochemistry* **9**, 2917 (1970).

[4] M. F. Holick, H. K. Schnoes, H. F. DeLuca, R. W. Gray, I. T. Boyle, and T. Suda, *Biochemistry* **11**, 4251 (1972).

[5] T. Suda, H. F. DeLuca, H. K. Schnoes, Y. Tanaka, and M. F. Holick, *Biochemistry* **9**, 4776 (1970).

70% with ammonium sulfate in order to precipitate the protein to which the vitamin D metabolites are bound. The precipitate is allowed to stand at 4° for 7 days and then collected by centrifugation at 25,000 rpm for 25 minutes in a Sharples AS-16-P centrifuge. The protein precipitate (1 part) is extracted with methanol (2 parts) and chloroform (1 part) with a portable mixer and allowed to stand for 24 hours at 4°. This is filtered through a Büchner funnel, and the protein precipitate is again reextracted for 4 hours with the same volume of methanol and chloroform and once again filtered. The filtrates are combined, and chloroform (2 parts) and tap water (2 parts) are added whereupon separation into two layers, i.e., methanol–water–protein and chloroform–lipid, occurs [in the event that the phases do not separate small amounts of saturated NaCl (0.1 part) is added to encourage phase separation]. The lower chloroform layer is removed and the upper layer reextracted with more chloroform (1 part). Tap water (0.5 part) is added to the combined chloroform extracts (1 part), mixed well, and then stored at 4° until two clear layers are apparent (this takes approximately 2–8 days). The chloroform phase is removed and the water layer is reextracted with 0.5 parts chloroform. The combined chloroform phases are concentrated under vacuum by flash evaporation with a rotary flash evaporator. The resulting black oily residue (1 part) is partitioned with 5 parts of Skellysolve B (petroleum ether, redistilled at 67–69°) and 6 parts of methanol–water (9:1) in a separatory funnel. (The upper phase contains less polar metabolites of vitamin D_3 and the lower phase contains 25-OH-D_3 and its more polar metabolites.) After separation of the phases, chloroform (5 parts) and tap water (2 parts) is added to the lower phase. The chloroform phase containing the polar metabolites is collected and the aqueous phase is reextracted with chloroform (2 parts). The combined chloroform extracts are then taken to dryness by flash evaporation.

In order to provide a means of locating the vitamin D metabolites in subsequent chromatography, an extract containing tritium-labeled vitamin D_3 metabolites is prepared as follows. [1,2-^{3}H]Vitamin D_3 (2.5 μg) is dosed intravenously to each of 50 chickens which had been maintained on a vitamin D-deficient diet (see table) for 27 days. After administration they are fasted and 20 hours later the blood is collected by decapitation. The resulting plasma (1 part) is extracted with methanol–chloroform (2:1) and the tritiated labeled polar metabolites are recovered in a similar fashion as described above. The yellow residue is dissolved in 1.5 parts chloroform–Skellysolve B (65:35) and applied to a 3 $\times$ 70-cm glass column packed with 110 g of Sephadex LH-20 (a hydroxy propyl ether derivative of Sephadex G-25, Pharmacia Corp., Piscataway, N.J.), with 3 cm of Celite on top of the bed. One hundred 18-ml fractions are col-

PRACTICAL RACHITOGENIC DIET FOR CHICKS

Component	Percent
Soy protein	25.6
Sucrose	52.55
DL-Methionine	0.6
Glycine	0.4
Cellulose	3.0
Cottonseed oil	4.0
$CaHPO_4$	0.52
KH_2PO_4	1.55
$CaCO_3$	2.28
NaCl	0.8
Salt mixture[a]	6.6
Vitamin mixture[b]	2.1

[a] The trace salt mixture contains the following (g/kg): $MgSO_4 \cdot 7\ H_2O$, 745; $MnSO_4$, 365; KIO_3, 7.7; $CuSO_4 \cdot 5\ H_2O$, 2.2; ZnO, 5.1; $NaMoO_4 \cdot 2\ H_2O$, 2.9; Na_2SeO_3, 0.18; and ferric citrate · 6 H_2O, 200.

[b] The vitamin mixture contains the following (g/kg): Thiamine–HCl, 2.85; riboflavin, 4.28; calcium pantothenate, 9.5; nicotinamide, 23.7; pyridoxine, 3.8; biotin, 0.14; inositol, 474; folic acid, 0.95; B_{12}, 0.01; and sucrose, 477.78. A fat-soluble vitamin mixture is added dropwise to the diet as a cottonseed oil solution which supplies 100 μg β-carotene, 900 μg α-tocopherol, and 120 μg 2-methyl-1,4-napthoquinone weekly per chick.

lected and 100 μl of each fraction are taken for counting in 15 ml of the scintillation counting solution [2 g of PPO 2,5-diphenyloxazole and 100 mg of POPOP 1,4-bis-2-(4-methyl-5-phenyloxyazolyl)benzene/liter of toluene]. The radioactive profile of the column effluent should appear as in Fig. 2. The peak Va fractions (tubes 42–54) which mainly contain

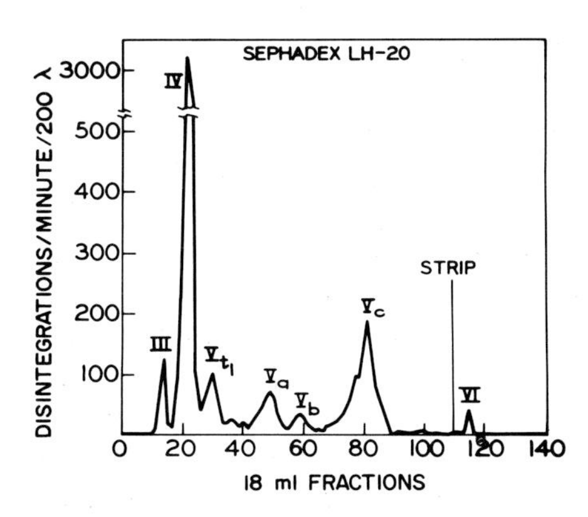

FIG. 2. Sephadex LH-20 column (3 × 70 cm packed in chloroform–Skellysolve B 65:35) profile of plasma lipid extract from chicks given [1,2-^{3}H]vitamin D_3 orally 24 hours earlier.

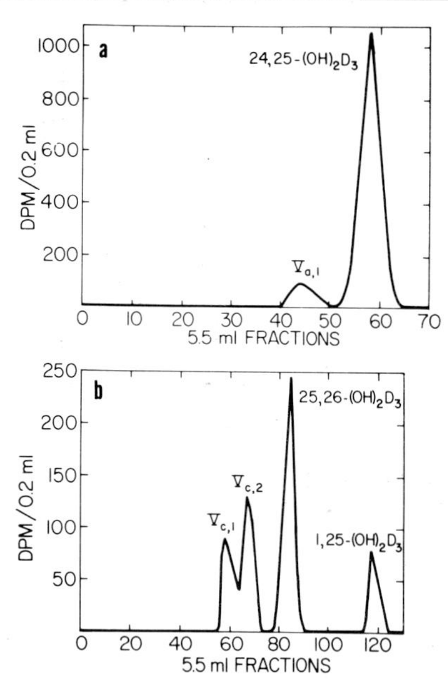

FIG. 3. Liquid–liquid partition chromatography of (a) peak Va and (b) peak Vc fractions from Fig. 2.

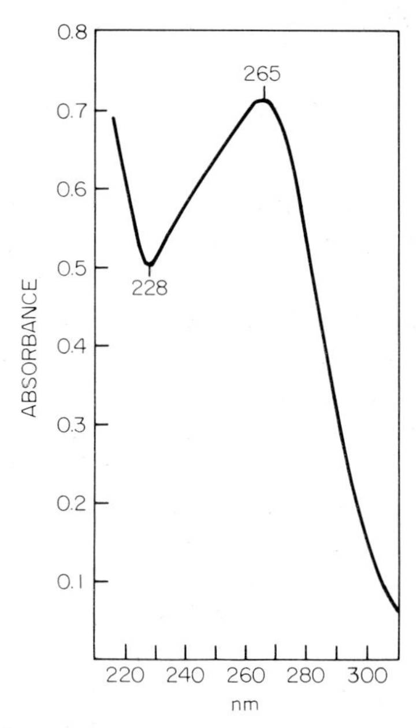

FIG. 4. Ultraviolet absorption spectrum for the 5,6-*cis*-triene system of 24,25-$(OH)_2D_3$, 25,26-$(OH)_2D_3$, and 1,25-$(OH)_2D_3$.

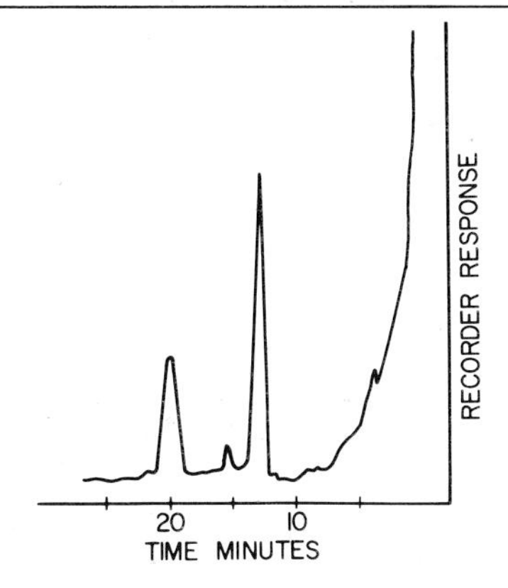

FIG. 5. Gas-liquid chromatographic profile of the tritrimethylsilyl ether derivative of 24,25-$(OH)_2D_3$. (Reproduced with the kind permission of the publisher, Holick *et al.*[4])

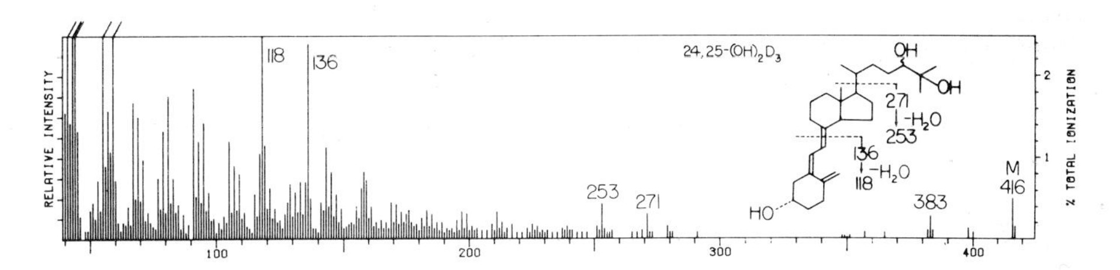

FIG. 6. Mass spectrum of 24,25-$(OH)_2D_3$.

24,25-dihydroxyvitamin D_3 [24,25-$(OH)_2D_3$] are combined by flash evaporation and rechromatographed on a Celite liquid–liquid partition column especially designed for this metabolite. The column is constructed as follows. Three-hundred milliliters of methanol–water (9:1) are equilibrated at 4° with 750 ml Skellysolve B–chloroform (8:2). The aqueous methanol phase (15 ml) is mixed with 20 g of Celite and dry-packed in a 1 × 60-cm column in 2-cm portions. The column is developed with the upper phase (mobile phase) at 4° and 5.5-ml fractions are collected. Figure 3a shows a typical radioactivity profile of the column effluent. The peak Va fraction from the Sephadex LH-20 column separates into two distinct fractions, peak Va_1 (as yet unidentified) and 24,25-$(OH)_2D_3$ (fractions 50–62). The contents of tubes 53–62 from the partition column are combined and applied in 0.1 ml of methanol to a 1 × 60-cm Sephadex LH-20 column which is packed and developed in the same solvent. The ultraviolet absorption spectrum of each fraction (1 ml) containing 24,25-$(OH)_2D_3$ should appear as in Fig. 4, and the purity of the metabolite is conveniently checked by gas-liquid chromatography (as the tritrimethylsilyl ether derivative) (Fig. 5), ultraviolet absorption spectrophotometry ($\lambda_{max} = 265$ nm, $\lambda_{min} = 228$ nm), and its mass spectrum [$m/e = 416$ (M^+) and fragments = 384, 273, 251, 136, and 118] (Fig. 6).

Isolation of 25,26-Dihydroxyvitamin D_3

The peak Vc fractions from the Sephadex LH-20 column as described above (tubes 70–90) are combined, dryed by flash evaporation, and redissolved in the mobile phase and chromatographed on the Celite liquid–liquid partition column as described above for 24,25-$(OH)_2D_3$. The heterogeneity of the peak V fraction is clearly demonstrated in Fig. 3. The peak Vc fraction is separated into at least 4 peaks on this column. Vc_1 and Vc_2 are unidentified while tubes 77–90 contain 25,26-dihydroxyvitamin D_3 [25,26-$(OH)_2D_3$] and tubes 115–125 contain 1,25-dihydroxyvitamin D_3 [1,25-$(OH)_2D_3$]}. Tubes 77–90 are combined, dried, and redissolved in 0.1 ml of methanol and applied to a 1×60-cm Sephadex LH-20 column which is developed in the same solvent. One-milliliter fractions are collected and the ultraviolet absorption spectrum for each of the tubes is taken. Tubes 32–36 contain the metabolite which demonstrates the typical ultraviolet absorption spectra $\lambda_{max} = 265$ nm and a $\lambda_{min} = 228$ nm for the 5,6-*cis*-triene (Fig. 4). Gas-liquid chromatography of the tritrimethylsilyl ether derivative of this metabolite shows the characteristic pyro- and isopyro form of the vitamin, and mass spectrum shows $M^+ = 416$ and fragments m/e = 398, 383, 367, 271, 253, 136, and 118, demonstrating the metabolites purity.

Preparation of 24,25-$(OH)_2D_3$ from Chicken Kidney Homogenates[4]

One-day-old white leghorn cockeral chicks obtained from Northern Hatcheries (Beaver Dam, Wisc.) are kept in cages at 38° and fed *ad libitum* for 2 weeks on a soy protein rachitogenic diet (see table). At the end of the second week the chicks are switched to a high calcium diet (3%) and given 0.50 μg of vitamin D_3 orally each day for an additional 12 days. The animals are sacrificed and their kidneys removed and placed in a buffer solution at pH 7.4 containing 14 m*M* Tris-acetate, 0.19 *M* sucrose, 1.87 m*M* magnesium acetate, and 5 m*M* succinate at 4°. The kidneys are teased apart to remove any extraneous connective tissue as well as testes. The kidney tissue (1 part) is homogenized with 4 parts of the buffer solution with three strokes in a Potter-Elvehjem homogenizer. The homogenate (1.5 ml) is placed in a 25-ml Erlenmeyer flask and then flushed for 1 minute with 100% oxygen. One-half microgram of [^{3}H]25-OH-D_3 (4,000 dpm/μg) is then added to the homogenate in 10 μl of 95% ethanol and placed in a water bath shaker at 37° for 30 minutes. At the end of 30 minutes, 3 ml of methanol (2 parts) is added to each of the flasks to stop the reaction. Flask contents are combined. To this is added chloroform (2 parts) and the phases are permitted to separate. The lower phase containing 24,25-$(OH)_2D_3$ is collected and the upper phase is reextracted with chloroform (1 part). The lower phases are combined, dried by flash evaporation, and the resulting yellow residue is dissolved in 1.5 ml of chloroform–Skellysolve B (65:35). The sample is applied to a 2 × 60-cm glass column packed with 60 g of Sephadex LH-20 and developed with the same solvent. Forty 18-ml fractions are collected and the 24,25-$(OH)_2D_3$ is found in fractions 25–28. Depending on the degree of purity desired, one can either stop at this point and use the material as authentic 24,25-$(OH)_2D_3$ containing some lipid contamination, or purify it further in the following manner: Fractions containing the 24,25-$(OH)_2D_3$ are combined, dried under nitrogen, redissolved in 50–100 μl of chloroform–Skellysolve B (1:1), and applied to a 1 × 60-cm glass column packed with 18 g of Sephadex LH-20 in chloroform–Skellysolve B (1:1) and eluted in the same solvent [60 fractions (5.5 ml) are collected]. Peak fractions (37–45) are combined, dried under nitrogen, redissolved in 50–100 μl of methanol, and applied to a 1 × 60-cm glass column of Sephadex LH-20, packed, and developed with methanol [30 fractions (1.95 ml) are collected]. Fractions 15–18 contain the metabolite and are used for mass spectrometry, ultraviolet absorption spectrophotometry, and gas-liquid chromatography to demonstrate its purity.

Preparation of 1,25-Dihydroxyvitamin D_3 from Chick Intestines[6]

Although very small amounts of 1,25-dihydroxyvitamin D_3 [1,25-$(OH)_2D_3$] are obtained from chicken intestine (0.01 μg/chick intestine), it is possible to generate small amounts of this material by this method. One-day-old chickens (the authors use white Leghorn chicks from Northern Hatcheries, Beaver Dam, Wisc.) are kept at 38° and fed *ad libitum* for 4 weeks on a purified soy protein rachitogenic diet (see table). At the end of the fourth week the chicks are starved for 12 hours and then each are given orally 2.5 μg of [1,2-^{3}H]vitamin D_3 (specific activity 3.2 × 10^5 dpm/μg) in 0.2 ml vegetable oil. Twenty-four hours later the chicks are killed by decapitation and the whole small intestine

[6] M. F. Holick, H. K. Schnoes, H. F. DeLuca, T. Suda, and R. J. Cousins, *Biochemisry* **10**, 2799 (1971).

is excised, flushed with chilled distilled water, and immediately frozen with solid CO_2. The chick intestines are homogenized in a blender with 2 parts methanol (w/v) for 1 minute. One part chloroform is then added to the homogenate and the mixture is allowed to stand at 4° for 1 day. The mixture is filtered and chloroform (1 part) and tap water (0.8 part) are added to the filtrate to effect phase separation. The lower phase is separated from the aqueous phase and the aqueous phase is reextracted with chloroform (1 part). The residue from the initial extraction is reextracted with methanol–chloroform (2:1) as described above. The lower phase is collected and combined with the other chloroform phases and mixed with tap water (1 part) (v/v). This mixture is allowed to stand at 4° for 24 hours. (If an emulsion occurs and cannot be broken up easily, then saturated sodium chloride is added to the mixture.) The chloroform phase is removed and the upper aqueous phase is extracted with chloroform (0.5 part). The combined chloroform phases are concentrated on a rotary flash evaporator and the yellow oily residue is partitioned between Skellysolve B and methanol–water (9:1) as described in the preparation of 24,25-$(OH)_2D_3$ from hog plasma. The resulting chloroform phases are combined, evaporated to dryness, dissolved in chloroform–Skellysolve B (65:35), and applied to a Sephadex LH-20 column slurried and developed in the same solvent. [Note: The size of the column used as well as the amount of solvent needed to dissolve the residue depends on the amount of residue present. For example, if one recovers 5 ml of yellow residue it is necessary to dissolve this in 5–6 ml of chloroform–Skellysolve B (65:35) and apply it to either a 2×60-cm or a 3×70-cm Sephadex LH-20 column. If one is making a very small preparation such that the residue is only a few milligrams, one needs only to dissolve it in 1 ml of chloroform–Skellysolve B (65:35) and apply it to a 1×60-cm Sephadex LH-20 column. Assuming that the void volume of the column is equal to the number of grams in the column, 1,25-$(OH)_2D_3$ comes out in approximately 12.5 void volumes.] The material recovered in this region is homogenous but contains lipid contaminants. If, however, the investigator wishes to purify it further he should follow the additional procedures described for the purification of 1,25-$(OH)_2D_3$ generated by chicken kidney homogenates.

Preparation of 1,25-$(OH)_2D_3$ from Chicken Kidney Homogenates[7]

Preparation of 1,25-$(OH)_2D_3$ from chicken kidney homogenates is followed in identical fashion as described for the production of 24,25-

[7] I. T. Boyle, L. Miravet, R. W. Gray, M. F. Holick, and H. F. DeLuca, *Endocrinology* **90**, 605 (1972).

$(OH)_2D_3$ from chicken kidney homogenates with one exception. The chickens are fed a rachitogenic vitamin D-deficient soy protein diet (see table) for 4 weeks instead of the 3% calcium diet supplemented with vitamin D. The homogenates are combined and worked up in identical fashion and the residue is dissolved in chloroform–Skellysolve B (65:35) and applied to a 60 g Sephadex LH-20 column (2 × 60 cm) which is developed with the same solvent. The 1-hydroxyl metabolite is found in 12.5 void volumes and can be considered as homogeneous 1,25-$(OH)_2D_3$ with contaminating lipids. If further purification is desired the sample is dried under N_2, dissolved in 50–100 μl of chloroform:Skellysolve B (65:35), and applied to a 1 × 150-cm glass column packed with Bio-Beads SX-8 (a polystyrene resin produced by Bio-Rad Corp., Richmond, Calif.) to a height of 148 cm in the same solvent. Fractions (2.5 ml) are collected and the metabolite is recovered in fractions 24–28. The combined fractions are dried under N_2, redissolved in 50–100 μl of methanol, and applied to a 1 × 150-cm glass column packed to a height of 148 cm with Sephadex LH-20 in methanol. Eighty fractions (2.5 ml) are collected and the elution position of the metabolite is shown in Fig. 7. If the purity of the metabolite is unsatisfactory as demonstrated by its ultraviolet absorption spectrum, and mass spectrum, one can apply the sample in 100 μl chloroform–Skellysolve B (1:1) to a 1 × 60-cm Sephadex LH-20 column packed in the same solvent. The metabolite is found in 25 void volumes (approximately 500 ml). The fractions containing the metabolite are combined, dried under nitrogen, and once again applied to the methanol Sephadex LH-20 column. 1,25-$(OH)_2D_3$ has an ultraviolet absorption spectrum characteristic for the 5,6-*cis*-triene system, demonstrating a λ_{max} = 265 nm and λ_{min} = 228 nm (Fig. 4). The mass spectrum demonstrates *m/e* 416 (M^+) and fragments at 398, 380, 287, 269, 152, and 134 (Fig. 8).

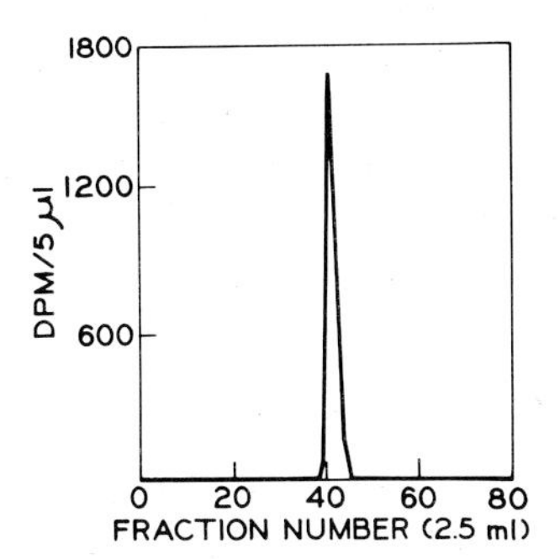

FIG. 7. Sephadex LH-20 column (1 × 150 cm packed in MeOH) profile of 1,25-$(OH)_2D_3$. (Reproduced with the kind permission of the publisher, Holick *et al.*[6])

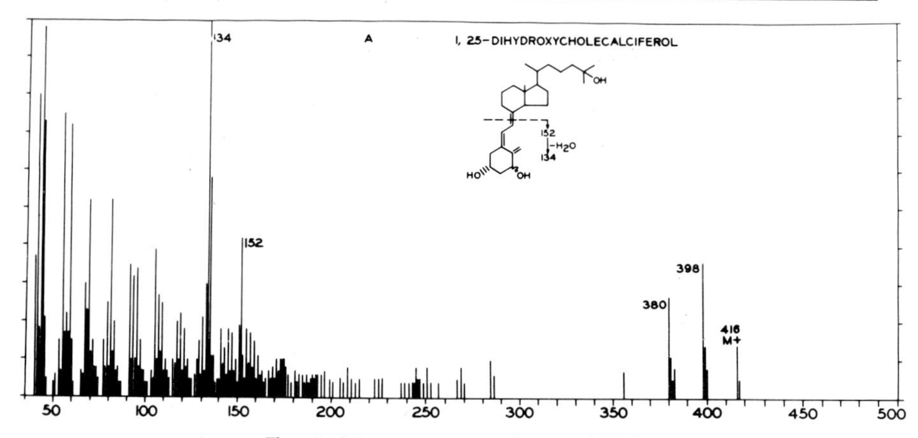

FIG. 8. Mass spectrum of 1,25-$(OH)_2D_3$.

Chemical Synthesis of 1,25-Dihydroxyvitamin D_3[8]

Of the four known biologically active metabolites of vitamin D_3, virtually every metabolite identified to date has now been synthesized chemically. The 25-hydroxy derivative of vitamin D_3 has been prepared from the photochemical isomerization of cholesta-5,7-diene-3β-25-diol and has been adequately described in a previous volume.[2]

Of great interest is the synthesis of 1,25-$(OH)_2D_3$, which is thought to be the hormonal form of vitamin D responsible for intestinal calcium transport and bone calcium mobilization. This metabolite is prepared from the photoisomerization of 1α,3β,25-triacetoxycholesta-5,7-diene (Fig. 9).

Protection of Δ^5. Homocholanic acid *i*-methyl ether (23.83 g) (1) is dissolved in 150 ml of methanol, and diazomethane is added dropwise until the solution remains slightly yellow. This solution is evaporated *in vacuo* to yield a syrupy residue and then dissolved in 240 ml of acetone and 25 ml of 1 *N* perchloric acid. The reaction is stirred at room temperature and after 3 hours a precipitate begins to form. The reaction is continued for another 8 hours and then 150 ml of water is added and the mixture cooled on ice. The precipitate is filtered by suction and washed with water. The product is crystallized in ethanol–chloroform to give 21.25 g of methyl-3β-hydroxy-25-homochol-5-enate [m.p. 114–115°; NMR ($CDCl_3$) δ 3.5 (m, C-3-H), 3.7 (s, $COOCH_3$), 5.4 (m, C-6-H); mass spectrum *m/e* 402 (M^+) and 387, 384, 369, 317, and 291].

Compound 2 is acetylated by dissolving 21.48 g in 35 ml acetic anhy-

[8] E. J. Semmler, M. F. Holick, H. K. Schnoes, and H. F. DeLuca, *Tetrahedron Lett.* **40,** 4147 (1972).

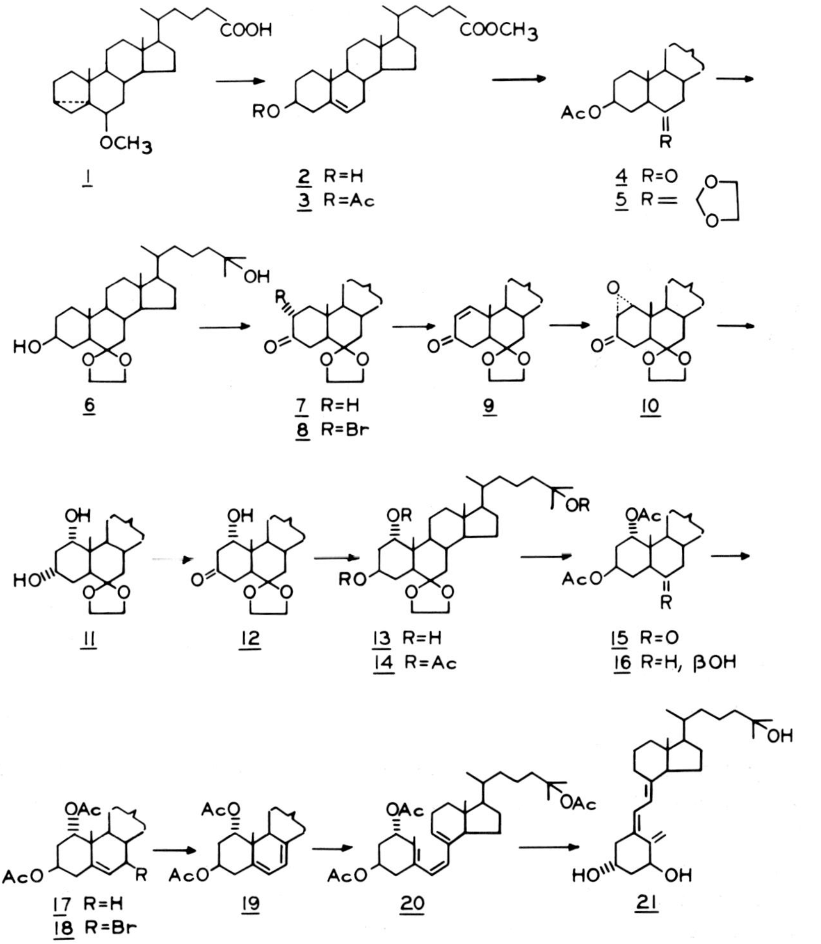

FIG. 9. Synthesis of 1α,25-$(OH)_2D_3$. (Reproduced with the kind permission of the publisher, Semmler *et al.*[8])

dride containing 4 drops of pyridine. The reaction is refluxed for 1 hour and permitted to cool. Thirty milliliters of water is added and the mixture is filtered by suction. Another 20 ml of water is then added to the filtrate to give additional product. The solids are combined and crystallized from methanol–chloroform to yield 22 g of methyl-5β-acetoxy-25-homochol-5-enoate (3) [m.p. 111–113°; NMR ($CDCl_3$) δ 2.0 (s, —$OCOCH_3$), 3.65 (S, $COOCH_3$), 4.6 (m, C-3-H), 5.4 (m, C-6-H); mass spectrum *m/e* 384 (M^+-60), 263, 145].

Compound 3 (14.98 g) is dissolved in 250 ml of diethyl ether and

3 drops of fuming nitric acid is added. The solution is cooled in an ice bath and 90 ml of fuming nitric acid is added over a 30-minute period. The reaction is stirred for 1 hour and then diluted with diethyl ether. The mixture is washed with 2 $\times$ 150-ml and 1 $\times$ 230-ml portions of 5% sodium bicarbonate, and the phases are allowed to separate. The upper (ether) phase is washed with a saturated NaCl solution (8 $\times$ 200 ml). If mass spectrometry shows that the reaction is incomplete, then the ether solution is evaporated to dryness and the nitration is repeated as described above. The crude product is dissolved in 300 ml of acetic acid plus 30 ml of water. Thirty grams of zinc dust is added over a half-hour period. The solution is refluxed and an additional 10 g of zinc dust is added. After refluxing for 3 hours the solution is cooled, diluted with water and diethyl ether, and reextracted once with water. The ether phase is concentrated to about 70 ml with a flash evaporator. An equal quantity of ethanol is added along with additional water to the hot solution. On cooling (7.96 g), white crystals of methyl-3β-acetoxy-6-oxo-25,(5α) homocholanate are obtained [m.p. 140–145°; NMR ($CDCl_3$) δ 2.0 (s, —$OCOCH_3$), 3.7 (s, $COOCH_3$), 4.7 (m, C-3-H); IR ($CHCl_3$) 1735, 1710, 1260 cm^{-1}; mass spectrum m/e 460 (M^+ $C_{28}H_{44}O_5$) and 400, 385, and 284].

The ketal is formed by dissolving 7.6 g of compound in 270 ml of distilled Skellysolve B. One-hundred milligrams of p-toluene sulfonic acid monohydrate and 30 ml of ethylene glycol are added. The solution is refluxed for 20 hours with a Dean-Stark trap to remove water. After cooling and the addition of 0.3 g of sodium acetate, the Skellysolve B layer is decanted and diluted with 100 ml of diethyl ether. The organic phase is extracted twice with 100-ml portions of 2% sodium acetate and twice with water. The ethylene glycol layer is added to the sodium acetate washings and the mixture extracted with 200 ml of diethyl ether. The ether layer is washed twice with water and combined with the organic extracts. On evaporation, 7.43 g methyl-3β-acetoxy-6-ethylenedioxy-25,(5α) homocholanate (5) is obtained. This is chromatographed on neutral alumina AG-7 minus 200 mesh (100 g in a 3.0-cm-diameter column) prepared and eluted with diethyl acetate (25-ml fractions are collected). Compound 5 is found in fractions 4–10 [NMR ($CDCl_3$) δ 2.0 (s, —$OCOCH_3$), 4.7 (m, C-3-H); IR ($CHCl_3$) 1735, 1710, 1260 cm^{-1}, mass spectrum m/e 460 (M^+), 400, 385, 384].

Addition of the 25-Hydroxyl Group. The Grignard reagent is prepared by adding 2.39 g of magnesium shavings (which have been dried overnight at 115°) in a three-necked round-bottom flask. To the flask fitted with a dropping funnel, and a reflux condenser 5.95 ml of iodomethane in 15 ml of diethyl ether is added over a half-hour period. After refluxing for 20 minutes almost all of the magnesium is dissolved. Compound 5 (7.33 g) is dissolved in 25 ml of diethyl ether and added over a half-hour period to the Grignard reagent. A white solid begins to form immediately. At the end of the addition, the dropping funnel is rinsed with diethyl ether. The reaction mixture is refluxed for 3 hours and then poured into 200 ml of 20% ammonium chloride and diluted with additional diethyl ether. The ether layer is separated and washed with 100 ml of 2% sodium acetate. The 20% ammonium chloride layer is washed with diethyl ether (4 × 100 ml). The combined ether fractions are washed once more with 100 ml of 2% sodium acetate and then twice with 100 ml of water. The ether phase is dried *in vacuo,* and the 25-hydroxy compound is crystallized from ethyl acetate–heptane to give 6.35 g of 6-ethylenedioxy-5α-cholestane-3β,25-diol (6) [m.p. 174–177°; NMR ($CDCl_3$) δ 1.2 [s, $C(CH_3)_2OH$], 3.6 (m, C-3-H), 3.9 (m, ketal); mass spectrum *m/e* 462 (M^+), 307, 183, 99].

Addition of the 1α-Hydroxyl Function. The dihydroxy compound (6) (6.35 g) is dissolved in 20 ml of pyridine and then added to ice-cold chromium trioxide pyridine complex prepared by dissolving 9.1 g chromium trioxide in 90 ml of pyridine. An additional 45 ml of pyridine is used to aid in the transfer. The reaction mixture is stirred for 11 hours at room temperature after which it is diluted with an equal volume of ethyl acetate and filtered through a 50-g Celite column (6 cm diameter) using 500 ml of ethyl acetate as eluate. The recovered effluent is then filtered through a column (4 cm diameter) containing 100 g of neutral alumina activity 1, slurried, and eluted with ethyl acetate and 800 ml of effluent is collected. The solvent is evaporated and the solid is taken up in ethyl acetate and filtered through a column (3 cm diameter) containing 100 g of neutral alumina as described above. On evaporation of the solvent, 5.47 g of 6-ethylenedioxy-25-hydroxy-5α-cholestan-3-one (7) [NMR ($CHCl_3$) δ 1.2 [s, —$C(CH_3)_2OH$], 2.3 (m, C-2-H and C-4-H), 3.9 (m, ketal); IR ($CHCl_3$) 3600, 3430, 1710, 910 cm^{-1}; mass spectrum *m/e* 460 (M^+), 307, 99] is obtained.

Compound 7 (2.5 g) is then dissolved in 50 ml of tetrahydrofuran, and 0.65 g of acetamide is added. The solution is warmed to 70° and 1 drop of HBr and 3 drops of acetic acid are added. Bromine (0.87 g) dissolved in 2 ml of CCl_4 is added dropwise, allowing the solution to decolorize between successive additions. The reaction is rapidly cooled on ice and filtered through a sintered glass filter using 25 ml of ethyl acetate to aid in transfer. The filtrate is purified by passing it through a column (2 cm diameter) containing 25 g of neutral alumina and using ethyl acetate as the eluent. A total of 250 ml is collected and evaporated to dryness to yield a crude product containing 2-bromo-6-ethylenedioxy-

25-hydroxy-5α-cholestan-3-one (8) and starting material. This material is dissolved in 15 ml of collidine and flushed with nitrogen for a half hour and then refluxed for 2 hours under nitrogen. The mixture is cooled, taken up in diethyl ether, extracted twice with water, and the ether layer evaporated, leaving approximately 5 ml of a dark oily liquid. This residue is dissolved in 5 ml of chloroform–Skellysolve B (1:1) and applied to 300 g of Sephadex LH-20 (3 × 70-cm glass column), slurried, and developed in the same solvent. Ten-milliliter fractions are collected, and fractions 29–56 contain the desired unsaturated ketone. The starting material is also recovered and recycled to yield more of the unsaturated ketone. The products are combined, dissolved in ether, and cooled. A precipitate consisting mainly of the bromo compound is collected and recycled to yield more of the Δ^1 compound. The supernatant is concentrated, dissolved in chloroform–Skellysolve B(1:1), applied to a 200-g Bio-Beads SX-8 column, slurried, and developed in the same solvent. A total of 4.93 g of 6-ethylenedioxy-25-hydroxy- 5α-cholest-1-en-3-one (9) is recovered between 290–340 ml [NMR ($CDCl_3$) δ 1.2 [s, —$C(CH_3)_2OH$], 3.9 (m, ketal), 5.8 (d, $J = 10$; C-2-H), 7.05 (d, $J = 10$; C-1-H); mass spectrum m/e 458 (M^+), 307, 99].

Compound 9 (4.38 g) is dissolved in 95 ml of dioxane, and 23 ml of 1 N sodium hydroxide is added. Epoxidation is accomplished by the addition of 8 ml of 30% hydrogen peroxide over a period of 1 hour. After 8 hours another 1 ml of hydrogen peroxide is added and the reaction is continued for another 12 hours. The reaction mixture is diluted with water and diethyl ether; the ether layer is washed with water and the combined aqueous layers are reextracted with diethyl ether. The combined ether extracts are combined to yield 2.63 g of 1α,2α-epoxy-6-ethylenedioxy-25-hydroxy-5α-cholestan-3-one (10).

The crude epoxide (2.6 g) is dissolved in 42 ml of diethyl ether, and 1.2 g of lithium aluminum hydride dispersed in 30 ml of diethyl ether is added over a 30-minute period. The resulting mixture is refluxed for 6 hours (after 4 hours, an additional 0.5 g of lithium aluminum hydride is added). The reaction mixture is cooled and ethyl acetate is added to decompose any excess hydride. The solution is diluted with diethyl ether and saturated aqueous potassium sodium tartarate. The ether layer is washed once more with tartarate and then water. The first aqueous tartarate phase is extracted with diethyl ether and twice with chloroform. The ether and chloroform layers are combined, evaporated to dryness, and dissolved in 5 ml of chloroform–Skellysolve B (1:1) for application to a Sephadex LH-20 column (300 g) as previously described. The trihydroxy compound recovered between 1500 and 1830 ml of effluent is dried and applied to a silicic acid column, slurried in Skellysolve B, and eluted with a linear gradient of 25% acetate and Skellysolve B to 100% ethyl acetate. The 3α-hydroxy compound (11) elutes between 590 and 720 ml and precedes the 3β-hydroxy compound (13). Because the 3β-hydroxy compound is the desired product, one can recycle the 3α compound in the following manner to obtain more (13). Compound 11 (0.177 g) is dissolved in 11 ml of *t*-butanol and 27 ml of pyridine. *N*-Bromosuccinimide (0.086 g) is added. The reaction mixture is stirred in the dark at room temperature for 24½ hours. The solution is then diluted with ether and washed 3 times with water. The ether layer is dried over anhydrous sodium sulfate, evaporated to dryness, and applied to a 40 g silicic acid column, slurried with Skellysolve B, and eluted with a linear gradient beginning with 25% ethyl acetate : Skellysolve B, and ending with 100% ethyl acetate. Eighty-four milligrams of the 3-ketone (compound 12) elutes between 280 and 360 ml. This is dissolved in 5 ml of isopropyl alcohol and reduced with an excess of sodium borohydride for 1½ hours at room temperature. Diethyl ether is added and the solution is washed 3 times with water and dried over anhydrous sodium sulfate. The ether extract is dried on a flash evaporator and applied to a 40-g silicic acid column and eluted with a linear gradient beginning with 75% ethyl acetate : Skellysolve B to 100% ethyl acetate. The desired 3β-hydroxy compound (13) elutes between 328 and 392 ml.

Removal of the 6-Ketal and Generation of the Δ^5. Dihydroxy compound 13 (110 mg) is dissolved in 5 ml of acetic anhydride and 0.3 ml of pyridine and heated at 85° for 24 hours. The reaction mixture is cooled, diluted with diethyl ether, and washed twice with water. The ether layer is extracted 4 times with sodium bicarbonate followed by a final extraction with water. The organic layer is evaporated to dryness, dissolved in 3 ml of chloroform–Skellysolve B (1:1), and applied to a Sephadex LH-20 column (3 × 70 cm containing 110 g), which is slurried and developed with the same solvent. 6-Ethylenedioxy-5α-cholestan-1α,3β,25-triacetate (14) (110 mg), which elutes between 120 and 170 ml, is recovered after flash evaporation of the solvent, and dissolved in 7.5 ml of methanol. To the solution is added 7.5 ml of 95% ethanol containing 10 mg *p*-toluenesulfonic acid and the solution is stirred at 60° for 4 hours. The reaction is diluted with aqueous bicarbonate and extracted 3 times with diethyl ether. The ether phases are dried by flash evaporation and applied to a Sephadex LH-20 column (2 × 30 cm containing 10 g) in chloroform–Skellysolve B (1:1) (9.8 ml fractions are collected). One-hundred milligrams of the 6-keto compound (15) is recovered in fractions 2, 3, and 4 [NMR ($CDCl_3$) δ 1.4 [s, —$C(CH_3)_2OH$], 1.96, 2.03, 2.12 (Singlets 1,3,25 acetates) 5.0 (m, C-1-H and C-3-H); mass spectrum *m/e* 560 (M^+) 500, 440, 380].

The 6-ketone (90 mg) is dissolved in 15 ml of isopropyl alcohol and reduced with 20 mg of sodium borohydride in 2 ml of isopropanol. Four hours later at 25° the reaction is diluted with water (pH 4) and extracted 3 times with diethyl ether. The ether phases are combined, washed with aqueous sodium bicarbonate, and dried over anhydrous sodium sulfate. The product is dried, dissolved in chloroform–Skellysolve B (1:1), and applied to a 10 g Sephadex LH-20 column as described above. 1α,3β,25-Triacetoxy-5α-cholestan-6β-ol (16) is recovered (83 mg) in tubes 2–6.

Twenty milligrams of this (16) is dissolved in 0.8 ml of pyridine and brought to 4° in an ice bath. Phosphorus oxychloride (25 μl) is added dropwise while vigorously stirring the reaction over a period of 30 seconds. Two minutes later the reaction is brought to 25° and stirred for an additional 30 minutes. (The reaction is followed to completion by gas-liquid chromatography, the dehydrated product precedes the 6-OH compound.) The same procedure is repeated for the remaining 50 mg of compound 16. The reaction mixture is diluted with ice water (pH 4) and extracted 5 times with diethyl ether. The ether phases are combined, dried, and redissolved in chloroform–Skellysolve B (1:1) and applied to a 10 g Sephadex LH-20 column as described above. Fractions 2–5 contain 50 mg of 5-cholestene-1α,3β,25-triacetate (17) [NMR ($CDCl_3$) δ 1.43 [s, $C(CH_3)_2OH$], 1.98, 2.05, 2.10 (singlets, 1,3,25-acetates), 5.1 (m, C-1 and C-3-H), 5.6 (m, C-6-H); mass spectrum m/e 544 (M^+), 484, 424, 364].

Synthesis and Photoisomerization of 1α,3β,25-Triacetoxycholesta-5,7-diene (19). Fifty milligrams of 1α,3β,25-triacetoxycholesterol (17) is dissolved in 1.5 ml of benzene–Skellysolve B (1:1) and warmed to 72° in an oil bath. *N,N'*-Dimethyldibromohydantoin (13.5 mg) is added and the reaction continued for 10 minutes. After 10 minutes the reaction mixture containing a white flocculent precipitate is placed on ice for 2 minutes and then filtered. The filtrate is collected and the precipitate is washed twice with cold Skellysolve B. The filtrates are combined, dried under nitrogen, and dissolved in 0.4 ml of dry xylene. The 7-bromo derivative (18) is added dropwise over a period of 3 minutes to a solution of 0.1 ml trimethyl phosphite and 0.3 ml of xylene maintained at 135°. The reaction is kept at this temperature for 90 minutes and then the solvents are evaporated under nitrogen at 65°. The residue is dissolved in Skellysolve B and applied to a 1 × 10-cm glass column containing a mixture of 3 g of silver nitrate impregnated silicic acid and 1 g of Celite. The column is eluted batchwise with 40 ml of Skellysolve B, 50 ml of 2%, 60 ml of 6%, and 150 ml of 10% diethyl ether-Skellysolve B. The desired 5,7-diene (19) is recovered in the 10% diethyl ether-Skellysolve B effluent. The product showed the characteristic ultraviolet absorption spectra of λ_{max} = 254, 281, 272 nm for the 5,7-diene chromophore and the mass spectrum showed *m/e* 544 (M^+).

The 5,7-diene (19) (2.2 mg) is dissolved in 300 ml of diethyl ether, irradiated for 45 seconds in a jacket around a double-walled water-cooled quartz immersion well. The Hannovia high pressure quartz mercury vapor lamp model 654A is ignited for 3 minutes before being placed in the immersion well. During the irradiation the ether is flushed continually with a stream of nitrogen and stirred vigorously. Immediately after the irradiation the ether is evaporated and the residue applied to a 3-g silver nitrate impregnated silicic acid column containing 1 g of Celite in 4% diethyl ether in Skellysolve B. The column is eluted batchwise with 50 ml of 4%, 200 ml of 8%, and 50 ml of 12% diethyl ether–Skellysolve B. The 1α,3β-25-triacetoxyprevitamin D_3 (20) is recovered in the 8% effluent.

The previtamin D_3 (20) (0.1 mg) is saponified and converted to the 5,6-*cis*-triene in 3 ml of 95% ethanol containing 0.15 ml of saturated KOH. Saponification and equilibration is carried out under reflux conditions for 2 hours and is constantly flushed with nitrogen. The sample is extracted with chloroform–water and the aqueous phase is reextraced 3 times with chloroform. Chloroform phases are combined and dried by flash evaporation. The sample is dissolved in 0.05 ml of chloroform–Skellysolve B (1:1) and applied to a 20-g Sephadex LH-20 column (1 $\times$ 60 cm), slurried, and developed with the same solvent. Ten-milliliter fractions are collected. The product 1α,25-dihydroxyvitamin D_3 (21) is recovered in fractions 45–55. The product showed the characteristic ultraviolet absorption spectrum λ_{max} = 265 and λ_{min} = 228 and the mass spectrum showed *m/e* 416 (M^+), 398, 380, 152, 134. The biological activity of this synthetic product is identical to biosynthetic material.

Chromatographic Methods

Sephadex LH-20 Chromatography.[9] During the past decade the most widely used method for separating vitamin D compounds was the silicic acid column method; this has been described adequately in a previous volume.[2] In 1971 a powerful new chromatographic system involving Sephadex LH-20 (a hydroxypropylether derivative of Sephadex G-25, Pharmacia Corp., Piscataway, N.J.) packed in and developed with a mixed solvent of chloroform–Skellysolve B was introduced. This system offers many advantages over silicic acid absorption chromatography. These columns are simple to develop since only one solvent is needed

[9] M. F. Holick and H. F. DeLuca, *J. Lipid Res.* **12,** 460 (1971).

throughout the chromatography and the volume of solvent used is about 30% of that used in silicic acid chromatography. Besides being easy to prepare, the same column may be used repeatedly and the recoveries from the column approximate 100%. Our chief use of Sephadex LH-20 chromatography is in the resolution of vitamin D metabolites more polar than 25-OH-D_3 and in the purification of many of the intermediates in the synthesis of 1α,25-$(OH)_2D_3$, since many of these intermediates are unstable on either alumina or silicic acid.

Unlike silicic acid which must be activated before use, Sephadex LH-20 only needs to be slurried in the appropriate solvent and equilibrated for about 8–12 hours. The slurry is poured into a glass column containing 30 ml of solvent. The stopcock is opened at the same time the slurry is poured and allowed to settle by gravity with free solvent flow. The column is washed with at least 50 ml of solvent before the sample is applied. Air pressure (1–2 lb) may be applied to the column if the flow rate is less than 0.5 ml/minute. Figure 10 illustrates the effect

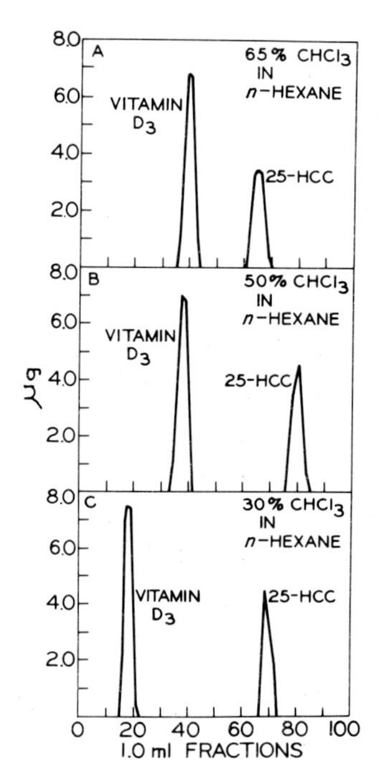

FIG. 10. Separation of vitamin D_3 and 25-OH-D_3 on Sephadex LH-20 in various solvent mixtures. Column 1.1×60 cm containing 20 g of Sephadex LH-20 was used. Solvent $CHCl_3$–Skellysolve B (A) 65:35, (B) 1:1, (C) 3:7. (Reproduced with the kind permission of the publisher, Holick and DeLuca.[9])

of different solvent mixtures of chloroform in Skellysolve B on the Sephadex LH-20 column separation of 25-OH-D_3 from vitamin D_3. The retention volume of 25-OH-D_3 relative to vitamin D_3 increases as the percentage of chloroform in Skellysolve B decreases. However, it must also be noted that the gel does not swell as well as the proportion of chloroform to Skellysolve B decreases. Chloroform–Skellysolve B (1:1) is a suitable solvent for Sephadex LH-20 and is the method of choice for purifying vitamin D, 25-OH-D_3, 24,25-$(OH)_2D_3$, 1,25-$(OH)_2D_3$, as well as intermediates in the chemical synthesis of 1α,25-$(OH)_2D_3$. As an approximation it can be assumed that the number of grams of Sephadex LH-20 in the column is equal to the void volume of the column. A solvent system of chloroform–Skellysolve B (65:35) is especially useful for the study of dihydroxyvitamin D metabolites. Figure 11 illustrates the capability of this chromatographic system in the separation of the various dihydroxy metabolites of vitamin D_3 in rat plasma as compared to the silicic acid system. For the resolution of metabolites more polar than

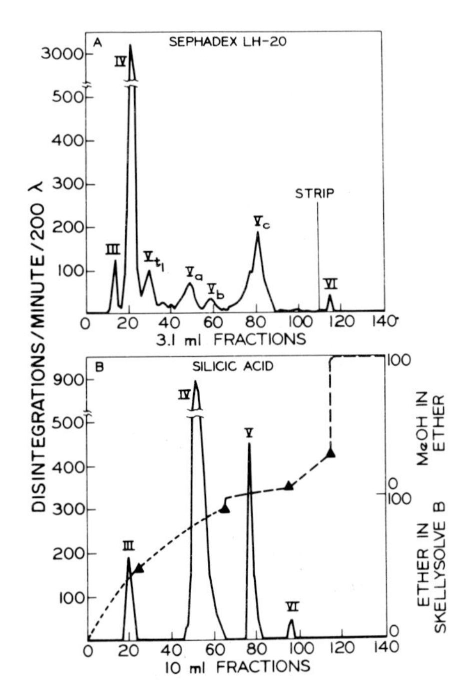

FIG. 11. Comparison of the chromatography of plasma lipid extract from rats that received 2.5 μg [1,2-^{3}H]vitamin D_3 24 hours earlier. (A) Sephadex LH-20 (20 g) in $CHCl_3$–Skellysolve B (65:35), (B) silicic acid (25 g). (Reproduced with the kind permission of the publisher, Holick and DeLuca.[9])

1,25-$(OH)_2D_3$ a solvent system of methanol–chloroform–Skellysolve B (2:75:23) is employed as demonstrated in Fig. 12.

Silver Nitrate Impregnated Silicic Acid Chromatography. Recently silver nitrate impregnated silicic acid has been used successfully in the purification of dihydrotachysterol from other reduction products of vitamin D_3 and in the purification of the 1α,3β,25-triacetaoxycholesta-5,7-diene from the 4,6-diene and the 1α,3β,25-triacetoxyprevitamin D_3 from the other irradiation products. Because this system separates compounds based on its degree of unsaturation this method works best when the hydroxyl functions on the compound are protected by an acetate or some other ester group.

Silver nitrate impregnated silicic acid is prepared by slurring 1 part silicic acid to 2 parts of 50% (w/v) aqueous silver nitrate at 100° for 1 hour. The product is collected by Büchner funnel filtration and dried at 110° for 24 hours. Because of the product's granularity, which normally slows the flow rate of the column considerably, 0.2 parts of Celite is mixed with silver nitrate impregnated silicic acid before slurring it with the appropriate solvent for chromatography purposes.

Chromatography of 1α,3β,25-Triacetoxy-cholesta-5,7-diene. Four milligrams of 1α,3β,25-triacetoxycholesta-5,7-diene is successfully separated from its 4,6-diene isomer on a 5-g silver nitrate impregnated silicic acid column containing 1 g of Celite. The column is eluted batchwise with 40 ml of Skellysolve B, 50 ml of 2%, 60 ml of 6%, and 150 ml of 10% diethyl ether-Skellysolve B. The desired 5,7-diene is recovered in 10% diethyl ether–Skellysolve B effluent.

Chromatography of the Irradiation Mixture from 1α,3β,25-Triacetoxycholesta-5,7-diene and Recovery of the Corresponding Pre D. After ultraviolet irradiation of the 5,7-diene the corresponding product is dried and applied to a 5-g silver nitrate impregnated silicic acid column con-

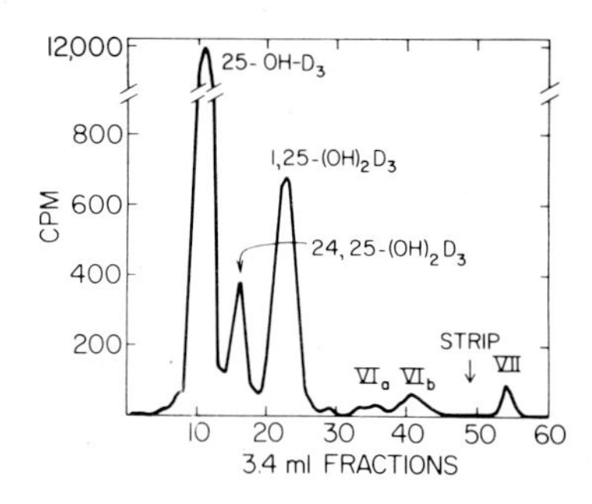

FIG. 12. Sephadex LH-20 column (1 × 60 cm packed and developed in $CHCl_3$–Skellysolve B–MeOH, 75:23:2) profile of a lipid plasma extract from rats that had received 0.25 μg [26,27-^{3}H]25-OH-D_3 48 hours earlier.

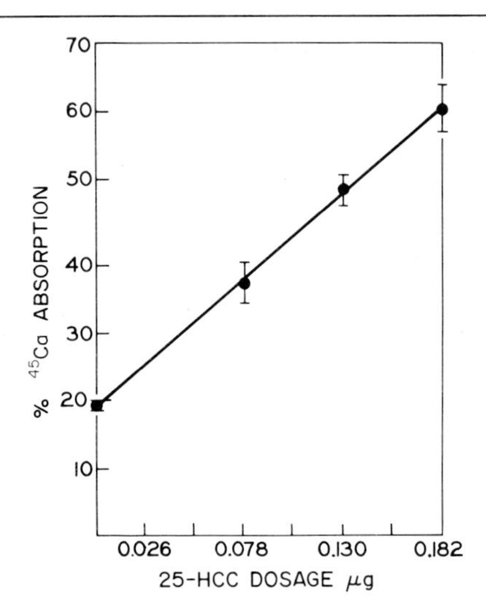

FIG. 13. Intestinal calcium absorption test in chicks. A single dose of 25-OH-D_3 (25-HCC) was given to chicks and 24 hours later the absorption of ^{45}Ca from an *in situ* ligated loop was measured as described in the text.

taining 1 g of Celite in 4% diethyl ether–Skellysolve B. The column is eluted batchwise with 50 ml of 4%, 200 ml of 8%, and 50 ml of 12% diethyl ether-Skellysolve B. The 1α,3β,25-triacetoxyprevitamin D_3 is recovered in the 8% effluent.

Chick Bioassay Methods

In addition to the methods already described in a previous volume,[2] two methods of measuring intestinal calcium absorption using chicks have been widely used. One involves the injection of calcium-45 into the duodenum and subsequent measurement of the rate of appearance of calcium-45 in the blood. This method suffers from two disadvantages. First it lacks precision and second the calcium-45 level in the blood is subject to modification by bone and other tissues.

In the author's laboratory, a measurement of disappearance of calcium-45 from ligated loops of intestine *in situ* has been used with considerable success.[10] The precision of the method is illustrated by Fig. 13. Day-old chicks (white leghorn or similar breed) are maintained in a 38° incubator and fed the rachitogenic diet (see table) for 3 weeks. Chicks weighing between 170 and 190 g are used. The vitamin D preparation

[10] J. Omdahl, M. Holick, T. Suda, Y. Tanaka, and H. F. DeLuca, *Biochemistry* **10,** 2935 (1971).

may be given orally in either vegetable oil or propylene glycol or intravenously in less than 0.05 ml of 95% ethanol. If vitamin D_3 is being tested, it must be given either 24 or 48 hours before the test. 25-OH-D_3 must be given 12–24 hours before and 1,25-$(OH)_2D_3$ given 6 hours before the test. All chicks are fasted 15 hours prior to the test. They are anesthetized with chloral hydrate (35 mg/100 g). A 10-cm portion of the duodenal loop is ligated and its lumen rinsed with 5 ml of a phosphate-free bicarbonate solution (120 m*M* NaCl, 4.9 m*M* KCl, and 9.2 m*M* $NaHCO_3$, pH 7.0, adjusted with CO_2) after which the proximal ligature is tied off. The distal ligature is tightened around a blunted 18 gauge needle through which 0.2 ml of phosphate-free calcium-45 bicarbonate solution (0.5 μCi of ^{45}Ca/ml; 10 m*M* $CaCl_2$) is injected into the lumen of the duodenal loop. Following withdrawal of the needle the ligature is secured and the loop replaced in the peritoneal cavity for 20 minutes at which time the animal is killed. The loop is rapidly excised and subsequently dry-ashed in porcelain crucibles at 600° for 48 hours. The ashed product is dissolved in 2 *N* HCl, neutralized with 2 *N* Tris, and 0.05 ml aliquots spotted on filter paper disks (2.3 cm diameter). Radioactivity is estimated by placing the paper disk in a 20-ml scintillation counting vial, adding 10 ml of toluene counting solution,[11] and counting in a liquid scintillation spectrometer. Results are expressed as percent ^{45}Ca absorption and are calculated using the following formula: $\% \ ^{45}Ca_{abs} = [1 - {}^{45}Ca_R/{}^{45}Ca_A]$ (100), where $^{45}Ca_R$ is the amount of ^{45}Ca remaining in the duodenal loop following the *in situ* incubation and $^{45}Ca_A$ is the amount of ^{45}Ca initially added to the loop preparation.

The *in situ* method for detection of changes in calcium absorption is quite sensitive (Fig. 13). As little as 195 pmoles of 25-OH-D_3 gives a significant increase in calcium absorption. The retrograde movement of calcium (i.e., blood to intestine) ranges between 0.7 and 1.4% of the calcium placed in the intestine, a factor which adds negligible error to the calculations.

Competitive Protein Binding Assay for Vitamin D_3 and 25-OH-D_3

With the availability of [^{3}H]vitamin D_3 and [^{3}H]25-OH-D_3 (New England Nuclear, Boston, Mass. and Amersham Searle, Chicago, Ill.) and the recognition of specific binding proteins, competitive binding assays have been developed for vitamin D_3 and 25-OH-D_3 which are sensitive and easily executed. Haddad and associates[12] utilize a kidney supernatant

[11] P. F. Neville and H. F. DeLuca, *Biochemistry* **5**, 2201 (1966).
[12] J. G. Haddad, Jr. and K. J. Chyu, *Biochim. Biophys. Acta* **248**, 471 (1971).

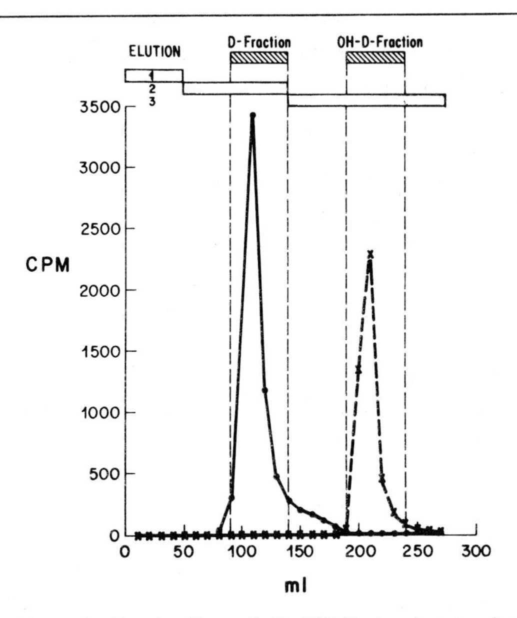

FIG. 14. Separation of vitamin D_3 and 25-OH-D_3 by batch elution from silicic acid columns as described in the text. (Reproduced with the kind permission of the publisher, Belsey *et al.*[13])

protein while Belsey *et al.*[13] use the binding proteins in the serum of vitamin D-deficient rats. These methods are particularly useful as a potential diagnostic tool in measuring vitamin D and 25-OH-D_3 levels in human blood. The following method has recently been successfully applied although certain improvements are already in progress.[13] It is essential to realize that the vitamin D_2 series binds with different affinity to the proteins than does the vitamin D_3 series. Great care must be exercised in interpretation of results when vitamin D_3 or 25-OH-D_3 tracers are used in the measurements of the vitamin D_2 and 25-OH-D_2 levels in blood.

Assay Procedure. A blood serum sample (0.1–0.2 ml) with 0.001 μCi of [^{3}H]vitamin D_3 and 0.001 μCi of [^{3}H]25-OH-D_3 (New England Nuclear) added to monitor recovery is subjected to chloroform–methanol extraction. Vitamin D_3 and 25-OH-D_3 are separated by chromatography on silicic acid using batch elution with (1) 10%, (2) 30%, and (3) 80% diethyl ether in Skellysolve B. The solvents are evaporated under nitrogen and the vitamin D_3 and 25-OH-D_3 fractions (Fig. 14) are redissolved in 0.1 ml of absolute ethanol. Diluted lipoprotein (3.5 ml) is added to

[13] R. Belsey, H. F. DeLuca, and J. T. Potts, Jr., *J. Clin. Endocrinol. Metab.* **33,** 554 (1971).

each sample, mixed well, and a 1.0-ml aliquot counted for recovery. Diluted lipoprotein (2.5 ml) (see below) is added to each standard (in 0.1 ml absolute ethanol). One milliliter of diluted lipoprotein with 0.00875 μCi [^{3}H]vitamin D_3 is added to each sample and standard. Four 0.8-ml aliquots are put into 10 × 75 glass test tubes. Five microliters of vitamin D-deficient rat serum is added to two tubes (final dilution 1:1,000). After 10 days at 4°, the lipoprotein is precipitated with heparin and $MnCl_2$. After standing 15 minutes in the cold, the mixture is centrifuged for 15 minutes at 4°. The supernatant is decanted and counted.

Preparation of Lipoprotein. β-Lipoprotein can be prepared from human plasma by selective precipitation with heparin and $MnCl_2$. Forty microliters of 5% heparin and 50 μl of 1 *M* $MnCl_2$/ml of plasma is added and mixed well. After standing in the cold for 30 minutes, the mixture is centrifuged at 3,000 rpm for 1 hour at 4°. The precipitate is washed with distilled water and redissolved with 0.3 sodium citrate. The volume of the solution was adjusted to the original plasma volume with 0.01 barbital–acetate buffer, pH 8.6. For use in the assay, the lipoprotein solution is further diluted with the same buffer.

Standard Curves. Increasing amounts of nonradioactive vitamin D_3 or 25-OH-D_3 produces progressive displacement of the [^{3}H]vitamin D_3 from the "bound" fraction (Fig. 15). It is possible to detect 1–2 ng/ml and comfortably measure up to 400 ng/ml of both vitamin D_3 and 25-OH-D_3. Controls with no added binding protein ("damage" controls) indicate that the phase-separation technique precipitates 95% of the added [^{3}H]vitamin D_3. The standard curves are reproducible from assay to assay with little change in either the detection limits or the shape of the curve.

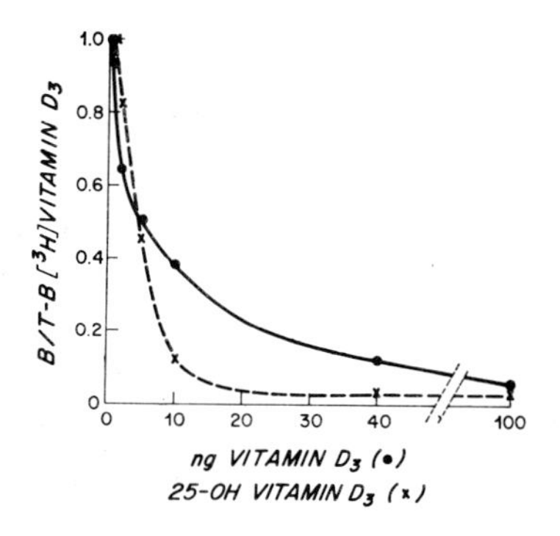

FIG. 15. Standard curve for competitive binding assay of either 25-OH-D or vitamin D as described in the competitive binding assay in the text. (Reproduced with the kind permission of the publisher, Belsey *et al.*[13])

[47] Methods for Determining the Conversion of L-Thyroxine (T_4) to L-Triiodothyronine (T_3)[1]

By MARTIN I. SURKS and JACK H. OPPENHEIMER

I. T_4 to T_3 Conversion in the Rat

A. Approach

The conversion of T_4 to T_3 in the rat is determined by measuring the rate of change of radioactive T_3 in the whole animal after the injection of tracer T_4.[1a] The rate of change of tracer T_3 in the body is equivalent to the net effect of conversion of T_4 to T_3 and the irreversible metabolism of the T_3 which is generated. Thus,

$$\frac{d(T_3)^*}{dt} = k(T_4)^* - \lambda_3(T_3)^* \tag{1}$$

where $(T_4)^*$ and $(T_3)^*$ represent the total body pools of radioactive T_4 and T_3 at time t after the injection of radioactive T_4; k, the fractional rate of conversion of radioactive T_4 to T_3; and λ_3, the irreversible fractional removal rate of T_3. The irreversible fractional removal rate of T_4 is described by

$$\frac{d(T_4)^*}{dt} = -\lambda_4(T_4)^* \tag{2}$$

where λ_4 is the fractional rate of removal of T_4.

Solution of Eqs. (1) and (2) with rearrangement of terms results in the following expression for k:

$$k = \frac{(T_3^*/T_4^*)(\lambda_3 - \lambda_4)}{1 - e^{-(\lambda_3-\lambda_4)t}} \tag{3}$$

Since the 3′- and 5′-iodine atoms are randomly labeled in commercially available phenolic-ring labeled T_4 preparations, for each molecule of radioactive T_4 monodeiodinated to T_3, a molecule of nonradioactive T_4 must also be converted to T_3. Evidence supporting this conclusion has been presented.[1a] Thus, the fractional conversion of nonradioactive T_4 to

[1] Supported by USPHS Grant 9 RO1 AM 15421-12 and US Army Contract DA-49-193-MD-2967. M.I.S. is a Research Career Development Awardee from the USPHS (KO4 AM 19502-O1A1).

[1a] H. L. Schwartz, M. I. Surks, and J. H. Oppenheimer, *J. Clin. Invest.* **50,** 1124 (1971).

T_3 is equivalent to $2k$, and a conversion ratio (CR) which represents the percentage contribution of T_4 to T_3 conversion to the net turnover of T_4 can be defined

$$CR = \frac{2k(100)}{\lambda_4} \tag{4}$$

Substituting from Eq. (3), the final expression for the conversion ratio is

$$CR = \frac{200(T_3^*/T_4^*)(\lambda_3/\lambda_4 - 1)}{1 - e^{-(\lambda_3-\lambda_4)t}} \tag{5}$$

Thus, measurement of T_4 to T_3 conversion in the rat requires the experimental determination of λ_3, λ_4, and T_3^*/T_4^* at a time t after injection of radioactive T_4.

B. Experimental Procedures

1. Measurement of $(T_3)^*/(T_4)^*$

L-Thyroxine labeled with $[^{125}I]$iodine ($[^{125}I]T_4$) is obtained commercially (Tetramet-^{125}I, Abbott Laboratories, North Chicago, Illinois) and purified by descending paper chromatography on Whatman 3 MM paper in a *tert*-amyl alcohol–2 *N* ammonia–hexane (5:6:1) solvent system (TAA). Thirty to fifty microcuries of the purified $[^{125}I]T_4$ in 4% albumin is injected intravenously via the tail vein of rats weighing between 100 to 150 g. One-half a milligram of sodium iodide is injected subcutaneously daily in order to prevent thyroidal reutilization of radioactive iodide released from hormonal degradation.. The rats are killed 48–96 hours after injection of $[^{125}I]T_4$. The intestinal contents are removed and the entire carcass is then homogenized with 1–2 volumes of water. Two 5-g portions of the carcass homogenate are repeatedly extracted with 5-ml volumes of 95% ethanol until at least 85–90% of the ^{125}I counts have been removed. The counting rate of the combined ethanol extract is determined and T_3 labeled with ^{131}I ($[^{131}I]T_3$) is added to the extract in amounts approximately equal to 3–5% of the ^{125}I radioactivity. The $[^{131}I]T_3$ is obtained from the same commercial source and purified by paper chromatography as described above for $[^{125}I]T_4$. The added $[^{131}I]T_3$ is employed as a radioactive marker for this hormone, allows corrections for losses during subsequent analytic procedures, and enables the measurement of the $[^{125}I]T_3$-to-$[^{131}I]T_3$ ratio as a criterion for the identification of any T_3 derived from the injected $[^{125}I]T_4$.

The $[^{125}I]T_3$ derived by conversion from the injected $[^{125}I]T_4$ and the $[^{131}I]T_3$ added to the carcass extract are then purified by paper and thin-layer chromatography until a constant isotopic ratio in the T_3 area is achieved. The ethanolic extract is concentrated *in vacuo* at a temperature less than 50° and applied along with carrier T_4 and T_3 to Whatman 3 MM paper. Descending chromatography is carried out for 16–20 hours in a TAA solvent system. The T_3 area is visualized under ultraviolet light, cut out, and eluted with methanol–ammonia (95:5). The eluate from the T_3 area is concentrated as indicated above and rechromatographed in the same solvent system. This procedure is repeated for a total of three cycles of paper chromatography. The eluate of the T_3 area from the third chromatogram is then applied to thin-layer silica gel sheets (Eastman Kodak Company) and ascending chromatography is performed in a toluene–acetic acid–water (2:2:1) solvent system.[2] The thin-layer chromatogram is used to remove $[^{125}I]$tetraiodothyroacetic acid which is generally found in the carcass extract. In the thin-layer system, this compound has an R_f of approximately 0.6, whereas the iodothyronines and iodide remain at the point of application. After the chromatogram is developed, the area of application is eluted with methanol–ammonia (95:5), and the extract is concentrated *in vacuo*. This extract is then divided into three equal portions, each of which is chromatographed on paper in a different solvent system: (1) TAA, descending; (2) *n*-butanol–*p*-dioxane–2 *N* ammonia (4:1:5), descending; and (3) *n*-butanol–ethanol–0.5 *N* ammonia (5:1:2), ascending. After 16–20 hours, the chromatograms are dried and the iodothyronines visualized by staining with diazotized sulfanilic acid.[3] The T_3 area of these chromatograms is cut into 0.5 cm sections and the counting rates of $[^{131}I]T_3$ and $[^{125}I]T_3$ are determined in a scintillation spectrometer. These counting rates are corrected for isotopic decay for the time which has elapsed since the initial extraction of the carcass homogenate. The ratio of the ^{125}I to ^{131}I across the T_3 peak is then calculated. The recovery of the $[^{131}I]T_3$ added to the ethanolic extract of the carcass is generally between 50–60%.

Since losses during the analytical procedures of T_3 carrying either radioactive label are proportional, the percent of the total ^{125}I counts in the carcass present as T_3 is equivalent to

$$\frac{r \times [^{131}I]T_3 \times 100}{^{125}I} \tag{6}$$

where r represents the $^{125}I/^{131}I$ ratio in the T_3 area of the final chromato-

[2] M. I. Surks, S. Weinbach, and E. M. Volpert, *Endocrinology* **82,** 1156 (1968).

[3] R. Pitt-Rivers and H. L. Schwartz, *in* "Chromatographic and Electrophoretic Techniques" (I. Smith, ed.), 3rd ed., Vol. I, p. 244. Heinemann, London 1969.

grams; $[^{131}I]T_3$, the cpm of this compound added to the ethanolic carcass extracts; and ^{125}I, the total cpm of this isotope in the original carcass extract.

The fraction of the carcass extract ^{125}I which represents $[^{125}I]T_4$ is determined indirectly. Since $[^{125}I]$iodide is the only major component of the carcass extracts aside from the iodothyronines, the fraction of the carcass extract radioactivity representing $[^{125}I]$iodide is assessed by chromatography of a small aliquot of the original carcass extract in the TAA solvent system. Thus, the fraction of radioactivity in the extract representing T_4 is calculated as the difference between the total extract radioactivity and the $[^{125}I]T_3$ and $[^{125}I]$iodide which are determined directly. The radioactive T_3^*/T_4^* ratio is then readily calculated and substituted into Eqs. (3) and (5).

2. Measurement of λ_3 *and* λ_4

In the animals injected with $[^{125}I]T_4$, the residual body T_4 at the time the animals are killed is calculated from the product of the percent T_4 in the carcass (see above) and the total carcass radioactivity which is determined from a weighed portion of the carcass homogenate. The fractional turnover rate of T_4, λ_4, is then calculated from these measures using the standard isotopic decay equation.

The fractional turnover rate of T_3, λ_3, is determined in a separate group of litter mates paired to the first group insofar as diet and sodium iodide injections are concerned. They are injected intravenously with 20–30 μCi of purified $[^{131}I]$ or $[^{125}I]T_3$ and killed 17 hours after the injection of labeled hormone. At this time, only negligible quantities of tissue- and plasma-labeled iodoproteins are present.[4] After the animals are killed, homogenates of the total carcass are prepared as described above and the counting rate of weighed portions of the homogenates is determined. The residual body T_3 is calculated from the difference of the counting rate of the total carcass homogenate and the fraction of the carcass radioactivity due to radioiodide. The latter is determined from the product of the plasma radioiodide concentration as determined by rapid paper electrophoresis[5] and an iodide space of 48.3 ml/100 g body weight.[6] With the residual body T_3 at 17 hours known, the fractional disappearance of T_3, λ_3, is calculated in the same fashion as λ_4.

[4] M. I. Surks and J. H. Oppenheimer, *J. Clin. Invest.* **48,** 685 (1969).

[5] S. A. Berson and R. S. Yalow, *Ann. N.Y. Acad. Sci.* **70,** 56 (1957).

[6] J. H. Oppenheimer, H. L. Schwartz, H. C. Shapiro, G. Bernstein, and M. I. Surks, *J. Clin. Invest.* **49,** 1016 (1970).

3. *Controls*

Since some T_3 may be generated from T_4 during the preparation of the carcass homogenates, extraction procedures, and chromatographic purification, appropriate control studies are carried out to determine the extent of *in vitro* monodeiodination of $[^{125}I]T_4$. Littermates of the experimental rats are injected with $[^{125}I]T_4$ and killed within 2 minutes after injection of isotopic hormone. Homogenates of their carcasses are prepared. Extraction of the carcass homogenates and purification of the labeled T_3 is carried out in a manner identical to that described above. In our hands, the (T_3^*/T_4^*) ratio of the control rats is 0.0127 ± 0.0022 (SEM), while that of the experimental rats is 0.1000 ± 0.0038.[1] The (T_3^*/T_4^*) ratio used for calculation of conversion ratio is the difference between the ratio in experimental animals and the mean (T_3^*/T_4^*) ratio of the control animals. The mean corrected (T_3^*/T_4^*) ratio is 0.0873 ± 0.0038. Substitution of this ratio, λ_3, λ_4, and t into Eq. (5) results in a mean conversion ratio of $16.9 \pm 1.0\%$ in our laboratory.

II. T_4 to T_3 Conversion in Human Subjects

A. Approach

In human beings, the fractional rate of conversion of T_4 to T_3 is determined by measuring simultaneously the turnover rates of T_4 and T_3 in subjects without endogenous thyroid function who are maintained in the euthyroid state by treatment with synthetic L-thyroxine. Since the only source of T_3 in these subjects is from the metabolism of the administered T_4, the T_3 turnover is equivalent to T_3 converted from T_4. Thus, the T_4 to T_3 conversion ratio (CR), which represents the percentage of the T_4 turnover, converted to T_3 is readily calculated from Eq. (7).

$$CR = \frac{100 \times T_3 \text{ turnover, } \mu M/\text{day}}{T_4 \text{ turnover, } \mu M/\text{day}} \tag{7}$$

Although the ideal subjects for these determinations are patients who are completely athyreotic, we have found no difference in conversion ratio between these subjects and patients who had severe primary hypothyroidism who were euthyroid (undetectable serum thyrotropin concentrations) during treatment with synthetic L-thyroxine.

Iodothyronine turnover rates are calculated as the product of the metabolic clearance rates and mean plasma concentrations of each hormone.

B. Experimental Procedures

1. Measurements of Iodothyronine Concentrations

Serum T_4 concentrations are measured both by displacement analysis[7] and iodine determinations.[8] The average T_4 value of 4–6 plasma samples obtained during each turnover study is taken as the mean plasma T_4 concentration.

The concentration of T_3 is determined by radioimmunoassay after separation of the hormone from the plasma proteins on small Sephadex columns.[9] The Sephadex columns are prepared in the barrels of 3.0-ml syringes (Beckton-Dickinson, Rutherford, New Jersey). They are packed with 3.0 ml of a 10% suspension of Sephadex G-25 fine (Pharmacia) in 0.1 *N* NaOH. After packing, the columns are washed with 2.0 ml 0.1 *N* NaOH and the bottoms are closed by screwing on the disposable needles which have previously been filled with clay. Standard curves are prepared by applying to the columns mixtures of 0.1 ml gravimetrically prepared T_3 solutions [50–1000 pg/0.1 ml 0.3% bovine serum albumin (BSA) in 0.075 *M* barbital buffer, pH 8.6, 0.1 ml $[^{125}I]T_3$ (approximately 75 pg), and 0.1 ml 0.1 *N* NaOH]. For the analysis of plasma, 0.02–0.5 ml plasma is substituted for the T_3 standards. All the $[^{125}I]T_3$ in the assay is added before fractionation on the Sephadex columns. The various mixtures are transferred to the closed Sephadex columns with Pasteur pipets and the tubes in which the mixtures are prepared are washed with 0.1 *N* NaOH. After the washes are added, the columns are opened. When the samples have percolated through the Sephadex, the columns are washed twice with 1.5 ml 0.075 *M* barbital, pH 8.6 (barbital buffer). This procedure results in the elution of the plasma proteins and the equilibration of the columns at pH 8.6. The iodothyronines which are bound to the Sephadex (92–98% of the applied hormones) are then eluted with 0.3% BSA in barbital buffer (BSA buffer). Four-tenths milliliter BSA is first applied and allowed to drain into the column. Four 0.5-ml aliquots of BSA buffer are then separately applied and the eluates from these applications collected as a single pool (approximately 2.0 ml). The pooled eluates are mixed and two 0.9-ml portions are pipetted into disposable 12 × 77-mm plastic test tubes (Falcon Plastics, Los Angeles, California). The tubes are placed in an ice bath and 0.2 ml of a rabbit anti-T_3 anti-

[7] B. E. P. Murphy and C. J. Pattee, *J. Clin. Endocrinol. Metab.* **24,** 187 (1964.)

[8] V. J. Pileggi, N. D. Lee, O. J. Golub, and R. J. Henry, *J. Clin. Endocrinol. Metab.* **21,** 1272 (1961).

[9] M. I. Surks, A. R. Schadlow, and J. H. Oppenheimer, *J. Clin. Invest.* **51,** 3104 (1972).

body solution is then added. After mixing, the tubes are incubated at 4° for 16 hours. At that time, the antibody bound T_3 (B-T_3) and free T_3 (F-T_3) are separated by the addition of 1.0 ml of a dextran-coated charcoal suspension at 4°. This charcoal suspension is prepared by mixing equal volumes in barbital buffer of 1% activated charcoal (Norit A, Fisher Chemical Company, New York, N.Y.) and 1% dextran, clinical grade, molecular weight 70,000–90,000 (Nutritional Biochemical Corporation, Cleveland, Ohio). After 15 minutes, the charcoal and supernatant solution are separated by centrifugation. The counting rates of both the supernatant containing B-T_3 and the charcoal (F-T_3) are then measured in an automatic gamma spectrometer.

Antibodies to T_3 are raised in rabbits or guinea pigs by injection of T_3–BSA conjugates which are prepared by the method of Gharib *et al.*[10] Fifty milligrams of BSA (Sigma Chemical Company, St. Louis, Missouri) is dissolved in 30 ml of water. After the solution is filtered, 30 mg of Morpho CDI [1-cyclohexyl-3-(2-morpholinyl-4-ethyl)carbodiimide metho-*p*-toluene sulfonate] (Aldrich Chemical Company, Milwaukee, Wisconsin) is added. Twenty milligrams of L-triiodothyronine (free acid form) (Sigma Chemical Company, St. Louis, Missouri) dissolved in 7.5 ml dimethylformamide is then added dropwise with continuous stirring. The pH is continually monitored and kept at 5.5 by addition of several drops of either 0.02 *M* HCl or 0.02 *N* NaOH. The mixture is stirred for 10 minutes and an additional 10 mg of the carbodiimide is added. The resulting solution is then stirred for 20 hours at room temperature in the dark. The reaction mixture is then dialyzed against running tap water for 96 hours and lyophilized. As determined from the recovery of tracer amounts of $[^{125}I]T_3$ added before conjugation, the resulting T_3:BSA molar ratios of the conjugates which were synthesized varied between 7 and 10.1. One-half milligram of conjugate is then homogenized in 0.5 ml complete Freund's adjuvant and injected subcutaneously into either rabbits or guinea pigs every 14 days. Blood is collected by cardiac puncture one week after the fourth injection. In our studies, this procedure resulted in the production in all rabbits and in four of eight guinea pigs of anti-T_3 antibodies which were suitable for radioimmunoassay. A rabbit serum was used as the T_3 antibody at a final dilution of 1:8000 in BSA buffer.

Since the recovery of T_3 from the Sephadex columns is not uniform (65–80% in the 2.0 ml pooled BSA eluate, 30–35% per assay tube) it is necessary to measure precisely the specific activity of the added tracer

[10] H. Gharib, R. J. Ryan, W. E. Mayberry, and T. Hockert, *J. Clin. Endocrinol. Metab.* 33, 509 (1971).

so that appropriate corrections can be made for the mass of T_3 in the added $[^{125}I]T_3$. This is accomplished by a radioimmunoassay procedure. Mixtures of T_3 standards (50–400 pg in 0.1 ml BSA buffer) are added to 0.1 ml $[^{125}I]T_3$ (50 pg T_3 as estimated from the specific activity of the $[^{125}I]T_3$ furnished by the manufacturer). Additional tubes contain 0.2 and 0.3 ml of the $[^{125}I]$tracer solution alone without nonradioactive T_3. T_3 antibody is added and the volume of all tubes adjusted to 1.1 ml with BSA buffer. After incubation, the bound and free T_3 are separated as described above. A standard curve is plotted (B/F ratios vs. added *nonradioactive* T_3) and the mass of T_3 in the $[^{125}I]T_3$ solution is determined from the B/F ratios of the tubes containing two- and threefold volumes of tracer solution as outlined by Yalow and Berson.[11]

The two B/F ratios for each BSA column eluate are plotted against the *total* T_3 in each assay tube and a standard curve is drawn. The total T_3 in each tube is calculated as the product of the recovery of $[^{125}I]T_3$ in each assay tube (30–35%) (cpm B-T_3 + cpm F-T_3/cpm added $[^{125}I]T_3$) and the mass of T_3 (standard + $[^{125}I]T_3$) applied to each column. Total T_3 content in assay tubes containing serum is determined from their B/F ratios and the standard curve in the usual fashion. The contribution of T_3 in the $[^{125}I]T_3$ added to each sample is calculated as the product:recovery of $[^{125}I]T_3$ in each tube and the mass of T_3 in the added tracer. The total T_3 is then corrected for T_3 in the tracer. The resulting figure corrected for recovery ($[^{125}I]T_3$) and sample volume is equivalent to the concentration of T_3 in serum.

As in the case of T_4, T_3 concentrations should be measured in 4–8 plasma samples obtained during the course of the turnover studies and the average of these values employed as the mean plasma T_3 concentration for calculation of hormone turnover.

2. Measurement of T_4 and T_3 Metabolic Clearance Rates

A combined dose of 20 μCi $[^{125}I]T_4$ and 40 μCi $[^{131}I]T_3$ is prepared in 1% human serum albumin and injected intravenously. Plasma samples are obtained every 2 hours for the first 12 hours after injection and at 24- to 48-hour intervals thereafter. The collection of these samples is facilitated by the prior placement of an indwelling intravenous catheter. All plasma samples are treated with trichloracetic acid in order to precipitate the plasma proteins and remove inorganic radioiodide.[6] For the measurement of T_3 metabolic clearance rate, the plasma nonextractable radioiodine is also measured by extraction with ethanol as previously

[11] R. S. Yalow and S. A. Berson, *J. Clin. Invest.* **49**, 1157 (1960).

described.[6] The plasma radioiodothyronine concentration is calculated as the difference between the TCA precipitable radioactivity and the nonextractable radioactivity.

The metabolic clearance rate of T_4 may be calculated by conventional single compartment kinetics. In the case of the metabolic clearance rate of T_3, however, we have found that this approach results in a significant overestimation of the clearance rate of this hormone.[12] Thus, the T_3 metabolic clearance rate is calculated by the noncompartmental (integral) approach originally proposed by Tait[13] for the steroid hormones and recently applied to the thyroid hormones.[6] Thus,

$$\mathrm{MCR} = \frac{100}{\int_0^\infty c\,dt} \tag{8}$$

where c is percent of the injected dose/liter plasma and t is time (days). This integral expression can be solved graphically as follows. The concentration of radioactive hormone, c, is plotted on semilogarithmic paper and a time, t_m, with its corresponding radioactive iodothyronine concentration, c_m, is selected during the initial phase of the constant fractional removal rate of iodothyronine from plasma.[6] This time, t_m, is generally within one to two days after injection of labeled hormone. The integral expression can then be divided into two components, from $t = 0$ to t_m and from t_m to ∞.

$$\int_0^\infty c\,dt = \int_0^{t_m} c\,dt + \int_{t_m}^\infty c_m e^{-\lambda(t-t_m)}\,dt \tag{9}$$

The first integral $\left(\int_0^{t_m} c\,dt\right)$ is calculated graphically from a plot of radioiodothyronine concentration against time on Cartesian coordinates. The second integral can be calculated readily since

$$\int_{t_m}^\infty c_m e^{-\lambda(t-t_m)}\,dt = \frac{c_m}{\lambda} \tag{10}$$

In practice, although the T_4 metabolic clearance rate calculated from single compartment kinetics does not differ significantly from the estimate derived from noncompartmental kinetics, we generally apply the integral approach described above for both iodothyronines. The metabolic clearance rate data are expressed as liters per day.

[12] M. I. Surks, A. R. Schadlow, J. M. Stock, and J. H. Oppenheimer, *J. Clin. Invest.* **52**, 805 (1973).
[13] J. F. Tait, *J. Clin. Endocrinol. Metab.* **23**, 1285 (1963).

3. Calculation of the T_4 to T_3 Conversion Rate

The mean plasma hormone concentration and metabolic clearance rate for both T_4 and T_3 determined in the fashion described above are substituted in Eq. (7) and the T_4 to T_3 conversion rate is calculated. The mean T_4 to T_3 conversion rate in our laboratory is $42.6 \pm 3.1\%$ (range 31–51%) for seven subjects.

Author Index

Numbers in parentheses are reference numbers and indicate that an author's work is referred to, although his name is not cited in the text.

A

B

C

D

E

F

G

H

I

J

K

L

M

N

O

P

R

S

T

U

V

W

Y

Z

Subject Index

A

B

C

D

E

F

G

I

K

L

M

N

O

P

R

A 4
B 5
C 6
D 7
E 8
F 9
G 0
H 1
I 2
J 3